Role of Mutation Breeding in Floriculture Industry

S. K. Datta

Role of Mutation Breeding in Floriculture Industry

Springer

S. K. Datta
Former Scientist
National Botanical Research Institute
Lucknow, India

ISBN 978-981-99-5674-6 ISBN 978-981-99-5675-3 (eBook)
https://doi.org/10.1007/978-981-99-5675-3

This Springer imprint is published by the registered company Springer Nature Singapore Pte Ltd.
The registered company address is: 152 Beach Road, #21-01/04 Gateway East, Singapore 189721, Singapore

Paper in this product is recyclable.

Preface

Breeding of new ornamental varieties is one of the most rewarding professions in the world. All present-day ornamental varieties have evolved through bud sports, open-pollinated interspecific crosses, open-pollinated intraspecific crosses, planned crosses, natural chromosomal changes, chromosomal manipulations, induced mutations, management of chimera, in vitro mutation, molecular approaches, somaclonal variations, etc. Crop improvement is a never-ending project and there is always a search for new techniques.

It is more than 80 years since mutagens were identified as the source of inducing genetic variability in living systems. Since then scientific community is deeply engaged to apply mutation techniques in crop improvement. The widespread use of induced mutagenesis in plant breeding programs throughout the world has generated thousands of novel crop varieties in hundreds of crop species. Induced mutation techniques have successfully produced and commercialized worldwide quite a large number of new promising varieties in different ornamental plants.

I was involved with induced mutagenesis work since 1971 for crop (vegetable) improvement at Bose Institute, Kolkata, for my Ph.D. work. For almost 30 years (1977–2007) I was associated with CSIR-National Botanical Research Institute, Lucknow, India, in a multidisciplinary research program on induced mutagenesis on ornamental crops. National Botanical Research Institute, Lucknow, is one of the pioneer institutions where commendable work has been done on induced mutagenesis and has been most successful to produce more than 76 new promising mutant varieties in different ornamentals. Voluminous knowledge has been generated which extends different basic aspects to advance application for the success of classical mutagenesis. The scientific and technological achievements in induced mutagenesis by global experts are highly appreciable.

The book covers almost all aspects of induced mutagenesis on ornamental plants. Vast literature/knowledge generated on induced mutation in ornamental plants, worldwide, is spread in the form of books, book chapters, bulletins, and scientific journals. An attempt has been made in this book to put together nearly all information to develop a complete documentation of the results of the research conducted by different researchers over the last more than 80 years as a practical guidebook. The main objective of the book is to give a coherent and concise account of all work done on induced mutagenesis with an emphasis on recent developments. The document has been prepared only from published information as a practical document. All important work of scientists on different ornamentals has been cited. Sincere efforts have been made to highlight the original work reports as reported by individual scientists. The publication of this book is planned to reveal multifarious activities done on different aspects of mutation so that achievements made so far can be used judiciously for designing new experiments in the future. It is my sincere endeavor to present a complete and comprehensive scenario of mutation breeding research carried out on different ornamentals through this book. The book also provides a detailed review account of many aspects of current interest and progress in mutation research. The number of experiments conducted on mutation breeding, worldwide, on various ornamental plants is so voluminous that it is difficult to collect all by one scientist. Efforts have been made to highlight maximum important work on each ornamental so that one can get all about experimental designs, mutagens, mutants, etc.

The book deals with all the important and relevant aspects of mutation technique and provides an authoritative review account in the field of ornamental crops. Mutation work on some important ornamentals (Bougainvillea, Chrysanthemum, Gladiolus, Rose, and Tuberose) have been mentioned in individual chapters, and all other ornamentals have been arranged systematically in one chapter. All efforts have been made to give extensive coverage of literature dealing with different ornamentals to cover the whole field of floriculture crops. The chapter on the chrysanthemum needs a special focal point as it covers the entire mutation technology as it stands today. The author considers the chrysanthemum a model plant for mutation work. I am sure that the information in the book will be of great help to researchers, teachers, students, and breeders in planning future strategies for the development of new ornamental varieties. The book will be a very useful and important source of reference for practical mutation breeding in vegetatively propagated crops.

I sincerely acknowledge and profusely thank all my professional colleagues in mutation breeding for publishing a good amount of literature on mutation activities. I thank all scientific societies, book publishers, and journal editors who have published all this research work on ornamental mutagenesis. The book will be an excellent guide for updated information on the application of mutagenesis on ornamentals.

All important and useful references have been cited on ornamental species-wise and basic aspects of mutagenesis. Though meticulous care has been taken in reviewing while dealing with such a voluminous work, some mistakes/errors might have crept in, despite my best efforts and the non-availability of original publications.

I wish to record my sincere thanks to all the Directors, CSIR-NBRI, Lucknow, India, for providing all working facilities, favorable suggestions, and constant encouragement. I sincerely thank my teacher, the Late Prof. Dr. R.K. Base, D.Sc., Cytogenetic Laboratory, Department of Botany, Bose Institute, Kolkata, India, under whose guidance I did my Ph.D. thesis work. I sincerely acknowledge my long association at National Botanical Research Institute, Lucknow, India, where I did all my research activities on different ornamental crops in multidisciplinary aspects. I especially thank my students and colleagues and convey my deepest feelings of gratitude to all of them. I shall be failing my duty if I do not record the blessings of my parents, whose unfathomable love and blessings kept me ever up and doing. It would be remiss if I don't acknowledge the warmth of my wife who always stood by me steadfastly. She is a charming, understanding, sociable lady who took care of me and my family with effusive affection. She admirably adjusted me to pursue my research interest with unison devotion. I record my sincere gratitude to my wife Dr. Kalyani Datta and son Shouvik Dutt who extended full cooperation in many invisible ways during my research.

I warmly appreciate the cooperation and approval of Springer Nature for excellently publishing the book.

I am confident that the book will be widely accepted by students, teachers, and researchers in the field of crop improvement. I hope that this book will provide valuable data and also be a reference material for future research activities in this area.

Lucknow, India S. K. Datta

Introduction

Floriculture has developed as a highly profitable commercial area worldwide. Ornamental crops are marketed in many diversified forms and therefore the scope of business is maximum. A massive amount of diversity has been created in commercial ornamental species. Habitually breeders prepare ornamental varieties for the floriculture market either from existing promising and attractive cultivars or selecting casual chance variants or systematically breeding varieties for definite traits of commercial interest. Different conventional and advanced techniques are used for the development of new varieties. A massive amount of literature has been accumulated on general and crop-specific technology for the development of new varieties.

The book provides an authoritative re-evaluation review account of mutation technology, and one can get all the required information about its present status for the development of new ornamental varieties. Efforts have been made to highlight the different technological components and their fine points as it stands today. Breeders will be sensible of the ability of different early approaches to select the proper procedure which is most suitable. The most important and relevant work results of different researchers have been documented along with the author's practical knowledge. Induced mutation breeding aims to improve the characteristics of target ornamental plants to make them commercially profitable. New varieties are always in need and essential in the floriculture industry and the development of new varieties is a never-ending activity. The induced mutation is one of the most important approaches for the creation of new varieties through genetic manipulations. The author has published a series of papers highlighting technological advancement for the development of new and novel ornamental varieties. Mutation studies at CSIR-National Botanical Research Institute, Lucknow, India, on several important floricultural crops facilitated the correct utilization of the technique for the creation of new and novel cultivars of commercial importance.

The effects of mutagens on seed and other propagules of vegetatively propagated plants have been studied concerning different cytomorphological, anatomical, palynological, and biochemical parameters to standardize the test protocols for routine testing of mutagen effects on plants. The book covers mutation breeding work on almost all ornamentals. The author was deeply involved in mutation work on different ornamentals for more than 30 years. An attempt has been made to highlight details of major ornamentals and their mutagens treatment procedure. Different mutation breeding technical and practical applications have been discussed. Mutation breeding work carried out on some important ornamentals has been discussed in separate chapters. For these ornamentals only the initial and most important work has been cited. An attempt has been made to prepare the manuscript in such a way that one can get all details of technical information on individual crops. Classical-induced mutagenesis has successfully produced quite a large number of new varieties in different ornamental plants. It is not wise to mention all scientific papers related to the application of induced mutagenesis for the development of new varieties in all ornamental plants. A total number of new mutant ornamental varieties developed throughout the world are available in IAEA, Vienna, Mutant Database. Mutation studies on some important ornamentals (Bougainvillea, Chrysanthemum, Rose, Tuberose) have been mentioned crop-wise. Work on other ornamentals has been presented in one chapter. Attempts have been made to review the work on the application of mutagens, mutagen doses, and results starting from the early stage to the present.

It is more than 80 years since induced mutation was identified as a technique to induce genetic variability. Since then scientists are deeply engaged to develop new varieties in different crop plants for commercial exploitation. Worldwide activities have generated a massive amount of literature. Extensive research on induced mutagenesis has already been done, and a wide range of basic knowledge on multidisciplinary aspects of ornamental crops has been generated. It is now needful to assess the present research status on mutation as a potential substitute for breeding after the extensive research activities conducted by different research institutions. Induced mutation has proved to create further genetic variability of desirable characters. Efforts have been made to highlight the most important past research results. There is a need to introspect and review our approach and policies carefully and implement whatever correction is needed to correct the exploitation of mutation technology in floriculture. A proper strategy or correct approach has to be followed to use mutation technology for the development of new ornamental varieties within a reasonable time. The knowledge generated so far has been reviewed in this book, which will work as a knowledge base to prepare guidelines for future planning of successful application of mutation technology. Many popular articles, papers, review articles, book chapters, and books are published each year. Despite significant technological advances, we are still repeating some of our experiments as routine activities and just carrying a stereotype model year after year. We should prepare need-based planning considering ongoing and on-coming research areas and market demand.

The book will present a factual evaluation account of many aspects of the current mode of application and progress in the field of improvement of ornamental plants using mutation technology. This is a correct time to focus attention on the recapitulation of the past work, rise and fall, actuality and constraints of achievements, and promises of routine application of mutation technology for the creation of new varieties. Without going to argue, an attempt has been made to highlight the progress story of the improvement of ornamental crops as it stands today. At this juncture, there is no need for further thorough critical assessment of the earlier works. Floriculture scientists, breeders, and nurserymen will be aware of the capability and the constraint of various early approaches and should choose the strategy which is most appropriate as well as economic for reaching their aims. Although a huge number of research papers and review articles have been published highlighting different aspects of induced mutagenesis, the success of mutation has not been accurately documented. An attempt has been made to highlight different technical information, which is most important for the successful application of induced mutagenesis in ornamental crops. The author made an attempt to evaluate the entire multidisciplinary mutation technology work on different ornamental plants and tried to develop a complete documentation of the results of the research conducted by different researchers. The book has been prepared from published data and authors' knowledge as a review document.

About the Book

The book covers the vast literature/knowledge generated on induced mutation in ornamental plants worldwide. It will provide an authoritative review of many aspects related to induced mutagenesis in the improvement of ornamental crops and the floriculture industry. The book discusses mutation work on important ornamentals like Bougainvillea, Chrysanthemum, Gladiolus, Rose, and Tuberose in individual chapters. The entire mutagenesis work on other ornamentals (approx. 143) is arranged in a separate chapter titled "Other Ornamentals." The book not only caters to classical mutation breeding but also addresses relevant modern aspects. It serves as an up-to-date comprehensive practical guide for the proper application of mutagenesis to improve ornamental crops. Each ornamental features important and useful reference lists have been cited in all ornamentals. There is no recent book on Induced Mutations that covers mutation work on all ornamental plants with recent experimental results. An attempt has been made to compile almost all induced mutation work on all ornamentals. The book consists of 14 chapters providing holistic information on almost all important components related to vegetatively propagated ornamentals.

Contents

About the Author

S. K. Datta Ph.D., D.Sc., is an internationally acclaimed expert in floriculture and mutation breeding. Dr. Datta was engaged in both basic and applied research for the improvement of floriculture for over more than 35 years at CSIR-NBRI, Lucknow, India. His main field of research was induced mutations and improvement of ornamental plants. For mutation breeding work he applied both Physical (X-rays, Gamma rays) and Chemical (EMS, MMS, Colchicine) mutagens and developed more than 100 new varieties in different ornamentals. He has published a series of research papers (approx. 338) on different aspects related to floriculture and induced mutagenesis. He has published seven books, edited four books and eight bulletins on different aspects related to induced mutagenesis and floriculture. He has just published the book *Induced Mutation Breeding*. Dr. Datta visited Humboldt Universitat Zu Berlin, Berlin, for 2 months (under CSIR-DAAD Scientists Exchange Programme) and Korea for 1 month (INSA-KOSEF Scientists Exchange Programme). The International Atomic Energy Agency (IAEA), Vienna, selected Dr. Datta as an "Expert on Mission" for the evaluation of mutation breeding projects sponsored by IAEA to the Philippines and Jakarta, Indonesia, for a project evaluation mission. He organized international training program as supervisor on induced mutagenesis sponsored by IAEA, Vienna, at the Central Research Institute for Horticulture, Cipanas, Indonesia. Dr. Datta presented papers in different international symposium: at VIth International Congress of SABRAO held at Tsukuba, Japan; International Nuclear Conference held at Putra World Trade Centre, Kuala

Lumpur, Malaysia; International Symposium on "Underutilized Plant Species for food, nutrition, income and sustainable development" held at Arusha, Tanzania; FAO/IAEA International Symposium on the contribution of Plant Mutation Breeding to Crop Improvement, held at International Atomic Energy Agency, Vienna, Austria—four times.

Part I
Introduction

Introduction to Mutation

1

Abstract

The induced mutation is now recognized as well standardized valuable tools for the breeder. All types of mutagens (physical and chemical) have been applied to ornamentals crops. All experimental results from various ornamentals have been assessed individually or in a combined fashion, along with an indication of experiment-to-experiment variability. The book highlights views on the scientific and technological path that may allow future researchers to leapfrog to achieve the goal within a reasonable time.

Keywords

Mutagens · Mutant variety · Technological advancement

The induced mutation is now recognized as well standardized valuable tools for the breeder. A single gene or a few genes can be changed without altering the genetic makeup of a specific genotype through induced mutation. This approach is most suitable for creating further desirable change in outstanding cultivars and needs base mutations. Mutation events may develop in nature spontaneously and they can be induced. The rate of natural spontaneous mutation is very slow, i.e. 1×10^{-6} or 10^{-7} for alteration of a specific gene toward recessive in a single cell. Mutations may be induced with physical and chemical mutagens. Induced mutations using physical or chemical mutagens may increase the rate of the manifold. Most mutation events are from dominant to recessive (Brock 1979). Mutations can be differentiated into various types. It may be at the single gene level, genome level, and chromosome level. Breeders are more interested in recessive gene mutations than in large chromosome alterations. Foliage plant breeders are more interested in extranuclear mutations. The ultimate aim of the breeder is to create desirable mutant traits of commercial importance.

S. Datta, *Role of Mutation Breeding in Floriculture Industry*,
https://doi.org/10.1007/978-981-99-5675-3_1

The concept of induced mutation for crop improvement developed dates back to the beginning of the twentieth century. The first observation on artificial induction of genetic change was reported by Gager (1908), Muller (1927), and Stadler (1928a, b). High potential for bringing genetic improvement by induced mutation was realized after systematic studies by Hoffmann (1959), Hoffmann and Walther 1961, Gustafsson (1947), and Mackey (1956), and information accumulated on optimal treatment doses, treatment conditions, mutation frequency, and mutation spectra. De Mol obtained the first registered commercial X-ray mutant in Tulip in 1945. One successful story of mutation breeding is the "Horim" group of Chrysanthemum mutants in the Netherlands, which in 1979 took as much as 35% of the total Dutch market of 500 million plants (Broertjes et al. 1980).

A literature survey since the beginning to 2021 indicates that there is a step-by-step improvement in the technical procedure for the application of induced mutation for crop improvement and voluminous knowledge has been generated which broadens different basic aspects for successful and accurate application of the technique. It is more than 80 years since the mutation incident was identified as a technical source of inducing genetic variability for crop improvement. The literature review indicates an extensive multidisciplinary network approach by different researchers to enrich knowledge on induced mutations and to develop high-yielding varieties in different crops including ornamentals. Induced somatic mutation holds promise for the effective improvement of ornamental crops. Widespread use of induced mutations throughout the world has led to the official release of more than 3402 mutant varieties from more than 170 different plant species resulting in enormous positive economic impacts (Shu 2009). Induced mutation techniques have successfully produced and commercialized worldwide quite a large number of new promising varieties in different ornamentals.

Voluminous works have been done worldwide for the improvement of both seed and vegetatively propagated crops using this technology. A series of high-standard books have been published by the International Atomic Energy Agency (IAEA), Vienna, and a good number of books and edited books have been published by different publishers worldwide based on past and ongoing mutagenesis activities. The FAO/IAEA, Vienna has developed Mutant Variety Database (MVD) which provides information on plant mutant varieties (cultivars) released officially or commercially worldwide. A series of review papers have been published which cover almost all aspects of induced mutagenesis in general and ornamentals in particular (Broertjes and Van Harten 1978, 1988; Sigurbjornsson and Micke 1969; Gottschalk and Wolf 1983; Bhatia 1991; Datta 1988, 1992, 1997a, b, 2009a, b, 2012, 2014, 2017, 2020; Parry et al. 2009; Aida et al. 2018; Hernandez-Munoz et al. 1959, 2011, 2019; Micke et al. 1987, 1990; Maluszyuski et al. 1995; Schum 2003; Ahloowalia et al. 2004; Patade and Suprasanna 2008; Haq 2009; Martin et al. 2009; Kharkwal and Shu 2009; Nakagawa 2009; Lundquist 2009; D'Souza et al. 2009; Penna and Jain 2017; Jain and Suprasanna 2011; Suprasanna and Nakagawa 2012; Oladosu et al. 2015; Raina et al. 2016; Spencer-Lopes et al. 2018; Manzoor et al. 2019; García 2007, 2020; Yamaguchi et al. 2003; Melsen et al. 2021; Tutuncu et al. 2023).

Several excellent reviews and books describe the methods, the challenges, and the potential of mutation screening for crop improvement (Abbott and Atkin 1987; Van Harten 1998; Schum 2003; Schum and Preil 1998; Comai and Henikoff 2006; Datta 1990, 1997a, b, 2004; Datta and Chakraborty 2005; Datta and Teixeira da Silva 2006; Urrea and Ceballos 2005; Micke 1991; van Harten 1998; Wang 1991; Till et al. 2007a, b; Barkley and Wang 2008; Patil and Patill 2009; Jankowicz-Cieslak and Till 2016; Jankowicz-Cieslak et al. 2017; Oladosu et al. 2016; De 2017; Penna and Jain 2023; Ibrahim et al. 2018; Miri 2018; Hernandez-Munoz et al. 2019; Anne and Lim 2020; Kayalvizhi et al. 2020; Tulmann Neto et al. 2020; Ma et al. 2021; Melsen et al. 2021; Puripunyavanich et al. 2022; Werkissa and Takele 2022).

All types of mutagens (physical and chemical) have been applied to ornamentals crops. For practical mutation breeding both acute and chronic irradiations are used. For acute means short irradiation times which is recommended using a dose rate of 1–10 Gy/min. For prolonged or chronic irradiation gamma source is used. Both fast neutrons and thermal neutrons may be used if neutrons are available. Different treatment procedures like direct dose, split dose, recurrent irradiation, combined treatment, etc. have been applied to ornamentals. All types of radiation mutagens like ultraviolet light, X-ray, gamma ray, alpha and beta particles, protons and neutrons, and chemical mutagens like ethyl methanesulfonate (EMS), methyl methanesulfonate (MMS), diethyl sulfate (dES), ethyleneimine (EI), ethyl nitroso urethane (ENU), ethyl nitroso urea (ENH), methyl nitroso urea (MNH), Azides, etc. have been applied. It is not wise now to summarize all the works on different ornamentals. Worldwide researchers from floriculture/ornamental fields initiated mutation work to create new desirable cultivars through genetic variability. Details about mutagens have been mentioned in a separate chapter. For reviewing mutagenesis work, the author attempted to cover almost all important ornamentals.

The literature cited covers most of the important journals and books. One can get almost all the information about practical induced mutagenesis work on ornamental crops.

X -and gamma rays are widely used in routine mutation experiments. These mutagens have many benefits like simplicity in application, easy-to-treat bulk material, reproducibility, good penetration, accurate dosimetry, etc. The application of gamma rays in mutation work with ornamental plants is well recognized. A good amount of supportive data have been generated on the dose rate and mode of application of gamma rays (acute/chronic). Determination of a suitable dose of gamma radiations for the induction of somatic mutation is essential. Working doses of different ornamentals have been precisely determined and radiosensitivity related to all parameters has been determined and reported elaborately (Datta 2023a). Mutation technology has progressed significantly in its application mission in ornamental crops. Although a massive amount of literature has been accumulated on classical mutagenesis, the fruits of all past and recent research have not reached all researchers in the form of full technology. A huge amount of popular articles, papers, review articles, book chapters, and books have been published and there is the continuous addition of new articles. Now the majority of publications appear to be mechanical despite noteworthy technological progress. The author has taken the opportunity to argue about the current and future scientific approaches in some detail. It is now high time to prepare need-based planning considering ongoing and oncoming research areas.

The success of induced mutation depends upon the application of correct technological applications to fulfill the desired target. In this context, two recent publications of the author need to be pointed out. The book "Induced Mutation Breeding" (Datta 2023a) is an excellent reference collection of the research conducted by different researchers on induced mutagenesis, worldwide, for more than 80 years. The book highlights holistic information on almost all important components of mutation breeding. One can follow the references on crop and aspect basis since the start of mutation breeding work. Datta (2023b) in his review article highlighted the present status of the package of mutation technology for maximum commercial utilization. The knowledge generated so far on vegetatively propagated crops, especially on ornamental crops, has been critically assessed and prepared a complete technology package for future need-based planning of successful and accurate application of mutation techniques in the crop improvement program. Unyielding optimism and unrealistic expectations do not help to reach the goal. All experimental results from various ornamentals have been assessed individually or in a combined fashion, along with an indication of experiment-to-experiment variability. Both theoretical and practical implications of the mutation methodology have been discussed and presented some relevant future research needs. The ground reality of improvement work for the creation of desirable varieties for commercial exploitation has been highlighted and properly acknowledged the reality and limitations of achievements. The book highlights views on the scientific and technological path that may allow future researchers to leapfrog to achieve the goal within a reasonable time.

References

Abbott AJ, Atkin RK (eds) (1987) Improving vegetatively propagated crops. Academic Press, London. xvii + 416 pages

Ahloowalia BS, Maluszynski M, Nichterlein K (2004) Global impact of mutation-derived varieties. Euphytica 135:187–204

Aida R, Ohmiya A, Onozaki T (2018) Current research in ornamental plant breeding. Breed Sci 68(1):1. https://doi.org/10.1270/jsbbs.68.1

Anne S, Lim JH (2020) Mutation breeding using gamma irradiation in the development of ornamental plants: a review. Flower Res J 28(3):102–115. https://doi.org/10.11623/frj.2020.28.3.01

Barkley NA, Wang ML (2008) Application of TILLING and EcoTILLING as reverse genetic approaches to elucidate the function of genes in plants and animals. Curr Genomics 9(4):212–226

Bhatia CR (1991) Economic impact of mutant varieties in India. In: Plant mutation breeding for crop improvement. IAEA, Vienna, pp 33–45

Brock RD (1979) Mutation plant breeding for seed protein improvement. In: Seed protein improvement in cereals and grin legumes. Proc. Symp. IAEA/FAO/GSF, Neuherberg, BRD, 1978, IAEA, Vienna, pp 43–45

Broertjes C, Van Harten AM (1978) Application of mutation breeding methods in the improvement of vegetatively propagated crops. Elsevier, Amsterdam

Broertjes C, Van Harten AM (1988) Applied mutation breeding for vegetatively propagated crops. Elsevier, Amsterdam

Broertjes C, Koene P, van der Veen JWH (1980) A mutant of a mutant of a mutant of a . . .: irradiation of progressive radiation-induced mutants in a mutation-breeding programme with Chrysanthemum morifolium Ram. Euphytica 29(3):525–530

Comai L, Henikoff S (2006) TILLING: practical single-nucleotide mutation discovery. Plant J 45: 684–694

D'Souza SF, Reddy KS, Badigannavar AM et al (2009) Mutation breeding in oilseeds and grain legumes in India: accomplishments and socio-economic impact. In: Shu QY (ed) Induced plant mutations in the genomic era. FAO of the United Nations, Rome, pp 55–57

Datta SK (1988) Chrysanthemum cultivars evolved by induced mutations at National Botanical Research Institute, Lucknow. The Chrysanthemum 44(1):72–75

Datta SK (1990) Role of mutation breeding in floriculture. In: Proc Symp. Pl. Mut. Breed. for crop improvement, I., Vienna 18–22, June 1989, pp 273–281

Datta SK (1992) Induction and analysis of somatic mutations in vegetatively propagated ornamental. D. Sc. Thesis, Kanpur University, Kanpur, India

Datta SK (1997a) Ornamental plants—role of mutation. Daya Publishing House, Delhi, p 219

Datta SK (1997b) Role of mutation breeding for improvement of vegetatively propagated ornamentals. In: Siddiqui BA, Khan S (eds) Plant breeding advances and in vitro culture. Dept. of Botany Aligarh Muslim University/CBS Publishers and Distributors, Aligarh/New Delhi, pp 144–158

Datta SK (2004) Bougainvillea research at National Botanical Research Institute, Lucknow. J Ornam Hortic 7(1):1–14

Datta SK (2009a) A report on 36 years of practical work on crop improvement through induced mutagenesis. In: Shu QY (ed) Induced plant mutations in the genomics era. Joint FAO/IAEA Division of Nuclear Techniques in Food and Agriculture, International Atomic Energy Agency, Vienna, pp 253–256

Datta SK (2009b) Role of classical mutagenesis for development of new ornamental varieties. In: Shu QY (ed) Induced plant mutations in the genomics era. Joint FAO/IAEA Division of Nuclear Techniques in Food and Agriculture, International Atomic Energy Agency, Vienna, pp 300–302

Datta SK (2012) Success story of induced mutagenesis for the development of new ornamental varieties. In: Bioremediation, biodiversity and bioavailability 6 (special issue I). Global Science Books, Invited Review, pp 15–26

Datta SK (2014) Induced mutagenesis: basic knowledge for technological success. In: Tomlekova NB, Kozgar ML, Wani MR (eds) Mutagenesis: exploring genetic diversity of crops. Wageningen Academic Publishers, pp 95–137

Datta SK (2017) Improvement through induced mutagenesis: ornamental crops. In: Malik CP, Wani SH, Bhati-Kushwaha H, Kaur R (eds) Advanced technologies for crop improvement and agricultural productivity. Agrobios, Jodhpur, Chapter 4, pp 49–86

Datta SK (2020) Induced mutations: technological advancement for the development of new ornamental varieties. Nucleus 63:119–129

Datta SK (2023a) Induced mutation breeding. Springer. ISBN 978-981-19-9488-3. https://doi.org/10.1007/978-981-19-9489-0

Datta SK (2023b) Technology package for induced mutagenesis. J Biol Nat 15(1):70–88. Article no. JOBAN.11268 ISSN: 2395-5376 (P), ISSN: 2395-5384 (O), (NLM ID: 101679666)

Datta SK, Chakraborty D (2005) Classical mutation breeding and molecular methods for genetic improvement of ornamentals. In: Datta SK (ed) Role of classical mutation breeding in crop improvement. Daya Publishing House, Delhi, pp 260–303

Datta SK, Teixeira da Silva JA (2006) Role of induced mutagenesis for development of new flower colour and type in ornamentals. In: Floriculture, ornamental and plant biotechnology, vol I. Global Science Book, pp 640–645

De LC (2017) Improvement of ornamental plants a review. Int J Hortic 7(22):180–204
Gager CS (1908) Effects of the rays of radium on plants. Mem N Y Bot Gard 4:278
García JL (2007) Mutagenesis in the improvement of ornamental plants. Revista Chapingo. Serie horticultura, versión On-line ISSN 2007-4034 versión impresa ISSN 1027-152X Rev. Chapingo Ser Hortic 25(3). Chapingo sep./dic. 2019 Epub 29-Mayo-2020. https://doi.org/10.5154/r.rchsh.2018.12.022
García JL (2020) Mutagenesis in the improvement of ornamental plants. versión on-line ISSN 2007-4034 versión impresa ISSN 1027-152X. Rev. Chapingo Ser Hortic 25(3) Chapingo sep./dic. 2019 Epub 29-Mayo-2020
Gottschalk W, Wolf G (1983) Induced mutations in plant breeding. In: Monographs on theoretical and applied genetics, vol 7. Springer, Berlin
Gustafsson Å (1947) Mutations in agricultural plants. Hereditas 24:3–93
Haq MA (2009) Development of mutant varieties of crop plants at NIAB and the impact on agricultural production in Pakistan. In: Shu QY (ed) Induced plant mutations in the genomic era. Food and Agriculture Organization of the United Nations, Rome, pp 61–64
Hernández-Muñoz S, Pedraza-Santos ME, López PA, Gómez-Sanabria JM, Morales-Hoffmann W (1959) Neuere Moglichkeiten der Mutationszuchtung. Z Pflanzenz 41:371–394
Hernández-Muñoz S, Pedraza-Santos ME, López PA, Gómez-Sanabria JM, Morales-Jain MS, Suprasanna P (2011) Induced mutations for enhancing nutrition and food production. Gene Conserv 10:201–215
Hernandez-Munoz S, Pedraza-Santos ME, Lopez PA, Gomez-Sanabria JM, Morales-Garcia JL (2019) Mutagenesis in the improvement of ornamental plants. Rev Chapingo Ser Hortic 25(3):151. https://doi.org/10.5154/r.rchsh.2018.12.022
Hoffmann W (1959) Neuere Moglichkeiten der Mutationszuchtung. Z Pflanzenz 41:371–394
Hoffmann W, Walther F (1961) Die Wirkung von Mehrfachbestrahlungen auf die Mutabilitat eines Ein-Korn-Ramsches. Z Pflanzenz 45:361–388
Ibrahim R, Ahmad Z, Salleh S, Hassan AA, Ariffin S (2018) Mutation breeding in ornamentals. In: Handbook of plant breeding book series (HBPB), vol 11
Jain MS, Suprasanna P (2011) Induced mutations for enhancing nutrition and food production. Gene Conserve 10:201–215
Jankowicz-Cieslak J, Till BJ (2016) Chemical mutagenesis of seed and vegetatively propagated plants using EMS. Curr Protoc Plant Biol 1:617–635. https://doi.org/10.1002/cppb.20040
Jankowicz-Cieslak J, Mba C, Till BJ (eds) (2017) Chapter 1. Mutagenesis for crop breeding and functional genomics. In: Biotechnologies for plant mutation breeding. https://doi.org/10.1007/978-3-319-45021-6_1
Kayalvizhi K, Ramesh Kumar A, Sankari A, Anand M (2020) Induction of mutation in flower crops—a review. Int J Curr Microbiol App Sci 9(6):1320–1329
Kharkwal MC, Shu QY (2009) The role of induced mutations in world food security. In: Shu QY (ed) Induced plant mutations in the genomic era. FAO of the United Nations, Rome, pp 33–38
Lundquist U (2009) Eight years of Scandinavian barley mutation genetics and breeding. In: Shu QY (ed) Induced plant mutations in the genomic era. FAO of the United Nations, Rome, pp 39–43
Ma L, Kong F, Sun K, Wang T, Tao G (2021) From classical radiation to modern radiation: past, present, and future of radiation mutation breeding. Front Public Health Sec Radiat Health 9:768071. https://doi.org/10.3389/fpubh.2021.768071
MacKey J (1956) Mutation breeding in Europe. In: Genetics in plant breeding, Brookhaven Symp in Biol, vol 9, pp 141–152
Maluszyuski M, Ahloowalia BS, Sigurbjornsson B (1995) Application of in vivo and in vitro mutation techniques for crop improvement. Euphytica 85:303–315
Manzoor A, Ahmad T, Bashir MA, Hafiz IA, Silvestri C (2019) Studies on colchicine induced chromosome doubling for enhancement of quality traits in ornamental plants. Plants 8(7):194. https://doi.org/10.3390/plants8070194

Martin B, Ramiro M, Martinex-Zapater JM et al (2009) A high-density collection of EMS-induced mutations for TILLING in *Landsberg erecta* genetic background of *Arabidopsis*. BMC Plant Biol 9:147

Melsen K, van de Wouw M, Contreras R (2021) Mutation breeding in ornamentals. HortScience 56(10):1154–1165. https://doi.org/10.21273/HORTSCI16001-21

Micke A (1991) Plant mutation breeding: its future role, the methodology needed, training, and the research priorities. Introduction. In: IAEA (ed) Nuclear techniques and in vitro culture for plant improvement. IAEA, Vienna, pp 473–474

Micke A, Donini B, Maluszynski M (1987) Induced mutations for crop improvement—a review. Trop Agric (Trinidad) 64:259–278

Micke A, Donini B, Maluszynski M (1990) Induced mutations for crop improvement. Mutation Breed. Rev., FAO/IAEA, Vienna No. 7, pp 1–41

Miri SM (2018) Mutation technique and its applications in the breeding of ornamental plants. In: Conference: 2nd international & 3rd national congress on flower and ornamental plants, at Mahallat, Iran, October 2018

Muller HJ (1927) Artificial transmutation of the gene. Science 66:84–87

Nakagawa H (2009) Induced mutations in plant breeding and biological research in Japan. In: Shu QY (ed) Induced plant mutations in the genomics era. Proc of an International Joint FAO/IAEA Symp. FAO of the United Nations, Rome, pp 48–54

Oladosu Y, Rafii MY, Abdullah N et al (2015) Genetic variability and diversity of mutant rice revealed by quantitative traits and molecular markers. Agrociencia 49(3):249–266

Oladosu Y, Rafii MY, Abdullah N, Hussin G, Ramli A, Rahim HA, Miah G, Usman MG (2016) Principle and application of plant mutagenesis in crop improvement: a review. Biotechnol Biotechnol Equip 30:1–16. https://doi.org/10.1080/13102818.2015.1087333

Parry MAJ, Madgwick PJ, Bayon C, Tearall K, Hernandez-Lopez A, Baudo M, Rakszegi M, Hamada W, Al-Yassin A, Ouabbou H, Labhilili M, Phillips AL (2009) Mutation discovery for crop improvement. J Exp Bot 60(10):2817–2825. https://doi.org/10.1093/jxb/erp189

Patade VY, Suprasanna P (2008) Radiation-induced *in vitro* mutagenesis for sugarcane, improvement. Sugar Tech 10(1):14–19

Patil SD, Patill HE (2009) Improvement of major ornamental crops through mutation breeding. Int J Agric Sci 5(2):628–632

Penna S, Jain M (2017) Mutant resources and mutagenomics in crop plants. Emir J Food Agric 29(9):651–657. https://doi.org/10.9755/ejfa.2017.v29.i9.86

Penna S, Jain SM (2023) Mutation breeding for sustainable food production and climate resilience. Springer, Singapore. ISBN: 9811697191, 9789811697197

Puripunyavanich V, Maikaeo L, Limtiyayothin M, Orpong P (2022) New frontier of plant breeding using gamma irradiation and biotechnology. In: Kumar B, Debut A (eds) Green chemistry. https://doi.org/10.5772/intechopen.104667

Raina A, Laskar RA, Khursheed S et al (2016) Role of mutation breeding in crop improvement—past, present, and future. Asian Res J Agric 2(2):1–13. Article no. ARJA.29334

Schum A (2003) Mutation breeding in ornamentals: an efficient breeding method? In: Forkmann G et al (eds) Proc. 21st IS on classical/molecular breeding. Acta Hort., 612. ISHS, pp 47–60

Schum A, Preil W (1998) Induced mutations in ornamental plants. In: Jain SM, Brar DS, Ahloowalia BS (eds) Somaclonal variation and induced mutations in crop improvement. Kluwer Academic Publishers, Dordrecht, pp 333–366

Shu QY (2009) A summary of the international symposium on induced mutations in plants. In: Shu QY (ed) Induced plant mutations in the genomics era. Joint FAO/IAEA Division of Nuclear Techniques in Food and Agriculture, IAEA, Vienna, pp 15–18

Sigurbjornsson B, Micke A (1969) Progress in mutation breeding. In: Induced mutations in plants (Proc. Symp. Pullman, 1969). IAEA, Vienna, pp 673–698

Spencer-Lopes MM, Forster BP, Jankuloski L (eds) (2018) Manual on mutation breeding. 3rd ed. pp xvii + 299 pp. ISBN: 9789251305263. http://www.fao.org/3/I9285EN/i9285en.pdf

Stadler LJ (1928a) Mutations in barley induced by x-rays and radium. Science 68:186–187

Stadler LJ (1928b) Genetic effects of x-rays on maize. Proc Natl Acad Sci U S A 14:69–75
Suprasanna P, Nakagawa H (2012) Mutation breeding of vegetatively propagated crops. In: Plant mutation breeding and biotechnology. FAO of the United Nations, Rome, pp 347–358
Till BJ, Comai L, Henikoff S (2007a) TILLING and ECOTILLING for crop improvement. In: Genomics-assisted crop improvement, vol 1. Genomics approaches and platforms. Springer, Germany, pp 333–349
Till BJ, Cooper J, Tai TH, Colowit P, Greene EA, Henikoff S, Comai L (2007b) Discovery of chemically induced mutations in rice by TILLING. BMC Plant Biol 7:19
Tulmann Neto A, Latado RR, Tsai SM, Derbyshire MT, Yemma AF, Scarpare Felho JA, Cera Volo L, Rossti AC, Namekata T, Pompeu Jr J, Figueiicedo JO, Pio R, Tobias Domngues E, Santos PC, Boliani A (2020) Mutation breeding in vivo and in vitro vegetatively propagated crops. https://inis.iaea.org/collection/NCLCollectionStore/_Public/32/022/32022680.pdf
Tutuncu M, Kantoglu YT, Kunter B, Mendi YY (2023) Induced mutations for developing new ornamental varieties. In: Mutation breeding for sustainable food production and climate resilience. Springer, Singapore. https://doi.org/10.1007/978-981-16-9720-3_22
Urrea AI, Ceballos SM (2005) Empleo de las radiaciones gamma en la inducción de variabilidad genética en *Heliconia psittacorum*. Actual Biol 27(82):17–23. https://aprendeenlinea.udea.edu.co/revistas/index.php/actbio/article/view/329425
van Harten AM (1998) Mutation breeding: theory and practical applications. Cambridge University Press, Cambridge
Wang LQ (1991) Induced mutation for crop improvement in China. In: IAEA (ed) Plant mutation breeding for crop improvement. IAEA, Vienna, pp 9–32
Werkissa Y, Takele M (2022) Mutation breeding and its importance in modern plant breeding. J Plant Sci 10(2):64–70. https://doi.org/10.11648/j.jps.20221002.13
Yamaguchi H et al (2003) Mutation induced with ion beam irradiation in rose. Nucl Instrum Methods Phys Res B206:561–564

2 Introduction to Floriculture

Abstract

Floriculture has become a very important industry in many countries as a result of science-based techniques and a steady supply of improved plant materials. Ornamental plants with their wide range of diversity in flowers and foliage are highly demand in the floriculture trade. The diversity of ornamental plants has multifarious use in the floriculture industry.

Keywords

Floriculture · Ornamental diversity · Science based techniques · Novel varieties

Floriculture scientists and nursery experts have done painstaking exercises to list the names of plants that have been recommended for their utilization in one form or another in the floriculture industry. Ornamental diversity represents all kinds of plants and they are used in the floriculture industry for trees, ornamental shrubs, loose flowers, cut flowers, foliage plants, etc. Ornamental/floricultural crops have been classified into different groups: bulbous ornamental crops (categorized as the bulb, tuber, rhizome, tuberous root, etc.); flowering bulbous; foliage bulbous; annuals (most important popular ornamental plants: seasonal (annual) plants—summer seasons, rainy seasons, winter seasons, annuals suitable for pots, annuals suitable for cut flowers); annual creeper; houseplants; ferns suitable for houseplants; palms; ornamental climbers; bonsai; hedge plants; edge plants; topiary plants; ornamental trees; ornamental shrubs; cacti and succulents; etc. Plants belonging to each category have been listed very systematically.

Floriculture is by far the most diverse sector of horticulture. Ornamental plants and flowers are associated with our civilization since time immemorial. Floriculture has now been established as an independent sector of agriculture being important from aesthetic, social, and economic points of view. It has the potential for

S. Datta, *Role of Mutation Breeding in Floriculture Industry*,
https://doi.org/10.1007/978-981-99-5675-3_2

generating employment opportunities around the year. Floriculture is a highly specialized sector where the contribution of traditional and advanced techniques is equal. The advancement of scientific approaches helped floriculture to develop into a global industry. More than 140 countries are connected in the floriculture business. It is an age-old farming activity. Production commodities in floriculture have been diversified in the recent past. Floriculture deals with the cultivation, processing, and marketing of all above-mentioned ornamental categories of plants. Ornamental plants are highly appreciated for their special and unmatched qualities such as attractive floral and leaf characteristics and are widely cultivated worldwide. Ornamental plants with their wide range of diversity in flowers and foliage are highly demand in the floriculture trade. The diversity of ornamental plants has multifarious use in the floriculture industry. More and more people are involved in the production and purchasing of floriculture products. There are millions of people involved in the floriculture industry worldwide. This industry is growing very fast and it has metamorphosed into a viable commercial enterprise all over the world. The demand for floriculture products is throughout the year and their demand significantly increases on special occasions like Christmas, Valentine's Day, Easter, Mother's Day, etc. Today, floriculture has been recognized as the most profitable profession with a much higher potential for returns per unit area than most fields and even some other horticultural crops (Aida et al. 2018; Anonymous. 2018; Burchi et al. 1996; Datta 2006, 2015, 2019a, b, c; Gupta and Agnihotri 2017; International Trade Center (ITC) 2019; Singh 2017; Khan 2018; Malhotra 2017; Melsen et al. 2021; Xia et al. 2006; Vahoniya et al. 2018; Van Uffelen and Groot 2010).

Floriculture has become a very important industry in many countries as a result of science-based techniques and a steady supply of improved plant materials. In ornamental crops, there is a regular development of new varieties through conventional cross-breeding and selection by nurserymen, amateur growers, etc. Although classical breeding is still a powerful tool in the breeder's hands, the available gene pool for new traits is limited. There are many limitations of classical breeding for crop improvement and therefore the application of such a technique (induced mutagenesis) is a necessity. Floriculture has a significant role in driving the economy of many countries. This economic area can be more profitable using nuclear techniques. Crop yield can be increased over traditional varieties and the development of new varieties will support the floriculture industry. The new varieties will fulfill the demand in terms of higher yield, improved flower quality, good agronomic performance, adaptability to climate change variation, etc. New varieties developed through mutation breeding may be adapted and resilient to climate change with improved other commercial qualities as per the need of consumers. The farming community will be benefitted from the heat tolerance varieties even under the changing weather (Abrol and Baweja 2019; Anonymous. 2018; Chawla et al. 2016; Khan 2018; Kuzichev and Kuzicheva 2016; Papademetriou and Dadlani 1998; Vahoniya et al. 2018; Van Uffelen and Groot 2010).

For the improvement and development of new varieties using induced mutation or any other techniques, knowledge of propagation techniques of all categories of ornamentals is very important. All cultural practices have been described very precisely. To start floriculture activities/business and improvement through induced mutagenesis one should have basic knowledge of different cultural practices of horticultural operations. All combinations of the agronomical trial using cultural practices have been worked out for optimization of production in all ornamental crops. Methods of propagation help treatment procedure and dose determination in induced mutagenesis.

Cut flower: In the floriculture industry, important cut flowers are roses, carnations, chrysanthemums, orchids, gerbera, Lilium, anthurium, gladiolus, narcissus, bird of paradise, heliconia, ranunculus, tulip, calla lily, etc.

Loose flowers comprise rose, chrysanthemum, marigold, jasmine, tuberose, gaillardia, crossandra, barleria, chandni, kaner, hibiscus, spider lily, and eranthemum.

Some examples of potted plants are aglaonema, aralia, azalea, begonia, calathea, chlorophytum, croton, dieffenbachia, dracaena, ferns, ficus, kalanchoe, maranta, money plant, etc.

A large number of bulbous plants, such as gladiolus, tuberose, amaryllis, dahlia, lilies, freesia, tulip, calla lily, etc., are multiplied and marketed.

The demand for natural floral extracts like perfumes from flowers is increasing by the day. Some flowers, such as rose, jasmine, screwpine (kewra), and tuberose are used for the extraction of essential oils for the preparation of perfumes or attar.

Propagation: Different floricultural crops are propagated/multiplied by different means like sexual propagation, asexual propagation, and micropropagation. Ornamentals are usually propagated in two ways: A. Seeds and B. Vegetative parts of the plant. Techniques for asexual/vegetative propagation of both woody and herbaceous and bulbous ornamental plants have been standardized through stem cuttings (softwood cuttings, hardwood cuttings), leaf cuttings, ground layering, air layering, divisions, bulbs, and corms, tubers, and rhizomes, grafting, budding, pinching, disbudding and dis-shooting, etc.

Vegetative propagation: Vegetative propagation is opted in different ornamentals due to several reasons like plants where seeds are not viable or do not produce seeds (*Acalypha*, *Eranthemum*); some plants are highly heterogeneous due to cross-pollination and are propagated only by vegetative means. Ornamental crops are propagated by all types of vegetative propagules like bulbs, tubers, corms/cormels, rhizomes, suckers, soft cuttings (rooted or unrooted), hardwood cuttings, axillary buds, budding eyes, grafts, other plant parts or whole plants, explants (for in vitro experiments—shoot apex, leaf, petals, pedicels, etc.), etc. Sometimes, it is difficult to correctly classify a few ornamentals due to their propagation methods. It has been classified into more classes by different growers. However, an attempt has been made to highlight the position of different ornamentals as per available literature.

Bulbous ornamental plants: Bulbous ornamentals are categorized as follows:

Bulbs (underground modified stem): Amaryllis (Amaryllidaceae), Begonia sp. (Begoniaceae), *Crinum bulbispermum* Syn. *C. longifolium* (Amaryllidaceae), daffodils, Freesia (*F. refrecta*—Iridaceae), *Gerbera jamesonii* (Compositae), Glorosia superba (Liliaceae), Hyacinths, Hymenocallis, Hemerocallis, *Haemanthes multiflorus*, Cooperanthes, *Iris reticulata*, tuberose, Zephyranthes, tuberose, etc. are mainly grown from bulbs. Some other ornamentals are also propagated by the same methods: Anemone, Arum, Colchicum, Crocus, Cyclamen, Endymion, Fritillaria, Galanthus, Ipheion, Leucojum, Lilium, Muscari, Narcissus, Ornithogalum, Scilla, Sternbergia, and Tulipa.

Corms/cormels: Gladiolus, Freesia, Crocus, liatris, crocosmia.

Rhizomes and tubers: Calathea, Caladium, Alstroemeria, Anthurium, Alocasia, Alpinia, Hedychium, Heliconia, Gloriosa, Canna (Cannaceae), Dahlia, ranunculus, zantedeschia etc.

Runner or stolon: Chlorophytum, Episcia, etc.

Offsets (branches ending in a rosette of leaves): Sansevieria and *Agave americana.*

Root suckers: *Millingtonia hortensis, Clerodendron splendens, Quisqualis indica,* etc.

Enlarged hypocotyl: Cyclamen, gloxinia, tuberous begonia.

Air layer and ground layer: *Gustavia augusta, Brownea ariza, B. grandiceps, Ficus elastica and Ficus Krishna,* and *F. benjamina* are easily propagated from the air layer.

Flowering bulbous: Achimenes, Alstroemeria, Amaryllis, Anthurium, Canna, Cooperia, Crinum, Dahlia, Eucharis, Eurycles, Gladiolus, Gloriosa, Haemanthus, Hedychium, Hemerocallis, Hippeastrum, Hymenocallis, Nelumbo, Nymphaea, Pancratium, Polyanthes, Sinningia, Zantedeschia, Zephyranthes, etc.

Foliage bulbous: Alocasia, Asparagus, Caldium, Calathea, Kaemferia, Sansevieria, etc.

Cuttings: Both woody and herbaceous plants are propagated by cuttings of stems, leaves, and roots. A plant can be propagated by more than one method of cutting.

Stem Cuttings: A wide range of ornamentals are propagated by stem cuttings and the process is named differentially in different species as per practice. It may be tip cutting, stem cuttings, or both. Stem cuttings may be herbaceous (non-woody—coleus, chrysanthemum, and dahlia), softwood, semi-hardwood, and hardwood—African Violet, Acalypha, Aglaeonema, Begonia, Beloperone, Bougainvillea, *Brassaia actinophylla*, Christmas cactus, Cissus, Coleus, Crassula, Croton, Cordyline terminalis, Dieffenbachia, Dracaena, *Ficus elastic*, *Ficus benjamina*, Fittonia, Geranium, Gliricidia, Heder, Helxine, *Hoya carnosa*, Impatiens, Maranta, Monstera, Nepthitis, Peperomia, Philodendron, Pothos, Pilea cadierea, Plectranthus, Plumerias, Podocarpus, Poinsettia, Selaginella, *Spathodia campanulata* (also by seeds), etc.

Leaf Cuttings: The process is the same as stem cuttings. A leaf blade or leaf with a petiole may be used. One of many new small plants forms at the base of the leaf. African violet, Begonia, Cactus (a few varieties), Crassula, Kalanchoe, Peperomia, Plectranthus, Sansevieria, Sedum, etc. may be propagated by this method.

Leaf vein cuttings: Rex begonia, Sinningla, Smithianthas (Temple Bells), etc.

Budding: Budding is the most suitable and commonly used method for the propagation of roses. The technique is simple but requires proper practice, precaution, and care. Different steps are followed in budding. The first step is a selection of rootstock and variety to be budded. Budwood is collected from a healthy branch with 3–4 eyes. The eye at the axile of the leaf is removed with a sharp knife (budding knife) along with a piece of bark 1.5 cm above and below the eye. Either an I-shaped or T-shaped sharp cut is made at the lower portion of the stock. The eye is very carefully kept into the T-shaped or I-shaped cut. The eye is tied with narrow tape-like alkathene fiber. The eye sprouts and develop into a plant on rootstock.

Pinching, disbudding, and dis-shooting: These are different cultural operations of vegetatively propagated crops—to induce the growth of vegetative laterals, to increase branch number, to increase the number of flowering stems, to develop a desired number of flowers, etc.

Customers' desire for the beauty and appealing characteristics of ornamental plants regulates the prospect of the floriculture industry. The creation of new character in any ornamental species is motivated by the tastes and understanding of the beauty of different consumer preferences. Creation and commercialization of new flowers and planting materials increase the farmer's economic status. The new quality of attraction for ornamentals is determined by a new color, mixed color design, new flower type, flower size, flower yield, hardy flower stem, long vase life, fragrance, leaf variegations, plant type, disease resistance, etc.

Ornamentals are a very special group of plants where the scope of alterations in new forms is wide in comparison to another group of plants. Different economic characteristics of ornamental groups of plants have been highlighted in Fig. 2.1a, b. Creation of novelty in any such desirable characters has good commercial value. In most ornamental plants especially chrysanthemums, gladiolus, gerbera, etc. there is always demand for new appealing and resistant to different environmental stress cultivars (Datta 2020).

Every country has its selective ornamentals in general for commercial exploitation and selected ornamentals for mutation breeding experiments.

CSIR-National Botanical Research Institute (NBRI), India is working extensively on induced mutagenesis for the last 35 years using gamma radiation for the development of new and novel ornamental varieties. Mutation technique has been standardized very precisely on different ornamental crops (bougainvillea, chrysanthemum, gladiolus, rose, tuberose, *Lantana depressa*, etc.) and generated voluminous literature/knowledge on various parameters which cover methods of exposure, LD_{50} dose, working dose, radiosensitivity, mechanism of induction of mutations, isolation of mutant tissues, management of chimera, in vitro mutagenesis, multiplication, assessment of mutants, commercialization, etc. (Datta 2006, 2015, 2019c, 2020). NBRI exploited the following methods for the development of new varieties: spontaneous mutations, hybrid seed technology, classical breeding, chromosome manipulations, induced mutagenesis, management of chimera, in vitro mutagenesis, etc. Chrysanthemum was selected as a model material for the application of all techniques.

Literature/knowledge generated at CSIR-NBRI and different research institutes/universities throughout the world is spread in the form of books, bulletins, catalogs, scientific journals, newsletters, newspapers, popular magazines, etc. All are not easily accessible. The book will provide all technical and scientific information

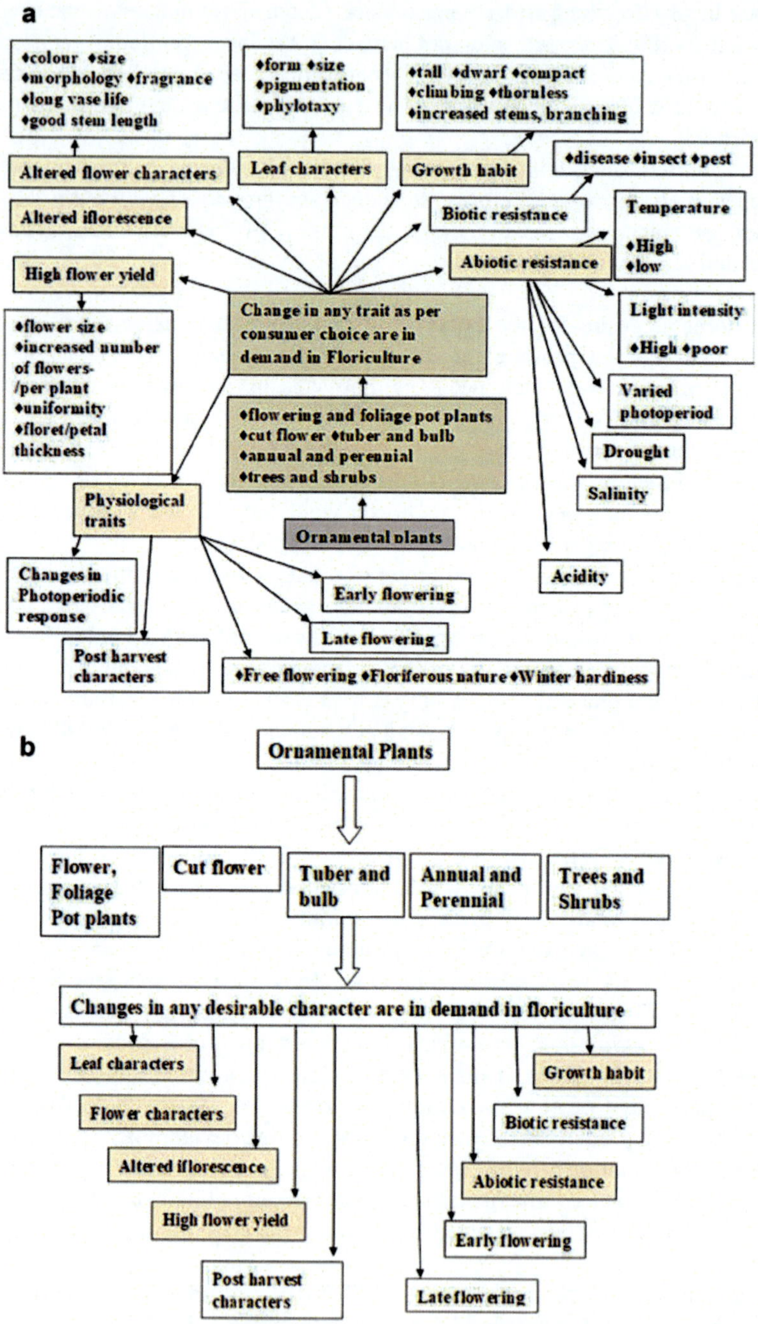

Fig. 2.1 (a, b) Different economic characters of ornamental plants suitable for improvement through induced mutations

generated at NBRI along with important publications on different ornamentals by other research institutions/universities. Readers will get all mutation work activities for the development of new ornamental varieties.

References

Abrol A, Baweja HS (2019) Floriculture worldwide production, trade, consumption pattern, market opportunities, and challenges. https://medium.com/@preetisharma_51610

Aida R, Ohmiya A, Onozaki T (2018) Current research in ornamental plant breeding. Breed Sci 68(1):1. https://doi.org/10.1270/jsbbs.68.1

Anonymous (2018) A review of production statistics for the cut flower and foliage sector. Agriculture and Horticulture Development Board (AHDB)

Burchi G, Mercuri A, De Benedetti L, Giovannini A (1996) Transformation methods apply to ornamental plants. Plant Tissue Cult Biotechnol 2:94–104

Chawla SL, Patil S, Ahlawat TR, Agnihotri R (2016) Present status, constraints, and future potential of floriculture in India. In: Patel NL, Chawla SL, Ahlawat TR (eds) Commercial horticulture. New India Publishing Agency, New Delhi, pp 29–38

Datta SK (2006) Floriculture worldwide trade. In: Bhattacherjee SK (ed) Advances in ornamental horticulture. Pointer Publishers, Jaipur, pp 1–19

Datta SK (2015) Indian floriculture; role of CSIR. Regency Publications, A Division of Astral International (P) Ltd., New Delhi. 432 pages

Datta SK (2019a) Floriculture work at CSIR-National Botanical Research Institute, Lucknow. Sci Cult 85(7–8):274–283

Datta SK (2019b) Present status of research on floriculture in India. LS Int J Life Sci 8(2):71–93

Datta SK (2019c) Need-based tissue culture in floriculture: a success story. J Plant Sci Res 35(2): 245–254

Datta SK (2020) Induced mutations: technological advancement for the development of new ornamental varieties. Nucleus 63:119–129

Gupta YC, Agnihotri R (2017) Vistas in the breeding of roses. In: Malhotra SK, Ram L (eds) Floriculture and landscape gardening. Central Institute of Horticulture, India, pp 152–156

International Trade Center (ITC) (2019) International Trade Center. https://www.trademap.org/Country_SelProductCountry_TS.aspx?nvpm=3%7c484%7c%7c%7c%7c06%7c%7c%7c2%7c1%7c1%7c1%7c2%7c1%7c2%7c1%7c1. Accessed 22 May 2019

Khan H (2018) Top 10 flower-producing countries in the world. www.worldblaze.in

Kuzichev OB, Kuzicheva NY (2016) Innovative processes in floriculture: current status, problems and prospects. Indian J Sci Technol 9(16). https://doi.org/10.17485/ijst/2016/v9i16/89804. ISSN (Print): 0974–6846. ISSN (Online): 0974–5645

Malhotra SK (2017) Emerging floriculture industry in India. In: Malhotra SK, Ram L (eds) Floriculture and landscape gardening. Central Institute of Horticulture, India, pp 32–40

Melsen K, Mark van de Wouw M, van de and Contreras, R. (2021) Mutation breeding in ornamentals. Hortscience 56:1154. https://doi.org/10.21273/HORTSCI16001-21

Papademetriou MK, Dadlani N (1998) Cut flowers in Asia. RAP Publication, Asia, p 85

Singh HP (2017) Landscape gardening for ecological and aesthetic gains. In: Malhotra SK, Ram L (eds) Floriculture and landscape gardening. Central Institute of Horticulture, India, pp 1–10

Vahoniya D, Panigrahy SR, Patel D, Patel J (2018) Status of floriculture in India: with special focus to marketing. Int J Pure Appl Biosci 6(2):1431–1438

Van Uffelen RLM, Groot NSPD (2010) Floriculture worldwide; production, trade, and consumption patterns show market opportunities and challenges. RuudvanUffelen@wurnl

Xia Y, Deng X, Zhou P, Shima K, da Silva JAT (2006) The world floriculture industry: dynamics of production and markets (Chapter 35). In: Teixeira da Silva JA, Floriculture, ornamental and plant biotechnology IV: advances and topical issues (eds) , vol IV, 1st edn. Global Science Book

World Status of Mutant Varieties of Different Ornamentals

3

Abstract

Mutant varieties registered during the period 1949–2020 have been scrutinized. Attempts have been made to analyze and highlight the entire registered document year-wise, country-wise, and cropwise.

Keywords

Ornamental crops · IAEA Mutant Database · Mutant varieties · World status

It is more than 80 years since the mission-induced mutation started for crop improvement through the development of new varieties. It is now very important to evaluate the results related to the development of new varieties worldwide. A literature survey indicates that there has been a noteworthy rise in the development of new varieties in different crop plants. The author was also a scientific partner in this mission to develop new and novel varieties in different ornamentals through induced mutagenesis. To conclude fulfillment of the mutation technology mission at this stage, the total evaluation of new ornamental varieties developed worldwide is very important. After over 80 years, mutation technology must be grown-up enough and it is high time to evaluate the achievements in terms of the development of new varieties. It will provide scope to reflect the achievements of the technology in the last phase. Despite an early lead and substantial development of new varieties, there seems to be a lack of appreciation of the technology. The author has taken the opportunity to argue about the current and future approaches of the technology as it is very relevant to the floriculture industry. Considering the length of the period of mutation research there must be a flow of knowledge and feedback from the scientific community down to the industry level. Despite such impressive advancement, doubts are often amplified whether the benefits of all this progress could benefit the floriculture sector through the development of new varieties.

S. Datta, *Role of Mutation Breeding in Floriculture Industry*,
https://doi.org/10.1007/978-981-99-5675-3_3

The success of any technology depends upon its correct application. Unyielding optimism and unrealistic expectations do not help the technology produce the right product. At this stage, it is very important to develop a clear picture of the contribution of mutation technology in the development of new ornamental varieties. Results from multilocational experimental designs have been evaluated individually or in a combined form, along with a suggestive crop-to-crop experimental variability.

Both theoretical and practical implications of the technology have been discussed and some relevant future research needs have been proposed. Maximum mutation-related experiments on all important ornamentals have been evaluated to understand the achievements and impacts of past research. These results will elicit a productive response among the floriculture scientists. While highlighting several achievements of early efforts, a sincere attempt has been made to acknowledge sensitive basic weaknesses. The ground reality and limitations of achievements of mutation work for the development of new varieties should be realized. It is very significant to understand the nitty-gritty of the creation of new variety related to some specific topic-wise research results for designing new activity of futurology. The author has taken the opportunity of discussing the new scientific activity in some detail as it is so relevant to the current focal theme. It is very important to highlight some specific topic-wise research results related to crop improvement. A massive amount of available literature has been consulted and judicially analyzed to prepare this document. Literature published during the entire period was surveyed. Without going through reference-wise work details, the entire work done by concentrated efforts of research scientists from different countries has been highlighted.

Analysis of different publications and the IAEA Vienna Mutant Database motivated author to evaluate the achievements and impact of past and ongoing research processes toward the development of new varieties. There has been a leap forward to generate knowledge on multidisciplinary aspects. Mutant varieties developed and documented worldwide have been reported earlier (Ahloowalia et al. 2004; Anonymous 1977, 1994; IAEA Vienna Mutant Database; Datta 1988, 1990, 1997a, b, 2000a, b, 2010; Datta and Mitra 1999; Gupta and Datta 2006; Melsen et al. 2021; Schum 2003; Schum and Preil 1998). Mutant varieties registered during the period 1949–2020 have been scrutinized. Attempts have been made to analyze and highlight the entire registered document year-wise, country-wise, and cropwise.

Countries that registered their ornamental mutant varieties are Brazil, Belgium, Canada, China, France, Germany, Hungary, India, Japan, Korea, Malaysia, Netherlands, Philippines, Poland, Russia, Slovakia, Thailand, United States, Vietnam, etc. Several mutant varieties registered during the period 1949–2020 have been shown in Figs. 3.1, 3.2, 3.3, and 3.4. There is no specific pattern of the number of new varieties developed for each year. Both increases and decreases in the number of new varieties were observed. A maximum number of new varieties were registered between 1985 and 1986. Development of new varieties declined after 2006.

Mutants have been developed in a wide range of ornamental crops. Several mutant varieties and their percentage values (in parenthesis) of a total number of mutants of individual ornamentals have been shown in Figs. 3.5, 3.6, 3.7, 3.8, 3.9,

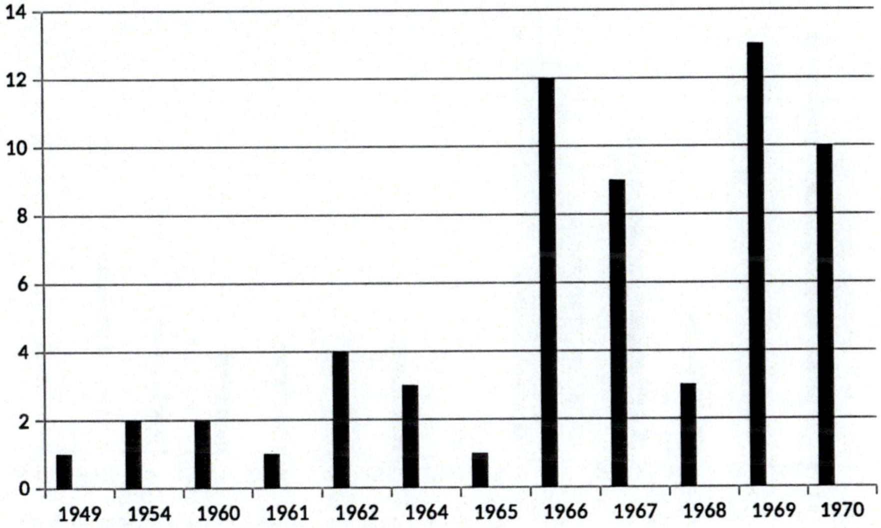

Fig. 3.1 Number of mutant varieties registered during the period 1949–1985 (Mutant Variety Database, IAEA, Vienna)

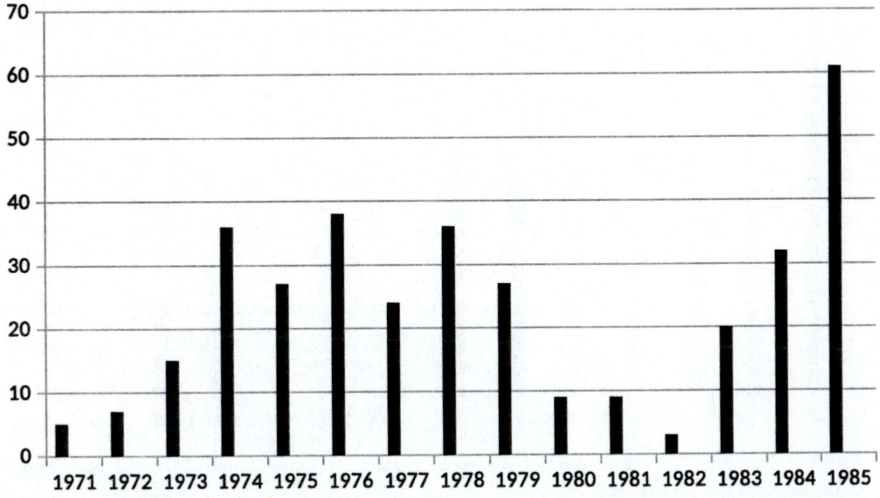

Fig. 3.2 Number of mutant varieties registered during the period 1949–1985 (Mutant Variety Database, IAEA, Vienna)

and 3.10 and as follows: Abelia 1 (0.13), Abelmoschus manihot (Aibika) 4 (0.54), Achimenes 10 (1.36), Agrostis 3 (0.40), Alopecurus pratensis 1 (0.13), Alstroemeria 33 (4.48), Antirrhinum (Snapdragon) 5 (0.68), Aster 2 (0.27), Begonia 25 (3.40), Bidens (Beggarticks) 1 (0.13), Bougainvillea 17 (2.30), Caladium 1(0.13), Calathia 1 (0.13), Canna 8 (1.08), Chinese tallow tree 1 (0.13), Chrysanthemum 293 (39.80),

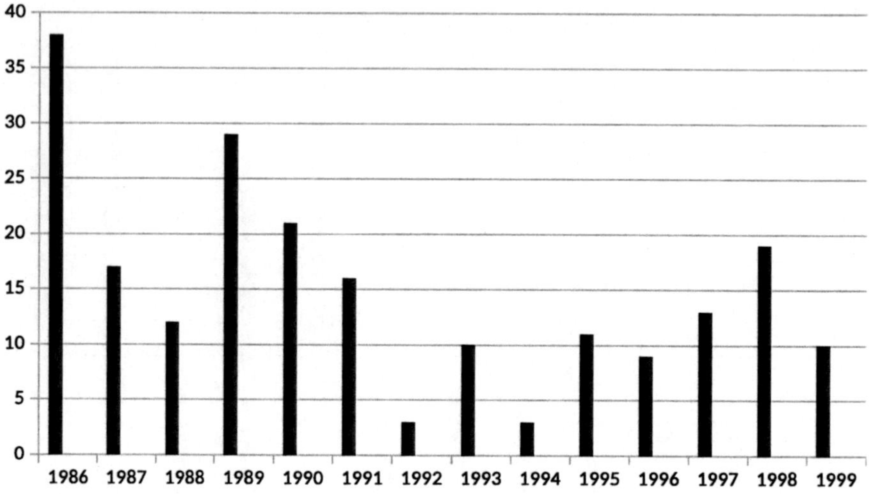

Fig. 3.3 Number of mutant varieties registered during the period 1986–2020 (Mutant Variety Database, IAEA, Vienna)

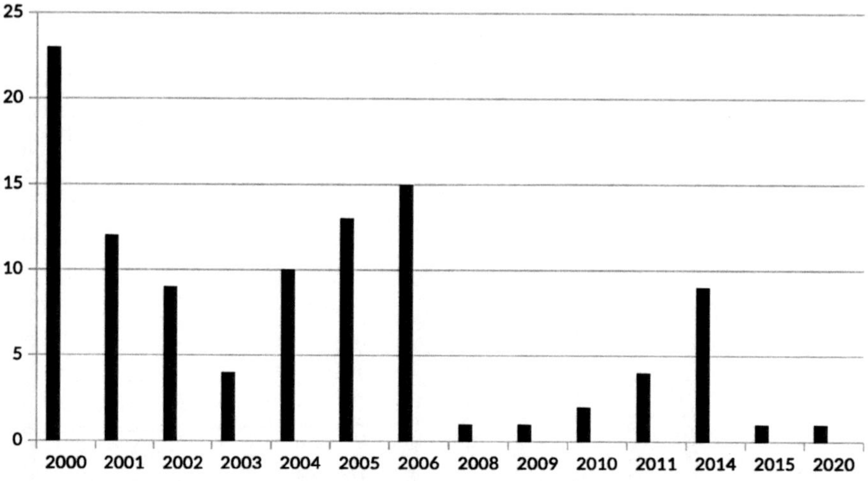

Fig. 3.4 Number of mutant varieties registered during the period 1986–2020 (Mutant Variety Database, IAEA, Vienna)

Clematis 2 (0.27), Cordyline 1 (0.13), Cyperus malaccensis 1 (0.13), Cytisus scoparius (Scotch broom) 9 (1.22), Dahlia 35 (4.76), Delphinium (Larkspur) 1 (0.13), Dianthus (Carnation) 28 (3.80), Dracaena 1 (0.13), Euphorbia fulgens 1 (0.13), Eustoma 3 (0.40), Festuca pratensis 3 (0.40), Freycinetia multiflora 1 (0.13), Ficus 2 (0.27), Forsythia 2 (0.27), Gardenia jasminoides 1 (0.13), Gentiana 2 (0.27), Gerbera 1 (0.13), Gladiolus 4 (0.54), Glory bush (Tibouchina organensis) 4 (0.54), Guzmania 1 (0.13), Gypsophyla 3 (0.40), Hibiscus 11 (1.49), Hoya

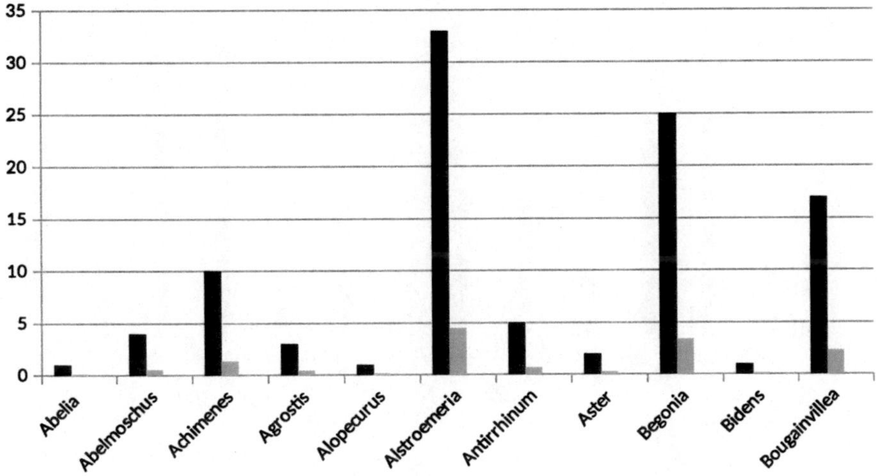

Fig. 3.5 Number of mutant varieties ■ and percentage values □ of individual ornamentals

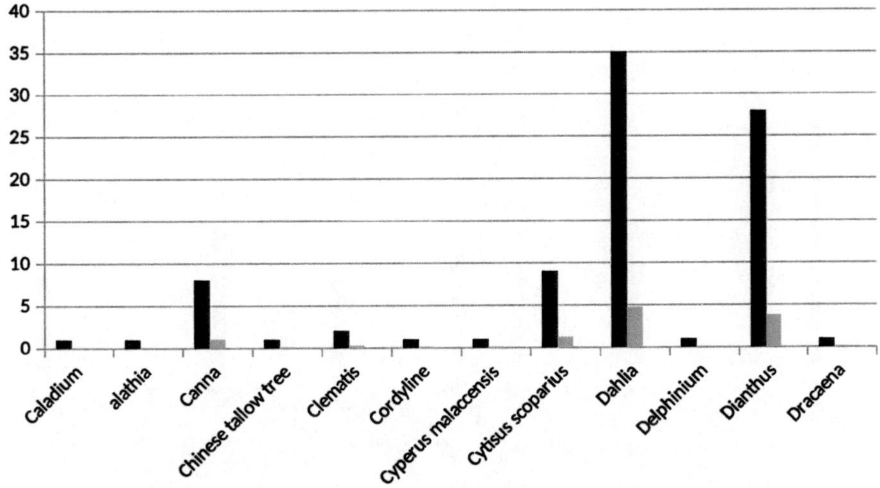

Fig. 3.6 Number of mutant varieties ■ and percentage values □ of individual ornamentals

4 (0.54), Hyacinth 1 (0.13), Iris 5 (0.68), Joysia japonica (Japanese lawngrass) 1 (0.13), Juncus effuses (Mat rush) 3 (0.40), Kalanchoe 4 (0.54), Lagerstroemia 2 (0.27), Lantana depressa 3 (0.40), Larkspur (Delphinium) 1 (0.13), Lily 5 (0.68), Lilium 1 (0.13), Limonium (Statice) 6 (0.81), Lotus (Nelumbo) 3 (0.40), Mat rush (Juncus) 1 (0.13), Melastoma 1 (0.13), Muraya 1 (0.13), Orchid (Oncidia) 8 (1.09), Panicum maximum (Guinea grass) 1 (0.13), Pelargonium (Geranium) 3 (0.40), Petasites (Butterbar) 1 (0.13), Petunia 1 (0.13), Portulaca 17 (2.30), Populus trochocrpa 1 (0.13), Rhododendron (Azalea) 15 (2.03), Rose 67 (9.10), Saintpaulia

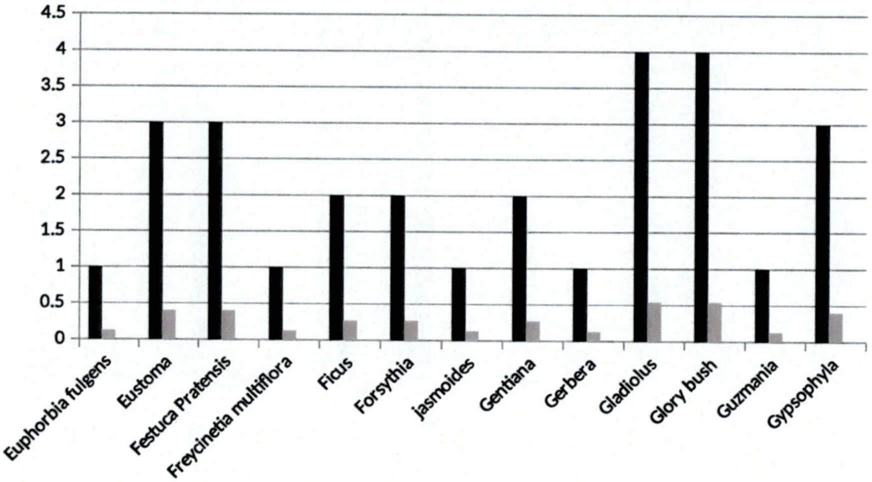

Fig. 3.7 Number of mutant varieties ■ and percentage values □ of individual ornamentals

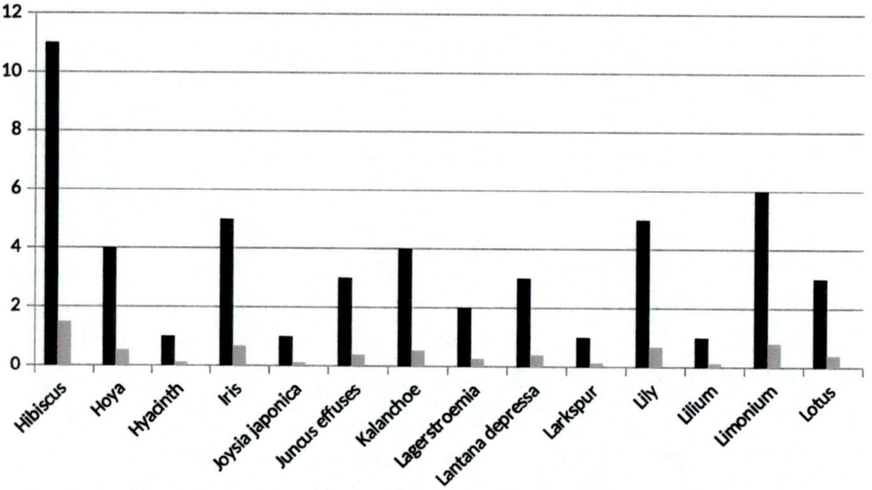

Fig. 3.8 Number of mutant varieties ■ and percentage values □ of individual ornamentals

African violet 1 (0.13), Schefflera 1 (0.13), Sonerila picta 1 (0.13), Stenotaphrum secundatum Kuntze (St. Augustine grass) 2 (0.27), Streptocarpus 30 (4.07), Syringa vulgaris (Lilac) 1 (0.13), Tibouchiana organensis (Glory bush) 1 (0.13), Torenia 3 (0.40), Trifolium repens (White clover) 1 (0.13), Tuberose 2 (0.27), Tulip 10 (1.36), Weigela 3 (0.40), Zoysia matrella (Manila grass) 1 (0.13), Zoysia japonica (Japanese lawngrass) 1 (0.13).

Among all the ornamental varieties, maximum mutant varieties were developed in chrysanthemums from all the countries (Figs. 3.11 and 3.12). Countrywide

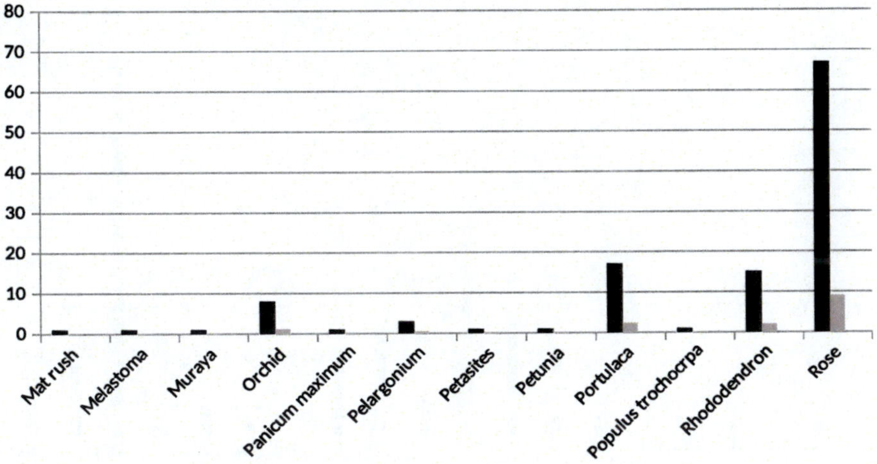

Fig. 3.9 Number of mutant varieties ▬ and percentage values ▭ of individual ornamentals

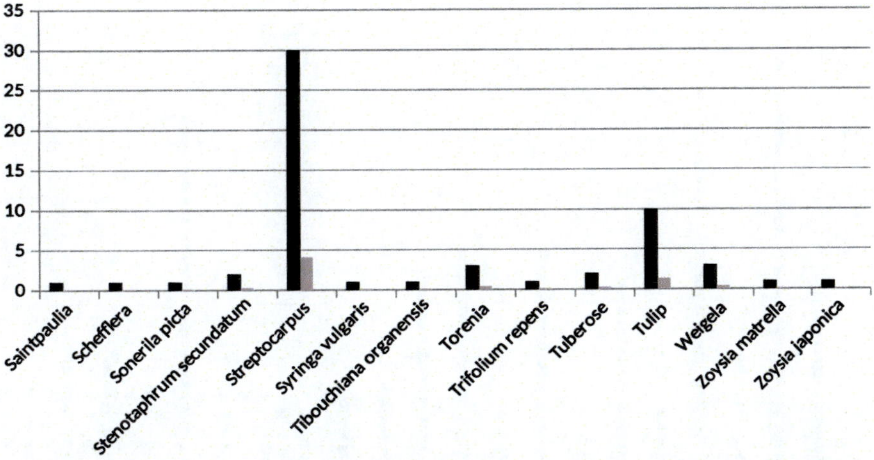

Fig. 3.10 Number of mutant varieties ▬ and percentage values ▭ of individual ornamentals

registration of total mutant varieties has been shown in Fig. 3.13. Netherlands registered maximum mutant varieties followed by Japan, India, Germany, and China. Cropwise number of total mutants and percentage values reported from different countries have been shown in Figs. 3.14, 3.15, 3.16, 3.17, 3.18, 3.19 and 3.20. Worldwide researchers were motivated to initiate research on such sensitive topics. Collaborative research among researchers having different skills and experiences helps to achieve the target results faster. But much collaborative work on induced mutagenesis toward the development of mutant varieties could not find. Recent publications trend indicates that the majority of experimental designs are

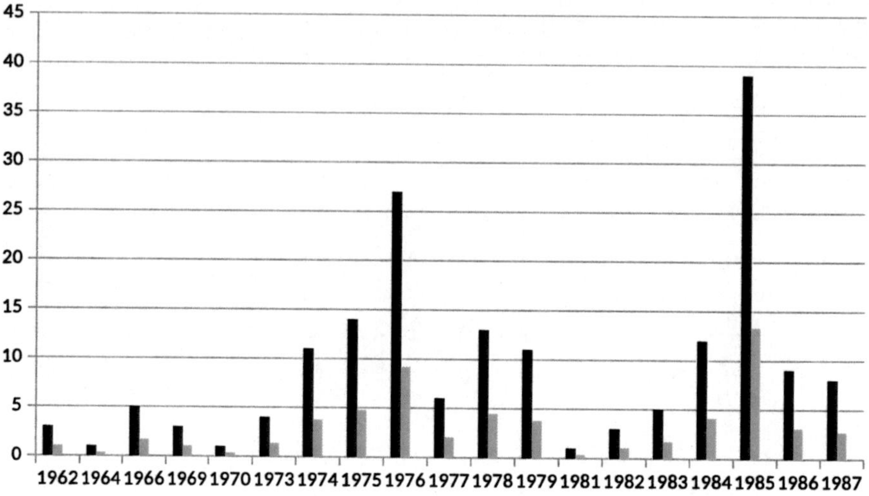

Fig. 3.11 Year wise number ■ and percentage □ of chrysanthemum mutants

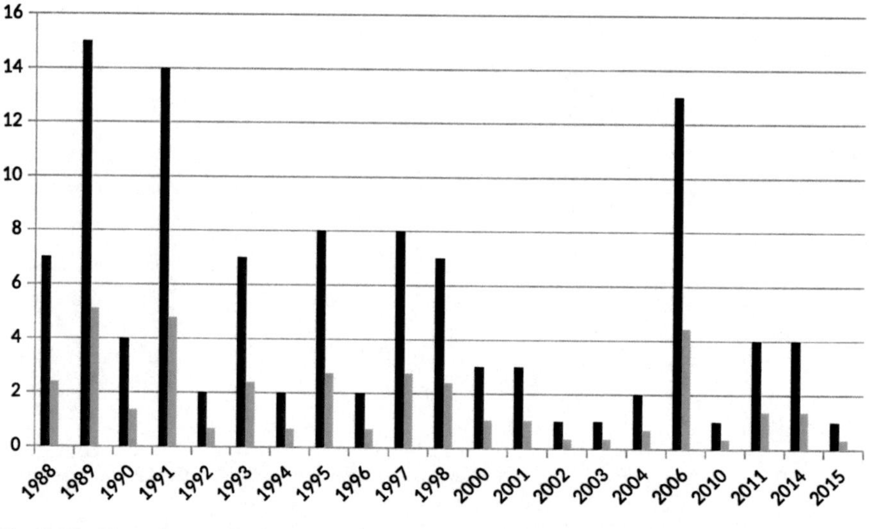

Fig. 3.12 Year wise number ■ and percentage □ of chrysanthemum mutants

almost routine research covering similar parameters. These publications are important for individual scientists but do not enrich mutation technology.

The author was deeply engaged (35 years) to utilize gamma radiation for the improvement of ornamental crops. The mutation breeding work carried out by the author on ornamentals will be highlighted which will give the complete picture of the role of induced mutations in the development of new ornamental varieties and how it can support the floriculture industry. Work has enriched very significantly different

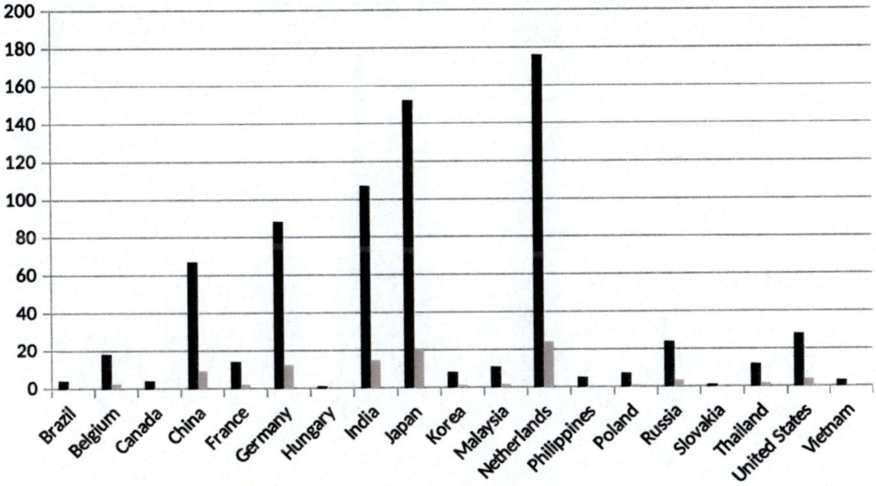

Fig. 3.13 Country wise registration of number ■ of mutants and percentage □ values

parameters of mutation technology and all mutation technology-related data have been generated from a large number of ornamentals (amaryllis, bougainvillea, canna, chrysanthemum, gerbera, gladiolus, hibiscus, mesembryanthemum, Narcissus tazetta, Perennial portulaca, Polyanthus tuberose, rose, Tagetes erecta, *Lantana depressa*, etc.). Efforts were most successful to produce more than 87 new promising mutant varieties in different ornamentals which are highly praiseworthy (Fig. 3.21).

As the author is a man of floriculture, all experiments were designed as per the requirements of the floriculture trade (Datta 2015). Entire mutation activities were designed in two directions. Fifty percent work was designed to generate basic knowledge/information and another 50% was applied. No experiment was done at random just to develop and increase the number of mutants. All the work in the laboratory was "Target Oriented" and "Need Base." Entire work results will be the focal point of discussion to give the complete picture of the present status of both basic and applied induced mutagenesis in ornamentals.

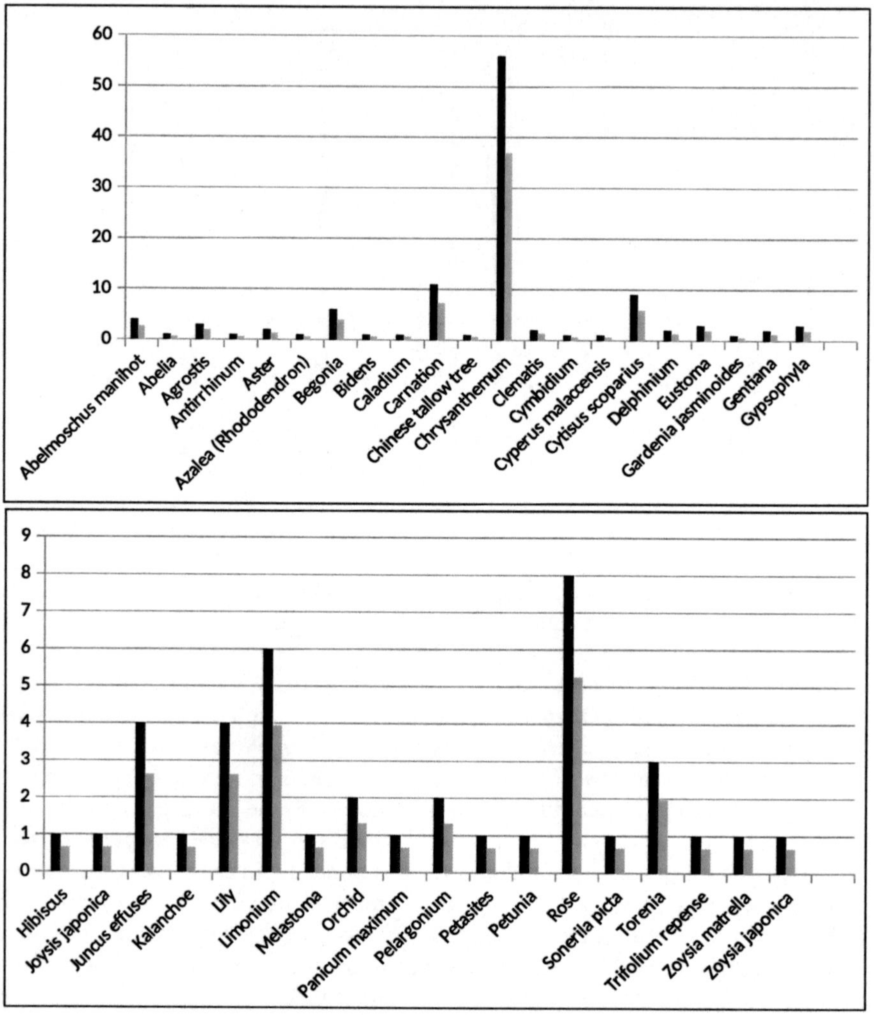

Fig. 3.14 Cropwise number of mutants and their percentage value reported from Japan

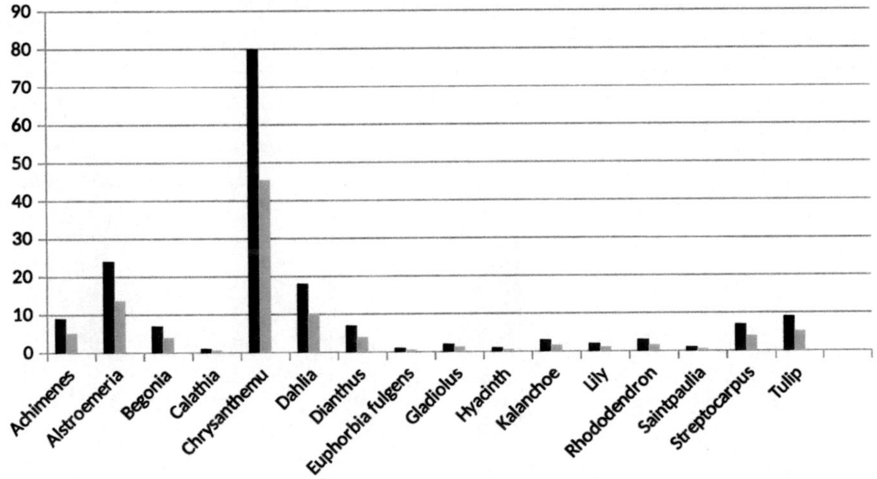

Fig. 3.15 Cropwise number ▬ of mutants and percentage value ▭ reported from the Netherlands

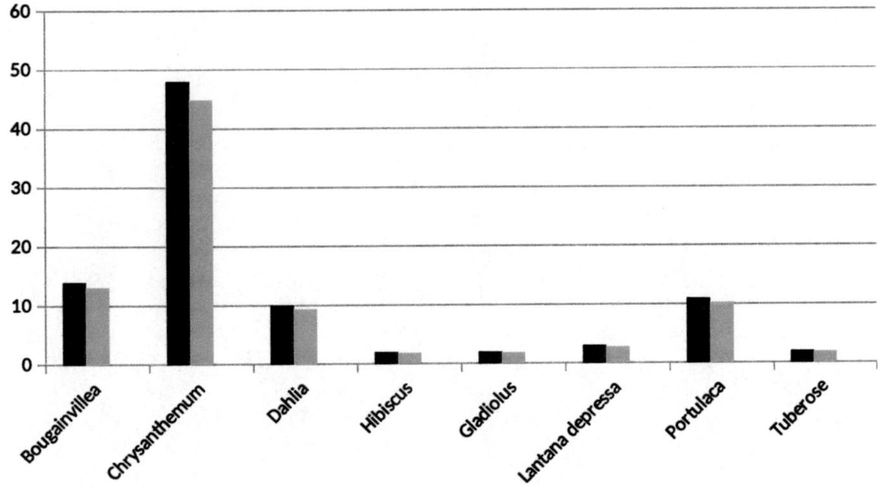

Fig. 3.16 Cropwise number ▬ of mutants and percentage value ▭ reported from India

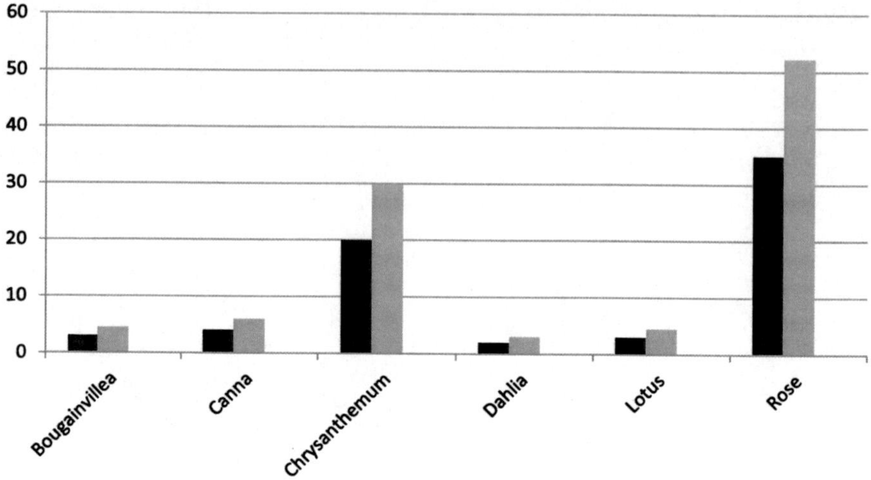

Fig. 3.17 Cropwise number ▬ of mutants and percentage value ▭ reported from China

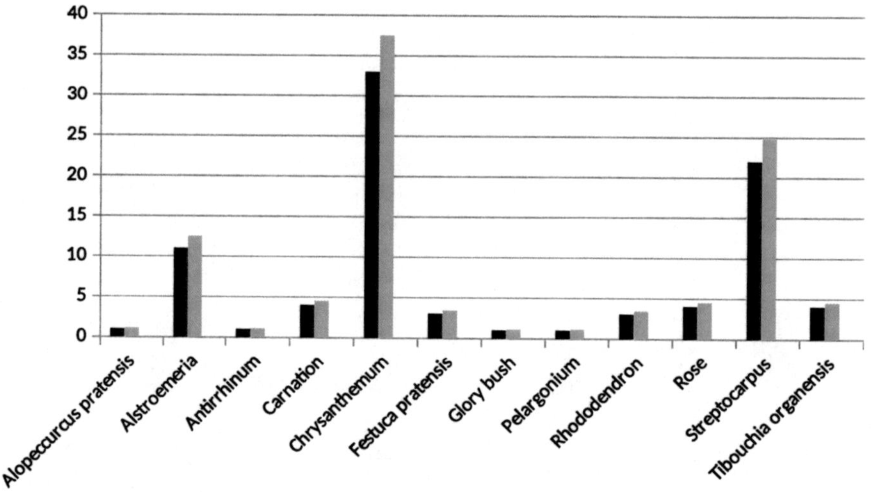

Fig. 3.18 Cropwise number ▬ of mutants and percentage value ▭ reported from Germany

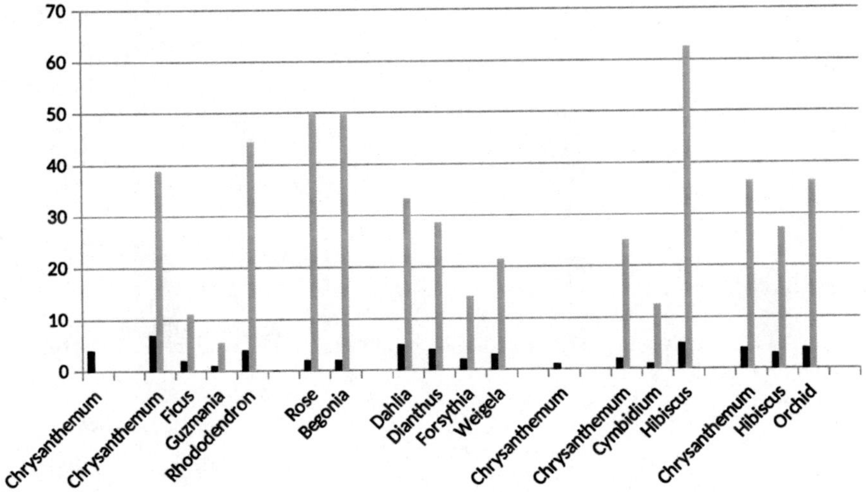

Fig. 3.19 Cropwise number ■ of mutants and percentage value ▭ reported from different countries

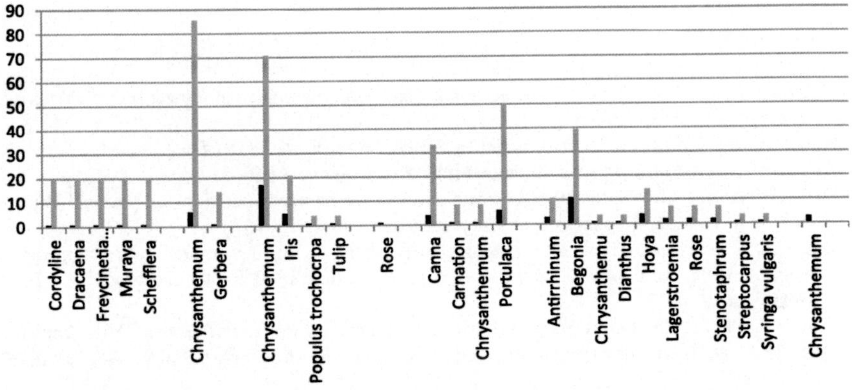

Fig. 3.20 Cropwise number ■ of mutants and percentage value ▭ reported from different countries

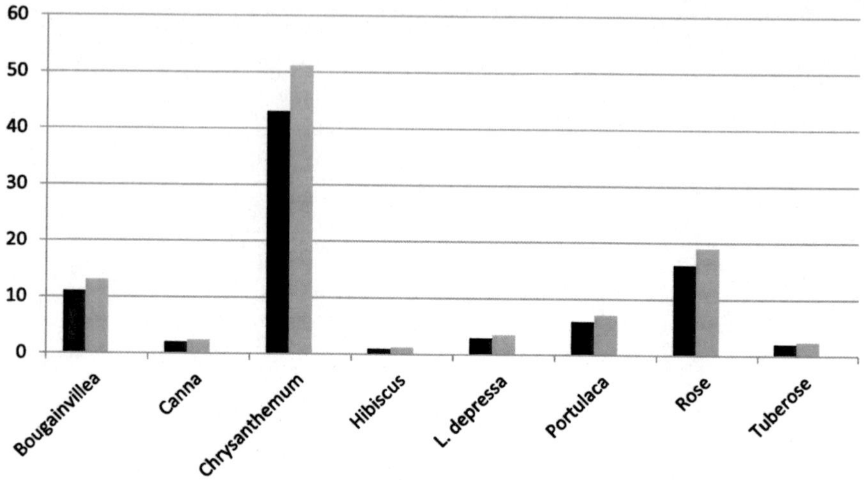

Fig. 3.21 Number ▬ of mutants and their percentage value ▭ of different ornamentals developed by author (CSIR-NBRI, Lucknow, India)

References

Ahloowalia BS, Maluszynski M, Nichterlein K (2004) Global impact of mutation-derived varieties. Euphytica 135:187–204

Anonymous (1977) List of mutant varieties. Mutat Breed Newsl 9(1):14–17

Anonymous (1994) List of new mutants. Mutat Breed Newsl 39:14–33

Datta SK (1988) Chrysanthemum cultivars evolved by induced mutations at National Botanical Research Institute, Lucknow. The Chrysanthemum 44(1):72–75

Datta SK (1990) Role of mutation breeding in floriculture. In: Plant mutation breeding for crop improvement, vol I. Proc of a Symp., Vienna, 18–22 June 1990, jointly organized by IAEA and FAO, pp 273–281

Datta SK (1997a) Ornamental plants—role of mutation. Daya Publishing House, Delhi, total, p 219

Datta SK (1997b) Mutation studies on garden roses: a review. Proc Indian Natl Sci Acad B63(1&2): 107–126

Datta SK (2000a) Mutation studies on garden chrysanthemum—a review. In: Singh SP (ed) Scientific horticulture, vol 7. Scientific Publisher, Jodhpur, pp 159–199

Datta SK (2000b) New chrysanthemum cultivars. In: Indian horticulture, pp 16–18

Datta SK (2010) Induced mutagenesis for the development of new varieties in vegetatively propagated ornamentals. In: NAARRI International Conference—2010 (NIC-2010), isotope technologies and applications—new horizons, December 13–15, 2010, Mumbai, organized by National Association for Applications of Radioisotopes & Radiation in industry (NAARRI), Mumbai, sponsored by Department of Atomic Energy, Government of India, vol 1. Invited talks, pp 227–230

Datta SK (2015) Indian floriculture; role of CSIR. Regency Publications, A Division of Astral International(P) Ltd, New Delhi. 432 pages

Datta SK, Mitra R (1999) Commercial floriculture in India. Appl Bot Abstr, IBIS, NBRI 19(3): 202–221

Gupta VN, Datta SK (2006) Chrysanthemum cut flower varieties developed at NBRI, Lucknow for the floriculture industry. J Rural Technol 3(1):37–41

Melsen K, van de Wouw M, Contreras R (2021) Mutation breeding in ornamentals. HortScience 56(10):1154–1165. https://doi.org/10.21273/HORTSCI16001-21

Schum A (2003) Mutation breeding in ornamentals: an efficient breeding method? In: Forkmann G et al (eds) Proc. 21st IS on classical/molecular breeding. Acta Hort 612, ISHS, pp 47–60

Schum A, Preil W (1998) Induced mutations in ornamental plants. In: Jain SM, Brar DS, Ahloowalia BS (eds) Somaclonal variation and induced mutations in crop improvement. Kluwer Academic Publishers, Dordrecht, pp 333–366

Bud Sports/Spontaneous Mutations

4

Abstract

Gardeners, horticulturists, amateur growers, and researchers have identified bud sports from time to time and propagated as new cultivars in clonally propagated plants as a chimera. This chapter covers how sports have contributed extensively to the generation of diversity, especially within vegetatively propagated ornamental species.

Keywords

Bud sports · Chimera · New variety · Diversity

Bud sport is a natural process of genetic mutation. Any standing plant variety generates a new shoot with changed morphological features from the original mother plant. The new feature may be in leaf character, flower character, branch character, or plant stature. The new branch is propagated by vegetative means and established as a new cultivar. Natural genetic variation performs a very meaningful function to produce a new trait through genetic variation. Spontaneous mutations have been observed in natural populations for a long time. This natural mutation process made radical changes from the wild phase to the cultivated mode by maintaining their new chimeric nature through vegetative propagation. Nurserymen recognized plant mosaics long back when they were aroused as bud sports with novel phenotypes (Darwin 1868). These spontaneous changes in phenotype were observed by Darwin, who named them "sports" and subsequently reported by many others (Carrière 1865; Cramer 1907; Shamel and Pomeroy 1936).

Gardeners, horticulturists, amateur growers, and researchers have identified bud sports from time to time and propagated as new cultivars in clonally propagated plants (Crane and Lawrence 1952) as a chimera. It is estimated that spontaneous mutation rates in higher eukaryotes range from 0.1 to 100 per genome per sexual

S. Datta, *Role of Mutation Breeding in Floriculture Industry*,
https://doi.org/10.1007/978-981-99-5675-3_4

generation. Bud sport contributed a maximum number of new varieties in ornamental crops. The most notable changed characters in ornamentals are changed flower color and bract color, new flower form, leaf variegation, plant stature, etc. A maximum number of sports varieties have been developed in bougainvillea, chrysanthemum, and rose. Sports arise commonly in polyploid and heterozygous plants and plants which perpetuate clonally. Although the rate of spontaneous mutation is low, chances of getting useful and uncommon features are more in ornamental crops. Sports have contributed extensively to the generation of diversity, especially within vegetatively propagated ornamental species (Schum and Preil 1998).

Research and utilization of plant chimera are very novel and inventive fields. The development of chimeric plants and their significance in the development of new ornamental varieties is very interesting. Variegation is a wonderful episode in the plant kingdom and is found in many plant families. A plant is called a chimera when cells of more than one genotype grow alongside the same meristem, organ, or tissue. A plant may be a chimera for almost any trait. Nurserymen recognized plant mosaics long back when they developed as bud sports with novel phenotypes (Darwin 1868; Grant 1975; Neilson-Jones 1969; Swingle 1927; Szymkowiak and Sussex 1996; Tilney-Bassett 1963, 1986). Plant chimera may evolve by grafting, spontaneous mutation, induced mutation, sorting-out from variegated seedlings, mixed callus cultures, or protoplast fusion (Stewart and Dermen 1970; Kameya 1975; Eyerdom 1981; Norris and Smith 1981; Norris et al. 1983; Tilney-Bassett 1963; Peary et al. 1988).

A wide range of leaf variegations and flower color characteristics have been developed in many plants through chimera. Mutated genes have affected flower colors by anthocyanin variegation on many ornamentals. Chimerism has significantly changed floral characteristics through the reorganization of histogenic layers. In chimeric breeding, the position of the epidermis and its rearrangements play a critical role to develop unique flower color, morphology, and fragrance. All present-day commercially available important chimeras in floriculture have been developed through spontaneous or induced mutations. These include both the variegated leaf patterns and flower color sports.

In ornamental floriculture, one major drawback is that nurserymen do not disclose the parents of bud sports, and therefore scientific literature on many bud sports has not been enriched. There is no centralized data bank to trace out all bud sport varieties and their evolutionary details. It is very tiresome to prepare the record on the correct number of bud sport varieties and their details of origin due to the non-availability of information on every bud sport. An attempt has been made to mention the bud sport varieties of different ornamentals from available and accessible literature. Names of some bud sports along with parental variety (in parenthesis), where available, in different ornamental species, are mentioned as a ready reference. The list of chimeric varieties is very big. To mention a few of the chimeras which have variegated foliage/flower color: Antirrhinum (De Vries 1905; Baur 1924), Mirabilis (Correns 1910; Kanna 1933), Pharbitis (Imai 1927, 1931, 1934, 1935), Portulaca (Ikeno 1929), Delphinium (Demereo 1931), Impatiens (Kanna 1934), "William Sim" carnation (Melquist et al. 1954), "Indianapolis" chrysanthemums

(Stewart and Dermen 1970), poinsettia (Stewart and Arisumi 1966), African violet (Lineberger and Druckenbrod 1985), Ficus (Beardsell and Norden 2004), *Cornus alba* "Argenteo Marginata," *Vinca minor* "Variegata," *Ajuga reptans* "Burgundy Glow," and many selections of *Hosta*, *Diffenbachia*, *Peperomia*, *Chlorophytum*, *Saintpaulia*, *Pittosporum tobira*, variegated bromeliad, *Begonia* (Bigot 1981; Westerhof et al. 1984), succulent leaves of many aloes, *Paphiopedium*, *Alpinia zerumbet* (shell ginger), *Pisonia umbellifera "variegata," Ficus aspera*, *Hedera helix*, *Abutilon*, *Euonymus japonica*, *Acalypha wilkesiana* "Marginata," *Calathea makoyana*, aroids, *Callisia elegans*, *Tradescantia zebrine*, Silybum marianum, *Ficus elastica*, *Ficus benjamina* "Golden Princess," etc.

As mentioned, all types of characters may be mutated spontaneously in nature. The contribution of spontaneous mutation in some commercially important ornamentals has been reported long back (Wasscher 1956): chrysanthemum (30%), carnation (25%), rose (40%), and begonia (70%). The most successful cultivar and "mutant family" in chrysanthemum was the cultivar "Reagan", "Sei-Rosa" in Japan, and its family, with more than 20 mutant cultivars and 400 million stems (representing 35–40% of the total Dutch market) being sold annually from 1992 to 1993 (Van Harten 2002). Few more bud sports are "William Sim" carnation (Melquist et al. 1954), "Indianapolis" chrysanthemums (Stewart and Dermen 1970), many poinsettia cultivars (Stewart and Arisumi 1966), African violet (Lineberger and Druckenbrod 1985), Ficus (Beardsell and Norden 2004 and (Tilney-Bassett 1963), Variegated perianth in wild Anemone coronaria (Horovitz and Zohary 1966), Rhododendron (Heursel 1980).

Lilium speciosum is one of the most important wild species used for breeding Oriental hybrid lily cultivars. *L. speciosum* flowers have red tepals and dark red anthers. Spontaneous mutation developed white-flowered *L. speciosum* with two distinct phenotypes, one with white tepals and dark red anthers and the other with white tepals and yellow anthers (Suzuki et al. 2015). Morimoto et al. (2020) detected pinkish-red flower carnation developed spontaneously through cell layer rearrangement from the deep pink cultivar "Feminine Minami."

Foliage plants such as *Ananas comosus*, *Aglaonema*, *Ficus* spp., *Hedera helix*, etc. are examples that are often changed directly to present status by bud sports. Some very important chimeric foliage, floricultural and landscape plants are *Cornus alba* "Argenteo Marginata"; *Vinca minor* "Variegata"; Ajuga reptans "Burgundy Glow"; many selections of cultivated forms of *Hosta*, Diffenbachia, Peperomia, Chlorophutum, Saintpaulia, *Pittosporum tobira*, variegated bromeliad, *Begonia* (Bigot 1981; Westerhof et al. 1984), succulent leaves of many aloes, *Paphiopedium*, *Alpinia zerumbet* (shell ginger), *Pisonia umbellifera "variegata," Ficus aspera*, *Saccharum officinarum*, a cultivar of *Coprosma* X *kirkii*, *Hedera helix*, *Abutilon*, cultivars of *Euonymus japonica*, *Acalypha wilkesiana* "Marginata," *Calathea makoyana*, the numerous species of cultivated aroids, *Callisia elegans*, *Tradescantia zebrine*, Silybum marianum, *Ficus elastica*, *Ficus benjamina* "Golden Princess," etc. are but a few of the chimeras which have variegated foliage. Bromeliad is known as "variegata" when it has two or more different colors. Over 60% of cultivated bromeliads have bands, dots, lines, and streaks and can therefore be considered variegated. However, the term is accepted in horticulture, when applied to bromeliads that have lines, streaks, and longitudinal bands of contrasting colors, especially those that show differences in pigmentation between the green chlorophyll-containing tissues and albino tissues.

Bougainvillea is a very high-ranking ornamental where the sequence of new bract color and chlorophyll-variegated mutant varieties has been developed through bud sports in *B. glabra* Choisy, *B. peruviana* Humboldt and Bonpland, and *B. spectabilis* Willdenow and *B. x buttiana* Holttum and Standley (Datta 2021). "Mrs. Butt" is an outstanding mutant that comprises high genetic heterozygosis-bearing genes of many ancestors. It is a miraculous variety that has developed a series of new bract color and chlorophyll-variegated varieties including double-bracted variety through bud sports like "Scarlet queen," "Surekha," "Scarlet queen variegate," "Versicolour," "Alick Lancaster," "Louise wathen," "Enid Lancaster," "Bhabha," "Louise wathen mediopicta," "Mrs. Butt variegatta," "Magenta Queen," "Purple King," "Purple Prince," "Rao," "Kuvempu," "Vellayani," "Mrs. McClean," "Yellow Queen," "Rosevilles delight," "Archana," "Mary Baring," "Golden Glow," "Lady Mary Baring," "Gangaswamy," "Gangamma," etc. (Banerji and Datta 1993; Datta 2021). The author evaluated 212 varieties of bougainvillea and found 31.60% of varieties developed through bud sport (Datta 2021).

Following are more new chlorophyll-variegated and new bract color bougainvillea cultivars developed through bud sports along with their parental varieties (in parenthesis) obtained from different sources "Abhimanya" ("Arjuna"), "Afterglow," "Agnihotri," "Alic Lancaster" ("Scarlet Queen"), "Anindita" (Refulgens), "Archana" (Roseville's Delight), "Bhabha" (Louise Wathen), "Brilliant Variegata," "Carmencita," "Cindrella," "Coconut Ice," "Daphene Mason," "Daya" (Partha), "Dona Rosita delight," "Doubloom," "Enid Lancaster" ("Louise Wathen"), "Easter Parade," "Fantasy" ("Peincess Margaret Rose"), "Gangamma" ("Lady Mary Baring"), "Gangaswamy" ("Lady Mary Baring"), "Gem," "Glabra variegatea" ("*B. glabra*"), "Godrej Cherry Blossom" ("Cherry Blossom"), "Golden Glow," "Golden Queen," "Hawaiian Beauty" ("Hawaiian White"), "Helen Coppinger" "Jane Stansfeld" (Refulgens), "Jawaharlal Nehru" (*B. spectabilis*), "Jayalakshmi Variegata" ("Jayalakshmi"), "Killie Campbell Variegata," "Kuvempu" ("Rao," "Lady Hudson of Ceylon Variegata" ("Lady Hudson of Ceylon"), "Laxminarayana" (*B. spectabilis*), "L.N. Birla" ("Maharaja of Mysore"), "Los Banos Beauty," "Louise Wathen" ("Mrs. Butt" or "Scarlet Queen"), "Louise Wathen Mediopicta" ("Louise Wathen"), "Louise Wathen Variegata" ("Louise Wathen"), "Magenta Queen" ("Mrs. Butt"), "Mahara," "Mahatma Gandhi Variegated" ("Mahatma Gandhi"), "Manohar Chandra Variegata" ("Manohar Chandra"), "Marietta" ("Mahara"), "Marigowda" ("Gladys Hepburn"), "Mary Palmer" ("Mrs. H.C. Buck"), "Mary Baring" ("Mrs. McClean"), "Mataji Agnihotri" ("Mary Palmer"), "Maude Chettebursh," "Meera Sport" ("Meera"), "Midget," "Mrs. Butt Magenta" ("Mrs. Butt"), "Mrs. Butt Scarlet" ("Mrs. Butt"), "Mrs. G.S. Randhwa" ("Killie Campbell"), "Mrs. McClean" ("Mrs. Butt"), "Mrs. McClean Nirmal" ("Mrs. McClean"), "Mrs. Eusenia," "Munivenkatappa" (Mary Palmer"), "Nirmal" ("Mrs. McClean"), "Odisee" ("Mary Palmer"), "Parthasarathy" ("Partha"), "Partha

Variegata" ("Partha"), "Preeti" ("Partha"), "Penelope," "Phoenix," "Purple Prince" ("Mrs. Butt"), "Purple Queen" ("Mrs. Butt"), "Raja Singhe," "Rao" ("Mrs. Butt"), "Refulgens variegata" ("Refulgen"), "Red Glory Improved" ("Red Glory"), Scarlet Glory" ("Padmi"), "Scarlet Queen Variegated" ("Scarlet Queen"), "Sharma" ("Mrs. Fraser"), "Shweta" ("Trinidad"), "Shubhra" ("Mary Palmer"), "Soundarya" ("Mrs. H.C. Buck"), "Spectabilis Variegated" (*B. spectabilis*), "Surekha" ("Scarlet Queen"), "Thimma" ("Mary Palmer"), "Vishakha" ("Mrs. H.C. Buck"), "Zakerina" ("Maharaja of Mysore"), "Zakir Hussain" ("Maharaja of Mysore"). "Purple King" ("Mrs. Butt"), "Shweta" ("Trinidad"), "Suvarna" ("Lady Hudson of Ceylon"), "Yellow Queen" ("Mrs. McClean") (Datta 2004, 2021; Singh et al. 1999; Sharma 1996; Banerji and Datta 1993; Marigowda and Gangaswamy 1974; Sharma and Yadav 1981; Srivastava 1977).

A good number of striped cultivars have been developed through bud sports: "Abhisarika," "Anant," "Anvil Sparks," "Careless Love," "Cynosure," "Chitrarekha," "Courageous Indira," "Calcutta 300," "Dattaji," "Dr. Noshir Wadia," "Festival Funfair," "Harry Wheatcroft," "Ico Trimurthi," "Jaslok," "Mahalaxmi," "Madhosh," "Modern Times," "Nav Sadabahar," "Narmada Lahari," "Orange Sparks," "Rare Edition," "Shoba," "Sahasradhara," "Sidhhartha," "Stars "N Strpes," "Strawberry Swirl," "Supriya," "Suvernarakha," "Tata Centenary," "Visveswarayya," "Vichitra," "Venu Vaishali," "Yashwant," etc. (Datta 2021).

Chrysanthemum is another important ornamental where a large number of new varieties have been developed through sports. It is very difficult to say the exact number of sport varieties in chrysanthemums as the frequency is very high and no authentic data is available. One report mentioned that approximately 30% of the chrysanthemum cultivars originated as sports (Wasscher 1956). Many of these varieties developed worldwide are very promising. Shamel and Pomeroy (1936) reported 400 varieties that originated through bud sports. A few excellent bud sport chrysanthemum varieties in India are "Kasturba Gandhi," "Sonar Bangla," "White Cloud," "Pink Cloud," "Sharda," "Queen of Tamluk," "R. Venkatraman," "William Turner," "J.S.Lloyd," "White Ball," etc. (Datta 1997).

Bud sports or spontaneous mutations have played an important role in the origin and evolution of garden gladioli. Many cultivars of gladioli are available in commerce and developed through spontaneous mutation (Anonymous 1972; Fisher and Lapins 1966). The first reports of such a mutation were GxColvillci albums and the bride both of which are white and mutated from bright scarlet Gx Colvillei. Color mutations in gladiolus are quite common, particularly in some varieties. Piendy, the most widely grown variety in the globe has produced several white mutants, namely Leading Lady, Silver Wing, Bingo, and Wanda while Lady Luck is a lighter colored pink than "Picandy." "Elizabeth the Queen" a beautiful lavender-colored variety has produced many white and white mutations as well as several dull white types. In all cases noted the new color is always lighter in color than the original verity. To propagate such (mutation) sports mutation must occur in corms and cormlets. More often they may be due to the expression of recessive mutants as has occurred in the case of recessive mutations for extending the blooming period (Anonymous 1972).

Rose: The origin of the moss rose was observed for the first time in 1696 as a mutant of *Rosa centifolia* (Hurst and Breeze 1922). Among 5819 rose cultivars marketed during 1937–1976, 865 were developed from bud mutations (Haenchen and Gelfert 1978). Bud Sport has developed a new rose variety not only with changed flower color but also with changed growth habits, i.e. climbing Hybrid Tea rose like "Crimson Glory," "Mrs. Sam McGredy," "Climbing Blue Moon," "Climbing Cinderella," "Climbing Fragrant Cloud," "Climbing Kronenbourg," "Climbing Mr. Lincoln," "Climbing Queen Elizabeth," "Climbing Zambra," "Climbing Guitare," "Climbing High Field," "Climbing Ladies" Choice," "Climbing Miss Harp," "Climbing Over the Rainbow," "Climbing Peace," "Climbing Rina Herholdt," "Climbing Sterling Silver," "Climbing Yellow Doll," "Climbing Zambra" etc. Reports support that 18% of the varieties in the Hybrid Tea group ("Mme Butterfly," "Lady Sylvia," "Rapture," "Better Times," "Jewel," "Royal Beauty," etc.) and about 54% of varieties Dwarf polyanthas group varieties ("Miss Edith Cavell," "Coral Cluster," "Juliana Rose," "Locarno," "Cameo," "Ideal," "Little Dorrit," etc.) have developed through bud sports. Bud sport resulted in the development of striped roses ("Careless Love" from "Red Radiance"; "Candy Stripe" from "Pink Peace"; "Banhar" from "Charlotte Armstrong"; "Harry Wheatcraft" from "Picadilly" etc.). Interestingly miniature rose varieties have been developed through this process ("Climbing Baby Darling," "Climbing Over the Rainbow," "Climbing Mary Marshall," "Climbing Yellow Doll," "Climbing Peace" etc.) (Osburn 1994; Gudin 2001). Sported varieties are equally stable as normal varieties. There are reports that such varieties persist for hundreds of years: "Winchester Cathedral," "Redoute," "Rose Marie," many sports of "Peace" ("Climbing Peace" and "Chicago Peace"), "Prairie Snowdrift," etc. Following rose varieties have also been developed through sports "Anand Rao," "Balwant," "Careless Love," "Chandralekha," "Chicago," "City of Lucknow," "Dazzling Flame," "Durgapur Delight," "Family Circle," "Harry Wheatcroft," "Hutton Village," "Janaki, Kanchani," "Nava Sadabahar," "Orange Sparks," "Pink Montezuma," "Priti," "Rose Bansal," "Sahasra Dhara," "Shanti," "Shirakawa Star," "Siddartha," "Tapti," "Tata Centenary," "White Queen Elizabeth," etc. (Narayana Gowda 1999; Datta 2018).

A good number of striped roses have been developed through bud sports. Many stripe roses are stable and some show variations. These varieties are "Abhisarika," "Anant" (1967), "Anvil Sparks" (1962, Coral red with golden yellow stripes and splashes), "Hassena"—white stripes on soft pink petals), "Anvil Sparks," "Careless Love" (1955, "Red Radiance"—pink with white stripes), "Cynosure," "Chitrarekha" (1982, "Chitralekha"—wine red stripes on pale pink), "Courageous Indira" ("Summer Holiday"—pink and white splashes on orange), "Calcutta 300" (1991, "Tajmahal"—white stripes on pink), "Dattaji" (1991, "Lusamba"—white and dull red stripes on dark red background), "Dr. Noshir Wadia" (1991, "Norma"—white stripes cross each other and form a mosaic structure on red petals)," Festival Funfair" (1986, "Fred Loads" —pale vermillion with creamy-white stripes), "Harry Wheatcroft" (1972, "Picadilly scarlet striped yellow), "Ico Trimurthi," "Jaslok" (1992, "Blue Ocean" —dark mauve stripes on light silvery mauve), "Mahalaxmi,"

"Madhosh," "Modern Times" (1956, "Better Times"—Small white stripes on red petals), "Nav Sadabahar" ("Sadabahar"—cream white striped on pink), "Narmada Lahari" (1976, "Dayanand"—white stripes on light pink), "Orange Sparks" (1969, "Cherry Brandy"—orange with gold stripes), "Rare Edition" (1983, "Kusum-luminous"—scarlet with white splashes and stripes), "Shoba" (1988, "Oto" hime—broad checkered lines of white and salmon over red), "Sahasradhara" (1981, "Century Two"—deep pink with white stripes), "Sidhhartha" (1973, "Christian Dior"—stripes and splashes of white on red petals), "Stars N Strpes," "Strawberry Swirl," "Supriya" (1984, "Princes Margaret of England"—pure pink with distinctive cream and white splashes and stripes), "Suvernarakha" (1981, "Soraya" —vermillion oranges red with stripes of golden yellow), "Tata Centenary" (1979, "Pigalle"—deep purple mauve splashed with pale yellow stripes), "Visveswarayya" (1987, "Tourmaline" —pink with white stripes), "Vichitra" (1985, "Gbriella" —vermillion red striped with apricot bronze), "Venu Vaishali" (1970, "Astree"—silvery pink steatched and veined white and yellow), "Yashwant" (1972, "Suspense"—stripes and splashes of white and yellow), etc.

Haenchen and Gelfert (1978) surveyed about 5819 rose varieties, marketed from 1937 until 1976, and found that 865 varieties developed from bud mutations, of which 289 were climbers. They noted differences in mutation frequency between classes and varieties. Several examples of "sports families" have been presented such as that of the cultivar Columbia (from 1917) with the greatest number of sports, namely 79 in 1978.

Somatic mutations have been recorded in bougainvillea, chrysanthemum, rose, dahlia, pelargonium, cosmos, etc. The chances to get such bud sports periodically are due to high heterozygosity. Therefore, bud sport is a continuous process and breeders should have keen observations of their germplasm stock to detect such natural changes. The available literature on spontaneous mutation is plenty and one can develop knowledge on different prospects of spontaneous mutation (Bull et al. 2007; Bradwell et al. 2013; Crane and Lawrence 1952; Crow and Kimura 1970; Cuevas et al. 2009, 2015; Dahiya et al. 1984; Doorenbos 1977; Drake 1991; Drake et al. 1998; Foster 2006; Foster and Wenseller 2006; Ganai and Johansson 2016; Gago et al. 2009; Kennedy et al. 2014; Kuchkarov and Ogurtsov 1987; Lynch 2010, 2011; Mergen 1963; Miller 1996; Ranel 1989; Rutger 1983; Sanjuán and Domingo-Calap 2016; Schmitt et al. 2012; Simmonds 1965; Schaapert et al. 1986; Nuraini et al. 2021).

References

Anonymous (1972) Gladiolus breeding. Splash, pp 45–47

Banerji BK, Datta SK (1993) 'Mrs. Butt'—bud sports and its near relatives. Sci Hortic 3:141–151

Baur E (1924) Untersuchungen das Wesen die Entstelung und die Verorhung ven Rassenunterschieden bei Antirrhinum majus. Bibliotheca Genetica 4:1–170

Beardsell D, Norden U (2004) Ficus rubiginosa 'Variegata', a chlorophyll-deficient chimera with mosaic patterns created by cell divisions from the outer meristematic layer. Ann Bot 94:51–58

Bigot C (1981) Multiplication vegetative in vitro de *Begonia* x *hiemalis* ('Rieger' et 'Schwabenland'). II. Conformite des plantes elevees en serre. Agronomie 1(6):441–447

Bradwell K, Combe M, Domingo-Calap P, Sanjuán R (2013) Correlation between mutation rate and genome size in riboviruses: mutation rate of bacteriophage Qβ. Genetics 195:243–251. PMID: 23852383

Bull JJ, Sanjuán R, Wilke CO (2007) Theory of lethal mutagenesis for viruses. J Virol 81:2930–2939. PMID: 17202214

Carrière EA (1865) Production et fixation des varietesdanslesvegetaux. Paris, 72p

Correns C (1910) Der Ubergang aus dem homozygotischen in cinen heterozygotischen Zustand in selben Individuum bei buntblottrigen und gestreiftbluhenden Mirabilis – Sippon. Ber Deutsch Bot Gesellsch 28:418–434

Cramer PJS (1907) Kritische Übersicht der bekanntenFälle von Knospenvariation. Natuurk. verh. v.d. Holl. Myd. Wetensch., Haarlem. VI, 474p

Crane MB, Lawrence WJC (1952) Genetics of garden plants. Macmillan, London

Crow JF, Kimura M (1970) An introduction in population genetics theory. Harper and Row, New York

Cuevas JM, González-Candelas F, Moya A, Sanjuán R (2009) The effect of ribavirin on the mutation rate and spectrum of hepatitis C virus *in vivo*. J Virol 83:5760–5764

Cuevas JM, Geller R, Garijo R, López-Aldeguer J, Sanjuán R (2015) The extremely high mutation rate of HIV-1 in vivo. PLoS Biol 13:e1002251. PMID: 26375597

Dahiya BS, Lather VS, Solanki IS, Kumar R (1984) Useful spontaneous mutants in chickpeas. Int Chickpea Newsl 11:4–7

Darwin C (1868) Variation of animals and plants under domestication, Parts I, II. Murray, London

Datta SK (1997) Mutation studies on garden roses: a review. Proc Indian Natl Sci Acad B63(1&2): 107–126

Datta SK (2004) Bougainvillea research at National Botanical Research Institute, Lucknow. J Ornam Hortic 7(1):1–14

Datta SK (2018) Breeding of new ornamental varieties: rose. Curr Sci 114(6):1194–1206

Datta SK (2021) Genetic diversity and improvement of bougainvillea. LS Int J Life Sci 10(2):61–79

De Vries H (1905) Species and varieties: their origin by mutation. Open Court Publishing, Chicago

Demereo M (1931) Behaviour of two mutable genes of Delphinium ajacis. J Genet 24:179–193

Doorenbos J (1977) Spontaneous mutation as a source of clonal variation on deciduous fruits. Acta Hort (ISHS) 75:13–18. http://www.actahort.org/books/75/75_1.htm

Drake JW (1991) A constant rate of spontaneous mutation in DNA-based microbes. Proc Natl Acad Sci U S A 88:7160–7164. PMID: 1831267

Drake JW, Charlesworth B, Charlesworth D, Crow JF (1998) Rates of spontaneous mutation. Genetics 148:1667–1686

Eyerdom H (1981) Flower color sports and variations in *Saintpaulia* hybrids. Afr Viol Mag:33–37

Fisher DV, Lapins KO (1966) Spur-type fruit varieties resulting from natural and induced mutation. In: Synge PM (ed) Fruit—present, and future. Meet. R. Hortic. Soc, pp 72–77

Foster PL (2006) Methods for determining spontaneous mutation rates. Methods Enzymol 409: 195–213

Foster KR, Wenseller T (2006) A general model for the evolution of mutualisms. J Evol Biol 19: 1283. https://doi.org/10.1111/j.1420-9101.2005.01073.x

Gago S, Elena SF, Flores R, Sanjuán R (2009) The extremely high mutation rate of a hammerhead viroid. Science 323:1308. PMID: 19265013

Ganai RA, Johansson E (2016) DNA replication—a matter of fidelity. Mol Cell 62(5):745–755

Grant V (1975) Mosaicism. In: Genetics of flowering plants. Columbia Univ. Press, New York, pp 278–299

Gudin S (2001) Rose breeding technologies. Acta Hortic 547:23–26

Haenchen E, Gelfert E (1978) Der Anteil der Mutationen an der Entstehung des Rosensortimentes unter besonderer Berücksichtigung des Zeitraumes der letzten 40 Jahre. Archiv für Gartenbau 26:231–241

Heursel J (1980) Sportvorming bij Rhododendron simsii Planch. (Azalea indica L.): een synthese. Landbouwtijdschrift 33:985–994

Horovitz A, Zohary D (1966) Spontaneous variegation for perianth color in wild Anemone coronaria. Heredity. 21:513–515. https://doi.org/10.1038/hdy.1966.49. Corpus ID: 41825958

Hurst CC, Breeze MSG (1922) Notes on the origin of the moss rose. J RHS:26–42. Cited in: Krüssmann G (1974) Rosen, Rosen, Rosen. Paul Parey, Berlin, Hamburg

Ikeno S (1929) Studien uber die Vererbung der Blutenfarhe bei portulaca grandiflora. III. Mitt. Mosaikfarbe. Jpn J Bot 4:189–217

Imai Y (1927) Vegetative and seminal variegations were observed in the Japanese morning glory, with special reference to its evolution under cultivation. J Coll Agric Tokyo Imp Univ 9:223–274

Imai Y (1931) Analysis of flower color in Pharbitis nil. J Genet 24:203–224

Imai Y (1934) On the mutable genes of Pharbitis with special reference to their bearing on the mechanism of bud-variation. J Coll Agric Tokyo Imp Univ 12:479–523

Imai Y (1935) Variegated flowers and their derivatives by bud variation. J Genet 30(1):1–13

Kameya T (1975) Culture of protoplasts from chimeral plant tissue of nature. Jpn J Genet 50(5): 417–420

Kanna B (1933) On the striped flowers of Mirabilis jalapa L. Jpn J Genet 8:165–178

Kanna B (1934) Genetic studies in Impatiens balsamina. J Coll Agric Tokyo Imp Univ 12:421–477

Kennedy SR, Schmitt MW, Fox EJ, Kohrn BF, Salk JJ, Ahn EH et al (2014) Detecting ultralow-frequency mutations by duplex sequencing. Nat Protoc 9:2586–2606. PMID: 25299156

Kuchkarov U, Ogurtsov KU (1987) Spontaneous mutants of mulberry. Shelk 6:3–4

Lineberger RD, Druckenbrod M (1985) Micropropagation of chimeral African violets. http://aggie-horticulture.tamu.edu/tisscult/chimeras/valprop/val.html

Lynch M (2010) Evolution of the mutation rate. Trends Genet 26:345–352. PMID: 20594608

Lynch M (2011) The lower bound to the evolution of mutation rates. Genome Biol Evol 3:1107–1118. PMID: 21821597

Marigowda MH, Gangaswamy (1974) Variegated bougainvillea. Lalbaugh J Mysore Hort Soc 19(3&4):27–29

Melquist GAL, Ober D, Sagawa Y (1954) Somatic mutations in the carnation, *Dianthus caryophyllus* L. Proc Natl Acad Sci U S A 40:432–436

Mergen F (1963) Evaluation of spontaneous, chemical, and radiation-induced mutations in the Pinaceae. BNL-6896, Brookhaven Natl. Lab., Upton, Long Isl., N.Y. 27 pp

Miller D (1996) Configurations revisited. Strateg Manag J 17:505. https://doi.org/10.1002/(SICI)1097-0266(199607)17:7<505::AID-SMJ852>3.0.CO;2-I

Morimoto H, Narumi-Kawasaki T, Takamura T, Fukai S (2020) Flower color mutation is caused by spontaneous cell layer displacement in carnation (*Dianthus caryophyllus*). Plant Sci 299(2020): 110598. https://doi.org/10.1016/j.plantsci.2020.110598

Narayana Gowda JV (1999) Recent advances in rose breeding. Indian Rose Annu XV:42–43

Neilson-Jones W (1969) Plant chimeras, 2nd edn. Methuen & Co. LTD, London

Norris R, Smith RH (1981) Regeneration of variegated African violet (*Saintpaulia ionantha* WENDL.) leaf chimeras in culture (Abstr.). Plant Physiol 67:117

Norris R, Smith RH, Vaughn KC (1983) Plant chimeras are used to establish de novo origin of shoots. Science 220:75–76

Nuraini L, Tatsuzawa F, Ochiai M, Suzuki K, Nakatsuka T (2021) Two independent spontaneous mutations related to anthocyanin-less flower coloration in *Matthiola incana* cultivars. Hortic J 90(1):85–96. https://doi.org/10.2503/hortj.UTD-212

Osburn W (1994) Watch for mini mutations. American Rose, pp 16–18

Peary JS, Lineberger RD, Malinich TJ, Wertz MK (1988) Stability of leaf variegation in *Saintpaulia ionantha* during in vitro propagation and during the chimeral separation of a pinwheel flowering form. Am J Bot 75(5):603–608

Ranel C (1989) Mutation research/fundamental and molecular mechanisms of mutagenesis. Mutat Res 212(1):33–42

Rutger JN (1983) Applications of induced and spontaneous mutation in rice breeding and genetics. Adv Agron 36:383–413
Sanjuán R, Domingo-Calap P (2016) Mechanisms of viral mutation. Cell Mol Life Sci 73:4433–4448. PMID: 27392606
Schaapert M, Danforth BN, Glickmanx BW (1986) Mechanisms of spontaneous mutagenesis: an analysis of the spectrum of spontaneous mutation in the Escherichia coli lacI gene Roe1. J Mol Biol 189:273–284
Schmitt MW, Kennedy SR, Salk JJ, Fox EJ, Hiatt JB, Loeb LA (2012) Detection of ultra-rare mutations by next-generation sequencing. Proc Natl Acad Sci U S A 109:14508–14513. PMID: 22853953
Schum A, Preil W (1998) Induced mutations in ornamental plants. In: Jain SM, Brar S, Ahloowalia (eds) Somaclonal variation and induced mutations in crop improvement. Kluwer Academic Publishers, Dordrecht, pp 333–366
Shamel AD, Pomeroy CS (1936) Bud mutations in horticultural plants. J Hered 27:487–494
Sharma SC (1996) Bougainvilleas in India. Economic Botany Information Service, N.B.R.I., Lucknow
Sharma SC, Yadav SN (1981) Variegated Bougainvillea. Lal Bagh J Mysore Hort Soc 26(3):35–38
Simmonds NW (1965) Chimeral potato mutants. J Hered 56:139–144
Singh B, Panwar RS, Voleti SR, Sharma VK, Thakur S (1999) The International Bougainvillea Check List. Division of Floriculture and Landscaping, IARI, New Delhi
Srivastava GS (1977) A new variegated multibracted Bougainvillea. Delhi Garden Magazine
Stewart RN, Arisumi T (1966) Genetic and histogenic determination of pink bract colour in poinsettia. J Hered 57:216–220
Stewart RN, Derman H (1970) The origin of adventitious buds in Chrysanthemum. Am J Bot 57:734–735
Suzuki K, Tasaki K, Yamagishi M (2015) Two distinct spontaneous mutations involved in white flower development in Lilium speciosum. Mol Breed 35:193. https://doi.org/10.1007/s11032-015-0389-z
Swingle CF (1927) Graft hybrids in plants. J Hered 18:73–94
Szymkowiak EJ, Sussex IM (1996) What chimeras can tell us about plant development. Annu Rev Plant Physiol Plant Mol Biol 47:351–376
Tilney-Bassett RAE (1963) The structure of periclinal chimeras. Heredity 18:265–285
Tilney-Bassett RAE (1986) Plant Chimeras. Edward Arnold, London, 199 pp
Van Harten AM (2002) Mutation breeding of vegetatively propagated ornamentals. In: Vainstein A (ed) Breeding for ornamentals: classical and molecular approaches
Wasscher J (1956) The importance of sports in some florist's flowers. Euphytica 5:163–170
Westerhof J, Hakkart FA, Joke Versluijs MA (1984) Variations in two Begonia X Hiemalis clones after in vitro propagation. Sci Hortic 24:67–74

Check for updates

Adventitious Bud Technique in Mutation Breeding

5

Abstract

Experimental results motivated scientists to use the adventitious bud technique for creating solid mutants as the plant develops from a single cell. Several ornamental plants have been identified where adventitious buds develop from different plant parts. This chapter reports the results of experiments conducted by several scientists indicating that chimera formation can be prevented/decreased and the probability of obtaining a complete mutant is to a greater extent.

Keywords

Adventitious buds · Chimera · Solid mutant · New variety

The development of chimera is the main stumbling block to getting solid mutants in vegetatively propagated ornamentals. After mutagen treatment, there is always the development of chimera in a multicellular organism. Such unpleasing conditions can be bypassed by developing a complete shoot/plant from only one cell. Irradiation of a single cell may result in either formation of normal or solid mutant and more mutation spectrum and mutant plants (Broertjes 1967, 1968a, b). The concept of mono-cell culture and the application of tissue culture for chimera management were not matured at an early stage. Efforts were made to find out the experimental material (propagule) which develops from a single cell or few cells to avoid such chimera formation. Experimental results motivated scientists to use the adventitious bud technique for creating solid mutants as the plant develops from a single cell (Broertjes 1982; Broertjes et al. 1968, 1976; Broertjes and Keen 1980; Broertjes and Van Harten 1985, 1988; Duron 1992; Haccius and Hausner 1974; Mikkelsen and Sink 1978; Parliman and Stushnoff 1979; Roest and Bokelmann 1975; Stangler 1956). Results of experiments conducted by several scientists indicated that chimera

S. Datta, *Role of Mutation Breeding in Floriculture Industry*,
https://doi.org/10.1007/978-981-99-5675-3_5

formation can be prevented/decreased and the probability of obtaining a complete mutant is to a greater extent.

Several ornamental plants have been identified where adventitious buds develop from different plant parts like leaves (achimenes, begonia, carnation, chrysanthemum, dendrophthoe, kalanchoe, lachenalias, saintpaulia, streptocarpus, etc.—Naylor and Johnson 1937; Sparrow et al. 1960; Broertjes 1968a, b, 1969a, b, 1972; Wongpiyasatid et al. 2007; Messeguer et al. 1993; Park et al. 2005; Broertjes and Ballego 1967; Wilden 1934; Nag and Johri 1970; Broertjes and Leffring 1972; Cook 1930; Marston 1962, 1964); bulb scales (Amaryllis, daffodil, hyacinth, iris, Lilium, nerine, tuip, etc.—Luyten 1935; Traub 1935; Alkema 1971a, b; Sytsema 1977); stem-segment (gypsophila, torenia—Ahroni et al. 1997; Takeuchi et al. 1985); roots (phlox), hypocotylary buds (euphorbia, Lilium, linaria—Broertjes et al. 1968). A definite number of cells involved in the formation of adventitious shoots have not been worked out in all cases. Broadly, the development of non-chimeral mutants may help to anticipate the number of cells involved in the development of adventitious buds. The involvement of only one epidermal cell has been noted in the development of adventitious buds in Saintpaulia, Begonia, and Streptocarpus (Sparrow et al. 1960; Naylor and Johnson 1937; Broertjes 1968a).

Few early works on this topic need to acknowledge which generated meaningful basic knowledge. Maximum adventitious solid mutants were obtained in streptocarpus, achimenes, and kalanchoe after irradiating half-leaves with X-rays (Broertjes 1969a, b, 1972; Brown 1971; Broertjes and Leffring 1972). Irradiation of detached begonia leaves produced hundreds of solid mutants (Doorenbos and Karper 1975; Mikkelsen et al. 1975). Role of adventitious buds when develop from more than one cell in the development of mutants has been studied in chrysanthemum, carnation, poinsettia, gerbera, carnation, gypsophila, etc. (Stewart and Dermen 1970a, b, c; Park et al. 2005, 2007; Watad et al. 1996; Love 1972; Breutmann 1976; Ahroni et al. 1997; Messeguer et al. 1993; Zalewska 2010). Tymoszuk and Zalewska (2014) and Zuker et al. (1997) developed in vitro adventitious shoots from ligulate florets of *Chrysanthemum ×grandiflorum* (Ramat.) (some varieties). The subject has been enriched by many other early works (Bain 1940; Broertjes 1979, 1982; Broertjes and Alkema 1970; De Nettancourt et al. 1971; Broertjes and Keen 1980; Broertjes and Lock 1984; Bergann and Bergann 1982; Deutch 1974; Duron 1992; Heide 1965; Holm 1925; Norris et al. 1983; Smith and Norris 1983; Lineberger and Druckenbrod 1985; Eyerdom 1981; Broertjes and Van Harten 1985; Marcotrigiano and Stewart 1984; Ando et al. 1986; Marcotrigiano 1986; Smith and Norris 1983; Shen et al. 1990; Stewart and Dermen 1970a, b, c; Sripichitt et al. 1988; Vainstein et al. 1992; Yang and Schmidt 1994; Pinet-Leblay et al. 1992).

Plants having the characteristic of developing adventitious buds from a single cell are very useful in mutation breeding. The theory of inducing non-chimeral mutants from adventitious buds was very desirable in induced mutagenesis but this technique has not been exploited in routine practical work. Adventitious bud development in some ornamentals is cited as a ready reference.

African violet (Saintpaulia): Earlier Norris et al. (1983) reported that the adventitious buds in African violet are of multicellular origin. Broertjes and Van Harten

(1985) from their experiments confirmed that the adventitious buds are of single-cell origin. Sports developed in edged and Geneva-edged types of saintpaulia cultivars during vegetative propagation. The sports were believed to be periclinal chimeras considering their frequency and definite trend in origin. The mechanism of persisting chimeric structures during vegetative propagation has been discussed about the origin of adventitious shoots (Ando et al. 1986).

Wongpiyasatid et al. (2007) treated leaf cuttings *S. ionantha* (African violet) cv. Optima Hawaii (purple flower) with gamma rays (0, 10, 20, 40, 60, 80, and 100 Gy) to induce mutation. Treated leaves were planted in plastic trays containing peat moss medium and were placed in a shaded greenhouse. Several adventitious plantlets per irradiated leaf were developed. The number of plantlets per leaf decreased as radiation doses increased. There was no survival of leaves at doses higher than 80 Gy.

Begonia: Auxins, cytokinins, and other growth regulators have been tested to manipulate the formation of adventitious buds in Begonia (Heide 1965). Callus formation took place at the basal portion of the leaf petiole of Rieger Begonia cv. "Aphrodite Peach." Roots formed from cells of the internal portion of the callus and also from the parenchymal cells of the petiole. Shoots formed from cells on the surface of the enlarging callus. Histological observations on adventitious shoot formation have been discussed about mutation breeding (Mikkelsen and Sink 1978).

Carnation (*Dianthus caryophyllus* L.): Leaves, basal segments of the flower, and petals of *Dianthus caryophyllus* L. cultivar "Scania" were cultured on MS medium containing various concentrations of BA and NAA to study the adventitious shoot regeneration. Petals and floral segments exhibited a high morphogenetic capacity. Basal segments were the only part of the leaf that generated shoots when cultured in a medium supplemented with 0.01 mg L^{-1} NAA and l mg L^{-1} BA (Messeguer et al. 1993).

Vainstein et al. (1992) developed an efficient method for adventitious shoot regeneration from carnation petals. Ninety percent of the petals developed up to seven shoots.

Watad et al. (1996) standardized adventitious shoot regeneration from stem explants of carnation (*Dianthus caryophyllus* L., cultivar "White Sim") using three different culture procedures (agar-gelled medium, liquid-shaken medium, and an interfacial membrane raft floating on liquid medium). Explants derived from the first upper internode exhibited higher adventitious shoot formation than those from the second internode in all culture procedures. A comparison of the overall efficiencies of shoot regeneration for each culture procedure showed that maximum shoot regeneration on the raft was twice and three times that obtained with agar-gelled and liquid-shaken media, respectively.

Chrysanthemum: Stewart and Dermen (1970a, b, c) studied the genetic potential for the color of each of the three layers in the apical meristem of their shoots in the sports "Indianapolis" family of chrysanthemums. Adventitious buds were developed from the stems of each cultivar by excising all normal lateral buds. The position of pigment within cells and tissues was examined from fresh petals. Studies showed 12 of the 16 "Indianapolis" cultivars to be periclinal chimeras. Adventitious shoots often originated from two or more cells. The most frequent type of sporting resulted from the loss in mitosis of a chromosome carrying a suppressor for the formation of yellow chromoplasts, giving a yellow sector or shoot. Sectors resulting from the rearrangement of layers in the periclinal chimeras were less frequent than the sectors from chromosome loss. Roest and Bokelmann (1975) examined adventitious bud formation in two (cultivars "Super Yellow" and "Bravo") and then 11 cultivars in vitro under the influence of the explant length, growth regulators, sugar, vitamins, and minerals. The significance of this method for vegetative propagation, storage, and mutation breeding has been discussed. Park et al. (2005) made a comparative study on adventitious shoot regeneration from three explants (petals, stems, and leaves) of chrysanthemum (CVS. Orlando, Klondike, and Pink Pixie Time). Shoots were regenerated on a modified MS medium (MS salts, B5 vitamins, and 0.8% agar) supplemented with various combinations of benzyladenine, and kinetin IAA. The development of the adventitious shoot was highly influenced by plant growth regulators, sucrose, and dark period and Orlando developed the greatest number of shoots per explant. All regenerated plants showed the same morphological characteristics of vegetative organs compared to those of the stock plant. Park et al. (2007) studied adventitious shoot regenerations of chrysanthemum cultivars ("Klondike," "Pink Pixie Time," and "Orlando") through in vitro culture in presence of different concentrations and combinations of IAA, BAP, and kinetin. For the experiment two stage groups of petals (Stage I, 2–3 days before ray flowers were fully open and Stage II, 7–8 days after tubular flowers opened). Stage I was found to be more organogenic. Tymoszuk and Zalewska (2014) determined the effect of various factors on the number and length of shoots regenerating in vitro from ligulate florets after culturing in an MS medium supplemented with cytokinin and auxin in different combinations. There was determined the effect of various factors on the number and length of shoots regenerating in vitro from ligulate florets. No differences were observed in the number and length of shoots regenerating on ligulate florets inoculated on solid or in liquid MS medium with 8.88 μM dm^{-3} BAP and 2.69 μM dm^{-3} NAA. The subculture of regenerating ligulate florets from the solid into a liquid medium increases the number of regenerating shoots and stimulates their elongation growth; however, these shoots are deformed.

Dendrophthoe falcata: Nag and Johri (1970) studied the effects of cytokinins to induce shoot buds on leaves obtained from embryonal tissue (diploid) or endosperm tissue (triploid). No buds developed in absence of cytokinins. The buds develop either by division of an epidermal cell which finally is organized into a shoot meristem, or the epidermal cell first produces a callus which subsequently gives rise to shoot buds. Injury to the leaf plays a major role in the distribution and number of shoot buds formed. Normal leaves on the plant do not regenerate buds even in the presence of a cytokinin.

Gypsophila paniculata: For regeneration of adventitious shoot, three upper internodes of the stem of Gypsophila paniculata L. cultivar Arbel were cultured on MS media supplemented with different cytokinins (thidiazuron, benzyladenine, kinetin, or zeatin) and an auxin (naphthaleneacetic acid). Thidiazuron was found to be the most efficient where up to 100% of the explants developed shoots. The highest percentage of shoot formation was observed in the stem explants originating from the first internode, with all cytokinins tested (Ahroni et al. 1997).

Kohleria eriantha: Parliman and Stushnoff (1979) studied growth in terms of root and adventitious bud formations after treating freshly cut leaves of *Kohleria eriantha* and *K.* × "Longwood" with gamma irradiation. *K. Eriantha* could not be propagated from leaf half cuttings and "Longwood" produced a small number of adventitious plantlets. Plantlet development increased in low and moderate doses. Effects of colchicine and detailed analysis of mutations could not determine the organogenesis of adventitious buds from "Longwood" leaf halves. Adventitious buds may have developed from callus at the petiole and leaf-vein bases. Mutant plantlets formed from adventitious buds on detached leaf halves of "Longwood" appeared to arise from single cells.

Torenia fournieri Lind.: Stem segments of *T. fournieri* Lind were cultured in presence of cytokinin to induce adventitious buds. A large number of buds were formed in the marginal regions of the cut ends of explants and only a few buds were initiated in the middle part of the explants. A significant increase in the number of buds was noted when a slight injury was made transversely in the central part of the explant. When a wounding treatment was given lengthwise to an explant, a large number of adventitious buds were formed over the entire surface of the explant compared to the control. These wounding treatments did not affect the uptake into explants and/or the distribution pattern of radioactive benzyladenine applied to the culture medium (Takeuchi et al. 1985).

Weigela: Shoot internodes of in vitro propagated Weigela plants were treated with 0.50% EMS solution in 3% DMSO for 90 min to induce mutations and adventitious buds. Out of 388 acclimatized plants, six solid mutants were detected. Two were homogeneous mutants and the others appeared on previous normal-looking plants. The possibility to improve the low percentage of mutants has been discussed (Duron 1992).

References

Ahroni A, Zuker A, Rozen Y et al (1997) An efficient method for adventitious shoot regeneration from stem-segment explants of gypsophila. Plant Cell Tissue Organ Cult 49:101–106

Alkema HY (1971a) Nieuwe vermeerderingsmethoden bij bolgewassen. Weekbl Bloembollencult 81(46):1211

Alkema HY (1971b) Veredeling van bolgewassen door bestraling. Weekbl Bloembollencult 81(48):1262–1263

Ando T, Akiyama Y, Yokoi M (1986) Flower colour sports in *Saintpaulia* cultivars. Sci Hortic 29(1/2):191–197

Bain HF (1940) Origin of adventitious shoots in decapitated cranberry seedlings. Bot Gaz (Chicago) 101:872–880

Bergann F, Bergann L (1982) Zur Entwicklungeschichte des Angiospermenblattes. Uber Periklinalchimaren bei Peperomia und ihre experimentelle Entmischung und Umlagerung. Biol Zentralbl 101(4):485–502

Breutmann B (1976) Vegetative vermehrung von *Gerbera jamesonii* Gartenbauliche Versuchsber. Rheinland 15:216–219

Broertjes C (1967) Progress in mutation breeding. Euratom Bull VI(2):45–51
Broertjes C (1968a) Dose rate effects in *Saintpaulia*. In: Mutations in plant breeding II (Proc. Panel Vienna, 1967). IAEA, Vienna, pp 63–71
Broertjes C (1968b) Mutation breeding of vegetatively propagated crops. In: Fifth Congr. of the European Asso. for Res. on Plant Breed, Milano, pp 141–165
Broertjes C (1969a) Mutation breeding of vegetatively propagated crops. Genet Agraria 23:39–165
Broertjes C (1969b) Induced mutations and breeding methods in vegetatively propagated species. Induced mutations in plants. IAEA, Vienna, pp 325–329
Broertjes C (1972) Mutation breeding of *Achimenes*. Euphytica 21:48–63
Broertjes C (1979) The improvement of *Chrysanthemum morifolium* Ram. by induced mutations. In: Proc Eucarpia Meeting on Chrysanthemum. Littlehampton 1978, pp 93–102
Broertjes C (1982) Interessante ontwikkelingen in sortiment *Streptocarpus*. Vakbl Bloemisterij 10: 36–37
Broertjes C, Alkema HY (1970) Mutation breeding in flower bulbs. In: First Int. Symp. On Flowerbulbs, Noordwijk/Lisse II, pp 407–411
Broertjes C, Ballego JM (1967) Mutation breeding of *Dahlia variabilis*. Euphytica 16:171–176
Broertjes C, Keen A (1980) Adventitious buds: do they develop from one cell? Euphytica 29:73–87
Broertjes C, Leffring L (1972) Mutation breeding of *Kalanchoe*. Euphytica 21:415–423
Broertjes C, Lock CAM (1984) The use of irradiated soil in establishing *in vitro* adventitious plantlets of *Chrysanthemum morifolium* Ram. Cv. Parliament. Acta Bot Neerl 33(3):375
Broertjes C, Van Harten AM (1985) Single cell origin of adventitious buds. Euphytica 34:93–95
Broertjes C, Van Harten AM (1988) Applied mutation breeding for vegetatively propagated crops. Elsevier, Amsterdam. 316 pp
Broertjes C, Haccius B, Weidlich S (1968) Adventitious bud formation on isolated leaves and its significance for mutation breeding. Euphytica 17:321–344
Broertjes C, Roest S, Bokelmann GS (1976) Mutation breeding of *Chrysanthemum morifolium* Ramat. Using in vivo and in vitro adventitious bud techniques. Euphytica 25:1–19
Brown AG (1971) The effect of supplementary light on *Streptocarpus* 'constant nymph'. John Innes Inst Annu Rep 62:50–55
Cook HH (1930) Propagation of Lachenalias by leaf cuttings. Card Chron 88:52
De Nettancourt D, Dijkuis P, Van gastel AJG et al (1971) The combined use of leaf irradiation and the adventitious bud technique for inducing and detecting polyploidy, marker mutations, and self-compatibility in clonal populations of *Nicotiana alata* Link and Otto. Euphytica 20(4): 508–520
Deutch B (1974) Bulblet formation in *Achimenes longiflora*. Physiol Plant 30(2):113–118
Doorenbos J, Karper JJ (1975) X-ray induced mutations in *Begonia* x hiemalis. Euphytica 24(1): 13–19
Duron M (1992) Induced mutations through EMS treatment after adventitious bud formation on shoot internodes of *Weigela* cv. Bristol Ruby. Acta Hort 320:113–118
Eyerdom H (1981) Flower color sports and variations in *Saintpaulia* hybrids. Afr Violet Mag 34(4): 32–37
Haccius B, Hausner G (1974) Adventivknospen und nicht-zygotische Embryonen—Grundlagen und Anwendung. Vortr. 22. Vortragatag. Ges. Arzneipflanzenforsch., Tubingen, pp 1–10
Heide OM (1965) Interaction of temperature, auxins, and kinins in the regeneration ability of begonia leaf-cuttings. Physiol Plant 18:891–920
Holm Th (1925) On the development of buds upon roots and leaves. Ann Bot 29
Lineberger RD, Druckenbrod M (1985) Chimeral nature of the pinwheel flowering African violets (*Saintpaulia*; Gesneriaceae). Am J Bot 72:1204–1212
Love JE (1972) Somatic mutation induction in poinsettia and sweet potato. In: Constantin MJ (ed) Mutat. Breed. Workshop, Knoxville. Univ of Tennessee, Knoxville, TN
Luyten L (1935) Vegetative propagation of *Hippeastrum*. Yearb Am Amaryllis Soc 2:115–122
Marcotrigiano M (1986) Origin of adventitious shoots regenerated from cultured tobacco leaf tissue. Am J Bot 73(11):1541–1547

Marcotrigiano M, Stewart RN (1984) All variegated plants are not chimeras. Science 223(4635): 505

Marston ME (1962) The propagation of plants from leaf cuttings, with special reference to *Streptocarpus*. In: XVIth Intern. Hort. Congr., Brussels, pp 33–40

Marston ME (1964) The morphology of a *Streptocarpus*-hybrid and its regeneration from leaf cuttings. Sci Hortic 17:114–120

Messeguer J, Arconada MC, Mele E (1993) Adventitious shoot regeneration in carnation (*Dianthus caryophyllus* L.). Sci Hortic 54:153–163

Mikkelsen EP, Sink KC (1978) Histology of adventitious shoot and root formation on leaf-petiole cuttings of *Begonia* x hiemalis FOTSCH 'Aphrodite peach'. Sci Hortic 8:179–192

Mikkelsen JC, Ryan J, Constantin MJ (1975) Mutation breeding of Rieger's elatior begonias. Am Hortic 54(3):18–21

Nag KK, Johri BM (1970) Effects of cytokinins and injury on the formation of shoots by leaves of Dendrophthoe foliate. Planta 90:360–364

Naylor E, Johnson B (1937) A histological study of vegetative reproduction in *Saintpaulia ionantha*. Am J Bot 24:673–678

Norris RE, Smith RH, Vaughu RC (1983) Plant chimeras are used to establish the development of shoots. Science 220:75–76

Park SH, Kim GH, Jeong BR (2005) Adventitious shoot regeneration in *Chrysanthemum* as affected by plant growth regulators, sucrose, and dark period. J Kor Soc Hort Sci 46(5):335–340

Park SH, Kim GH, Jeong BR (2007) Adventitious shoot regeneration from cultured petal explants of *Chrysanthemum*. Hortic Environ Biotechnol 48(6):387–392

Parliman B, Stushnoff C (1979) Mutant induction through adventitious buds of *Kohleria*. Euphytica 28(2):521–530

Pinet-Leblay C, Turpin FX, Chevreau E (1992) Effect of gamma and ultraviolet irradiation on adventitious regeneration from in vitro cultured pear leaves. Euphytica 62:225–233

Roest S, Bokelmann GS (1975) Vegetative propagation of *Chrysanthemum morifolium* Ram. in vitro. Sci Hortic 3:317–330

Shen XS, Wan JZ, Luol WY et al (1990) Preliminary results of using *in vitro* axillary and adventitious buds in mutation breeding of Chinese gooseberry. Euphytica 49:77–82

Smith RH, Norris RE (1983) *In vitro* propagation of African violet chimeras. HortScience 18(4): 436–437

Sparrow AH, Sparrow RC, Schairer LA (1960) The use of X-rays to induce somatic mutations in *Saintpaulia*. Afr Violet Mag 13(4):32–37

Sripichitt P, Nawata E, Shigenaga S (1988) Radiation-induced mutation by using *in vitro* adventitious bud technique in red pepper (*Capsicum annum* L. cv. Yatsufusa) analysis of the variant appeared in the M1 generation. Jpn J Plant Breed 38:141–150

Stangler BB (1956) Origin and development of adventitious roots in stem cuttings of *Chrysanthemum*, carnation, and rose, vol 342. Cornell Exp. Stn., Ithaca, NY, pp 3–24

Stewart RN, Dermen H (1970a) Somatic genetic analysis of the apical layers of chimera sports in *Chrysanthemum* by experimental production of adventitious shoots. Am J Bot 57:1061–1071

Stewart RN, Dermen H (1970b) Determination of the number and mitotic activity of shoot apical initial cells by analysis of mericlinal chimeras. Am J Bot 57:816–826

Stewart RN, Dermen H (1970c) The origin of adventitious buds in *Chrysanthemum*. Am J Bot 57: 734–735

Sytsema W (1977) Schubvermeerdering van Nerine. Vakbl Bloemisterij 32(6/7):21

Takeuchi N, Tanimoto S, Harada H (1985) Effects of wounding on adventitious bud formation in *Torenia* stem segments cultured in vitro. J Exp Bot 36:841–847

Traub HP (1935) Propagation of Amaryllis by stem cuttings. Yearb Am Amaryllis Soc 2:123–126

Tymoszuk A, Zalewska M (2014) *In vitro* adventitious shoots regeneration from ligulate florets in the aspect of an application in chrysanthemum breeding. Acta Sci Pol Hortorum Cultus 13(2): 45–58

Vainstein A, Fisher M, Ziv M (1992) Shoot regeneration from petals as a basis for genetic variation and transformation. Acta Hortic 314:39–45

Watad AA, Ahroni A, Zuker A et al (1996) Adventitious shoot formation from carnation stem segments: a comparison of different culture procedures. Sci Hortic 65:313–320

Wilden CE (1934) Propagation of dahlias by leaf bud cuttings. Q Bull Mich Agric Exp Str 16:253–254

Wongpiyasatid A, Thinnok T, Taychasinpitak T et al (2007) Effects of acute gamma irradiation on adventitious plantlet regeneration and mutation from leaf cuttings of African violet (*Saintpaulia ionantha*). Kasetsart J (Nat Sci) 41(4):633–640

Yang H, Schmidt H (1994) Selection of a mutant from adventitious shoots formed in X-ray-treated cherry leaves and differentiation of standard and mutant with RAPDs. Euphytica 77:89–92

Zalewska M (2010) *In vitro* adventitious bud techniques as a tool in the creation of new chrysanthemum cultivars. In: Datta SK, Chakrabarty D (eds) Floriculture. Role of tissue culture and molecular techniques. Pointer Publishers, Jaipur, p 196

Zuker A, Ahroni A, Shejtman H, Vainstein A (1997) Adventitious shoot regeneration from leaf explants of *Gypsophila paniculata* L. Plant Cell Rep 16:775–778

Experiments and Results

6

Abstract

The chapter covers many important steps and technical procedures for induced mutagenesis experiments. The aspects which have been covered are mutagens, materials, treatment procedure and observations, optimum working dose (LD_{50} dose/radiosensitivity), treatment plan (split dose, prolonged radiation irradiation), etc.

Keywords

Experiment · Propagules · Working dose · Treatment procedure

This chapter covers many important steps and technical procedures for induced mutagenesis experiments. The aspects which have been covered are mutagens, materials, treatment procedure and observations, optimum working dose (LD_{50} dose/radiosensitivity), treatment plan (split dose, prolonged radiation irradiation), etc.

6.1 Mutagens

For experimental purposes mostly two types of mutagens, namely physical and chemical mutagens are used.

Physical mutagens: Several types of radiation like ultraviolet light, X-ray, gamma rays, alpha and beta particles, protons, and neutrons are used. Physical mutagens are mostly used in all kinds of plant parts for mutation experiments. One can choose any type of ionizing radiation. But it depends mainly upon the availability of a source of radiation. In practice, mostly X-rays and gamma rays are used. The mutation breeding experiment was based primarily on X-rays. But now gamma rays are

S. Datta, *Role of Mutation Breeding in Floriculture Industry*,
https://doi.org/10.1007/978-981-99-5675-3_6

mainly used for the experiment. If neutrons are available, fast neutrons are preferable to thermal neutrons, although both are useful. For prolonged (chronic) irradiation gamma sources are used for large populations of plants and voluminous materials.

Chemical mutagens: Different groups of chemical compounds such as alkylating agents, antibiotics, and others have been determined for their mutagenic properties. Alkylating agents are among the chemicals more frequently used. The most commonly used chemical mutagens are ethyl methanesulfonate (EMS), methyl methanesulfonate (MMS), diethyl sulfate (dES), ethyleneimine (EI), ethyl nitroso urethane (ENU), ethyl nitroso urea (ENH), methyl nitroso urea (MNH), azides, etc. The alkyl group of an alkylating agent reacts with DNA resulting in a point mutation by changing the nucleotide sequence. Alkylating chemicals may also induce chromosome mutations. Not much is known about the use of antibiotics like streptomycin, netromycin, etc. Several other chemical mutagens like ethidium bromide and several nitroso compounds like ethyl nitroso urea, methyl nitroso urethane, sodium azide (NaN3), etc. have been identified for point mutations. Chemical mutagens are not promising in vegetatively propagated crops. However, chemicals like ethidium bromide and nitroso compounds occasionally affect specifically extranuclear DNA in vegetatively propagated crops to induce plastome mutants with ornamental value, like green/white variegation in foliage plants such as Dracaena, Hedera, and others (Pohlheim 1981; Polheim 1981). The technology for the application of chemical mutagenesis in vegetatively propagated plants is inadequate due to a lack of success in preliminary experiments. Poor uptake and penetration of the chemical compounds, bulky material, and size and composition of the material negatively affect the reproducibility of the experiments in vivo (Bowen 1965; Moes 1966; Nybom 1961). In most cases, buds on scions, stem cuttings, tubers, bulbs or rhizomes and adventitious buds are treated, either by immersion of the plant part in the mutagenic solution or by applying the mutagens via a cotton plug, placed on the buds. Dryagina and Limberger (1974) suggested the technique of injecting perennial trees through a syringe.

6.2 Materials, Treatment Procedure, and Observations

To start a mutation breeding experiment one should be acquainted with a wide range of basic information. The category of the plant, i.e. whether it is a self-pollinator, a cross-pollinator, or a vegetative propagated should be familiar before starting the experiment.

As mentioned, ornamental plants are propagated by seeds and vegetative means. Several ornamental plants are propagated both by seeds and by vegetative organs. For example, chrysanthemums, gladiolus, rose, etc. can be propagated by seeds, but for commercial purposes, these are mostly propagated by vegetative organs. In vegetative propagated plants seeds, tubers, bulbs, cuttings, suckers, rhizomes, corms, cormels, budding eyes, in vitro raised plantlets, etc. with meristematic buds are used for multiplication and treated with mutagens (Fig. 6.1). For mutation experiment, plants, propagules, and mutagens are selected first. The seed size,

Fig. 6.1 Different types of propagules (A & B. Seeds; C. Cuttings; D-H. Bulbs; I-K. Cuttings & Suckers; L & M. Tissue raised plantlets) of different ornamentals used for mutagenesis experiments

seed coat thickness, and moisture content are to be determined first in seed-propagated crops. One pilot experiment may be conducted to determine the working dose. After determining the working dose the large-scale experiment is started with the target crop. Dry seeds are normally treated with mutagens in mutation experiments with seed-propagated plants. Soaked seeds and seedlings are also treated with mutagens. However, the results of the soaked seeds and seedling experiment are less in ornamental crops. Determination of propagule size is most important in vegetatively propagated crops.

Literature surveys and knowledge of all cultural practices for the cultivation procedure of experimental material are very important. Cultural practices play a notable role in the detection and management of mutants. All experiments should be carried out during the ideal planting season of the crop. It has been endorsed that chances to obtain mutation are more within a large population. But it is wise to have definite knowledge/information on obtainable uniform and trustworthy planting material, accessibility of funds, manpower, infrastructure, etc. Before designing the experiment one should have definite breeding objectives and proper planning to reach the goal must be sketched. For experimental design, three replications and the randomized block design are mostly followed. Field experiment is always favored, otherwise in some cases, experiments may be conducted in pots.

Treated materials are grown at suitable conditions and observations are recorded on different parameters to assess the effects of mutagen/s. At the beginning following steps are followed to collect the supporting data for determining LD_{50} dose.

Germination: Percent seed germination reduces after treatment with different doses of mutagens and with an increase in exposure. The LD_{50} dose is determined from 50% germination. Germination time is also delayed in all cases. There are reports on the stimulating effects of mutagens on germination and growth of treated materials in certain cases.

Sprouting: In vegetatively propagated crops sprouting of propagules is reduced after treatment. The criteria for determination of LD_{50} is the same as seed germination.

Seedling height: Seedling height is reduced with an increase in mutagen doses. Seedling height may be recorded 15 days after germination. The period may vary according to experimental material.

Morphological abnormalities: A very striking effect of mutagen is the manifestation of abnormal plant growth involving various plant parts. The abnormalities are mostly observed in initial growth as primary effects of mutagens. The manifestation of abnormal plant growth is not genus, species, and dose-specific but the frequency of abnormalities varies with the genus, species, and dose. The leaf abnormalities mostly include changes in shape, size, margin, apex, fission, and fusion of leaves. The most dominant type of leaf abnormality is the fission of a leaf from the leaf tip to various depths on the lamina and the fusion of two or more leaves. The hampered development and irregular distribution of chlorophyll in the leaves are another abnormality that is observed very frequently. Asymmetrical development of the wing and displacement of main veins result in the development of leaves with very peculiar morphology (Datta 2006; Datta and Basu 1977, 1978; Datta and Gupta 1980, 1981a, b, c). The percentage of abnormal plants and abnormal leaves per plant increase with the increase in mutagen doses mostly in all the cases but the frequency varies from plant to plant. Normally the percentage of morphological abnormalities decreases with the aging of the plant. Progeny tests of the abnormal plants show only normal plants in the second and subsequent generations indicating no immediate genetic background of those aberrations.

Stem abnormalities: The stem abnormality is mostly fasciation of the stem and the frequency of abnormalities varies with the dose and genus. Other stem abnormalities may be in stem color, stem stature, etc.

Anatomical abnormalities: The mutagen effect is found mainly in leaf stomata index and size. Both increases and decreases in stomata number per unit area and size are recorded in treated populations. Variegated leaves develop in some treated populations. The chloroplasts of the guard cells of the variegated leaves (chlorophyll deficient) are not of definite shape but they are of diffused type due to hampered development (Datta and Gupta 1980). Histological study of abnormal plant parts indicates that abnormal plant growth is mostly associated with a change in the course of vascular bundles (Datta 1986).

Flowering behavior: Mutagen treatment affects flowering behavior, i.e. delay/ early in flower bud formation, color showing, and full bloom stage. The floral abnormalities are observed mostly in flower shape and size and petal/floret structure.

Chromosomal observations—Root/shoot tip mitosis: Root/shoot tip mitosis shows normal cell division in control materials. Various kinds of chromosomal abnormalities are observed in mutagen-treated materials. Chromosomal abnormalities generally observed are micronuclei, early or late separation, bridges, fragments, laggards, exclusions, clumping, disturbed polarity, etc. The percentage of cells with chromosomal aberrations increases with an increase in doses. The mitotic index is generally illustrated in the form of a relative division rate (RDR, Hoda et al. 1991). The RDR is calculated at each dose of mutagen. The values in all the cases are commonly negative indicating inhibition of mitotic division. The severity of the mitotic inhibition can be estimated from the increasing negative value of RDR at each dose.

Meiotic abnormalities: The behavior of chromosomes during meiotic cell division is also studied as an indicator of mutagen effects. Chromosomal abnormalities observed are bridges, fragments, laggards, delayed separation of bivalents, non-disjunction and exclusion of chromosomes, formation of micronuclei, etc. In general, there is an increase in the frequency of cells with chromosomal aberrations with an increase in mutagen doses.

Pollen grains: The sensitivity of mutagens can be estimated by studying pollen grain sterility/fertility. Pollen sterility increases after mutagen treatment and with an increase in exposure. Polar diameter and exine thickness may be reduced after treatment. Changes in pollen grain morphology are also a good symbol of mutagen effects (Datta 2015; Datta and Datta 1998, 2018).

6.3 Optimum Working Dose (LD_{50} Dose/Radiosensitivity)

For mutation work, the determination of the optimum working dose of mutagens is very important. The working dose of radiation depends upon the radiosensitivity, plant part, and mode of development of experimental material. The application of random doses is not the correct step for mutation technology. It does not give the real picture of the mutation effect on any specific crop. It also does not reflect the mutagen dose relationship with experimental material. It is a loss of time without any beneficial results. But unfortunately, the majority of experiments are conducted on a random basis and interpretations on mutagen/s effects are drawn based on such work. Subsequent researchers follow these reports and repeat the same experiments. Experimenting with optimum doses gives the correct image of the mutation effect. The working dose may also speak softly LD_{50} dose. LD_{50} dose of any mutagen for any genus, species, or cultivar depends upon the sensitivity of experimental material to mutagens. Radiosensitivity is a very significant essential technical factor for the success of the mutation technique. Radiosensitivity depends upon many factors and considering its vital role in mutation experiments, the author worked extensively on this aspect and reported earlier (Datta 2019, 2023). A wide range of components related to a biological, physiological, and chemical category like genotype, stage of cellular development, chromosome number, and size, DNA content, interphase chromosome volume, interphase nuclear volume, age of the tissue, storage after irradiation, oxygen, chemicals, water, temperature, ionization density, combined treatment, etc. can modify the effects of radiation in plants. The water content of the experimental material very much affects the radiosensitivity. Water content is more in tubers, bulbs, or cuttings than in dry dormant seeds. Determination of LD_{50} dose for ornamental crops is very meaningful as different ornamentals are propagated by different propagules as mentioned earlier. Sensitivity may vary not only among the species but also in commercial varieties within the same species. Radiosensitivity, i.e. mode of action of radiations can be assayed by several factors like growth inhibition (reduced germination, reduced height of seedling), lethality, chromosome breakage/aberrations, vegetative and floral abnormalities, pollen grain sterility/fertility, mutation frequency, and spectrum. Considering its important role in induced mutations, the author published several review articles and book chapters and one can get all basic information related to radiosensitivity from these publications (Datta 2006, 2014, 2017, 2019, 2020, 2023).

It is desirable to conduct one pilot experiment by treating experimental material with a range of doses. To summarize, LD_{50} dose can be determined by selecting several parameters depending upon the propagules used for plantation: (1) Percent germination (for seed-propagated plant, calculated by the decrease of germination rate). (2) Percent sprouting (for bulb/corm/cuttings/rhizomes/budwood propagated plant). (3) 15-day seedling height. (4) 15-day sprout height. (5) Percent morphological abnormalities (generally leaf and stem character) on 30 days or the date may be fixed considering the growth nature of the experimental material. (6) Percent floral abnormalities. (7) Chromosomal abnormalities during root/shoot tip mitosis. (8) Meiotic abnormalities. (9) Percent pollen grain sterility/fertility. (10) Percent survival (calculated by the decrease of survival rate). (11) Percent mutation.

Varietal differences in radiosensitivity studied among most of the ornamental species and varieties indicate that a genotype-dependent mechanism is connected to the damage or repair of mutagen-induced damage within the organism. It is found that all plant materials are differentially sensitive to mutagens. Not only different genera and species but also different cultivars, different parts of the same plant, and even different tissues are sensitive differentially (Banerjee and Datta 1993; Datta 1985, 1987, 1990, 1992; Gupta et al. 1991; Mandal and Basu 1977, 1978, 1980). The most important is to determine first the LD_{50} dose of every mutagen from the above experimental parameters about particular vegetation. Once the LD_{50} dose is determined the maximum permissible level of radiation exposure of a particular crop can be determined. Survival and capacity to regenerate decrease with increasing dose and mutation frequency increases with increasing dose (linearly, exponentially). One must choose some point between a low dose (100% survival; low mutation frequency) and a high dose (low survival; higher mutation frequency). After determining the LD_{50} dose one must choose some point between a low dose (100% survival; low mutation frequency) and a high dose (low survival; higher mutation frequency). For a large-scale experiment total of five doses (LD_{50} + 2 doses above LD_{50} + 2 doses below LD_{50}) will be ideal. The working dose of most of the common ornamentals has been precisely determined (Datta and Gupta 1981d; Datta 2023). Dose levels vary with many factors like dry and weight seed, seed size, seed coat, seed moisture, post-irradiation storage, etc. (Froese-Gertzen 1962; Froese-Gertzen et al. 1963, 1964; Briggs 1966; Konzak et al. 1972; Gaul et al. 1972; Datta 2019a, b).

Understanding early work reports are very helpful to select the working dose for new species and cultivars. It is very important to look into various elements before selecting mutagen and treatment. Knowledge of the ability of chemical mutagens to enter into plant tissue and hard-coated seeds is necessary. Both physical and chemical mutagens have encouraging effects on vegetatively propagated crops. Chemical mutagens are preferred as they usually cause single-base substitution. The quantum of work on ornamental crops is huge which helps proper manipulation of treatment conditions. Mutagenic effectiveness varies from mutagen to mutagen. Gamma radiation is largely used mutagen but there is an opinion that the effectiveness of gamma rays is lesser than that of EMS (Gautam et al. 1992; Girija and Dhanavel 2009; Kaul and Bhan 1977; Wani 2009). There are possibilities to increase the effectiveness of gamma and X-irradiations through combined treatment of chemicals and application of low doses for a longer period. Among all chemical mutagens, the use of EMS is most routine in ornamentals (Gautam et al. 1992, 1998; Girija and Dhanavel 2009; Yamaguchi et al. 2009; Gichner and Ehrenbard 1967; Schendel and Michaeli 1984; Talebi et al. 2012).

6.4 Role of Radiation Protection

Mutagen treatment (either physical or chemical) induces genetic and physiological damages and mutation. Efforts have been made to use of combined treatment of physical and chemical mutagens in mutation experiments. Experimental results have pointed out that mutagen effects can be changed and influenced by pre-and post-treatment of propagules with chemicals. Pre-irradiation seed treatment with colchicine resulted in the origin of uncommon mutations and mutation types. Chromosomal aberrations decrease when propagules are pre- or post-treated with ultraviolet light and increase if materials are pre-or post-treated with infrared. GA pretreatment decreased the frequency of X-ray-induced chromosomal aberrations. Combined treatment of gamma rays and EMS significantly increased fertility and mutation frequency doubled in M_2. Pretreatment of seeds with colchicine before irradiation created rare mutations and mutation types which were not easily found after treatment of seeds with X-rays alone (D'Amato and Gustafsson 1948). Pre-treated with colchicine, CO_2, or certain chemicals before irradiation increased survival, fertility, or mutation frequency (Gustafsson and Nybom 1949; Ehrenberg et al. 1952; Gaul 1958). A good amount of experiments were conducted to change radiation effects by treatment with chemical mutagens before or after exposure to ionizing radiations like treatment with colchicine and DES before and after X-irradiation in tomato (Bose and Maiti 1972), treatment with EMS, MMS, DES, and NEU after X- and gamma irradiation in Sorghum (Sree Ramulu 1971), treatment with EMS before exposure to gamma rays in *Vicia faba* (Hussein and Abdalla 1974), treatment with hydroxylamine before X-irradiation in *Lens esculenta* (Jana et al. 1974), treatment of colchicine and X-ray on plant growth, pollen grains, chromosomal aberrations, and ploidy level, fruit and seed yield of *Trichosanthus anguina* (Basu and Datta 1977, 1982; Datta and Basu 1977, 1978; Datta 1991, 1992), and combined treatment with beta and gamma rays in soyabean (Killion and Constantin 1971, 1974; Killion et al. 1971). It has been suggested that this modification of radiation-induced effects can be best exploited in crop improvement if it is possible to increase the frequency of all mutations with simultaneous reduction of radiation-induced chromosomal aberrations, physiological injury, and sterility. In crop plants, such modifications of the mutation process may open the way for directing and controlling the production of desirable mutants (Bose and Maiti 1972; Datta 2019a, b). The concept of combined treatment has not been exploited much in ornamentals to increase mutation frequency and spectrum.

6.5 Treatment Procedure (Split Dose, Acute and Chronic Irradiations)

Preparation for the experiment and design of the experiment is very important for the success of induction of mutation. Many early published literature highlighted the mode of application of radiation, i.e. in the form of split dose, prolonged radiation, chronic and acute radiation, recurrent irradiation, etc., and their advantages and disadvantages (Datta 2012, 2023). The split dose concept has not been judicially applied to ornamental crops. Such an application system will generate encouraging feedback regarding treatment procedures for different propagules of ornamental crops.

References

Banerjee BK, Datta SK (1993) Varietal differences in radiosensitivity of garden chrysanthemum. Nucleus 36(3):114–117

Basu RK, Datta SK (1977) Effects of X-rays and colchicine on pollen of *Trichosanthes anguina* L. (Cucurbitaceae). Grana 16:105–109

Basu RK, Datta SK (1982) Sensitivity of two species of *Trichosanthes* to X-ray and colchicine. J Cytol Genet 17:107–112

Bose S, Maiti SN (1972) Effects of pre-irradiation treatment with colchicines and DES on tomato. Nucleus 15(1):22–30

Bowen HJM (1965) 'Mutations in horticultural chrysanthemums', the use of induced mutations in plant breeding (rep. FAO/IAEA Tech. Meeting Rome 1964). Pergamon Press, Oxford, pp 695–700

Briggs RW (1966) Time and temperature tolerances of maize seeds. In: Mutations in plant breeding (proc. panel Vienna, 1966). IAEA, Vienna, pp 39–44

D'Amato F, Gustafsson A (1948) Studies on the experimental control of the mutation process. Hereditas 34:181–192

Datta SK (1985) Radiosensitivity of garden roses. J Nucl Agric Biol 14:133–135

Datta SK (1986) Histological interpretations of morphological aberrations. J Anat Morphol 3(2): 129–138
Datta SK (1987) Sensitivity of garden roses to gamma irradiation. Indian Rose Ann VI:121–126
Datta SK (1990) Induction and analysis of somatic mutations in garden chrysanthemum. Adv Hortic For 1(31):241–254
Datta SK (1991) Effects of X-rays and colchicine on fruit and seed yield of *Trichosanthes anguina* L. J Indian Bot Soc 70(I-IV):401–402
Datta SK (1992) Assessment of single and combined X-ray and colchicine treatment of *Trichosanthes anguina* L. J Nucl Agric Biol 21(4):293–298
Datta SK (2006) Parameters for detecting effects of ionizing radiations on plants. In: Tripathi RD, Kulshreshtha K, Agrawal M, Ahmad KJ, Varshney CK, Krupa SV, Pushpangadan P (eds) Plant responses to environmental stress. International Book Distributing Co, Lucknow, pp 257–265
Datta SK (2012) The success story of induced mutagenesis for the development of new ornamental varieties. In: Bioremediation, biodiversity, and bioavailability 6 (Special Issue I), pp 15–26; Global Science Books, Invited Review
Datta SK (2014) Induced mutagenesis: basic knowledge for technological success. In: Tomlekova NB, Kozgar ML, Wani MR (eds) Mutagenesis: exploring the genetic diversity of crops. Wageningen Academic Publishers, Wageningen, pp 95–137
Datta SK (2015) Indian floriculture: role of CSIR. Regency Publications, New Delhi, p 432
Datta SK (2017) Improvement through induced mutagenesis: ornamental crops. In: Malik CP, Wani SH, Kushwaha HB, Kaur R (eds) Advanced technologies for crop improvement and agricultural productivity, vol 4. Agrobios, Chater, Jodhpur, pp 49–86
Datta SK (2019) Determination of radiosensitivity: prerequisite factor for induced mutagenesis. In: Trivedi PC (ed) Harnessing plant biotechnology and physiology to stimulate agricultural growth. Agrobios, Jodhpur, pp 39–54
Datta SK (2019a) Floriculture work at CSIR-National Botanical Research Institute, Lucknow. Sci Cult 85(7–8):274–283
Datta SK (2019b) Present Status of Research on Floriculture in India. Int J Life Sci 8(2):71–93
Datta SK (2020) Induced mutations: technological advancement for the development of new ornamental varieties. Nucleus 63:119–129
Datta SK (2023) Induced mutation breeding. Springer, Singapore. https://doi.org/10.1007/978-981-19-9489-0; ISBN 978–981–19-9488-3
Datta SK, Basu RK (1977) Abnormal plant growth in M_1 and C_1 generation of two species of *Trichosanthes*. Trans Bose Res Inst 40(3):63–67
Datta SK, Basu RK (1978) Modifications of chromosomal aberrations and polyploidy by combined treatment with X-rays and colchicine in *Trichosanthes anguina*. Indian Biol X(1):59–64
Datta K, Datta SK (1998) Palynological interpretation of gamma-ray and colchicine induced mutation in chrysanthemum cultivars. Isr J Plant Sci 46:199–207
Datta K, Datta SK (2018) Pollen grain characters—a useful parameter for testing radiosensitivity and characterization of mutants. FAO/IAEA International Symposium on Plant Mutation Breeding and Biotechnology, Vienna; Abstract No. IAEA-CN-263-36
Datta SK, Gupta MN (1980) Effects of gamma irradiation on rooted cuttings of small flower chrysanthemum. New Bot VII:73–85
Datta SK, Gupta MN (1981a) Effects of gamma irradiation on rooted cuttings of Korean type Chrysanthemum cv. 'Nirmod'. Bangl J Bot 10(2):124–131
Datta SK, Gupta MN (1981b) Cytomorphological, palynological, and biochemical studies on control and gamma induced mutant of Chrysanthemum cultivar 'D-5'. Chrysanthemum 34(4): 193–200
Datta SK, Gupta MN (1981c) Cytomorphological, palynological, and biochemical studies on control and gamma induced mutant of Chrysanthemum cultivar 'E-13'. SABRAO J 134(2): 136–148
Datta SK, Gupta MN (1981d) Studies on Chrysanthemum cultivar 'Otome Zakura' and its mutants. Bot Prog 4:88–92

Dryagina IV, Limberger GE (1974) A new method for treating perennial fruit trees with chemical mutagens. Moscow Univ Biol Sci Bull 29(6):50–53
Ehrenberg L, Gustafsson A, Nybom N (1952) Effects of ionizing radiations in barley. Arkiv Bot 1: 557–568
Froese-Gertzen EE (1962) The action of the chemical mutagen, ethyl methanesulfonate on barley M. Sc Thesis. Washington State University, Pullman, WA, pp 1–72
Froese-Gertzen EE, Konzak CF, Foster R, Nilan RA (1963) Correlation between some chemical and biological reactions of ethyl methanesulfonate. Nature (London) 1998:447–448
Froese-Gertzen EE, Konzak CF, Nilan RA, Heiner RE (1964) The effect of ethyl methanesulfonate on the growth response, chromosome structure, and mutation rate in barley. Radiat Bot 4:61–69
Gaul H (1958) Present aspects of induced mutations in plant breeding. Euphytica 7:275–289
Gaul H, Frimmel G, Gichner T, Ulonska L (1972) Efficiency of mutagenesis. In: Induced mutations and plant improvement (Proc. Meeting Buenos Aires, 1970). IAEA, Vienna, pp 121–139
Gautam AS, Soodand KC, Richarria AK (1992) Mutagenic effectiveness and efficiency of gamma rays, ethyl methanesulphonate and their synergistic effects in blackgram (Vigna mungo L.). Cytologia 57:85–89
Gautam AS, Soodand KC, Mittal RK (1998) Mutagenic effectiveness and efficiency of gamma rays and ethyl methanesulphonate in rajmash (Phaseolus vulgaris L.). Legume Res 21(3&4): 217–220
Gichner T, Ehrenbard L (1967) The influence of post-treatment storage on the frequency of EMS-induced chromosomal aberrations in barley. Biol Plant Acad Sci Bohemoslov 8:256–259
Girija M, Dhanavel D (2009) Mutagenic effectiveness and efficiency of gamma rays, ethyl methane sulphonate and their combined treatments in cowpea (*Vigna unguiculata* L. Walp). J Mol Sci 4(2):68–75
Gupta MN, Nath P, Datta SK (1991) Studies on bud uptake and determination of LD_{50} dose of gamma rays of 35 rose cultivars. Indian Rose Ann IX:98–101
Gustafsson A, Nybom N (1949) Colchicine, x-rays and the mutation process. Hereditas 35:280–284
Hoda Q, Bose S, Sinha SP (1991) Vitamin C mediated minimization of Malathion and Rogor-induced mitoinhibition and clastogenic. Cytologia 56:389–397
Hussein HAS, Abdalla MF (1974) Egypt J Genet Cytol 3:246–256
Jana MK, Prasad AK, Moutschen JH (1974) Hydroxylamine induced chromosome aberrations in Lens esculenta and combined effects with ionizing radiation. Cytologia 39(4):655–663
Kaul MLH, Bhan AK (1977) Mutagenic effectiveness and efficiency of EMS, DES, and gamma-rays in rice. Theor Appl Genet 50(5):241–246
Killion DD, Constantin MJ (1971) Acute gamma irradiation of the wheat plant: effects of exposure, exposure rate, and developmental stage on survival, height, and grain yield. Radiat Bot 11:367–373
Killion DD, Constantin MJ (1974) Effects of separate and combined beta and gamma irradiation on the soybean plant. Radiat Bot 14(2):91–99
Killion DD, Constantin MJ, Siemer EG (1971) Acute gamma irradiation of the soybean plant: effects of exposure, exposure rate and developmental stage on growth and yield. Radiat Bot 11: 225–232
Konzak CF, Wickham IM, DeKock MJ (1972) Advances in methods of mutagen treatment. In: Induced mutations and plant improvement (Proc. Meeting Buenos Aires, 1970). IAEA, Vienna, pp 95–118
Mandal SK, Basu RK (1977) Differential radiosensitivity on a tissue level in *Vicia faba* and *Lens esculenta*. Indian Biol IX:33–37
Mandal SK, Basu RK (1978) Differential radiosensitivity in a tissue level in *Nigella sativa*. Nucleus 21:198–201
Mandal SK, Basu RK (1980) Differential sensitivity on a tissue level in *Delphinium ajacis*. Indian J. Exp Biol 18:940–943
Moes A (1966) Mutations induites chez lie glaieul (Gladiolus). Bull Rech Agron Gembloux 1:76–95

Nybom N (1961) The use of induced mutations for the improvement of vegetatively propagated plants, mutation, and plant breeding, NAS/NRC, vol 891, pp 252–294

Pohlheim F (1981) Induced mutations for the investigation of histogenic processes as the basis for an optimal mutant selection. In: Proc Int Sym on induced mutations as a tool for crop plant improvement. IAEA, Vienna, pp 489–495

Polheim F (1981) Genetischer Nachweis einer NMH-induzierten Plastommutation bei Saintpaulia ionantha H. Wendl Biol Rundsch 19:47–50

Schendel PF, Michaeli I (1984) A model for the mechanism of alkylation mutagenesis. Mutat Res 125:1–14

Sree Ramulu K (1971) Effects of ionizing radiations and chemical mutagens on chiasma frequency in Sorghum. Cytologia 36:543–551

Talebi AB, Talebi AB, Shahrokhifar B (2012) Ethyl methane sulphonate (EMS) induced mutagenesis in malaysian rice (cv. MR219) for lethal Dose determination. Am J Plant Sci 3(12):25363. https://doi.org/10.4236/ajps.2012.312202

Wani AA (2009) Mutagenic effectiveness and efficiency of gamma rays, ethyl methane sulphonate and their combination treatments in chickpea (*Cicer arietinum* L.). Asian J Plant Sci 8(4):318

Yamaguchi H, Hase Y, Tanaka A, Shikazono N, Degi K, Shimizu A et al (2009) Mutagenic effects of ion beam irradiation on rice. Breed Sci 59(2):169–177

Part II

Crop Wise Mutation Work

Bougainvillea

7

Abstract

Bougainvillea is one of the most important and popular perennial ornamental crops in floriculture for multipurpose use and a very suitable plant for multidisciplinary research work due to its numerous bract colors, leaf characters, and plant stature. The fascinating variegation design in leaves is added beauty to its attraction. It is an all-purpose plant that grows very well under tropical and subtropical conditions. Attempts have been made to highlight the progress report of the improvement of bougainvillea for the development of new varieties through induced mutation. This chapter covers almost all mutation work, mutagens and their optimal working dose, development of mutant varieties, etc.

Keywords

Bougainvillea · Multipurpose crop · Bract color mutation · Variegated mutations

Bougainvillea is one of the most important and popular perennial ornamental crops in floriculture for multipurpose use and a very suitable plant for multidisciplinary research work due to its numerous bract colors, leaf characters, and plant stature. The fascinating variegation design in leaves is added beauty to its attraction. It is an all-purpose plant that grows very well under tropical and subtropical conditions. The plant grows well in various climatological zones and also in almost all types of soil. Bougainvillea belongs to the family Nyctaginaceae and it is the only genus in the monotypic sub-tribe Bougainvilleineae (Gills 1976) and has originated in tropical and subtropical South America. Out of 10 species, three (*B. spectabilis* Wild., *B. glabra* Choisy, and *B. peruviana* Humb. and Bonp.) are of horticultural importance with very showy and colorful bracts (Heimerl 1900). Bougainvillea is divided into (1) single bracted and (2) double or multibracted groups based on bract nature. The majority of Bougainvillea belongs to a single bracted group with a wide range of

S. Datta, *Role of Mutation Breeding in Floriculture Industry*,
https://doi.org/10.1007/978-981-99-5675-3_7

bract color spectra. Bougainvilleas with their outstanding and charming diversity is used as the bush, climbers, specimen plants, hedges, topiaries, ground covers, standards, pergolas and trees, arches, pot culture, cut flowers, slopes, and mound, bonsai, hanging basket, cascade, etc. Maximum variability has been created through spontaneous mutations, natural hybridization, manmade hybridization, and induced mutations. Bougainvilleas due to their multifarious use have many beneficial components for the consumer which can be created, enhanced, or improved through the induced mutation technique. Efforts have been made to highlight the progress story of the improvement of bougainvillea for the development of new varieties through induced mutation.

Bougainvillea is mostly propagated by vegetative means, i.e. hardwood cuttings. The hardwood cuttings of pencil thickness are most suitable to propagate. It can also be propagated from shoot tip cuttings. Bougainvillea is propagated by other conventional methods like air layering, gootie, budding, grafting, etc. Generally, seed setting is not observed in all cultivars but a few cultivars set seeds occasionally depending on the climatic conditions and age of the plant. Micro-propagation technique has also been standardized for the difficult-to-root cultivar (Datta and Mandal 2012).

A series of experiments were conducted on different cultivars of bougainvillea to induce genetic variability. Abraham and Desai (1977) initiated mutation breeding experiments and successfully developed mutant varieties with chlorophyll-variegated leaves and bract color by applying gamma radiation. All the mutants had desired characters for commercial utilization. The mutants along with their parental varieties (in parenthesis) are "Jaya"—bract character ("Jayalaxmi"); "Jayalaxmi variegated"—leaves variegated ("Jayalaxmi"); "Lady Hudson of Ceylon Variegata"—leaves variegated ("Lady Hudson of Ceylon") and "Silvertop"—bract character ("Versicolour"). Hong and Shaode (1990) treated rooted and sprouted cuttings of ***Bougainvillea spectabilis*** with 3KR and 3.5KR radiation doses and detected mutations in colors in the VM1 generation. 74% mutations (scarlet and orange-yellow) were detected from 3KR and 1.64% mutations (orange-yellow and red-yellow) were detected from 3.5KR treatments. Finally, two mutants with long blossoming periods and attractive colors from these selections were established in pure form in VM3 and named 85–1 and 85–2 for commercial use. Deng and Liu (1990) applied 3 and 3.5KR gamma rays to the rooted cuttings of ***B. spectabilis*** and established a mutant with changed flower color. The experiment was conducted to study the effects of different doses (500, 1000, 1500, and 2000 rads) of gamma rays on bougainvillea cv. "Mahatma Gandhi." Sensitivity was determined based on different characteristics like morphological abnormalities, reduction in sprouting, survival of plants, plant height, number of branches, leaf number, size, etc. (Swaroop et al. 2015). Abd El-Mageid and Awad (2009) treated two cultivars of Bougainvillea, i.e., B. glabra (white) and B. buttiana Mrs. Butte (red) to gamma, ray (0.0, 2.0, 4.0, and 6.0 Krad.) and chemical mutagen (N-nitroso-N methyle Urea) NMU 0.000, 0.010, 0.015, 0.020, and 0.025% to induce variations. Mutations in different parameters like growth (compact plants and dwarfness), bracts color, delayed flowering, flower-bearing, leaf shape, chlorophyll mutation, flower cluster, length, and fasciata were detected. In M2, NMU was found to be more efficient than gamma radiation and the variation and the mutations in B. buttiana Mrs. Butte were higher than that of B. glabra. Banerji (2010) made a review of induced mutation work on four multibracted bougainvilleas ("Cherry Blossom," "Los Banos Beauty," "Mahara," and "Roseville's Delight") and one single bracted variety "Pixie" using gamma rays (0, 5, 10, 15 Gy) and EMS (0, 0.01, 0.02, and 0.03%). The effects of mutagens on different cytomorphological characters have been highlighted. From the entire experiments, a total of seven chlorophyll-variegated mutants ("Pallavi," "Los Banos Variegata," "Los Banos Variegata Silver Margin," "Mahara Variegata," "Mahara Variegata—Abnormal Leaves," "Los Banos Variegata-Jayanthi," and "Pixie Variegata") have been developed. le Cuckoo (2011) mentioned based on his experimental findings that the chemical mutation breeding method can obtain high frequency mutation and a wide spectrum of variation and is helpful for the variety improvement of bougainvillea. The methodology has been patented. Large quantities of cottage pieces of female parent bougainvillea are treated with sodium azide (NaN3) which is good for the growth of leaf color mutants and also develops new strains with stable heredity. Technology helps to increase the number and frequency of variants, reduction in treatment costs and breeding period, and to overcoming chimera formation. Stem cuttings of five varieties ("Mahara," "Lady Mary Baring," "Blondie," "Golden Glow," and "Vishakha") were exposed to 500, 1000, 1500, and 2000 rads of gamma rays and differential sensitivity to gamma rays was recorded among the cultivars. Mutations in bract color were detected in "Lady Mary Baring" (1000 rads) and "Golden Glow" (500 rad) and isolated by vegetative propagation. The bract color of the original "Lady Mary Baring" was orange group 29A and the mutant color was orange-red group 35A. The mutant had maximum plant height, branch length, number of leaves, and bract size in comparison to its parent. The bract color of "Golden Glow" was red group 39A and its mutant had red group 53A. Mutant showed reduced plant height, length of the branch, and size of leaf and bract (Swaroop and Jain 2016). Swaroop et al. (2015) made a preliminary experiment on the cultivar "Mahatma Gandhi" after treating 500, 1000, 1500, and 2000 rads of gamma rays and observed a sharp decrease in survival with an increase in doses and mentioned that light color variations in foliage/bract can be induced. Sodium azide was applied to *B. glabra* to study its mutagenic and genetic effects. Mutation rate and survival were determined and genetic effects of isozyme patterns were studied for various populations. Phenotypic mutation rates had a positive correlation with dose. Nine variants were assessed by peroxidase isozyme and the fingerprint showed that variants BN6 and BN7 were possible new strains with desirable character (leaf color and leaf shape) (Chen et al. 2012a). Radiosensitivity of *B. glabra* and *B. spectabilis* was studied based on biological effect in VM_1 after treatment with different dosages of Co^{60} γ ray. There was a negative correlation between dose and survival, but a positive between dose and variation rates. The radiosensitivity of *B. spectabilis* was superior to *B. glabra* (Chen et al. (2012b). Stem cuttings of the cultivar "Sawitree" were exposed to 10, 20, 30, 40, 50, and 60 Gy gamma rays, and the LD50 dose was determined 32 Gy and 7 Gy for 50% growth reduction. 20–40 Gy has been determined most suitable for induction of mutation. Mutations in leaf and bract characters have been reported (Judsri et al. 2016). Jain and Swaroop (2016) did experiments to induce mutations in bougainvillea and to characterize bud sports. The stem cuttings of five bougainvillea varieties ("Mahara," "Lady Mary Baring," "Blondie," "Golden Glow," and "Vishakha") were exposed to gamma rays (0, 500, 1000, 1500, and 2000 rads) and noted the radiation effects. Mutants and sports were isolated and multiplied for characterization. Delayed sprouting and variations in other characters were observed after treatment. 2000 rods of gamma rays had taken 85 days for sprouting in the Golden Glow cultivar. The results indicated that gamma rays were found more effective at 500 rads and 1000 rads of gamma rays compared to other treatments in cultivars. The mutants of "Lady Mary Baring," "Golden Glow," and Bud Sport of Dr. Bhabha significantly differed and had a considerable change in leaf and mostly in bract color according to the RHS color chart as well as morphological variation. The experiments also evidenced that mutants are stable. Anitha et al. (2017) treated hardwood cuttings of Bougainvillea spectabilis Willd. cv. Lalbagh with gamma rays (0.0, 7.5, 10.0 Gy) and EMS (0.8, 0.9, and 1.0%) for induction of mutations. Sprouting and survival of cuttings gradually reduced with increasing mutagenic treatments. Morphological mutants' viz., dwarfness, early flowering, and thornless stem were observed within the mutagen-treated plant population. Nine chlorophyll mutants were observed. Namely albina, albina green, xantha, chlorina, viridis, yellow viridis, striata, maculata, and variegated type. Chlorina mutation was more predominant followed by maculata and viridis. EMS (1.0%) induced more morphological mutants and chlorophyll mutants. The 5.0 Gy dosage of gamma radiation induced the highest mutations than other treatments. The highest mutation rates in terms of effectiveness and efficiency were recorded in EMS than gamma radiation in Bougainvillea cv. Lalbagh. Chang et al. (2019) studied the effects of EMS, sodium azide, and pingyangmycin on cuttings of *B. glabra* "Mrs. Eva" and *B. × buttiana* "Miss Manila," respectively. In 20 h EMS treatments, the half-lethal dose of "Mrs. Eva" and "Miss Manila's" cuttings was 0.22% and 0.62%, respectively. The shoot number and length of the longest branch of two *Bougainvillea* varieties showed a decreasing trend, while the chlorophyll and carotenoid content increased gradually with the increasing concentrations of EMS. With the increasing EMS concentration, the leaf hue of "Mrs. Eva" changed to a grayish tone and that "Miss Manila" changed to a red tone gradually. Almost all "Mrs. Eva" cuttings died in 20 h NaN_3 treatments, while the survival rate of "Miss Manila" cuttings was 82–92%. With the increase in NaN_3 concentration, the shoot number of "Miss Manila" decreased significantly. Twenty hours PYM treatments significantly improved the length of the longest branch of two *Bougainvillea* varieties in contrast to the control, but had no significant influence on the survival rate. Moreover, variant seedlings were found in EMS and PYM treatments, representing transformative leaf margins, wrinkled leaf surface, or double leaf tips.

A considerable amount of mutation work has been carried out on bougainvillea at CSIR-National Botanical Research Institute, Lucknow, India. Both physical (gamma rays) and chemical (ethyl methane sulfonate, EMS, and methyl methane sulfonate, MMS) mutagens were used (Datta 2004). Quite a large number of single bracted varieties belonging to different bract color groups (orange/yellow group, mauve/purple group, red/magenta/pink group, bicolored/multicolored group) were selected under mutation breeding program: “Blondie,” “Chitra,” “Dr. Harbhajan Singh,” “Dr. R.R. Pal,” “Elizabeth,” “Glabra,” “Golden Glow,” “Isabel Green Smith,” “Lady Mary Baring,” “Manohar Chandra,” “Mrs. Butt,” “Mrs. H.C. Buck,” “Partha,” “Palekar,” “Padmi,” “Penang,” “Poultoni Special,” “President,” “President Rosevelt,” “Purple Gem,” “Red Glory,” “Scarlet Glory,” “Sensation,” “Shubhra,” “Tetra Mrs. McClean,” “Wazid Ali Sha,” “Zinna Barat,” “Zulu Queen,” etc.

“Cherry Blossom,” “Los Banos Beauty,” “Mahara,” and “Rosevilles Delight” are four double-bracted cultivars. In addition, two chlorophyll-variegated mutants developed spontaneously. One is “Marieta” developed from “Mahara” and another “Archana” developed from “Rosevilles Delight.” Conventional breeding is not possible for improvement in double-bracted bougainvillea due to the absence of flowers. Mutation induction is the available technique for improvement. These cultivars were included in the mutation breeding program.

Some varieties were utilized several times for repeat experiments.

Stem cuttings were dipped in 0.01, 0.02, and 0.03% aqueous solution of EMS for 6 h and the cuttings were thoroughly washed in running water after treatment before planting. Stem cuttings were dipped in 0.1, 0.2, and 0.3% aqueous solution of MMS for 6 h and the cuttings were thoroughly washed in running water after treatment before planting. Stem cuttings (13 cm height) of both single and double-bracted varieties were exposed to 250, 500, 750, 1000, 1250, and 1500 rads of gamma rays (dose rate 36 s/Krad, ^{60}Co). Treated plants showed a reduction in sprouting, plant height, and survival and an increase in chromosomal aberrations in all the cultivars. Stimulation in all these characters was recorded in some cultivars at lower doses. Various forms of morphological abnormalities in leaves including chlorophyll variegations were observed in the treated population. It was observed that the abnormal leaves per plant decreased with the aging of the plants from 45 to 90 days. The abnormal cells were gradually eliminated due to diplontic selection and the plants appeared to be slowly returning to near normalcy. The percentage of abnormal leaves increased again in some plants after 90 days. It has been presumed that the affected cells in some unusual cases retained and participated in growth and development at later stages of growth. Differential sensitivity of different cultivars to mutagens was observed irrespective of the bract color group. Sensitivity has been determined based on several parameters like sprouting; survival; plant height; leaf, branch, and bract number and size; chromosomal aberrations, and mutation frequency. The most suitable dose of gamma rays for irradiation of stem cuttings has been standardized from 250 to 1250 rads. Studies indicated that radiosensitivity in Bougainvillea is a genotype-dependent mechanism. Mutations in leaf and bract color were recorded in many cultivars.

The most potentially beautiful chlorophyll-variegated mutant “Arjuna,” developed from the cultivar “Partha” after 250 rad treatment, has been commercialized (Gupta and Shukla 1974; Gupta and Nath 1977). Four gamma ray induced chlorophyll-variegated mutants have been developed in the double-bracted group. One mutant variety “Pallavi” has been developed from “Roseville’s Delight” from

10Gy treatment (Banerji et al. 1987). "Mahara Variegata" has been developed from "Mahara" (Datta and Banerji 1997). Two mutants have been developed from "Los Banos Beauty" and released for commercial exploitation in the name of "Los Banos Variegata" and "Jayanthi" (Anonymous 1982; Banerji and Datta 1987; Banerji et al. 1987; Datta 1992; Datta and Banerji 1990, 1997).

No new bract color could be induced in double-bracted varieties. It was interesting to note that although there was no new bract color mutation the development of induced chimera of existing bract color was recorded. The irradiated population of "Roseville's Delight" produced chimeric bract of "Mahara," "Cherry Blossom," and "Los Banos Beauty" type; "Cherry Blossom" produced "Mahara" and "Los Banos Beauty" type and "Los Banos Beauty" produced only "Mahara" type chimera. But surprisingly no change in bract color in the irradiated population of "Mahara" was recorded (Datta and Banerji 1997; Banerji and Data 1993). Bract color mutation in bougainvillea is infrequent. Only a few varieties showed bract color mutations. One branch from 750 rad gamma ray treatment showed bract mutation in the cultivar "Palekar." The mutant color has been established in pure form through cuttings. The original young bract color of "Palekar" is Current Red (821/3) changing to Salferino purple (26/1) on maturity. The mutant bract color at the young stage is orange (12/1) which on maturity becomes Azalia Pink (523/1). Leaves in the original cultivar are obtuse and flat while in the mutant these are elliptic and incurved (Sharma et al. 2002; Srivastava et al. 2002).

Reports on chemical mutagenesis to improve multibracted bougainvillea are very limited. Mature stem cuttings of the cultivar "Los Banos Variegata" were treated with 0.01–0.3% of EMS for 6 h. One plant from 0.02% treatment showed one mericlinal chimaeroid branch with variegated leaves in M_1V_1 generation. The size of chlorophyll variegation in leaves varied from a narrow streak on a leaf to a whole leaf. The growth of the axillary bud associated with variegated leaf was arrested by the influence of the apical dominance of the main apex of the mericlinal shoot. The chimeric branch has been isolated in pure form by air layering method and subsequently multiplied through repeated cuttings and named "Los Banos Variegata-Jayanthi" to release as a new variety (Jayanti et al. 2000; Jayanthi et al. 1999).

One gamma ray induced variant with all abnormal leaves has been detected and isolated in double-bracted cultivar "Mahara." The lamina of all the leaves showed unequal growth. This is a factual genetic deformity due to radiation (Datta et al. 2017). This type of permanent genetic abnormality has been detected for the first time in about 30 years of experiments with different types of mutagens and plants at CSIR-NBRI Lucknow. It may be mentioned that for such genetic abnormalities, it is not necessary to treat the plant with very high doses. The abnormal bougainvillea plant has been detected within LD_{50} dose in which previously hundreds of cuttings of bougainvillea have been irradiated.

Management of leaf chimera.

Chlorophyll variegations mostly appeared as mericlinal chimera, i.e. some leaves were green and others were variegated and detected in M_1V_1. Mericlinal chimeras are unstable and there is always a change in the frequency of normal and chlorophyll-variegated branches. Periclinal chimeras are stable plant sports that have resulted in unique features in many ornamentals. In the chimeric chlorophyll-variegated branch, the growth of the axillary bud associated with the variegated leaves in the chimeric branch is arrested by the influence of the apical dominance of the main apex of the mericlinal shoot. A simple technique has been standardized to convert the mericlinal branch to the periclinal chlorophyll-variegated branch. This technique is very easy but very important in mutation, especially for the management of chlorophyll-variegated branches. Considering its significance the technique has been reported earlier. To avoid apical dominance all the green leaves of the mericlinal branch were removed and the branch was made an arch to provide better conditions for the growth of the axillary buds of variegates leaves. Ultimately, new branches developed from the axillary buds of variegated leaves. Out of many new branches, some were purely periclinal where all the leaves were variegated and some were mericlinal. The same process was again repeated for the mericlinal branch. By this technique, new chlorophyll-variegated mutants have been isolated in pure form and commercialized in bougainvillea and other crops (Banerji and Datta 1987; Banerji et al. 1987; Datta 1992, 2023).

A literature survey indicates that a good number of varieties have been developed in different parts of India through the detection and isolation of bud sports, hybridization, and induced mutation. But unfortunately, no further systematic research work is being carried out at any research institute/university for further improvement of bougainvillea. Considering its importance more work should be initiated for the development of improved varieties through induced mutations which have now been proven to be a good source for inducing genetic variability.

References

Abd El-Mageid YA, Awad AE (2009) Studies on inducing mutations in bougainvillea. Zagazig J Agric Res ISSN:1110–0338

Abraham V, Desai BM (1977) Radiation induced variegation mutants in bougainvillea. Curr Sci 46(10):351–352

Anitha K, Surendranath R, Jawaharlal M, Ganga M (2017) Mutagenic effectiveness and efficiency of gamma rays and ethyl methane sulphonate on bougainvillea spectabilis willd. (cv Lalbagh). Int J Bioresour Stress Manag 8(2):247–256. https://doi.org/10.23910/IJBSM/2017.8.2.1634

Anonymous (1982) Mutation breeding. News Lett 20:18

Banerji BK (2010) Mutation breeding is an ideal and unique tool for the genetic improvement of bougainvillea. National symposium on lifestyle floriculture: challenges and opportunities. Session-2, crop improvement, biotechnology, and biodiversity, march 1921, 2010 at DR. Y.-S. Parmar University of Horticulture and Forestry, Nauni, Solan, H.P., India, Abstract No 2.1 (O):9–10

Banerji BK, Data SK (1993) 'Mrs. Butt'–bud sports and its near relatives. Sci Hortic 3:141–151

Banerji BK, Datta SK (1987) Gamma ray induced mutation in double bracted bougainvillea cv. "Los Banos Beauty". J Nucl Agric Biol 16:48–50

Banerji BK, Nath P, Datta SK (1987) Mutation breeding in double-bracted bougainvillea cv. "Roseville's Delight". J Nucl Agric Biol 16(1):48–50

Chang S, Huang S, Shisong XU, Yang G (2019) Chemical mutagenesis for the cuttings of *Bougainvillea glabra* 'Mrs. Eva' and *B.* × *buttiana* 'Miss Manila'. Chin J Trop Crops 40(2): 238–246

Chen T, Fan Y-W, Liu W (2012a) Study on the biological effect of bougainvillea induced by Co60-γ ray. J Henan Agric Sci 41:133
Chen T, Wang A, Liu Y et al (2012b) The study on the mutagenic effect of bougainvillea induced by NaN3. Chin Agric Sci Bull 25:191
Datta SK (1992) Mutation studies on double-bracted bougainvillea at National Botanical Research Institute (NBRI), Lucknow, India. Mutat Breed Newslett 39:8–9
Datta SK (2004) Bougainvillea research at National Botanical Research Institute, Lucknow. J Ornam Hortic 7(1):1–14
Datta SK (2023) Induced mutation breeding. Springer, Singapore. https://doi.org/10.1007/978-981-19-9489-0; ISBN 978–981–19-9488-3
Datta SK, Banerji BK (1990) "Los Banos Variegata"—a new double-bracted chlorophyll variegated bougainvillea induced by gamma rays. J Nucl Agric Biol 19(2):134–136
Datta SK, Banerji BK (1997) Improvement of double-bracted bougainvillea through gamma-ray induced mutation. J Plant Sci 64(4):395–400
Datta SK, Mandal AKA (2012) Standardization of in vitro multiplication of two difficult-to-root bougainvillea cultivars for commercial exploitation. Sci Cult 78(5–6):251–254
Datta SK, Jayanthi R, Janakiram T (2017) Bougainvillea. New India Publishing Agency, New Delhi
Deng H, Liu S (1990) The study of radiation-induced breeding in bougainvillea. J South China Agric Univ 13(1):89–93
Gills WT (1976) Bougainvilleas for cultivation (Nyctaginaceae). Baileya 20(1):34–41
Gupta MN, Nath PN (1977) Mutation breeding in bougainvillea. II. Further experiments with cvs. 'Partha' and 'President Rosevelt'. J Nucl Agric Biol 6(4):122–124
Gupta MN, Shukla R (1974) Mutation breeding in bougainvillea. Indian J Genet 34A:1296–1299
Heimerl A (1900) Denkschriften der Kaiserlichen Akadmiceder Wissenschaften, Mathematische, Naturwissens Chaftliche 70:97–124
Hong D, Shaode L (1990) The study of radiation-induced breeding in bougainvillea. J South China Agric Univ 13:89–93
Jain R, Swaroop K (2016) Mutation breeding and performance of induced mutants/bud sports of bougainvillea. Int J Hortic Floric Plant Sci 4(1):63–72
Jayanthi R, Datta SK, Verma JP (1999) Effect of gamma rays on single bracted bougainvillea. J Nucl Agric Biol 28(3):228–233
Jayanti R, Datta SK, Verma JP (2000) Chemical induced mutation in double bracted bougainvillea cv. 'Los Banos Beauty'. Indian Bougainvillea Ann 15:9–11
Judsri W, Taychasinpitak T, Jompuk P, Pipattanawong N, Puripunyavanich V (2016) Induced mutation by using gamma rays in bougainvillea cv. Sawitree. Agric Sci J 47(3):429–440
Cuckoo YZ le (2011) Chemical mutation breeding method for bougainvillea. Publication of CN102499071B. CN 2011104202762011–12–152,011-12-15. Application granted 2013-04-10
Sharma SC, Srivastava R, Datta SK et al (2002) Gamma-ray induced bract color mutation in single bracted bougainvillea 'Palekar'. J Nucl Agric Biol 31(3–4):206–208
Srivastava R, Datta SK, Sharma SC et al (2002) Gamma rays induced genetic variability in bougainvillea. J Nucl Agric Biol 31(1):28–36
Swaroop K, Jain R (2016) Mutation breeding and performance of induced mutants/bud sports of bougainvillea. Int J Hortic Floric Plant Sci 4(1):63–72
Swaroop K, Jain R, Janakiram T (2015) Effect of different doses of gamma rays for induction of mutation in bougainvillea cv Mahatma Gandhi. Indian J Agric Sci 85(9):1245–1247

Chrysanthemum

8

Abstract

Chrysanthemum with its large number of cultivars in respect of growth habit, size, color, and shape of bloom has attracted its admirers and enthusiasts all over the world for its use both as a commercial flower crop and as a popular exhibition flower. The chapter will provide a transparent picturesque application of mutation techniques for the creation of novel varieties in chrysanthemums. All important basic aspects and technical advancement related to induced mutagenesis work have been discussed. One can see all types of mutation work starting from classical to modern on this crop and can get a clear picture of technological advancement and its successful application for the development of new varieties. The knowledge generated so far on chrysanthemum will work as a model system for future need-based planning of successful and accurate application of mutation techniques in crop improvement programs.

Keywords

Chrysanthemum · Management of chimera · Induced mutagenesis · Colchi-mutation · Model plant

The autumn flowering chrysanthemum (C. *morifolium* Ramat.) is one of the highly profitable ornamental crops in the floriculture industry throughout the world. Chrysanthemum with its large number of cultivars in respect of growth habit, size, color, and shape of bloom has attracted its admirers and enthusiasts all over the world for its use both as a commercial flower crop and as a popular exhibition flower. The genus constitutes a large polyploid complex ranging from 2x to 22x, besides several aneuploids. The name *Chrysanthemum morifolium* was changed to *Dendranthema grandiflora* Tzueleu (Heywood and Humphries 1977; Kitamura 1978; Anderson 1987). But now it is mostly written *Chrysanthemum morifolium* Ramat. All the

S. Datta, *Role of Mutation Breeding in Floriculture Industry*,
https://doi.org/10.1007/978-981-99-5675-3_8

present-day colorful chrysanthemums are the result of extensive hybridization, spontaneous and induced mutations, and selections. Sports or spontaneous somatic mutations have played a very important role in the evolution of many Chrysanthemum cultivars. Wasscher (1956) has reported that the sporting process has contributed to about 30% of the Chrysanthemum cultivars. Some cultivars have given rise to a great number of sports such as "Sweetheart," "The Favourite" (Anonymous 1961; Bowen et al. 1962), and "Indianapolis" (Yoder 1967).

Chrysanthemum is propagated by seed, terminal cuttings, mature stem cuttings, suckers, and tissue culture.

8.1 Work Review

Most extensive work has been done on C. *morifolium*, Ramat. for its improvement through induced mutation by a huge number of scientists worldwide. The possibilities for creating different forms and improving the chrysanthemum are infinite due to its high heterozygosity. The chapter will cover most of the induced mutation breeding works carried out by the author and his contribution to floriculture. Most important mutation studies carried out at different other institutes have also been reviewed. An attempt has been made to discuss all important basic aspects and technical advancement related to induced mutagenesis work on chrysanthemums. It will not be an intelligent approach to consider all the published papers on mutagenesis for the improvement of chrysanthemums. The author was deeply involved in mutagenesis work for more than 30 years and such long concentrated efforts have so far resulted in several success stories on chrysanthemums. Voluminous knowledge has been generated on different aspects like radiosensitivity, selection of material, treatment period, suitable doses of γ-rays, colchicine treatment, recurrent irradiation, need-based and directive mutations, detection and isolation of mutants, the role of cultural practices on mutation, and commercial exploitation of mutants. All attempts have been made to concentrate on all minor details of applications of mutation technology as followed by the author and important work done at other institutions have been highlighted as and when required. The chapter will provide a transparent picturesque application of mutation techniques for the creation of novel and desired varieties in chrysanthemums.

Since the beginning, there is a step-by-step improvement in the technical procedure for the application of induced mutation for improvement of chrysanthemum and voluminous knowledge has developed for successful and accurate application of the technique. Considering the multifarious mutation activities, the chrysanthemum may be considered a model plant where one can see all types of mutation work starting from classical to modern. One can get a complete chronological picture of mutation technology from premature to the most advanced stage of chrysanthemum. Maximum mutant varieties have been developed in the chrysanthemum. Mutagenesis work on chrysanthemums generated maximum knowledge and its applied technical refinements have enriched the mutation technology maximum in comparison to other ornamentals. The experimental results and mutants obtained from different experiments in chrysanthemums are very interesting both from an academic and novelty point of view. Attempts have been made to pinpoint all minor technical details of mutagenesis work on chrysanthemums. The entire mutation breeding work has been represented in graphical and pictorial style so that one gets all about mutation technology from one crop. One can get almost all minor details of the mutation technology package in general and the chrysanthemum in particular. Author (Datta 2023) has reviewed in his contributed chapter the mutation work on chrysanthemum. He has mentioned that the *Chrysanthemum* is one of the most interesting ornamental groups of plant, where, perhaps, maximum work have been done worldwide. The chapter provides an authoritative review account of many aspects of current interest and progress in the field of induced mutagenesis for the improvement of chrysanthemum/ornamental crops. One can see all types of mutation work starting from classical to modern on this crop and can get a clear picture of technological advancement and its successful application for the development of new varieties.

The scope of application of mutation work for improvement is maximum in chrysanthemums in comparison to other ornamentals. The chrysanthemum (*C. morifolium* Ramat.) is classified into large flowered and small flowered categories and each category comprises different flower forms which increase wide popularity and high market demand. Chrysanthemum represents an interesting ornamental flower crop with a wide range of genetic diversity in flower size, flower type, flower color, plant stature, etc. The scope of selection parameters for experiments is maximum considering its flower form, color, and many other trade-related attributes (Fig. 8.1). Different approaches for mutation work on chrysanthemums have been shown in Fig. 8.2. The most interesting work on chrysanthemums is in vitro mutagenesis and management of induced and spontaneous chimeras. Experiments were conducted for the induction of mutations through direct irradiation, recurrent irradiation, and colchicine treatment. The scope of improvement of chrysanthemum is illimitable for its wide range of commercial characteristics like appealing flower color and shape, no pinch no stake dwarfness, out-of-season blooming, cut flowers (attractive color, long erect stem, uniform bloom opening, tough florets, long vase life, and healthy leaves), pot culture (dwarf and compactness, profuse branching, uniform spreading of branches, simultaneous blooming habit, attractive color and good color retention quality and healthy leaves), high yielding, garland purpose, exhibition type, chlorophyll variegation in leaves, showy decorative leaves, etc. (Fig. 8.3). A complete picture of induced mutagenesis experiments for the development of new varieties of chrysanthemum has been shown in Fig. 8.4.

A quick glimpse of the chronological advancement of mutation work on chrysanthemums is acknowledged in a nutshell. A wide range of physical mutagens like X-ray, gamma rays, fast neutrons, thermal neutrons, radioactive phosphorus, and microwave and chemical mutagens like ethyleneimine, ethyl methane sulfonate, and colchicine have been used for such studies.

Bowen et al. (1962) exposed rooted cuttings of seven cultivars ("Sweetheart," "Pearl Sweetheart," "Salmon Sweetheart," "Golden Sweetheart," "Orange

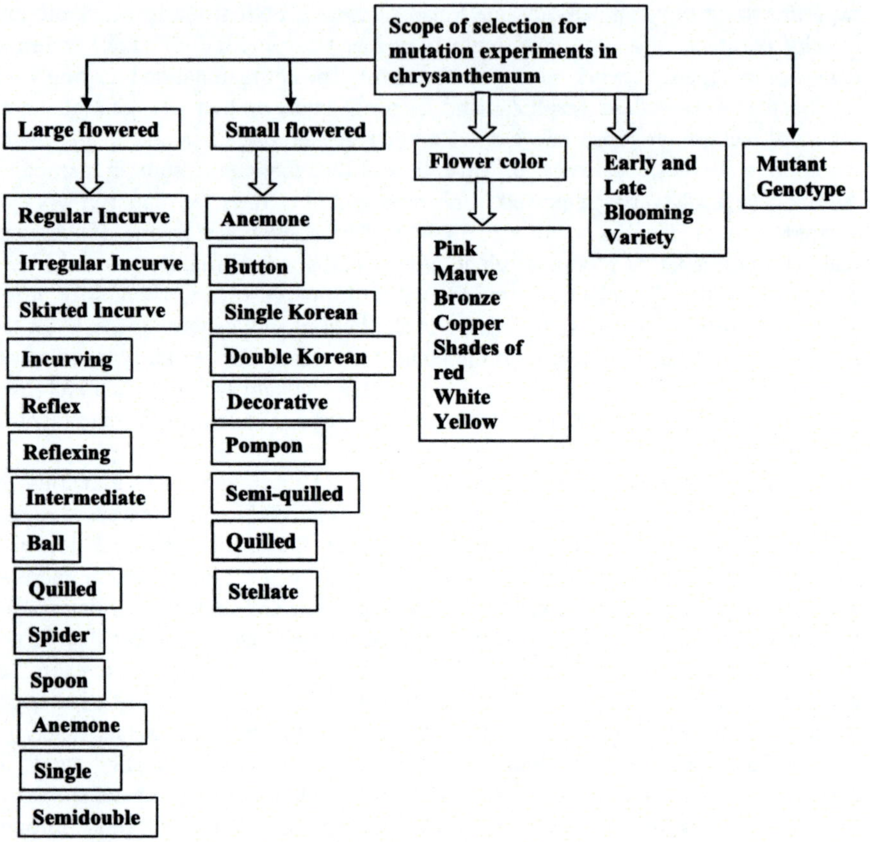

Fig. 8.1 Diversity in chrysanthemum and scope of selection for experiments

Sweetheart," "Red Sweetheart," and "Morcar Jewel") to 2.5–10 Krad gamma rays. LD_{50} ranged from 3.0 to 4.3 Krad. Treatment delayed flowering and induced color mutations in about 40% of treated plants and the mutation frequency varied from cultivar to cultivar. Dowrick and El-Bayoumi (1966a, b) irradiated the cultivar "New Princess" with X-rays (500–2000r) and gamma rays (1–4 Krad). Flower color changes were accompanied by a change in chromosome number and chromosome fragmentation. Most of the flower color changes resulted from the loss of pigmentation but a number with deeper colors were found. Mutation frequency was directly proportional to the dose. They have explained that mutations developed by the fragmentation of chromosomal chimeras or by the rearrangement of pre-existing chimeras. Broertjes (1966a, b) treated rooted cuttings of pot-grown Chrysanthemum cv. "Hortensien Rose" with X-rays, fast neutrons, thermal neutrons, and electrons. Electron was found to be ineffective. The optimum dose for X-rays was 1550 Rads. Higher mutation frequency was observed in both fast and thermal neutrons. The genetic constitutions of the treated material had an important role in mutation

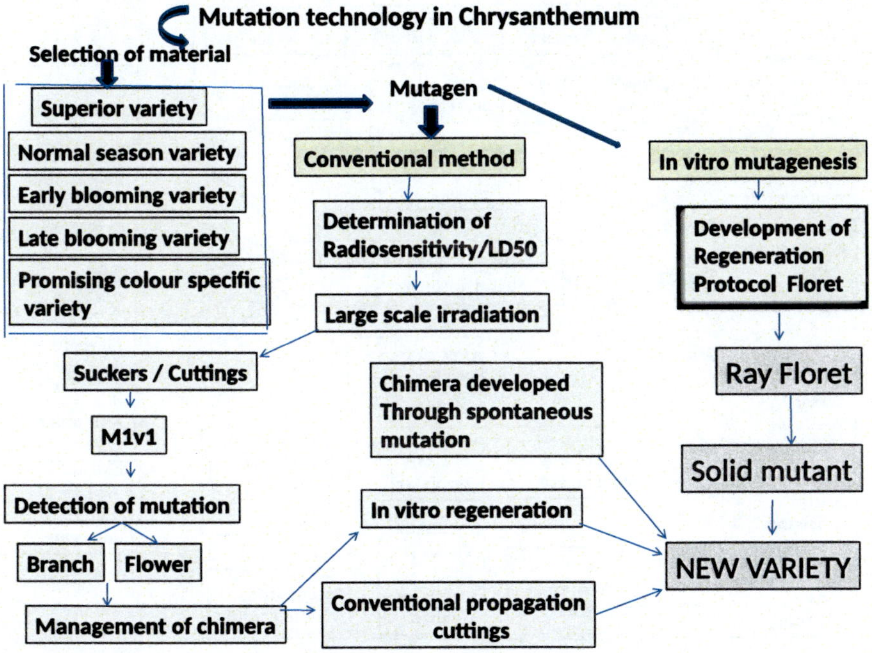

Fig. 8.2 Different steps of induced mutation technology for improvement of chrysanthemum

frequency and spectrum. Pink flower color varieties greatly developed maximum color variability. Broertjes et al. (1983) irradiated rooted cuttings of several cultivars and detected several low-temperature tolerant mutants but they were not stable. An adventitious bud technique was used to produce solid mutants from irradiated pedicel explant. Broertjes and Jong (1984) used rooted cuttings of four pollen-producing daisy Chrysanthemums and treated them with 17.5 Gy (1750 rad) X-rays and detected several male-sterile mutants identical to the control plants. Other mutants showed additional changes as reduced plant height or changes in the form, size, and color of the flower. Yamakawa and Sekiguchi (1968) applied high doses of gamma rays to inactivate all initial cells but one which resulted in increased mutant sector size and percentage of complete mutants. Matsubara (1982) and Matsubara et al. (1971) worked on both in vitro and in vivo techniques and isolated total mutants from sectorial chimera. Ichikawa et al. (1970) treated young plants of *C. morifolium* cv. Yellow Delaware and Delaware with ^{60}Co gamma rays in the gamma field chronically. The total exposures were 5, 10, and 20 kR. Many whole-type sports were produced after irradiation. Cytological examination of 19 mutant and 12 normal shoots produced after irradiation revealed that the somatic chromosome numbers ranged from 49 to 55 with an exceptionally doubled number of 110. More reduction of chromosome number was caused by higher exposures. The reduction of chromosome number was also correlated with the type of mutation

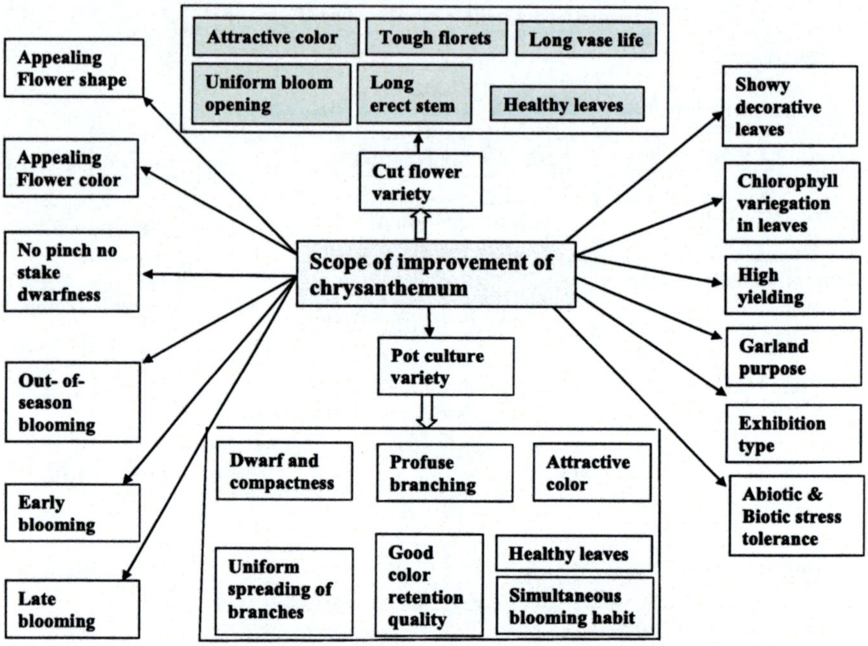

Fig. 8.3 Scope of improvement of chrysanthemum on various commercially important characters

and with the size of inflorescence. Jong (1978, 1984) and De Jong and Custers (1986) studied flowering characters (number of days to flowering, number of flowers per plant) at 12 °C, 13 °C, 15 °C, and 17 °C night temperatures in 79 F_1 populations from 15 parents of C. *morifolium*. Only a few F_1 plants flowered earlier than their parents; the majority flowered at the same time, later, or not at all. The number of flowers of the F_1 plants was higher than that of the parents. Considerable genetic variation for both flowering time and flower production was found in the experiment and recommended that in the selection of parents, special attention should be paid to earliness. If rapid flowering selections are sought for cultivation at lower temperatures at least one but preferably both parents should be early. Jung-Heliger and Horn (1980) treated cuttings, callus, and suspension cultures of five different clones with gamma rays (1.2–1.8 Krad) and one clone with EMS (0.5–1.5%), and treated plants were regenerated from stem cuttings, node cuttings and in vitro cultures. The frequency of flower color variants was low following treatment of cuttings, larger when plants were regenerated from single nodes, and largest following the treatment of suspension cultures. De Jong and Custers (1986) treated pedicel segments and petal epidermis of *C. morifolium* cv. "White Spider" with 8 GY X-ray doses and induced in vitro to regenerate adventitious buds either directly from the original tissue or indirectly via callus to induce genetic variability for yield. All treatments yielded variations in flower morphology, changes in flower color, flowering time, flower number, production variants, etc. Preil et al. (1983) irradiated

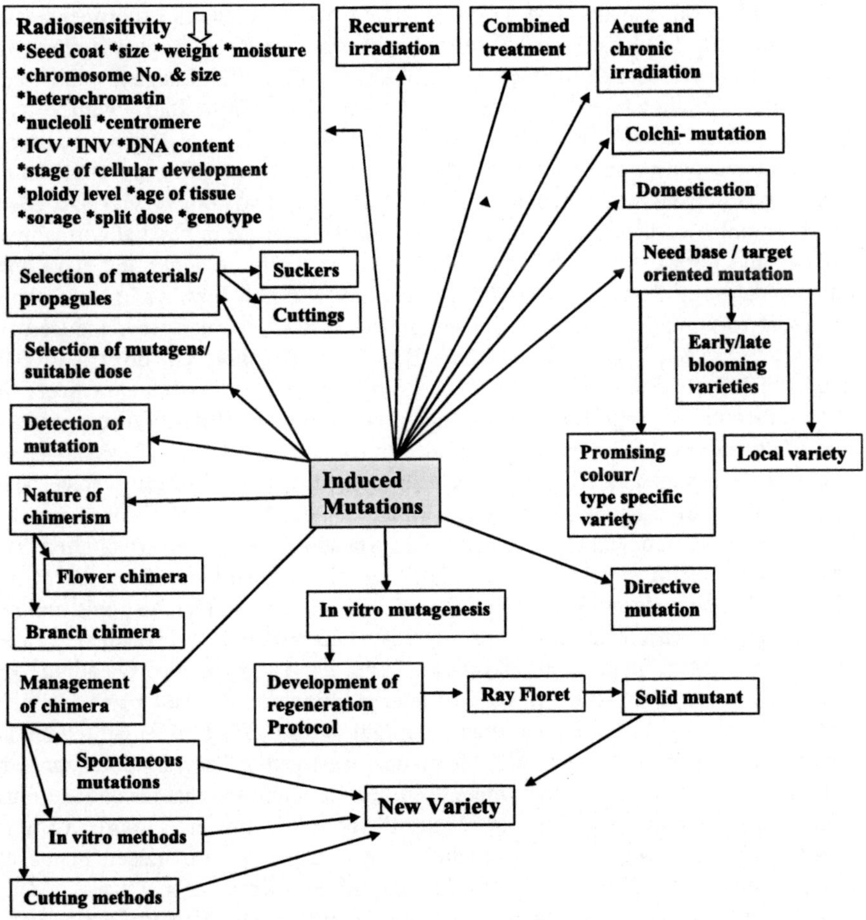

Fig. 8.4 Diagrammatic representation of induced mutation technology package showing different steps

single cells and small aggregates from suspension cultures of chrysanthemum, cv. "Puck" and then exposed for 170 days to in vitro stress temperatures (8 °C). Calli with embryo-like structures and initiated shoot differentiations were transferred to optimal temperatures (24 °C) at the end of the stress period. Regenerated plants were grown in a greenhouse under natural short-day conditions up to flowering at stress temperatures (10 °C/10 °C). Sixteen plants among 2112 chrysanthemums induced flowers at 10 °C. Most of the plantlets selected after in vitro temperature stress were not mutants. Two selected chrysanthemum clones flowered 7–10 days earlier than the original variety under summer greenhouse conditions as well as under 14 °C or 10 °C temperature regimes. These plants were considered real physiological mutants. Lema-Rumińska and Zalewska (2002) developed mutants through in vivo and in vitro mutagenesis using ionizing radiation and determined their ploidy level. Latado et al. (1996) did in vitro mutagenesis experiment with cv. "Pink repin." Dilta et al. (2003) irradiated rooted cuttings of chrysanthemum cultivars ("Ajay," "Baggi," "Bright Golden Anne," "Ellen Van Langen," "Glance," "Gulmohr," "Mountaineer," "Nanako," "Shyama," and "Snow Ball") with 2 krad gamma rays and studied radiosensitivity. The radiosensitivity was determined based on various vegetative and floral characteristics. There was a significant reduction in plant survival, plant height, growth, number of branches, leaf number, leaf size, plant spread, number of flowers, flower size, etc., and an increase in plant abnormalities (foliage and floral), days to bud formation, days to reach the harvesting stage and ratio of disc: ray florets per head in all cultivars with 2.0 k rads of gamma rays treatment as compared to control. The somatic color mutations were induced in "Ellen Van Langen," "Gulmohar," "Snow Ball," and "Shyama." Latado et al. (2004) applied EMS (0.77%—0.075 M) for 1 h and 45 min to immature pedicels of chrysanthemum cv. Ingrid (dark pink color) followed by the induction and production of adventitious buds in vitro to induce mutations. LD_{50} of EMS was close to 0.82% (v/v). Regenerated plants were evaluated at the flowering stage and 48 mutants with changed flower color (pink-salmon, light pink, bronze, white, yellow, and salmon color) were detected. Nagatomi et al. (2003) applied ion beams and developed six mutant varieties in chrysanthemum. Zalewska et al. (2007) did experiment with 11 ("Richmond" and its 10 radio mutants) cultivars by propagating in vitro with shoot tips and leaves as explants to investigate if the explant type used for micropropagation affects the genotype and phenotype of chrysanthemums. Few cultivars showed altered inflorescence color plants derived from adventitious buds. Those changes might be an effect of either chimeral structure or somaclonal variation of the plants investigated. The variation appeared only if non-meristematic explants were used. Results indicated that the adventitious buds technique might be useful in chrysanthemum breeding as a source of new variability. The experiment was conducted by treating in vitro raised plants of chrysanthemum cv. "Taihei" with 15, 30, and 60 Gy gamma rays at a rate of 0.5, 1, 2, and 5 Gy/h. The objective was to study the influence of total irradiation dose and dose rate on flower color mutation and nuclear DNA content as an index of radiation damage. The regeneration rate and frequency of flower color mutations were studied from the tissue culture of leaf explants from irradiated plants. From the analysis of mutation frequency, the effect of dose rate and total dose, and nuclear DNA content it has been interpreted that gamma ray irradiations of high total doses at low-dose rates efficiently induce mutations with less radiation damage in chrysanthemum (Yamaguchi et al. 2008). Padmadevi (2009) did both in vivo and in vitro mutagenesis experiments to induce genetic variability in chrysanthemums. Results enriched basic knowledge and developed new variants. Lema-Rumińska and Sliwinska (2009, 2015) demonstrated that somatic embryogenesis in chrysanthemums can be applied for separating periclinal chimera components for chimeric cultivars and for receiving an additional source of variation in the breeding of cultivars. Dwimahyani and Widiarsih (2010) exposed shoot explants of chrysanthemum cv. "Yellow Puma" to gamma rays (10, 15, and 20 Gy) and cultured in vitro for induction of variations. 20 Gy inhibited the growth of leaves and branches and

10 Gy promoted the multiplication rate. They recommended that in vitro mutagenesis using gamma irradiation with a dose of 10–15 Gy can be used for inducing genetic variability of chrysanthemum cv. Yellow Puma. Barakat et al. (2010) did in vitro mutagenesis experiments with cv. "Delistar White" by treating ray florets with gamma rays (0.5 and 1.0). Callus induction and shoot formation percentages were affected by gamma ray doses, whereas the variation between medium protocols and the variations due to the interaction among medium protocols and doses were statistically insignificant. The results indicated that the irradiation dose of 0.5 Gy was the most effective in inducing mutations in flower shape and the number of florets per flower-head. Genetic polymorphism among chrysanthemum mutants and their parent was determined using RAPD analysis. Mahure et al. (2010) exposed unrooted cuttings of the cultivar "Red Gold" to gamma rays (10, 20, and 30 Gy) to induce variability. Lower doses resulted in hormesis and induced encouraging novelties while the higher doses often induced a high degree of abnormalities and consequent mortality. New color mutants induced by gamma radiation were isolated and purified in M_2 generation. Misra et al. (2010) reported results of the effects of gamma radiation (100, 150, 200, and 250 Gy) on different morphological characters of chrysanthemum cv. "Pooja." Somatic mutation in flower color (yellow, pink, and white) and form (tubular floret) were detected in the chimeric form in V1M1. Momin et al. (2010) reported the evaluation reports of induced chrysanthemum mutants under the net house in VM3. The survival of plants of all the mutants was significantly reduced over the control ("Akola Local"). The mutants were derived from a 2.0 kR dose of gamma rays. In the VM3 generation, the early and dwarf mutants of 2.0 and 3.0 kR doses were found to be stable. Padmadevi et al. (2010a) did in vitro mutation experiment by treating the chrysanthemum cultivar "Ravi Kiran" with gamma rays (0.5 and 1.0 kR) and EMS (0.1%, 0.2%, and 0.3%). Mutagen effects were measured based on different morphological and floral characters, and different biochemical parameters (chlorophyll content, catalase, peroxidase, IAA oxidase, anthocyanin content, etc.). Padmadevi et al. (2010b) reported in vitro protocol for regeneration of florets and data was recorded at different developmental stages. They treated chrysanthemum cv. "Ravi Kiran" with gamma rays (0.5 and 1.0 kR) and EMS (0.1%, 0.2%, and 0.3%) separately and in combination. All steps for in vitro regeneration of treated explants have been reported. Desirable mutants were detected. Salleh et al. (2010) studied the early effects of acute gamma and ion beam irradiations on the vegetative growth of in vitro irradiated chrysanthemum in both low and high lands. Data were recorded on the number of leaves, plant height, and internode length of the irradiated plants and statistically analyzed. Verma et al. (2010a, b, c) treated rooted cuttings of small-flowered chrysanthemum cv. "Little Darling" with 0.0625%, 0.125%, and 0.25% EMS for 4 h and recorded dose effects on morphological and floral characters. Chimeric new floret color mutants were observed in a few plants after 0.0625% treatment in V1M1. Verma et al. (2010a, b, c) treated rooted cuttings of chrysanthemum cv. "Flirt" and the mutant cultivar "Purnima" with gamma rays (0, 1000, 1500, and 2000 Gy) and studied radiation effects on survival, plant growth, branch and leaf characters, floral characters, etc. Chimeric new flower color mutations were detected in both cultivars. Salleh et al.

(2012a, b) studied the comparative effectiveness of acute gamma and ion beam irradiation in generating flower color mutations on nodal explants of Chrysanthemum morifolium cv. Reagen Red. The nodal explants were irradiated with acute gamma (0, 10, 20, 30, 40, 50, 60, 70, 80, 90, 100, 110, and 120 Gy) and ion beam (0, 0.5, 1.0, 2.0, 3.0, 5.0, 8.0, 10, 15, 20, and 30 Gy). The optimal dose determined is 10–15.0 Gy for acute gamma radiations and 3.5–4.0 Gy for ion beam. Relative biological effectiveness for the ion beam was found 3.75 higher than the acute gamma. 77.8% flower color mutations were recorded at acute gamma rays and 42.3–58.3% at ion beam. Kaul et al. (2011) treated in vitro raised shoots of cv. "Snow Ball" with gamma rays (5, 10, 20, and 30 Gy) and found 10 Gy as the most effective dose to induce flower color mutations. Twenty RAPD primers were used to amplify DNA segments from the genomic DNA of the control and its ten variants, and the genetic similarity among them ranged from 0.06 to 0.79 revealing high genetic diversity. An attempt was made to find out the relative effectiveness of ion beam irradiation on ray florets and nodal explants of *C. morifolium* cv. "Reagan Red." Both propagules were treated with 0.5, 1.0, 2.0, 3.0, 5.0, 8.0, 10, 15, 20, and 30 Gy doses. Based on survival, plant morphology, and flowering characteristics the optimal dose (RD_{50}) was detected at 2.0 Gy for ray florets and 4.0 Gy for nodal explants. The ray florets were more sensitive and developed more mutations as compared to nodal explants (Salleh et al. 2012a, b). Verma et al. (2010a, b, c) reported details of in vitro techniques and how induced yellow chimeric floret was regenerated and established in pure form. Kapadiya (2014) did a mutation experiment with two varieties ("Jaya" and "Maghi") of chrysanthemum and induced variability in both vegetative and floral characters. Kapadiya et al. (2014) treated cuttings of chrysanthemum variety "Maghi" with three concentrations each of EMS and DES (0.02%, 0.03%, and 0.04%) and five doses of gamma rays (0.5, 1.0, 1.5, 2.0, and 2.5 kR) to exploit variability and evaluated for heritable effects on various parameters, viz. survival rate, morphological and flowering. All the mutagenic treatments delayed flowering up to 6–7 days, whereas flowering duration was significantly reduced. Flower-head diameter and the number of ray and disc florets were significantly increased with 0.5 and 1.0 kR gamma rays, respectively. Flower weight, the number of flowers, and flower yield per plant were highest at the lowest dose of γ-rays. Total abnormalities in vegetative and floral characters were higher in γ-rays as compared to EMS and DES. "Maghi" exhibited two foliage mutants but with no flower at 0.03% EMS and 1.0 kR gamma rays. Verma et al. (2012) treated cuttings of chrysanthemum cv. "Thai Chen Queen" with gamma rays and grown under polyhouse and detected chimeric mutation in floret color (red). The chimeric red ray floret explants were cultured in vitro and the regenerated plants maintained the true-to-type to the original red mutant color. Kumari et al. (2013) treated rooted cuttings of var. "Otome Pink" with gamma rays (0, 10, 15, and 20 Gy) and evaluated radiation effects on morphological, palynological, and anatomical characters. Plant survival, plant height, number of flower heads, stems per plant, stem diameter, and leaves per plant reduced after gamma irradiation. The delayed flowering and plant in the vegetative stage were observed at a 20 Gy gamma irradiation dose. Pollen fertility, the number of chloroplasts per guard cell, flower-head size, and fresh weight also decreased with increasing doses. Different types of leaf abnormalities, flower-head fasciation, and asymmetrical development of flower heads increased with an increased dose of gamma irradiation. Changes in flower color and shape were recorded after treatment in the form of chimeras. Two variants, one at 10 Gy having yellow color and the other at 15 Gy having quilled petals were obtained and further multiplied vegetatively. Singh and Bala (2015) treated terminal rooted cuttings of cv. "Bindiya" with gamma rays (0, 10, 20, and 30 Gy) and determined 30 Gy as the LD_{50} dose. Irradiation with doses 10 and 20 Gy induced flower color mutations which were isolated in pure form. Soliman et al. (2014) did in vitro experiment to induce mutations. White petals of chrysanthemum cv. "Youka" were cultured in vitro and the calli were irradiated with gamma rays (0, 10, 15, and 20 Gy). Plants developed from the irradiated calli were different from control plants in the number of leaves, leaf length, and width, number of flowers, flower diameter, petiole diameter, and petiole length after transplanting into the greenhouse for almost 70 days. 15 Gy produced three mutations in flower color and shape (tubular petals, yellow flowers, spooned-shaped ray florets, and flat-shaped florets). Semi-quantitative RT-PCR showed that most carotenoid-biosynthesis-related genes, except for violaxanthin deep oxidase (VDE) and lycopene ε-cyclase (LCYE), showed similar expression patterns in petals of the original "Youka" and its mutants (M2 and M3). The yellow mutants were maintained vegetatively and proved to be true-to-type in one successive generation. It has been concluded that gamma radiation with a 15 Gy dose can be used for in vitro induction of flower color and shape mutations of chrysanthemum cv. "Youka." Telem et al. (2015) treated four local varieties with gamma rays (10, 20, and 30 Gy) to induce desirable variation. Lower dose induced suitable novelties, while the higher dose induced a high degree of abnormalities. New color mutants were found distinct from parents through the RAPD technique. Patil et al. (2015, 2017) treated suckers of chrysanthemum cv. "Local Golden" with gamma rays (0.5, 1.0, 1.5, 2.0, 2.5, and 3.0 Krad) and radiation effects were studied on vegetative and flower traits. No leaf and floral abnormalities were observed in treated plants. No flower color mutation was detected but one plant showed a tubular flower shape mutant. Terminal cuttings (6–7 cm) of chrysanthemum cultivar "Jaya" were treated with EMS and DES (0.02%, 0.03%, and 0.04%) and gamma rays (5, 10, 15, 20, and 25 Gray). Both chemical mutagens and gamma rays changed the flowering period from late to early blooming at lower doses (0.02% EMS and DES, and 5, 10, and 15 Gy gamma rays). Gamma rays induced more variations in colors like light pink, yellow, and brick red with yellowish disc florets (Kapadiya et al. 2016). *Dendranthema indicum* var. *aromaticum* is a rare aromatic species of chrysanthemum. Miao et al. (2016) experimented to find out a suitable treatment combination of colchicine to induce polyploidy. Three propagules (shoot tips, pre-germinated seeds, and grin), five colchicine concentrations (100, 200, 500, 1000, and 2000 mg L^{-1}), and three treatment durations (12, 24, and 48 h) were the experimental design. The chromosomal analysis confirmed the development of 7 tetraploids and 301 chimeras. The treatment of grin seeds with 1000 mg L^{-1} colchicine for 24 h (14.5%) and shoot tips with 1000 mg L^{-1} colchicine for 7 days (40%) was suitable for induction of chromosome doubling. The tetraploid plants displayed much larger stomata with lower density, as well as a higher chloroplast count than the diploid paints. Moreover, tetraploid plants developed larger, thicker leaves, greater flower diameter, more epidermis hairs, and shorter plant height than diploid plants. An attempt has been made to find out the effects of microwave treatment in inducing genetic variability in chrysanthemum cv. "Alchimist." Leaf explants with or without callus were exposed to microwaves for different periods and in different milieus. Changes recorded morphological and floral characteristics like the development of longer shoots, inflorescence with greater floral diameter, prolonged bud opening, and changed flower shape and color. Results indicated that microwave treatment may be used in chrysanthemums for induction of mutation (Mile and Kulus 2018). Rooted cuttings of cv. Maghi were treated with 100, 150, 200, 250, 300, 350, 400, and 450 Grays gamma rays. Lower radiation doses induced a maximum percent of foliage (variegation) and flower color mutations and, also stomatal aberrations. Higher doses increased aberrations in morphological and anatomical traits (Bajpay and Dwivedi 2019). Shafiel et al. (2019) treated leaf explants of three chrysanthemum cultivars with different doses of gamma rays to induce desirable variations. From experimental results, 25 Gy was found most appropriate dose to induce mutations. 25 G ray produced a maximum change in purple petal color with a mutation rate of 54.56%. 25 G ray developed changes of 32.11% in the pink cultivar. Four new mutants were established for commercialization. The white "Bacardi" variety was irradiated by gamma radiation to create variation and to determine the Effective Mutation Dose (EMD-LD50). In vitro, cuttings were irradiated with gamma (0, 5, 10, 15, 20, 25, and 30 Gy) and subcultured. The EMD was calculated by linear regression as 20 Gy. Some changes were recorded in the leaves and flowers of the plants such as variable flowers, flowering time, differentiation on plant length, chimeric leaves, flower number per bunch, and ray floret differentiation. The changes in the ray florets were pink color, pink strips, and length changes. These useful mutant lines were selected from the treated population (Haspolat et al. 2019). Puripunyavanich et al. (2019) cultured in vitro leaf explants of seven Chrysanthemum varieties and healthy regenerated plantlets were treated with gamma rays (0, 20, 40, 60, 80, and 100 Gy). Irradiated healthy plantlets were transferred into natural environment. After 16 weeks in greenhouse, chrysanthemum no.D23 and no. D27 had no flower and grew slowly. Twenty gray irradiated chrysanthemum no.12-10 showed whole pink shade petal while the parent line was pink-orange shade flower. Chrysanthemum flower no.11-4 showed decrease yellow shade petal compared with pink-orange parent line and no.11-8 showed increase yellow shade petal on the middle of flower compared with pink-yellow parent line. Chrysanthemum no. M07-4 showed flower form mutation which produced numerous petals. Su et al. (2019) made a review on the recent achievements in conventional and modern molecular breeding methods and emerging omics technologies and discussed their future applications for improving the agronomic and horticultural characteristics of chrysanthemums. Ghormade et al. (2020) did mutagenesis experiments with three cultivars ("Raja Pandharpuri," "Brown," and "Shinaton") of chrysanthemum (rooted cuttings) using colchicine (0.01%, 0.02%, 0.03%) and EMS (0.01%, 0.05%, 0.1%,

0.5%, 1.0%, 1.5%). A significant reduction occurred in both colchicine and EMS-treated cuttings of chrysanthemum varieties in terms of the number of flowers per plant, yield of flowers per plant, number of ray florets per flower, peduncle length, peduncle diameter, the vase life of cut flowers, etc. Whereas the weight of a single flower and diameter of the flower were increased due to the colchicine and decreased due to EMS-treated population of chrysanthemum varieties over control. Miler et al. (2020) measured DNA content for early detection of somaclonal variants and chimeric mutants in chrysanthemums. One chrysanthemum cultivar belonging to the Lady group and 10 X-ray or gamma ray induced solid and periclinal chimeras. FCM analysis of DNA content in young plantlets can be indicative of the stability of inflorescence color in chrysanthemums, especially chimeric cultivars, and for mutant detection. In vitro grown micro shoots of cv. "Candid" were exposed to 10, 20, 30, and 40 Gy gamma irradiation to generate flower color mutations. Survival, leaf size, and the number of leaves on the plant after the eighth-week interval also decreased with the increasing trend of gamma irradiation dose but recorded a minimum decline in plants developed from shoots irradiated with a 10-Gy gamma irradiation dose. The minimum delay in the number of days to floral bud appearance took under 10 Gy, as compared to the control. The highest number of flower color mutants was recorded under 10 Gy (light pink, orange pink, white, and yellow) (Din et al. 2020). Din et al. (2021) treated in vitro developed micro shootlets of chrysanthemum cv. "Candid" with 10, 20, 30, and 40 Gy gamma irradiation and maximum flower color mutations (light pink, orange pink, white, and yellow) were observed in 10 Gy. Anitha et al. (2021) treated rooted cuttings of the chrysanthemum cultivar "Poornima White" with different doses of gamma rays and EMS and found alterations in morphological and flower characters. Haspolat (2022a, b) treated rooted cuttings of pot chrysanthemum variety "Brandevil" (brownish-red color) with gamma rays (0, 10, 20, 30, 40, 50 Gy) to induce new mutation. The effective mutagen dose was determined 27 Gy. Mutations were observed for flower numbers per plant, plant heights and widths, and shapes and colors of both flowers and leaves. The color changes varied from brownish-red to yellow and orange. Two different colors appeared in the same pot at some genotypes as well as form changes of flowers. Remarkable mutations of the selected mutants were multiplied by tissue culture. Chrysanthemum morifolium (Ramat.) Bacardi is a spray-type cut flower with white flowers and green disc florets. Bacardi is one of the most preferred cultivars by growers because of the strength of plants and the resistance to diseases. Haspalot et al. (2022) treated in vitro bud explants of the white Bacardi variety with gamma rays at 20 Gy (Gray) and subcultured them in vitro until the M1V4 period. Different changes were observed in the heights and flowers of the plants such as variable flowers, flowering time, differentiation on plant length, flower number per bunch, and ray floret differentiations. The changes of the ray florets were determined as color changes to pink and yellow. Mutation frequency was calculated by 1.1% of the population. Approximately 0.9% of useful mutant lines were determined from the selected mutants. Nasri et al. (2022) did in vitro mutagenesis experiment with four chrysanthemum cultivars ("Homa," "Fariba2," "Arina," and "Delkash") using ethyl methanesulfonate (EMS) (0%, 0.125%, 0.25%, and 0.5%). Genetic polymorphism among mutants and their parents was detected using ISSR and IRAP molecular markers. The explant survival rate decreased by increasing EMS concentration. A wide range of phenotypic leaf, flower color, and inflorescence variability was obtained in all the cultivars which were isolated by cuttings. The used ISSR and IRAP primers could classify mutants based on cultivar and somewhat based on used EMS concentration confirming their effectiveness for the discrimination of real variants that allow their earlier selection and reduction of the mutant population size. Palekar et al. (2022) treated rooted cuttings of three chrysanthemum varieties "V1-Pink Cloud," "V2-Devi," and "V3-Bidhan Agnisikha" with EMS (0.05%, 0.1%, 0.5%, 1.0%) and gamma rays (0.5 kR, 1.0 kR, 1.5 kR, and 2.0 kR). Both the mutagens had a significant influence on the vegetative and flowering characters. A significant reduction occurred in both EMS and gamma rays treated seedlings of chrysanthemum varieties in terms of plant height, branches per plant, and leaf area. The higher dose of EMS and gamma rays delayed the flower bud initiation, flower formation, and 50% flowering. Purente et al. (2020) applied different doses of EMS on seeds of *C. indicum* var. aromaticum to induce variability. Germination significantly decreased with increasing doses of EMS. Morphological mutants with changed leaf and stem characters were detected. The leaf size and height of mutant plants were significantly increased. Changes in anatomical characteristics of leaves and stems were observed in the mutant plants. The total lignin and cellulose contents of the mutant plant stem decreased significantly. EMS was found to be very suitable for the induction of mutation in the present material.

Extensive work has been carried out at CSIR-NBRI, Lucknow, India, and the author was deeply involved in these activities. The summary of the entire mutation work on the chrysanthemum at CSIR-NBRI, Lucknow, India has been highlighted in Fig. 8.5.

Improvement of chrysanthemum through induced mutagenesis was started at CSIR-NBRI during 1965–1970 initially with 46 large flowered and 4 small-flowered cultivars. Suckers/cuttings were treated with 1–4 Krad gamma rays and detected variations in radiosensitivity among the experimental cultivars. Somatic mutations in flower color were detected on 43 plants but no mutation could be found in the white and yellow cultivars. 1–2 Krad produced a maximum number of mutations and 22 new flower color mutations were established in pure form from this experiment (Gupta 1966, 1979; Gupta and Shukla 1971).

These experimental results motivated working group scientists to design experiments as per the need of the floriculture trade. All subsequent experiments and selection of varieties had a specific motive. During the period 1978–1995, 25 cultivars ("D-5," "Lalkila," "Lilith," "Nimrod," "Grape Bowl," "Kingsford Smith," "M-24," "Otome Zakura," "Flirt," "Megami," "Anupam," "Jaya," "Bhima," "Fish tail," "Lalima," "M-71," "Ajay," "Gouri," "Maghi," "Puja," "Basantika," "Cotton Ball," "Maghi," "Sharad Mala," and "Shyamal") were selected for improvement. The cultivars were selected from large flowered, small-flowered, flower color and type basis based on the market assessment (Gupta and Datta 1978; Datta and Gupta 1980, 1981a, 1984; Datta et al. 1985; Banerji and Datta 1990, 1993; Datta 1992; Shukla and Datta 1993; Datta and Banerji 1995a, b). Rooted suckers/

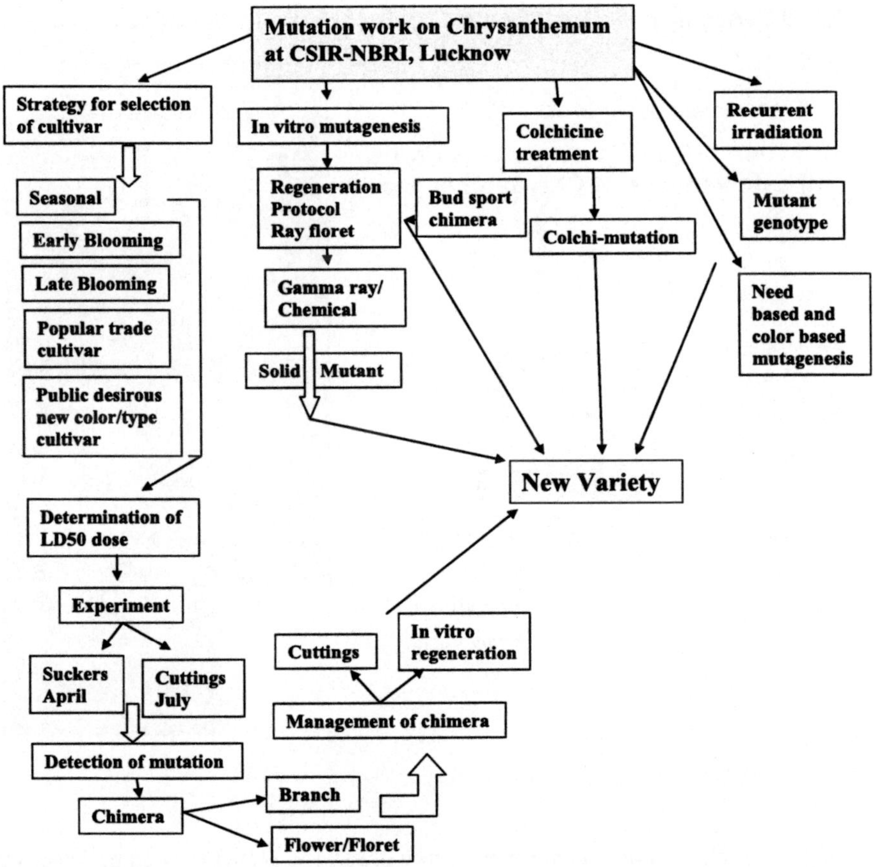

Fig. 8.5 Scheme of mutation work on chrysanthemum at CSIR-NBRI

cuttings were exposed to 1.5, 2.0, and 2.5 Krad gamma rays (^{60}Co). The treated population of all the cultivars showed increased chromosomal aberrations in root tip mitosis; reduction in survival, plant height, branch, leaf, and flower-head number per plant and different types of abnormalities in leaf and flower; delayed flowering; decrease in flower-head number; increased pollen sterility; decrease in polar axis and exine thickness, etc. However, few cultivars showed specific alterations. Chlorophyll variegation in leaves and stem fasciation was observed in some treated plants. Chlorophyll variegations in leaves provide additional beauty to the plants throughout the year. Chlorophyll variegation is very uncommon in popular chrysanthemum cultivars. The author from his mutation experiments with a huge number of cultivars detected chlorophyll variegations only in cultivars "Grape Bowl," "Lilith," and "Maghi" after gamma irradiation. A maximum number of chlorophyll-variegated plants and leaves were recorded in "Maghi." No chlorophyll-variegated cultivar

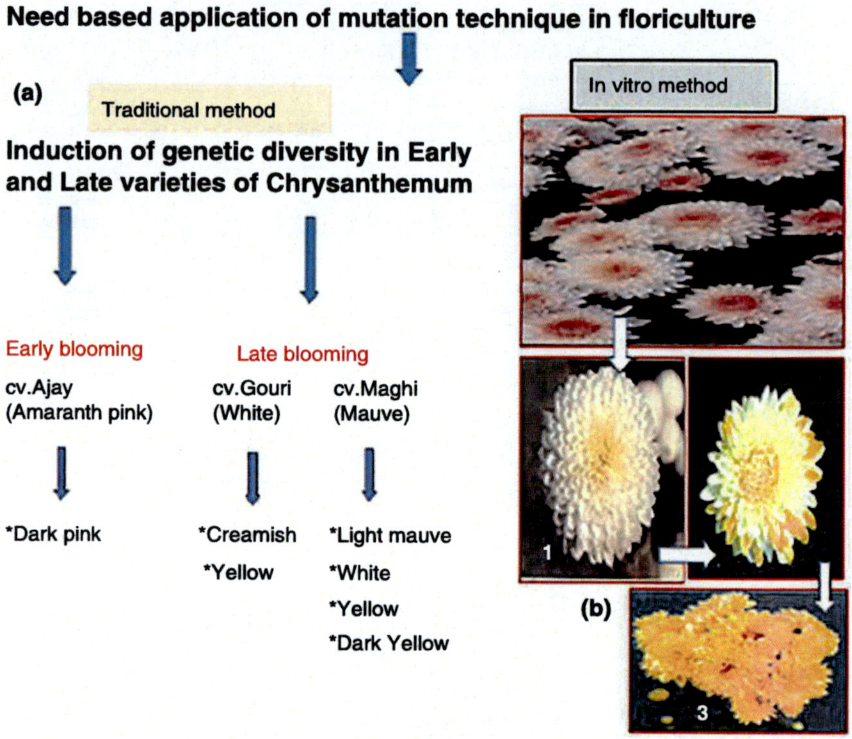

Fig. 8.6 Need based mutation on chrysanthemum

could be developed after gamma ray induced mutation in chrysanthemum (Datta and Gupta 1980; Gupta and Datta 1978; Shukla and Datta 1993).

In the floriculture trade, there is a wide range of attractive color variations in normal season blooming Chrysanthemum. Conversely, there is little range of color variations in early and late-blooming cultivars and consumers have very restricted choices. Conventional mutation techniques can be utilized to develop desired need base variety for trade. One early blooming cultivar "Ajay" and three late-blooming cultivars ("Gouri," "Maghi, and "Puja") were treated with gamma rays, and somatic mutations in flower color were induced in all the cultivars (Shukla and Datta 1993). This is a very interesting experiment on how the mutation technique has been successfully applied in chrysanthemums to create genetic variability in early and late-blooming varieties as per market demand (Fig. 8.6a).

The sensitivity of different floret colors to mutagens needs special elaboration for color mutation. Early color-wise reports mentioned that the maximal number of dominant genes for flower color is present in pink cultivars which cause recessive mutations and can be detected in M1V1 (Bowen et al. 1962; Jank 1957a, b; Bowen 1965; Broertjes 1966a, b; Dowrick and El-Bayoumi 1966a, b). Mutation can be induced in bronze color cultivars (Jank 1957a, b). Early attempts reported that white and yellow colors were immutable (Gupta 1966, 1979; Jank 1957a, b; Bowen 1965; Broertjes 1966a, b). Later experiments showed that yellow sectorial mutation was detected in the white cultivar "Lilith" and a complete Canary yellow mutant was induced in a white mutant cultivar "Himani" (Datta 1985b; Datta and Gupta 1980; Gupta 1979).

Several cultivars developed more than one flower color mutation after gamma irradiation. Cultivars like "Undaunted," "E-13," "Otome Zakura," "D-5," "Surekha," "Anupam," "Khumaini," "Kalyani Mauve," and "Lalima" are praiseworthy to acknowledge that series of mutants have been developed from the same cultivar (Gupta 1966, 1979; Datta and Banerji 1991, 1993, 1995a; Datta and Gupta 1980, 1981b, 1983). A series of mutants have been developed in the cultivar "Horim" by consecutive gamma irradiations (Broertjes et al. 1980).

The results pattern indicates that mutation is a chance process. The author did repeated experiments using color-specific cultivars with gamma rays and determined that the flower color (pigment composition) of experimental parental cultivars is a meaningful gauge for the development of new color mutation. Experiments resolved that white cultivars will develop yellow color, and red cultivars will develop either total yellow or a mixture of red and yellow. Yellow varieties will produce different shades of yellow or white or a mixture of yellow and white. There is a need to prepare a color chart based on the color (pigment composition) of the parent cultivar and the spectrum of induced color mutations. A color chart will be an indicator for selecting color-specific cultivars to develop specific color mutations as per the desire of the floriculture trade (directive mutation) (Datta 1990).

Mutant induction through the application of physical and chemical mutagens remains an important part of ornamental and floricultural plant breeding. It is a relatively easy-to-apply technology, in both ex vitro and in vitro systems, to induce novel characters in plant form, leaf, and flower color and shape. This is particularly relevant today where the consumer market is much more safety-orientated, and where the floriculture industry is always seeking novel characters to surprise the market. This chapter looks at classical mutation breeding as one of the important hallmarks of novel color, form, size, and shape breeding in ornamentals (Datta and Teixeira da Silva 2006).

For the mutagenesis experiment, it is very important to decide very precisely the sensitivity of materials to mutagens and LD_{50} dose. Radiosensitivity of almost all experimental cultivars was determined based on a large number of parameters like flower type, color, size, and shape; chromosome number, kayo-morphology, chromosomal abnormalities in the root meristem, Interphase Nuclear Volume (INV), Interphase Chromosome Volume (ICV), and 2c DNA content (Datta 2015). Experimental cultivars showed no definite relationship of radiosensitivity among the chromosomal aberrations, INV, ICV, DNA content, and other phenotypic characters. Some cultivars were moderately sensitive, others more sensitive and others still resistant to mutagens. Different cultivars were differently sensitive irrespective of flower color, size, and shape. Radiosensitivity of chrysanthemum has been considered as a genotype-dependent mechanism based on analysis of all sensitivity data (Datta 2015).

From the literature survey, it is difficult to select the LD_{50} dose of chrysanthemum as many researchers have reported suitable doses based on their experimental cultivars and radiation sources. Some cultivars survived 3000r X-rays treatment and the optimum dose set between 2000 and 4000r (Jank 1957b; Sheenan and Sagawa 1959). Fujii and Mabuchi (1961) reported an optimum dose between 2 and 4 Krad of gamma rays based on survival while Bowen et al. (1962) determined only 50% lethality after 4.3 Krad. 14 Krad of gamma rays was reported as a suitable dose by Dowrick and El-Bayoumi (1966a, b). Application of higher doses has also been reported by others like 25 Krad gamma rays (Cawse 1965), 10–12 kR gamma rays (Yamakawa and Sekiguchi 1968), and 8 Krad gamma rays (Broertjes 1966a, b). The application of such high doses was most likely for a low-dose rate (1 kr/day, 125–150 rad/ha) or due to the early stage of mutagenesis work on chrysanthemum. The author did repeat experiments with many cultivars and determined LD_{50} of chrysanthemum between 1.5 and 2.5 Krad gamma rays (Datta 1992, 1994, 2015).

8.2 Detection of Mutations

The creation of mutants through mutagen treatment is most important in the mutation breeding program. Detection of mutation/s in mutagen-treated populations and their isolation in pure form are most meaningful. Mutation in chrysanthemums and generally in vegetatively propagated plants is primarily noticed in M_1V1. The mutated cell expresses its mutant character if it gets a chance to express in M_1V_1. But mutations are also observed in M_1V_2 and later generations from normal looking at mutagen-treated plants (Brock 1966; Gupta and Jugran 1978; Das et al. 1974; Solanki and Sharma 2001; Sarker and Sharma 1988; Sneepe 1977; Usenbaev and Imankulova 1974; Buiatti and Tesi 1968). The mutated cells of the lower axillary buds remain in the dormant stage in the chrysanthemum and express their mutant character when included during vegetative propagation in M_1V_2. Several flower color mutants have been detected in M_1V_2, M_1V_3, and M_1V_4 in chrysanthemums (Datta 1992, 2000). Mutation frequency and spectrum of mutations vary with the cultivar and exposure to gamma rays.

8.3 Mutation in Flower Morphology

New flower form is always in good demand in the consumer market. Many attractive novel alterations in flower form have been induced in different ornamentals and commercialized (Heslot 1964; Broertjes 1966a, b). A few examples are the double flower type (Begonia—Broertjes and Van Harten 1988; Gladiolus—Lysikov 1990; Hyacinthus—Broertjes and Van Harten 1988), a dwarf mutant in *Impatiens platypetala* (Weigle and Butler 1983), semi-double or single type (carnation—Simard et al. 1992; Cassels et al. 1993; Portulaca—Gupta 1966, 1979), Hyacinthus (Broertjes and Van Harten 1988), double to single flower (Hibiscus—Banerji and Datta 1986, 1988), increase and decrease in petal number (rose—Broertjes and Van

Harten 1988), increased peduncles (Dahlia—Broertjes and Van Harten 1988), a variegated and dwarf mutant in Coleus (Love and Malone 1967), changed (broad, dentate/dissected margin) corolla (*Petunia nyctaginiflora*—Mahna and Garg 1989), etc. The author developed several interesting morphological mutants in chrysanthemums like "Cosmonaut"—anemone type flower shape mutant from Korean type cultivar (Datta and Gupta 1984), "Tulika"—paint brush type (semi-quilled) flower shape mutant developed from spoon type cultivar (Datta et al. 1985) and "Shabnam"—small appendage like structure developed at the tip of each floret (Datta 1990) through gamma irradiation and commercialized. The induction of tubular florets is one of the interesting observations in Chrysanthemum. Complete tubular floret mutants can be induced in those cultivars where there is a small tube at the base of each floret (Datta 2019). Several many other phenotypic mutants (anemone to Korean type, fringed ray florets, a fishtail shaped, etc.) have been developed in chrysanthemum (Gupta and Shukla 1971; Gupta 1979; Heslot 1964; Broertjes 1966a, b).

8.4 Color Mutations

Chrysanthemum cultivars with pink flowers have a maximum number of dominant genes for flower color. They have, therefore, greater possibilities of producing recessive mutations which can be detected in M_1V_1 (Bowen et al. 1962; Jank 1957a, b; Bowen 1965; Broertjes 1966a, b; Dowrick and El-Bayoumi 1966a, b). Jank (1957a, b) also reported the induction of flower color mutations in bronze cultivars. No mutations were detected in white-flowering chrysanthemums (Gupta 1966, 1979). Chrysanthemum cultivars with yellow flowers have been reported to be very stable (Jank 1957a, b; Bowen 1965; Broertjes 1966a, b; Gupta 1979). Datta and Gupta (1980) detected the yellow sector in the white-flowered cultivar "Lilith" after gamma irradiation. Datta (1985a) successfully induced Canary yellow flower color mutation after irradiation of a gamma ray induced white mutant ("Himani") of "E-13," a mauve-colored, pompon-type small-flowered chrysanthemum (Gupta 1979). More than one flower color mutation has been reported in different cultivars. Broertjes et al. (1980) have isolated several mutants in the chrysanthemum cultivar "Horim" by successive use of irradiation induced mutants in their mutation breeding program. Some of the cultivars like "Undaunted," "E-13" (Gupta 1966, 1979), "Otome Zakura" (Datta and Gupta 1981b), "D-5" (Datta and Gupta 1983), "Surekha," "Anupam," "Khumaini," "Kalyani Mauve," and "Lalima" (Datta and Banerji 1991, 1993, 1995b) are worth mentioning in this respect where series of mutants have been recorded from the same cultivar after gamma irradiation.

8.5 Colchicine Treatment

Colchicine is mostly used as a polyploidizing agent and no reports were available on the mutagenic effects of colchicine on chrysanthemum. The author treated rooted cuttings of four cultivars "D-5," "Flirt," "Lalkila," and "Sharad Bahar") by dipping (2.5 cm) into an aqueous solution of 0.0625% and 0.125% colchicine for 5 h. Somatic mutations in flower color could be detected in all the cultivars as chimera but no mutations could be isolated in the second generation except one mutation in cultivar "Sharad Bahar." The original color of "Sharad Bahar" was Purple whereas the mutant color was Terracotta Red. The mutant has been released as a new cultivar in the name of "Colchi Bahar." This is the first report of colchicine-induced somatic flower color mutation in Chrysanthemum (Datta 1985a, 1987, 1990; Datta and Gupta 1984, 1987).

8.6 Recurrent Gamma Irradiation

This treatment method has been recommended long back in mutation programs for different crop plants. Application of recurrent irradiation and assessment of its mode of action and achievements in ornamental crops is limited. The author did a comparative assessment between single irradiation and recurrent irradiation experiments on chrysanthemum and rose. The concept of recurrent irradiation was practically applied to increase genetic variability by irradiating plant material that had already been irradiated in one or more subsequent generations. Datta (1991) treated rooted cuttings of an early blooming cultivar "Sharad Bahar" with 1.5, 2.0, and 2.5 Krad of gamma rays, and all the plants were propagated bulk treatment-wise for the next vegetative generations and recurrent irradiation experiment was carried out in the second year. Reduction in survival, plant height, branch, and leaf number on the one hand, and increase in morphological abnormalities and somatic mutation frequency and spectrum, on the other hand, were more in recurrent irradiated populations. Experimental results indicate that recurrent irradiation methods may be utilized in mutation breeding programs as this method may provide an even greater range of genetic variability than a single mutagen treatment. It is confirmed that the physiological and genetic effects continue in the biological materials after irradiation. Normally the physiological effects fade away in the first generation with the consequence that it leaves behind some soft linkages within the chromosomes. This may cause a cumulative effect when gamma ray treatment is repeated in successive generations since the basic units affected are likely chromosomes and genes (Datta 1991). These data will help our understanding of the radiation effects in different forms and be useful in designing suitable radiation treatments for mutagenesis. This technical operandi may favorably be applied in routine mutagenesis programs for inducing novelties in floricultural crops.

8.7 Mutant Genotype

Literature on testing the sensitivity of different chrysanthemum cultivars to mutagens is very rich but on mutant genotype it is scanty. Broertjes et al. (1980) were successful to develop hundreds of mutants by successive use of radiation-induced mutants of Chrysanthemum cv. “Horim.” The author made a comparative test for the assessment of mutagen sensitivity on original cultivars and their respective induced mutants for induction of mutation (Datta et al. (1985). Three original cultivars “D-5” (magnolia purple, decorative), “E-13” (small-flowered, mauve, pompon type), and “Flirt” (small-flowered, red) and their gamma ray induced mutants were selected as materials. 1.5, 2.0, and 2.5 Krad of gamma rays were applied to the rooted cuttings of “D” and its three flower color mutants “Alankar” (Spanish Orange), “Agnisikha” (Erythrite Red), and “Shabnam” (flower shape mutant); “E-13” and its mutant “Himani” (white) and “Sheela” (canary yellow mutant of mutant “Himani”); and “Flirt” and its mutant “Man Bhawan” (reddish yellow). The frequency of mutations varied with the cultivars and exposure to gamma rays. The original cultivar “D-5” showed a reddish yellow flower color mutation. Out of three mutants of “D-5,” “Agnisikha” was found to be very stable and no mutation could be detected after irradiation. “Alankar” developed yellow flower mutation and Erythrite red flower color mutation in “Shabnam.” The original cultivar “E-13” was found to be very sensitive to gamma rays concerning flower color change. A wide range of new flower colors developed, i.e. erythrite red, white, brownish, and brownish-red. “Himani” developed only canary yellow flower color mutation. “Sheela” was found to be very resistant and no change in flower color could be induced after gamma irradiation. Yellow flower color mutation was induced in the original cultivar “Flirt.” “Man Bhawan” developed a light yellow flower color mutation (Datta 1992). This concept of induction of mutation is very important for the floriculture market to create further variability of mutant genotypes.

Castillo-Martínez et al. (2015) detected three mutants after gamma irradiation (30 Gy) and EMS (0.2% v/v for 30 min) treatment of the chrysanthemum mutant plant detected in the first stage. They detected three mutants with genotypes of slow development, dwarfs, and dotted leaves.

Mutagen-induced somatic recessive mutations were detected most often in $M1v_1$ in the chimeric form in the chrysanthemum. Repeat experimental results combined with cultural practices have established that mutations can be detected in M_1v_2, M_1v_3 and later vegetative generations form normal looking mutagen-treated plants of M_1v_1 (Gupta and Jugran 1978; Das et al. 1974; Datta 1990, 1992, 1994; Usenbaev and Imankulova 1974; Buiatti and Tesi 1968). Chances of recovering solid mutants are more in M_1v_2 and later generations. Detection of mutation in M_1v_2 has been justified by cultural practices followed for the propagation of chrysanthemum. Normally the mutated cell shows its mutant feature if it gets a chance to express in M_1v_1. The mutant cell of the lower axillary buds persists in the dormant stage and shows its mutant feature when incorporated during vegetative propagation of mature stem cuttings (Fig. 6.1k) in the M_1v_2 (Datta 2015).

Experiments were conducted to evaluate the comparative development of mutation events when gamma rays were applied on suckers (March/April) (Fig. 6.1j) and cuttings (July/August) (Fig. 6.1i). Treated propagules were grown and all cultural procedures were followed for healthy growth of the plants. Mutations in chrysanthemums commonly develop as chimera and the chimeric branches are propagated by cuttings to establish mutant tissue in pure form. Time of treatment was found to have a definite effect on mutation in the chrysanthemum. Suckers (March/April) treated plants were in good condition and more vigorous during the blooming period than cuttings (July/August) treated plants. The frequency of mutation was higher in the July/August experiment but they were not as healthy as those developed from the March/April experiment. Isolation of the mutant sector was easy from sucker-treated plants. This chimeric problem now can be solved by using the chimera management technique. Otherwise, July/August gamma ray treated plants can also be treated with some chemical mutagens to protect against the deleterious effects of gamma rays to encourage the healthy growth of plants. Both irradiation periods have advantages and disadvantages which can be gainfully utilized (Datta 1997).

8.8 Chimera Management and In Vitro Mutagenesis

Formation of a chimera is the main bottleneck in mutation work in vegetatively propagated plants. The mutation is always a single-cell incident. In multicellular organisms when multicellular apex is exposed to irradiation the newly developed mutated cell is exposed to diplontic selection. There is a struggle between the mutated cell and the neighboring normal cell. The mutated cell forms a group of cells and eventually forms a mutated cell layer through competition. The diplontic selection causes the development of a limited mutation spectrum and a small number of mutated plants. Such a process leads to the formation of chimera (Fig. 8.7). The size of the mutant area (chimera) varies from a narrow streak on a floret to an entire floret, from a portion of a flower to an entire flower, and from a portion of a branch to the entire branch (Fig. 8.8a–i) (Datta 1994). Isolation of such mutant tissue is one of the most significant processes in vegetatively propagated plants including chrysanthemums. Conventional multiplication methods (cuttings) have limitations to separate mutant tissues. Normally mutant tissues are isolated by the cutting propagation method when the entire branch is mutated. The majority of chrysanthemum mutants have been established by this method (Fig. 8.9). Conventional propagation technique cannot help for isolation of chimeric mutated individual floret. As a result, many such chimeric mutations are lost from most of the classical mutation breeding experiments. This motivated mutation breeder to standardize tissue culture technique for direct shoot regeneration from individual floret.

The concept of management of chimera developed and some isolated work has been reported in some ornamentals covering mostly theoretical presumptions. The author and his team made concentrated efforts to standardize the tissue culture method for direct shoot regeneration from chrysanthemum floret. *Chrysanthemum*

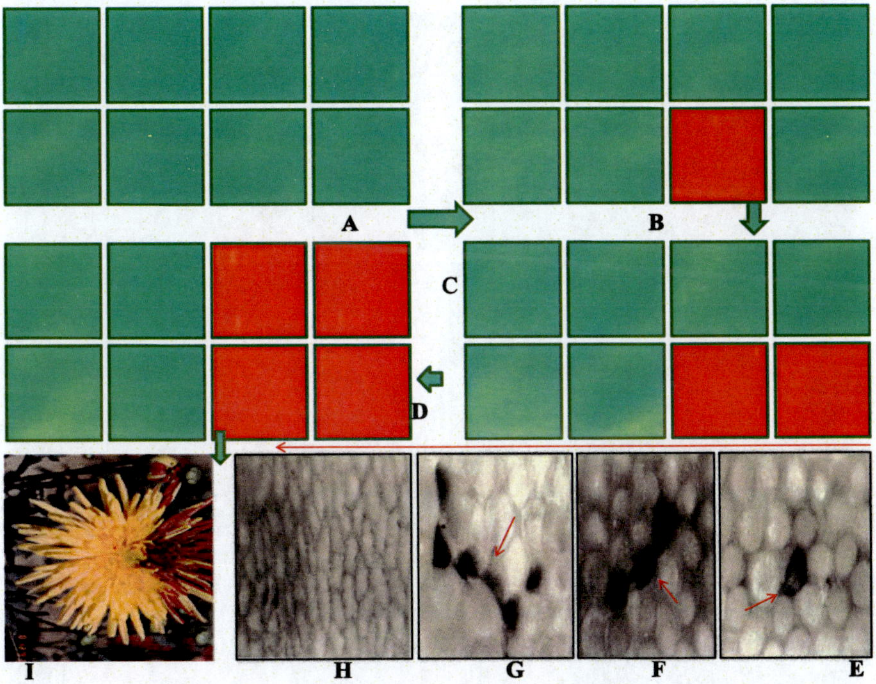

Fig. 8.7 (**a–d**) Schematic representation showing how mutant cell (*red*) develops and multiply. (**e–i**) Showing single cell mutation (**e**) and how it divides (**f** & **g**) and form chimera (**h** & **i**)

morifolium Ramat. was selected as the material for standardization of technique for the management of chimera and in vitro mutagenesis work. In vitro, protocol was utilized for the management of chimera developed through both spontaneous mutation and induced mutations. Perhaps this group has done the highest number of practical works and enriched the literature on the management of chimera and in vitro mutagenesis.

After mutagen treatment in vivo experimental plants are grown under field conditions. In in vivo experiments mostly population size is restricted due to the limitation of land and fund and also the availability of sufficient quality material. Unavoidable environmental conditions may also reduce the population size. In in vitro methods cells are uniformly treated with physical and chemical mutagens and are grown in a uniform cultural environment. One can conduct in vitro mutagenesis experiments with a large population, within limited space, and at any time of the year.

The main advantage of this technique is to overcome chimera formation in the M1V1 and the chances of getting solid mutant are more (Maliga 1984; Ahloowalia 1995; Chaleff 1983; Flick 1983; Maluszyuski et al. 1995; Novak 1991a, b; Smith 1985). Several experimental results have been reported in Saintpaulia and Pelargonium (Skirvin and Janick 1976; Grunewald 1983), Carnation (Johnson 1980; Simard

Fig. 8.8 (a–i) Chimera in chrysanthemum. (a–e) Flower chimera. (f–i) Branch chimera

Fig. 8.9 New mutant chrysanthemum varieties isolated by cuttings of mutant branch

et al. 1992; Cassels et al. 1993), chrysanthemum (Jung-Heliger and Horn 1980; Preil et al. 1983; Huitema et al. 1986, 1989; Jerzy 1990; Jerzy and Zalewska 1996; Dalsou and Short 1987; Schum and Preil 1998; Mishra et al. 2003; Datta and Mandal 2005), Eustoma grandiflorum (Nagatomi et al. 1996a, b, c); Gerbera and rose (Walther and Sauer 1985, 1986a, b, 1989; Laneri et al. 1990; Jerzy and Lubomski 1991; Jerzy and Zalewska 1992), Bryonopsis laciniosa (Caroline and Mallaiah 2011), *Gypsophila paniculata* L. (Barakat and El-Sammak 2011), Pear (Predieri 2001).

8.8.1 Induced Chimera

Rooted cuttings (13 cm height) of a large number of cultivars with different color combinations ("Maghi"—mauve; "Lilith"—white; "Purnima"—white; "Otome Zakura"—mauve; "Colchi Bahar"—Terracotta Red; "Sharad Bahar"—purple; "Puja"-red purple with flat spoon-shaped florets; "Lalima," "Flirt" "Sunil"—red; etc. were treated with gamma rays and grown in the field up to flowering. Chimeric mutations were detected in different cultivars like "Maghi" (two plants showed few chimeric florets with new colors white and yellow from 1.5 and 2.0 krad); "Lilith" (five plants showed yellow chimera from 1.5 and 2.0 krad); "Purnima" (sectorial yellow floret from 1.5 and 2.0 krad); "Colchi Bahar" (sectorial yellow floret from 1.5 and 2.0 krad); and "Puja" (two flowers with few florets with two mutations: one mutant floret color were yellow-orange with original flat florets and another mutant floret color was yellow-orange with tubular florets). Plants showing chimeric mutated florets were selected as materials for the management of chimera. Plantlets developed from all chimeric mutant florets through direct shoot organogenesis flowered true-to-explant floret color/shape in all the cultivars (Fig. 8.10a, b) (Chakrabarty et al. 2000; Mandal et al. 2000a, b; Dwivedi et al. 2000; Datta et al. 2001).

8.8.2 In Vitro Mutation

Chrysanthemum cultivars "Flirt," "Sunil," "Puja," "Lalima," "Maghi," and "Madam E Roger" were selected as experimental materials. Ray florets from field-grown plants were treated with gamma rays (500 and 1000 rad) after disinfection and cultured in vitro. Few plants from all regenerated plants showed solid somatic mutations in floret color and shape in all the cultivars: "Flirt" (produced two flower color mutations in 500 rad, Mutant 1—red ray floret with a yellow tip and Mutant 2—yellow ray floret with very fine red stripes); "Puja" (produced darker floret at 1000 rad); "Maghi" (produced tubular floret in 1000 rad); "Sunil" (developed red purple tubular floret mutant at 1000 rad); "Madam E Roger" (two plants produced chimeric yellow florets from 1000 rad); "Lalima" (produced two mutants in the 500 rad, both the mutants were yellow colored but one having flat spoon-shaped ray florets similar to the original cultivar, while the other having tubular florets). The

Fig. 8.10 Management of induced chimera in pure form (**a**–**k**, and **b**–**m**) and spontaneous chimera (**c**) and in vitro mutagenesis (**d**)

mutants were multiplied by cuttings (Fig. 8.10d) (Datta 2005a, b; Misra and Datta 2007).

8.8.3 Management of Sport Chimera

Development of new flower/floret/petal color/type sport (spontaneous mutation) are found frequently in different ornamentals including chrysanthemum. But it was difficult to isolate such chimeric florets. Now, novel in vitro technique is available for the management of such chimera. The original flower color of a large flowered chrysanthemum cultivar "Kasturba Gandhi" is white. Spontaneous mutation developed a few yellow color florets in the germplasm collection. Chimeric yellow florets were put through direct regeneration of shoot buds using a suitable tissue culture medium. All the plants regenerated either directly or through callus produced flowers true to the chimeric floret color, i.e. yellow (Fig. 8.10c) (Chakrabarty et al. 1999, 2000; Murashige and Skoog 1962).

This novel technique has created a sensation in mutation technology for developing new cultivars not only through the restoration of chimeric tissues but also through the development of solid mutants. The technique is effective for both induced mutation as well as natural mutation. This technique has practical importance not only for chrysanthemums but also for other ornamental breeding.

8.8.4 In Vitro Method for the Development of Need Base Variety

In vitro technique has been beautifully applied for the development of need base variety through directive mutation. It has been distinctly proved that the calculated and proper application of technology can create the most desirable and economically viable mutants in ornamentals. A very interesting example is cited as a ready reference. Chrysanthemum cv. "Otome Zakura" (pompon type, large flowered, pastel lilac flower color) is a highly favored variety in the consumer market. Considering the consumer choice and to increase the income of nurserymen mutation technique has been applied very judiciously to create desirable variations in the variety "Otome Zakura." In the first step, a complete white mutant variety ("Purnima") (Fig. 8.6b1) has been developed through gamma irradiation in "Otome Zakura." In the second step, the concept of directive mutation (white to yellow—as mentioned above) was applied to this white mutant ("Purnima) and developed yellow color new mutant variety (Fig. 8.6b3) through the management of chimera (Fig. 8.6b2).

Systematic efforts have been made to apply in vitro mutagenesis techniques to develop a specific characteristic associated with the mutation. In vitro mutation applying EMS developed stable NaCl-tolerant chrysanthemum variant through whole plant selection and callus line selection (Hossain et al. 2004, 2006a, b). For the management of single-cell mutation, effective somatic embryogenesis protocol has been standardized in chrysanthemum (Mandal and Datta 2005).

Chrysanthemum is highly heterozygous and develops floret chimera very frequently through spontaneous mutation. Bud sports/spontaneous mutations created many new chrysanthemums. According to an old report, 30% of the chrysanthemum cultivars originated as sports (Wasscher 1956). Some cultivars are very susceptible and developed a considerably good number of sports like "Sweetheart," "The Favourite," "Indianapolis," etc. (Anonymous 1961; Bowen et al. 1962; Yoder 1967). Spontaneous mutation has developed many outstanding cultivars in different countries. These varieties were generally noticed as branch chimeras and separated through cutting techniques. A large number of chimeric florets develop through spontaneous mutations in many ornamental plants every year and are lost. As an illustration, we can see the nature of such chimeric flowers in different ornamentals. Chimeric nature varies not only in different ornamentals but also within the same ornamental species. Such chimeric flowers are mostly reported by researchers, nurserymen, and amateur growers in different media including presently in Facebook (Fig. 8.11). The management of such mutated plant chimeras is an almost unexploited area in ornamental crops. Now in vitro approach can be applied in floriculture to separate and establish these spontaneous chimeras as new varieties. In vitro management of chimera will enrich the floriculture industry many times through the development of new varieties.

Fig. 8.11 Spontaneous (sport) developed different form of chimera in different ornamentals

8.9 Chlorophyll-Variegated Chimera and Its Management

Chlorophyll variegations, i.e. leaf variegations makes plants more decorative and favorite throughout the world. In the floriculture industry chlorophyll, and variegated ornamentals have huge demand due to their added beauty. Mutagenic treatments have developed broad categories of variegated plants in different ornamentals (Broertjes and Van Harten 1988; Datta 1997). In general, leaf variegations may develop in the meristematic tissue of a normal plant due to spontaneous or induced mutations. It is very tough to maintain true-to-type variegations through seed. Vegetative propagation (cuttings or divisions) is mostly applied to conserve and replicate the mother plant, but this method is not always reliable when dealing with variegates. There are always some changes in variegations in subsequent generations. The chimeric nature and its permanency depend upon the cell division pattern, frequency of cell division, and association sequence of mutated and non-mutated cell layers in the apical meristem. The genotypes of different cell layers of the tunica and corpus regions of the shoot apex have been described by using abbreviations L.I, L.II, and L.III (Fig. 8.12a) which represent the outermost layer, the next tunica layer in, and the corpus,

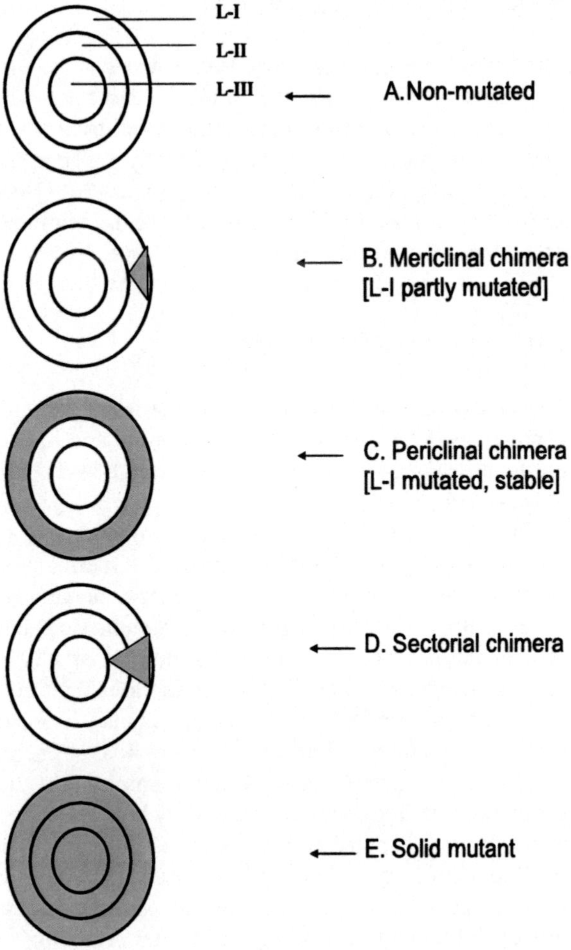

Fig. 8.12 Diagrammatic representation of cross-sections of shoots

respectively (Satina and Blakeslee 1941). Chlorophyll-variegated chimeric plants are classified mainly on the ground of position and corresponding ratio of mutated and non-mutated cells in the apical meristem. It shows layers of a non-mutated plant (Fig. 8.12), mericlinal chimera (Fig. 8.12b), periclinal chimera (Fig. 8.12c), sectorial chimera (Fig. 8.12d), and solid mutant (Fig. 8.12e) plant:

8.9.1 Mericlinal Chimeras

In such chimeras, the mutated cell does not cover the entire apical dome and is confined to one portion of the meristem. The new shoots/leaves which develop from these portions are chimeric and shoots/leaves which develop from other portions of the meristem are normal (Fig. 8.12b) (Stewart and Derman 1970).

8.9.2 Periclinal Chimeras

These chimeras are relatively stable and can be propagated true to their nature vegetatively. When the mutation takes place near the apical dome, the mutated cell may form an entire layer of the mutated cell by successive divisions. The entire new meristem contains one layer of genetically different mutant cells which are different from the remaining meristem cells (Fig. 8.12c) (Dermen 1947, 1953; Bush et al. 1976). Many periclinal chimeras have been identified and selected for commercial use due to their specific valuable phenotypes.

8.9.3 Sectorial Chimeras

In such cases, a portion of the apical meristem (extended through all the cell layers) is composed of mutated cells. Both normal and mutated shoots develop depending on the position of the mutated sector on the apical dome apex. These chimeras are unstable (Fig. 8.12d).

Out of three chimeric natures, the probability of multiplication of similar morphological characters is small in the case of mericlinal and sectorial chimeras. Due to comparative stability, periclinal chimeric plants are in good demand in trade. Horticulturists have selected many such chimeric plants for their special and delightful phenotypes. The frequency of normal and chlorophyll-variegated branches always changes on multiplication in mericlinal chimeras due to its unstable nature. The author has worked out very simple manageable techniques to convert mericlinal chimera to periclinal chimera. The mericlinal chimeric branch is composed of both green and variegated leaves. Apical dominance of the main apex of the mericlinal shoot restricts the growth of the axillary bud associated with variegated leaves. Two methods were adopted to bypass apical dominance and diplontic selection. In one method, all the green leaves of the mericlinal branch are detached and axillary buds of variegated leaves are encouraged to grow. In another approach, the chimeric branch is forcedly bent into an arch after removing all the green leaves to cater to better growth opportunities for the axillary buds. A new branch develops from the axillary bud of the variegated leaves that are either periclinal or mericlinal. Periclinal branches can be created as a new variety by cuttings. In the case of the mericlinal branch, the same process is repeated. Several new chlorophyll-variegated mutant varieties have been developed in bougainvillea and *Lantana depressa* following this technique (Datta and Banerji 1995b). This simple technical procedure will be very helpful for the management of chimera in different ornamental plants.

8.10 Development of Salt Resistant Chrysanthemum

Worldwide millions of hectares of land are affected with the problem of salinity and alkalinity, lying either unutilized or only partly utilized. Rectifications of the salinity problems are usually expensive and generally considered only temporary solutions.

Although the main focus is to develop salt resistant variety through induced mutation, conventional methods are also very important in this direction. Efforts are going on to develop techniques for selection of successful plant genotypes/species which can be grown under these saline conditions. No systematic efforts have been made for screening ornamental plants suitable for cultivation in salt affected soil to meet the increasing demand of floriculture trade. Few conventional approaches for development of salt resistant in chrysanthemum are mentioned.

Grafting is widely used to improve the salt tolerance of horticultural crops, but the mechanisms of grafted chrysanthemum responses to salt stress remained unclear. Li et al. (2022) showed that heterografted chrysanthemums with *Artemisia annua* as rootstock exhibited increased salt tolerance compared with self-grafted and self-rooted chrysanthemums. Thorough biochemical analysis revealed the mechanisms underlying the increased salt tolerance of heterografted chrysanthemums and makes it possible to have large-scale cultivation of chrysanthemums in saline areas. Chen et al. (2003) irrigated NaCl solution at different concentrations (0, 50, 100, 150, 200 mmol/L) during the growing period of *Chrysanthemum chanetii* and two geoecotypes (Nanjing and Hangzhou) of *Ch. indicum*. Salt tolerance was identified by measuring changes of chlorophyll, proline and soluble sugar, POD activity and water content. The chlorophyll content, POD activities and water contents of Nanjing *Ch. indicum* and *Ch. chanetii* maintained high levels under salt stress. Proline and soluble sugar accumulated significantly. These parameters were improved by applying 50 mmol/L NaCl and reduced dramatically when concentration was above 50 mmol/L in Hangzhou *Ch. indicum*. Salt tolerance of these three wild species of chrysanthemum decreased with Nanjing *Ch. indicum, Ch. chanetii* and Hangzhou *Ch. indicum*. Bañón et al. (2012) studied the effects of saline water (1.5 dS/m and 5 dS/m) and plant growth regulators (paclobutrazol at 25 ppm and ethephon at 500 ppm) irrigated potted native chrysanthemum (*Chrysanthemum coronarium*). Results indicate that *C. coronarium* was moderately tolerant to salinity, although the number of inflorescences was much reduced in the salt stressed plants. They have assayed salinity symptom and other related biochemical changes. Shatnawi et al. (2009) studied the growth and physiological traits of in vitro raised chrysanthemum under salt stress in vitro. Salinity was induced by incorporating different NaCl concentrations (0, 20, 40, 60, 80, 100, 150, 300 mM) into the media. Shoot length, number of shoots, number of leaves, and fresh and dry weights decreased with elevated salinity concentration. Chlorophyll and carotenoid content decreased with increasing salinity and proline increased with elevated salinity concentration.

Demand of floricultural crops/flowers is increasing day by day worldwide and to meet this demand CSIR-NBRI, India initiated multiform approach to utilize salt affected soil using different ornamental genus/species. A total of five practical approaches were attempted to select/develop salt resistant strains: (1) Screening of different ornamental plants suitable for growing under salt stress soil. (2) Development of salt resistant mutant varieties of different ornamentals through gamma irradiation for salt stress soil. (3) Development of salt resistant chrysanthemum through Gamma ray induced in vitro mutation of whole plant. (4) Development of salt resistant chrysanthemum through EMS-induced in vitro mutation of whole plant. (5) In vitro mutagenesis to create salt resistant strains of chrysanthemum through stepwise callus culture.

1. Screening of different ornamental plants suitable for growing under salt stress soil
2. Application of gamma rays to induce mutations.

Different ornamentals were screened to find out their salt tolerance capacity after growing under alkaline soil (Usar soils) of 8.10 pH range. Ornamental species tested were *Jasminum grandiflorum*, tuberose, *Rosa damascene*, *Tagetes erecta*, *Narcissus tazetta,* gladiolus, Chrysanthemum, etc. In another experiment, propagules of different ornamentals (all mentioned above) materials were treated with different doses of gamma rays (Cobalt 60, radiation source, 500, 1000, 1500 and 2000 rads) and treated samples were planted along with equal number of untreated (control) samples at both normal soil and alkaline soil. Performance of some species was very satisfactory in alkaline soil (Srivastava and Sharma 1976; Mishra et al. 1973; Rai et al. 2001; Singh 1970, 1971; Srivastava 1976; Rahi et al. 1998, 2001; Rahi and Datta 2000; Rahi and Singh 2011).

In vitro mutagenesis experiment conducted by author and his colleagues (Hossain 2005) for development of salt resistant mutant in chrysanthemum is highly noticeable for its practical approach. All the techniques can be used as model system in in vitro mutagenesis for developing trait specific (here salt resistant) mutation in chrysanthemum and other ornamentals.

Before entering into the actual in vitro mutation experiment, one pilot experiment (like conventional induced mutagenesis) was conducted to study the chromotoxic effects of salinity stress on chrysanthemum. The critical balance between the formation of active oxygen species (AOS) and quenching activity of the antioxidants is disturbed in plants under salt stress condition. Different physiological and biochemical aspects under such stress conditions have heretofore been well documented in different crops. Huge amount of experimental work on physiological basis of variability for salinity tolerance have been worked out on different crops based on cell and tissue culture techniques. The salt tolerance mechanism lies at the cellular level which reflects on the plant level. Roots are the most sensitive organ and affected first under salt stress. Such reports on biochemical analysis at cellular level are very limited in floricultural crops specially in chrysanthemum. There was no report on chromotoxic effects of salinity stress on chrysanthemum. Series of papers have been published regarding antioxidant behavior in callus line under salt stress condition in different crops but information regarding antioxidant activity in roots of NaCl-stressed plants are very scanty. For root induction, shoots (20–25 mm length) were excised from established cultures of *Chrysanthemum morifolium* Ramat cvs. "White Stafour" and "Kelvin Tito" and were grown in MS basal medium supplemented with sucrose (3%) and bactoagar (0.8%). After 7 days, shoots (with 2 mm long roots) were subcultured in liquid basal medium supplemented with different concentration of NaCl (0, 50, 75 and 100 mM). Chromosomal anomalies during root tip mitosis were measured in terms of mitotic inhibition and chromosomal abnormalities under salinity stress conditions. The percentage of dividing cells was reduced with increase in concentration and treatment period. Mitotic index was represented in the form of relative division rate (RDR). The RDR at each concentration and period was calculated and in all the cases the values were negative in both the varieties. Negative values of RDR indicate mitotic inhibition. Increase in the negative values of RDR was directly proportional to the severity of the mitotic inhibition. A wide range of chromosomal abnormalities (bridges, stickiness, fragments, early separation, exclusion, laggard, abnormal anaphase, micronuclei, clumping) were observed as shown in Figs. 8.13 and 8.14. Both the varieties showed increased chromosomal abnormalities with increase in NaCl concentration and treatment period. "Stafour" variety was found to be more sensitive to all NaCl concentrations in comparison to "Kelvin Tito." Root length decreased progressively in both the varieties with increasing NaCl concentration. Root tip necrosis was observed only in White Stafour variety after 2 weeks of 100 mM NaCl treatment (Fig. 8.15). SOD and GR activities were higher in "Kelvin Tito" with increase in NaCl concentration except GR at 100 mM NaCl treatment." Root tip proline content in cv. "Kelvin Tito" progressively increased with increasing concentration of NaCl. Accumulation of proline in roots under stress condition was clearly associated with the reduction in root growth and decrease in mitotic index with increase in NaCl concentration. A positive correlation was observed between the cytological behavior and biochemical changes under salinity stressed condition in the present experiment conducted with whole plant. Reduction in root length had good correlation with the increased proline content. Results signify that in vitro adaptation promotes induction of salt tolerance as it exhibited all meaningful cytological, physiological, and biochemical changes which are essential for acclimatization procedure in salinity tolerance. It was assumed that direct shoot culture is very effectual method to screen out salt tolerance as it is less prone to genetic variation. This was perhaps the first report on chromotoxic effects of NaCl in ornamentals as evident from RDR values and increased percentage of total aberration (Hossain et al. 2004). This pilot experimental result was used as base line data for designing next in vitro mutagenesis experiments.

3. Development of salt resistant chrysanthemum through gamma ray induced in vitro mutation of whole plant.

The main aim of the experiment was to induce genetic variability by in vitro mutagenesis with gamma rays and to develop stable NaCl-tolerant mutants in chrysanthemum through whole plant selection.

Ray florets of *Chrysanthemum morifolium* Ramat. cultivar "Shyamal" (a salt-sensitive variety) were exposed to gamma rays (5 and 10 Gy) after surface sterilization and cultured in vitro (MS medium) supplemented with 200 mM NaCl following standard procedure. Regenerated rooted plantlets were planted in earthen pots and pots were transferred to the field condition. After 1 week, control and treated plants were exposed to high levels of salinity stress by watering them with distilled water supplemented with 120 mM NaCl (300 mL per pot), three times per week up to the

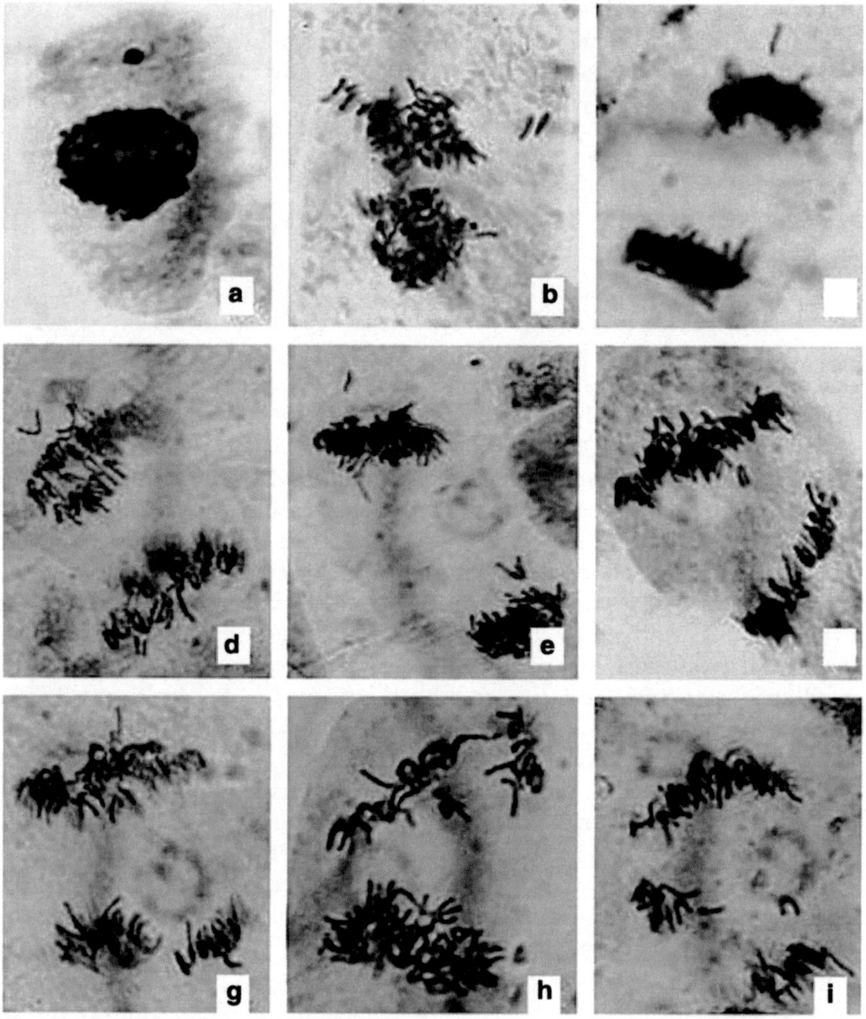

Fig. 8.13 Chromosomal aberrations induced by salt stress. (**a**) Micronuclei, (**b**) exclusion, (**c** and **d**) early separation, (**e** and **f**) late separation, and (**g–i**) abnormal anaphase. (C.f. Hossain et al. 2004, Plant Science)

flowering stage. Two best performing mutants R1 and R2 were selected from the treated populations based on plant growth, flower quality, and yield. The mutants were analyzed further morphologically and biochemically. Control plants subjected to equal NaCl stress were denoted as negative controls (NC), whereas control plants given only distilled water three times weekly, were considered as positive controls (PC). Performance of two selected mutants (R1 and R2) were assessed at the full bloom stage in terms of plant growth, flower characteristics, chlorophyll, carotenoid

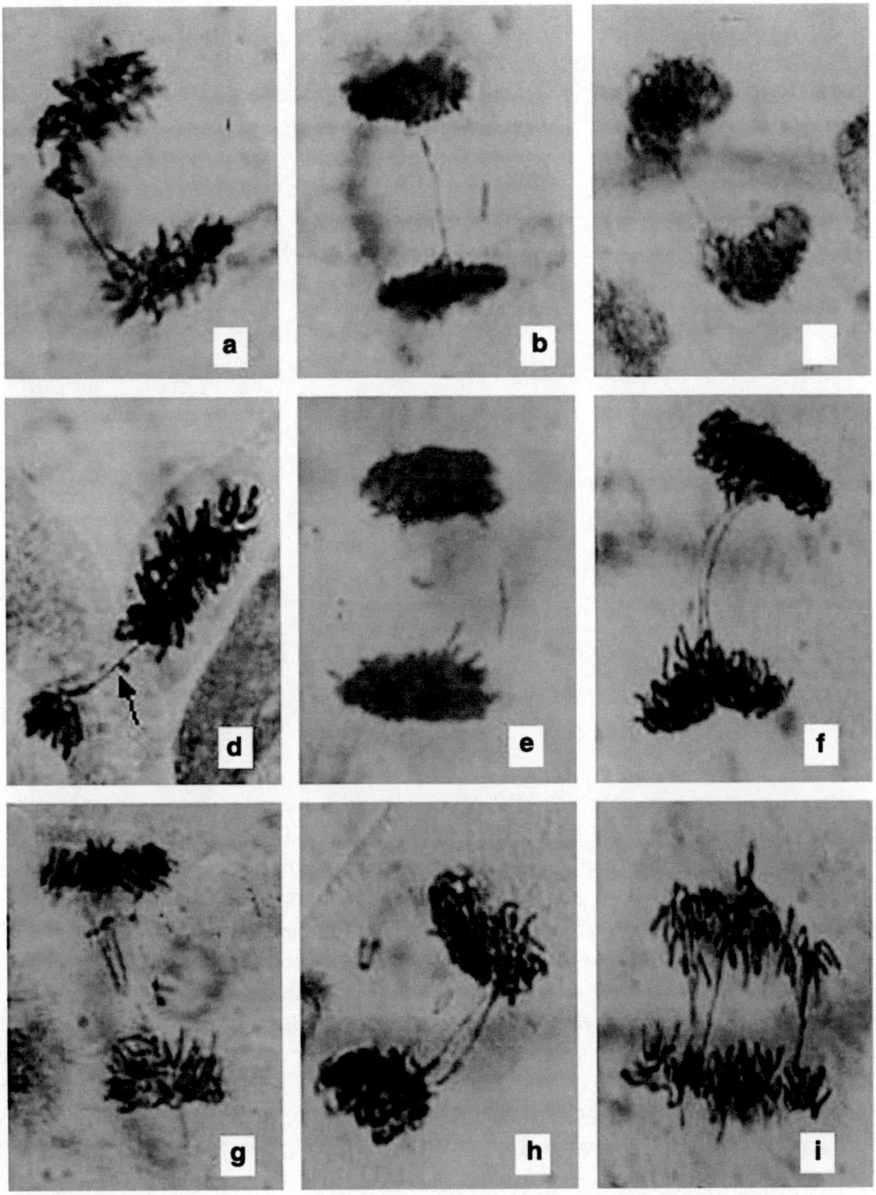

Fig. 8.14 Different types of chromosomal bridges, laggards, and fragments. (**a** and **c**) Single bridge; (**b**) bridge with laggard; (**d**) bridge with fragment; (**e**) laggard; (**f**) double bridges; (**g** and **h**) double bridges with laggards; (**i**) multiple bridges. (C.f. Hossain et al. 2004, Plant Science, S.K. Datta is also an author)

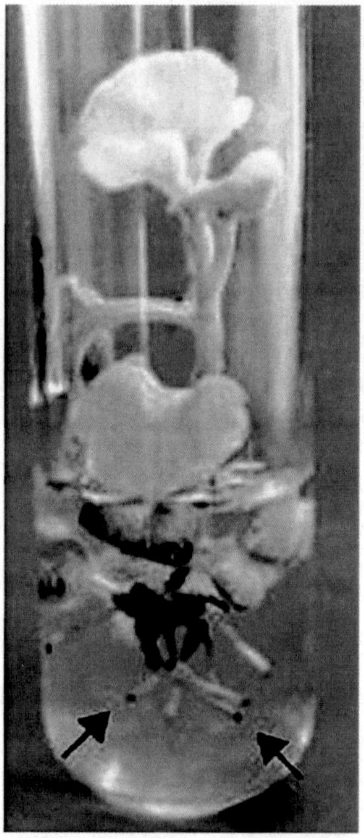

Fig. 8.15 Root tip necrosis (arrow marks) in cv. White Stafour. (C.f. Hossain et al. 2004, Plant Science, S.K. Datta is also an author)

contents, relative water content (RWC), membrane stability, and antioxidant enzyme activities (SOD, APX, MDAR, DHAR and GR). The R1 mutant, although showed a decrease in plant height under salinity stress, had flower quality and production comparable to those of PC plants, while R2 and NC plants showed poor performances in all aspects. A decrease in flower diameter was observed in the R1 mutant, but the magnitude of decrease was significantly greater in R2 and NC flowers. A change in flower shape was also observed in case of NC and R2 plants, while the R1 mutant maintained the regular shape, similar to PC flowers, even under NaCl treatment. NC plants produced irregular flowers, whereas R2 flower were mostly asymmetric in shape. Both R1 and R2 mutants and NC plants accumulated significantly high levels of proline under salinity stress, but R2 and NC plants produced flowers of inferior quality, whereas R1 flowers were as good as positive control flowers. The decrease in plant height in both the mutants and NC plants might be the results of accumulation of levels of endogenous free proline. Results indicated that the selected R1 mutant had an enhanced capacity for ROS scavenging, which might helped it to combat the salinity stress more easily. The R1 mutant developed by 5 Gy gamma irradiation can be considered as a salt-tolerant mutant showing all the positive characteristics of tolerance to NaCl stress (Fig. 8.16) (Hossain et al. 2006a).

Fig. 8.16 Plant regeneration and growth performances of two selected mutants (R1 and R2) NC plants of Chrysanthemum cv. "Shyamal" under NaCl stress conditions, along with PD plants, grown under unstressed conditions. (**a**) Direct shoot bud regeneration from gamma ray treated ray florets. (**b**) Growth and shoot proliferation. (**c**) In vitro rooting. (**d**) Acclimatization of in vitro grown plantlets in the hardening chamber. (**e**) Comparative growth performance of R1 and R2 and NC plant subjected to NaCl stress along with PC (from extreme left to right: Pc, R1, NC and R2, respectively). (**f**) Flower shape and size. Flower of PC (1), R1 (2), NC (3) and R2 (4–7) plants, respectively. Asymmetric shape of R2 flowers (4–7). (C.f. Hossain et al. 2006a, b, Functional Plant Biology, S.K. Datta is also an author)

4. Development of salt resistant chrysanthemum through EMS-induced in vitro mutation of whole plant.

The main aim of the experiment was to induce genetic variability by in vitro mutagenesis with EMS and to develop stable NaCl-tolerant mutants in chrysanthemum through whole plant selection.

Leaves of in vitro raised *Chrysanthemum mori*folium Ramat. cultivar "Regal Time" (a salt-sensitive variety) were cut into pieces and treated with EMS (0.025% and 0.05%) for 5 h and cultured in vitro. Control and EMS-treated shoots transferred separately to MS medium supplemented with 250 mM NaCl. Control shoots were maintained separately in MS medium having the same composition except for the NaCl. Plants showing better performance under in vitro salt stress conditions were selected along with control on the basis of growth performance. Better performed selected plants from 0.025% EMS and control were transferred to pots for further assessment. After hardening, both control and EMS-treated plants were subjected to salinity stress by watering them with 120 mM NaCl-supplemented distilled water (300 mL water in each pot), thrice weekly up to the flowering stage. Control plants were irrigated with only distilled water thrice weekly. Initially, seven variants (E1–E7) were selected from EMS-treated population on the basis of plant height, flowering time, number of flowers per plant, flower diameter, flower weight, number of ray florets, ray floret size and color, number and color of disc florets and finally the best variant (E2) was selected for further biochemical analysis. Better performance of E2 progeny under the same NaCl stress condition confirmed the genetic stability of the salt tolerance character. Salt tolerance was evaluated by the capacity of the plant to maintain both flower quality and yield under stress condition. Enhanced tolerance of E2 variant has been attributed to the increased activity of SOD, APX, DHAR, and, to a lesser extent, less membrane damage than in NaCl-treated control plants. Isoform analysis revealed that an increase in total SOD activity in the E2 variant was solely due to significant activation of the Cu/Zn isoform. Elevated levels of carotenoids and ascorbate in E2 leaves have been reflected in their higher free radical scavenging capacity (RSC) expressed in terms of DPPH (2,2-diphenyl-1-picryl-hydrazil) scavenging ability. The E2 variant developed through 0.025% EMS treatment might be considered as a NaCl-tolerant strain showing all positive characters toward NaCl stress (Fig. 8.17) (Hossain et al. 2006b).

5. In vitro mutagenesis to create salt resistant strains of chrysanthemum through stepwise callus culture.

The main aim of the experiment was to develop stable NaCl-tolerant mutants in chrysanthemum through stepwise selection of callus.

Ray florets were collected from flowers of field-grown *Chrysanthemum morifolium* Ramat. cv. "Maghi Yellow" (a salt-sensitive cultivar) and cultured in vitro for callus induction. The details of technical procedure have been reported (Hossain et al. 2007). The induced calli were subcultured for successive three times at 30 days interval. After three subcultures, the fast-growing green calli were

Fig. 8.17 Regeneration of a selected variant (E2) and control plants of C. morifolium Ramat. Cv. Regal Time, under salt stress conditions. (**a**) Leaf explant after 4 weeks of culture showing multiple shoot buds. (**b** and **c**) SEM photographs of leaf explants after 15 days (**b**) and 20 days (**c**) of culture, showing a distinct shoot bud development directly from the explant. (**d**) Proliferation and development of multiple shoots. (**e**) Comparative in vitro growth performance of EMS-treated (selected plants, regenerated from 0.025% EMS treatment (1) and control (3) plants under saline conditions (MS medium supplemented with 250 mM NaCl) along with control plants grown on NaCl-free MS medium. (**f**) Acclimatization of in vitro raised plantlets. (**g**) Comparative growth performance of E2 variant (2) and control (3) plants under NaCl-stressed condition. Control plant (1) grown without NaCl stress. (**h**) Flower size: Control flower (1), E2 variant (2), and NaCl-treated control (3) plant). (C.f. Hossain et al. 2006a, b, Plant Biology, S.K. Datta is also an author)

selected for NaCl treatment. The lethal dose of NaCl for callus survival was first determined and it was 150 nM. For development of NaCl-tolerant callus, two selection methods were adopted—direct selection (DS) and stepwise selection (SS). For DS method, approx. 500 mg fresh mass of callus pieces were shifted to the same callus induction medium but supplemented with different concentrations of NaCl (0, 50, 75, and 100 mM) and maintained for 9 consecutive months with regular subculture at 30 days interval. Calli cultured on MS medium devoid of NaCl were considered as control. In SS method, approx. 500 mg of callus was first shifted to 50 mM NaCl-supplemented medium and maintained for 3 months. Only surviving calli were transferred to 75 mM NaCl medium and maintained for another 3 months and finally in 100 mM (again for 3 months) with regular subculture at 30 days interval. All total of 60 replicates were maintained for each treatment. Few replicates were used up to 9 months for measuring growth and biochemical parameters and rest were maintained through regular subculture at 30 days interval. Well grown green calli of SS method were selected as 100 mM NaCl-tolerant line selected after 9 months long NaCl treatment and based on assessment of growth performance and antioxidant capacity. These selections showed all positive responses toward the salt stress. The selected lines were allowed to grow in NaCl-free medium for 3 months with regular subculture followed by retransfer to slightly higher saline medium (120 mM) to check the stability of NaCl-tolerance character. Regenerated shoots were shifted as per specific experimental design. Regenerated control shoots were transferred to MS medium supplemented with 250 mM Nacl and considered as negative control (NC). Similarly selected calli were transferred separately to MS medium supplemented with 250 mM NaCl and considered as S1 line. Another control line shoots were maintained in MS medium having the same compositions except for the NaCl and treated as positive control (PC). The selected callus line exhibited significant increase in superoxide dismutase, ascorbate peroxidase and glutathione reductase activities compared to control callus. Stability of salt tolerance character of the selected callus line was checked by growing the calli in NaCl-free medium for 3 consecutive months followed by re-exposure to higher salinity stress (120 mM NaCl). The regeneration potential of the NaCl-tolerant callus ranged from 20.8% to 0% against 62.4% to 0% in control callus line. Selected calli regenerated S1 plants under 250 mM stress condition exhibited significantly higher SOD and APX activities over both PC and NC plants. The NC plants showed stunted growth, delayed root initiation, and had lesser number of roots as compared to S1 plants. After 9 months long-term NaCl treatment callus of stepwise route showed significant increase in antioxidant enzymes activities (SOD, APX and GR) along with high endogenous free proline accumulation. Based on growth performance and antioxidant capacity, the S1 plants were considered as NaCl-tolerant line showing all positive adaptive features toward the salinity stress (Fig. 8.18) (Hossain et al. 2007).

Results obtained from the present experiment strongly indicate that stepwise increase in NaCl concentration from a relatively low level to cytotoxic level is a better way to isolate NaCl-tolerant callus line since direct transfer of callus to highest selection pressure (here it is 100 mM NaCl concentration) was found to be lethal. Among the different parameters responding to NaCl stress, rapid accumulation of free proline within the cell is the most significant one. In the present experiment with increase in NaCl stress there was an increase in endogenous free proline content, which was more prominent in case of stepwise route. Whenever, calli were transferred to a higher saline medium from a lower one, a sharp increase in endogenous free proline content was recorded. Interestingly, this accumulation is reversible, i.e. the withdrawal of NaCl from the medium (for 3 months) results in a sharp fall in proline content. According to Fukutaku and Yamada (1984) proline acts as a

Fig. 8.18 Callus growth and subsequent plant regeneration, and performance of such regenerants under in vitro NaCl-stressed condition in Ch. morifolium Ramat. cv. Maghi Yellow. (**a**) Green sectors (arrow marks) of callus developed on 50 mM NaCl-supplemented medium (SS). (**b**) Green sectors (arrow marks) of callus developed on 75 mM NaCl-supplemented medium (SS). (**c**) Green, well growing callus of selected line (100 mM NaCl-tolerant callus line developed through SS). (**d**) Dead callus (brown and soft tissue) of 100 mM NaCl medium (DS) after 9 months of treatment. (**e**) Plant regeneration from selected callus line. (**f**) Comparative growth performance of PC, S1, and NC plants under in vitro salinity stressed condition. (C.f. Hossain et al. 2007, Journal of Biotechnology, S.K. Datta is also an author)

reservoir of nitrogen and carbon sources for post stress growth. This may be the reason that when NaCl was withdrawn from the medium for 3 months, proline level fell down to control level. On retransfer to NaCl medium, high proline accumulated just to balance the osmoregulation. These results indicated that NaCl induced proline synthesis and accumulation are significant adaptive features of plant cells for their survival and growth in saline environment. On the contrary, accumulation of proline does not play a major role in combating salinity stress at whole plant level, as elevated proline accumulation does not give much protection to the NC plants. Both NC and S1 lines accumulated significantly high proline under salinity stress, but still NC line showed stunted growth whereas S1 plants attained much better height. The selected callus line exhibited several positive characters (better growth, higher antioxidant enzyme capacity and proline content) toward salinity stress. In interpreting the present experimental data, author (Hossain et al. 2007) proposed that salt resistance is much better correlated with antioxidant enzyme capacity than with proline content, as both in callus line selection (callus transferred directly to 100 mM NaCl medium) and NC plants, the increased proline level does not give them much protection to combat the unfavorable stress condition. These data are consistent with the notion that antioxidant enzymes are better marker for salt tolerance than proline content. Elevated antioxidant enzyme activity and better growth of S1 plants under NaCl stress indicated that efficient detoxification of both $O_2^{\bullet -}$ and H_2O_2 is required to combat salinity-induced oxidative stress in chrysanthemum. Based on growth performance and antioxidant capacity S1 plants could be considered as NaCl-tolerant line showing all positive adaptive features toward the salinity stress. Further study on agronomic performance of these S1 plants under saline soil condition need to be undertaken to check the genetic stability of the induced salt tolerance.

The entire works on development of salt resistant chrysanthemum have been shown diagrammatically (Fig. 8.19).

8.11 Acute and Chronic Irradiations

For experimental purposes, both acute and chronic irradiations are applied. Acute radiation treatment is applied from the gamma chamber and chronic dose from a gamma field or greenhouses. Acute irradiation with fairly hard X-rays or gamma rays is applied using a dose rate of 1–10 Gy/min. Irradiation dosages are changed by changing the distance between the source and plants. For chronic irradiation gamma fields or greenhouses have been established in several parts of the world (e.g. Brookhaven, USA and Ohmiya, Japan) (Ukai 1982). All earlier experimental plant varieties showed differential sensitivity to acute and chronic irradiations separately and in a combination of both methods. Chrysanthemum showed the highest mutation frequency and spectrum in chronic irradiation than in acute irradiation. Nagatomi and Degi (2009) reported ten times more mutation rate and solid mutants in combined treatment of acute and chronic treatment. For chrysanthemum cuttings, petal, and/or bud culture the optimum dose of chronic irradiation has been recommended almost 2.5 times more than acute irradiation. The subject has been enriched and several desirable mutant varieties have been developed in ornamentals by this technique (Broertjes 1971; Fujii 1962a, b; Fujii and Matsumura 1967; Broertjes 1972; Kovalchuk et al. 2000; Mergen and Thielges 1966; Nagatomi 1991, 1992, 2002; Nagatomi et al. 1993a, b, 1996a, b, c; 2000; Natarajan and Maric 1961; Richter and Singleton 1955).

8.12 Ion Beam Technology

In mutation breeding experiments ion bean technology has been considerably used as a powerful option for mutagens with different mutation spectra in comparison to other mutagens. Ion beams have much higher linear energy transfer than other radiations. Helium (He), carbon ©, neon (Ne), and argon (Ar) with 220 MeV C ions are most commonly used in these studies. Ion beams are graded as high-linear energy transfer (LET) and their utilization in mutation breeding experiments is

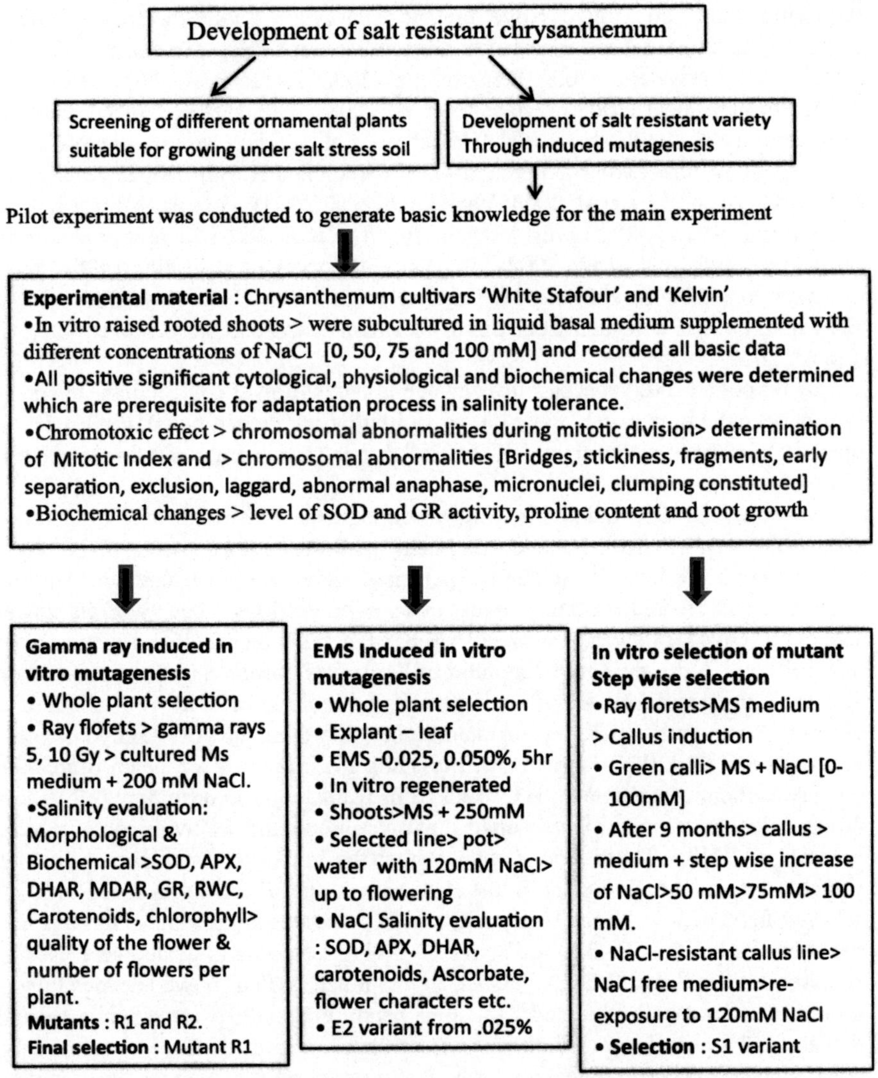

Fig. 8.19 Different methods for development of salt resistant chrysanthemum

restricted in a few counties like China, Japan, Korea, etc. (Du et al. 2017; Kim et al. 2016; Tanaka et al. 2010). Neutrons and heavy ions are used to induce mutations but their application is limited in ornamentals (Bolon et al. 2011, 2014; Broertjes 1976; Datta 2012; Love and Constantin 1965; Smith and Noyszewski 2018). An impressive number of ornamental mutant varieties have been developed in Japan applying high-energy ion beam radiations (Nakagawa 2009; Tanaka et al. 2010; Wu et al. 2005). Experimental results on the effects of ion beams on ornamentals (chrysanthemum) are limited. Novel mutant phenotypes, a wide spectrum of flower color and shape have been reported in chrysanthemums (Nagatomi et al. 1997a, b, 1998, 2003; Suzuki et al. 2005; Ikegami et al. 2006; Shirao et al. 2007; Ueno et al. 2005; Furutani et al. 2008; Watanabe et al. 2008; Wakita et al. 2009), dianthus (Sugiyama et al. 2008a, b), dahlia (Hamatani et al. 2001), Osteospermum (Iizuka et al. 2008), petunia (Miyazaki et al. 2002; Yoshida and Kusumi 2002; Yamaguchi et al. 2008), rose (Hara et al. 2003; Yamaguchi et al. 2003), Torenia (Miyazaki et al. 2006; Sasaki et al. 2008), Lotus, Verbena sp. (Abe et al. 2002; Kanaya et al. 2008), cyclamen (Sugiyama et al. 2008a, b), etc. (Nagatomi et al. 1995, 1996a, b, c; Okamura et al. 2002, 2003, 2006; Ikegami et al. 2005; Ueno et al. 2002, 2004; Hara et al. 2003; Yamaguchi et al. 2010; Oka-Kira et al. 2005; Miyazaki et al. 2006; Tanaka 2009).

Suzuki et al. (2005) successfully developed a new flower color mutant in spray-type chrysanthemum using heavy ion beam irradiation. Ikegami et al. (2006) developed a new chrysanthemum line "JCH1029" through ion beam irradiation to cultivar "Jinba."

Sato et al. (2006) irradiated in vitro raised explants with ion beams and studied effects on morphological characters and detected flower color mutants. Shirao et al. (2007) did an ion beam irradiation experiment with the Chrysanthemum cultivar "Sanyo-ohgon" to reduce lateral buds, or to improve flower size. Various flower shapes and days for flowering were observed in the mutation and 45 individuals were then selected. A decrease in the amount of DNA was observed in the mutants with increasing the dose of ion beam irradiation. It was indicated that the re-irradiation of the ion beam was effective in additional improvement of "Sanyo-ohgon." Matsumura et al. (2007) experimented to study the effects of ion beam irradiation on chrysanthemum cultivars ("H13" and "Shiroyamate"). Ion beam induced yellow flower color mutation in "Shiroyamate," and vermilion and white/red mutants were induced in "H13." These results suggest that the ion beam irradiation would be valuable for mutation induction in the chrysanthemum flower. Toyoda et al. (2007) selected light yellow "Jinba" chrysanthemum to induce deep yellow through ion beam irradiation (12C6+:320 MeV, 1 · 2 · 3 Gy). Leaves and petals were used as explants. Five mutants (2 Gy:2 mutants, 3 Gy:3 mutants) that showed deep yellow at flower bud time were obtained. But their petals gradually became white toward flowering. Transgenic chrysanthemum (tolerance to multi-diseases developed by recombinant DNA techniques) was included in ion beam breeding for the induction of further mutations. Leaf segments were treated with 220 MeV or 320 MeV carbon ion beams and cultured in vitro. Chrysanthemum Stunt Viroid (CSVd) and Tomato Spotted Wilt Virus (TSWV) 2). The tolerance of pac1 transgenic chrysanthemum to CSVd. Flower color mutants such as pale pink, dark pink, salmon, etc. were detected. Viruses belonging to the genus Tospovirus including TSWV are very serious pathogens not only in the chrysanthemum but on other horticultural plants. Ion beam breeding was effective to produce color variations, once the transgenic plants with the virus- and viroid-resistance have been acquired (Okamura et al. 2007). Furutani et al. (2008) induced flower color changes (vermilion, white/red, and several other colors) through ion beam irradiation in chrysanthemum cv. "H13" (red). Results suggested the valuable effects of beam irradiation on mutant induction in Chrysanthemum. Okamura et al. (2008) selected Agrobacterium-mediated transgenic chrysanthemum plants (Dendranthema grandiflora) expressing double-stranded RNA-specific ribonuclease gene (pac1) derived from Schizosaccharomyces pombe as materials for ion beam breeding. Leaf segments were exposed to 5–12 Gy of 320 MeV carbon ion beams and cultured in vitro. Regenerated transgenic chrysanthemum plants expressing the pac1 gene showed tolerance against Chrysanthemum Stunt Viroid (CSVd) and Tomato Spotted Wilt Virus (TSWV). New flower colors (pale pink, dark pink, salmon, white, yellow, etc.) mutants were also developed. Wakita et al. (2009) applied C-ion irradiation on spray-type chrysanthemum and detected new flower color mutations. Watanabe et al. (2008) developed a new mutant variety in chrysanthemum cv. "Jinba" through ion beam irradiation. Wakita et al. (2009) investigated the comparative effects of ion beams and gamma rays on mutation induction and nuclear DNA content as an index of radiation damage in C. morifolium. Yamaguchi et al. (2009) irradiated axillary buds with carbon ions at 2 Gy, helium ions at 10 Gy, and gamma rays at 80 Gy and found similar effects on survival. The main idea was to compare the effects of ion beam and gamma ray irradiation on mutation induction in axillary buds and to analyze the chimeric structure of the resulting mutants. The lower five nodes of the shoots elongating from the irradiated buds were cut one by one, and new shoots were allowed to grow from the axillary buds. This procedure was repeated twice, and flower color mutation was investigated. Flower color mutants emerged at a high frequency and there were no significant differences in the mutation frequency between the treatments. All the flower color mutants induced with gamma rays were periclinal chimeras. In contrast, some mutants obtained with ion beams had the same flower color as that of the plants derived from the roots. This suggests that they were solid mutants. Solid mutants were also obtained when irradiated with 5 Gy of helium ions, which had less effect on survival and mutation than other treatments. Matsumura et al. (2010) irradiated cultured ray florets and leaf explants of chrysanthemum cultivars, "Shiroyamate" and "H13" with $^{12}C^{5+}$ ion beam at doses of 1, 2, 4, and 8 Gy to induce mutations in ray florets color/shape. Shoot primordial formation on ray florets explants of "Shiroyamate" and shoot bud formation on leaf explants of "H13" was decreased at 8 Gy and 4 Gy irradiation. Yellow ray florets mutants from "Shiroyamate," various ray florets color mutants (dark red, light red, pink, pink spray), and a flower shape mutant (double-ray florets) from "H13" were induced by ion beam irradiation. A white mutant was obtained from a chimeric mutant.

Asami et al. (2010, 2011) developed new gene resources in chrysanthemums using ion beam irradiation. They have standardized easy screening techniques to assess the non-branching characteristics of the chrysanthemum line in the juvenile plantlet stage and in vitro stage. Matsumura et al. (2010) cultured ray florets and leaf explants of two chrysanthemum cultivars ("Shiroyamate" and "H13") and treated them with $^{12}C^{5+}$ ion beam at doses (1, 2, 4, and 8 Gy) to induce variability in ray florets. Shoot primordial formation in both the explants decreased at 8 Gy and 4 Gy treatment. The effective dose of the ion beam was less than 4 Gy in "Shiroyamate" and less than 2 Gy in "H13." Both varieties developed many floret colors (yellow, dark red, light red, pink, pink spray) and flower shape (double-ray florets) mutants. Results support that ion beam irradiation and tissue culture will be effective for inducing mutations in chrysanthemums. Tanaka et al. (2010) discussed the prospects of the application of ion beams in higher plants for the development of mutants. They developed flower color and flower form mutants in chrysanthemums by applying ion beams which are hardly produced by gamma rays or X-rays. Mutants were UV-B resistant, serrated petals and sepals, anthocyanins, etc. PCR and sequencing analyses showed that half of all mutants induced by ion beams possessed large DNA alterations, while the rest had point-like mutations. Both mutations induced by ion beams had a common feature that deletion of several bases was predominantly induced. It is plausible that ion beams induce a limited amount of large and irreparable DNA damage, resulting in the production of a null mutation that shows a new mutant phenotype. Yamaguchi et al. (2010) made a comparative study of ion beams and gamma rays on mutation induction and nuclear DNA content in *Chrysanthemum morifolium.* Leaf segments were treated with carbon ions (220 and 32 MeV), helium ions (100 MeV), and gamma rays. The number of plants with reduced nuclear DNA content increased with increasing irradiation doses of 320 MeV carbon ions, 100 MeV helium ions, and gamma rays. Irradiation treatment with 220 and 320 MeV carbon ions and gamma rays had a similar effect on mutation induction, while the effect of 100 MeV helium ions was not as great. 220 MeV carbon ion beam was found to be the most appropriate among the three types of ion beams because it gave a high mutation frequency with low damage to chromosomes. Shirao et al. (2013) selected four ion beam irradiated new desirable cultivars to detect the mutated zone by using PCR assays and successfully identified the mutants at the molecular level. Ueno et al. (2013) selected a white variety ("Jimba") for improvement of its some undesirable characteristics (lateral buds and delayed flowering under a low temperature) and successfully developed two new varieties ("Aladdin" and "Imagine") using ion beam irradiation. These two varieties were further irradiated and developed new cultivars with both desirable characteristics (few axillary buds and early flowering at a low temperature). Okamura et al. (2015) did tissue specific experiment with ion beam irradiation for inducing mutations. Somaclones regenerated from petal and leaf tissues and ion beam irradiated clones derived from the two tissues were analyzed for stem length and flower color mutation. 3 Gy of argon ion beams was found as an appropriate dose for irradiation because the resulting color mutants maintained an adequate stem length suitable for commercial use even with high mutation frequencies. Based on analysis of huge mutants/variants they concluded that somaclones from petal tissue had a higher mutation frequency than leaf somaclones; the effect of tissue source on the frequency of flower color mutations was significantly enhanced when combined with ion beam irradiation in that the mutation frequency of ion beam) the spectrum of carotenoid color mutations was remarkably increased by ion beam irradiation in that both leaf and petal ion beam irradiated clones had brown, red and yellow flowers in contrast to the somaclones that only produced yellow flowers. They have suggested that desirable flower color mutations can be induced by combining tissue source selection with ion beam irradiation. Sakamoto et al. (2016) applied C-ion beam irradiation and developed an early flowering mutant line at low temperatures in a spray-type chrysanthemum cultivar.

Tanokashira et al. (2015, 2016) selected a spray-mum variety ("Southern Chelsea") and induced mutations by treating cuttings with C ion (LET 23 keV/μ m) at doses of 2–5 Gy.

Tamaki et al. (2017) treated (C ion) a large number of cultivars comprising large-mum (17 cultivars), small-mum (19 cultivars), and a single cultivar of spray-mum and observed cultivar-specific color mutation, mutation frequencies, regardless of flower type. The mutation spectrum also varied according to the color of the starting cultivar. Mutated color spectrum has been compared with starting varieties and suggested that the selection of the color of experimental materials is very important to develop desirable color mutants.

Tamari et al. (2017) developed low-temperature-flowering chrysanthemum varieties "Ryujin" and "Touma" through ion beam irradiation. Suryawati et al. (2022) studied the effects of EMS to induce genetic variations on the callus of two chrysanthemum cultivars, "Jaguar Pink" and "Reagent Pink." Different EMS concentrations showed differential effects on survival callus, number of germinating calli, and number of callus with shoots on different varieties. The surviving callus from "Jaguar Pink" was higher than those of "Reagent Pink" and more calli from "Jaguar Pink" regenerated into shoots compared to "Reagent Pink." Based on the number of calli that produced shoots, the probability of mutants was found highly in "Jaguar Pink" and poorly in "Reagent Pink." The lethal concentration of 50% (LC_{50}) of the chrysanthemum callus "Reagent Pink" and "Jaguar Pink" was 0.4% of the EMS solution.

8.13 Selection of Chemical or Ionizing Radiation Mutagens

The book has provided an overview of the most frequently used mutagens for a wide range of ornamental genera under improvement program through induced mutagenesis. The benefit and drawbacks of each mutagen and technical procedure have been discussed. The potential and the limitations of various approaches have also been precisely discussed. It is clear from experimental results that every mutagen has differential mutagenic effectiveness and efficiency in terms of the creation of maximum beneficial changes. This helps the selection of mutagens and their effective doses. There are logical differences of opinion on the use of chemical and physical mutagens. This aspect has been discussed by many scientists at length which covers mainly the effectiveness and efficiency of each mutagen. Many chemicals were found to induce appreciable frequency and spectrum of mutations. But the main difficulty with chemical mutagen is the poor penetration into the propagules and the injurious effect of hydrolysis products. Another relevant problem with chemical mutagens is the bulk treatment of materials. This has restricted the application of chemical mutagens in vegetatively propagated crops (Van Harten 1998). It has been observed that mutagen effects depend upon the ploidy nature of the material. For example, mutations were induced in diploid roses but not in tetraploid roses by EMS. X-ray was very favorable to inducing mutations in both diploid and tetraploid species (Dommergues 1962; Heslot 1964). Chemical mutagens are chosen in many materials as they generally cause single-base substitutions (Greene et al. 2003; Shikazono et al. 2005). Physical mutagens may induce deletions at the chromosome level (Kodym and Afza 2003; Kodym et al. 2012; Koornneet et al. 1982; Oladosu et al. 2016). Physical mutagens provide harmonious treatments for experimental materials. But the limitation is accessibility to radiation sources (X-ray machines, gamma sources, particle accelerators, or nuclear reactors). Some other advantages of physical mutagens are easy post-treatment handling of treated materials, easy to handle delicate materials, application of split dose, recurrent dose, etc. (Oladosu et al. 2016; Predieri and Divrgilio 2007; Schum 2003).

As mentioned, each chemical and physical mutagens have differential mutagenic efficiencies. Studies demonstrated that EMS is more efficient than gamma rays (Gautam et al. 1992, 1998; Girija and Dhanavel 2009; Jayakumar and Selvaraj 2003; Kaul and Bhan 1977; Konzak et al. 1965; Mangaiyarkarasi et al. 2014; Solanki and Sharma 1994; Velu et al. 2007; Wani 2009). The efficiency of individual mutagen can be increased by combined treatment. Voluminous experimental results indicate that gamma ray is the most widely used ionizing radiation in mutagenesis. 63% of mutant varieties developed using a physical mutagen registered at the Mutant Variety Database (Joint FAO/IAEA) resulted from exposure to gamma rays.

8.14 Annual Chrysanthemum

The popularity of any ornamental in the floriculture market depends upon customer choice. Annual chrysanthemum (*C. carinatum*) is propagated by seeds and sold in the market along with other seasonal flowers. It has no special demand because of the less appealing form, size, and color of its flower. For the induction of genetic variability, only a few researchers started mutation work and applied X- and gamma radiations. Seeds were exposed to a 15,000r X-ray dose and also an acute X-ray dose of 15 Kr. Treated populations were examined in X_1, X_2, X_3, and X_4 generations. Several interesting phenotypic variations were recorded in flowers like tubular petals, double-type flower heads, and dissected type rays in which the strap-shaped petals were split into two along their entire length, small ray florets, apetalous type, etc. Results indicated that induced mutagenesis has notable scope for selecting and synthesizing new desirable varieties in annual chrysanthemum (Jain et al. 1961; Rana 1964a, b, c, d, 1965a, b).

If we look at all the above-mentioned experiments, conventional mutation technology is being applied in improvement programs using mostly either one or more physical and or chemical mutagen/s separately or in combinations. All experimental results contributed valuable technical guidelines for continuous improvement and refinement of mutation technology for betterment. There is no shortage of technical details and ideas of mutation technology at its present condition. There is a critical shortage of proper application of technique. This may be a painful reality but the fact is real. Lack of conceptual clarity is also seen in many initial experimental missions. Extensive research activities developed a complete package of mutation technology for maximum commercial utilization. The present mutation breeding technology package covers both traditional and biotechnological components (in vitro management) (Fig. 8.4). One should start mutation work with the full package. At this stage, classical mutagenesis combined with the management of chimera and in vitro mutagenesis is the most promising and standardized technique for developing new and novel ornamental varieties. The knowledge generated so far on vegetatively propagated crops, especially on ornamental crops will work as a model system for future need-based planning of successful and accurate application of mutation techniques in crop improvement programs.

References

Abe T, Katsumoto Y, Miyazaki K, Yoshida S, Kusumi T (2002) Isolation of sterile mutants of Verbena hybrida using heavy-ion beam irradiation. RIKEN Accel Prog Rep 35:129

Ahloowalia B (1995) *In vitro* mutagenesis for the improvement of vegetatively propagated plants. In: Induced mutations and molecular techniques for crop improvement. Proceedings of a symposium, Vienna, 19-23 June, jointly organized by IAEA and FAO, pp 531–541

Anderson YO (1987) Reclamation of the genus Chrysanthemum L. HortScience 22(2):313

Anitha G, Shiragur M, Patil BC, Nishani S, Seetharamu GK, Ramanagouda SH, Naika MBN (2021) Mutation studies in chrysanthemum cultivar Poornima white. J Pharmacogn Phytochem 10(1): 1235–1239

Anonymous (1961) Chrysanthemums. Bull Minist Agric Fish Food Lond 2(6)

Asami I, Hasegawa T, Yamada M (2010) Development of an easy screening method to evaluate the non-branching characteristics of the chrysanthemum line in the juvenile plantlet stage and in vitro stage. Res Bull Aichi Agric Res Ctr 42:1–6

Asami I, Tsuji T, Hasegawa T, Fukuta S, Kuroyanagi S, Hase Y, Yoshihara R, Narumi I (2011) Producing new gene resources in chrysanthemums using ion-beam irradiation. JAEA Takasaki Ann Rep 2009:68

Bajpay A, Dwivedi DH (2019) Gamma ray induced foliage variegation and anatomical aberrations in *Chrysanthemum* (*Dendranthema grandiflora* T.) cv. Maghi. J Pharmacogn Phytochem 8(4): 871–874

Banerji BK, Datta SK (1986) Induction of single flower mutant in Hibiscus cv. 'Alipur Beauty'. J Nucl Agric Biol 15(4):237–240

Banerji BK, Datta SK (1988) 'Anjali'—a new gamma ray induced single flower mutant of Hibiscus. J Nucl Agric Biol 17(2):113–114

Banerji BK, Datta SK (1990) Induction of somatic mutations in chrysanthemum cultivar 'Anupam'. J Nucl Agric Biol 19:252–256

Banerji BK, Datta SK (1993) Varietal differences in radiosensitivity of garden chrysanthemum. Nucleus 36(3):114–117

Bañón S, Conesa E, Valdés R, Miralles J, Martínez JJ, Sánchez Blanco MJ (2012) Effects of saline irrigation on phytoregulator-treated chrysanthemum plants. Acta Hortic 937:307–312. https://doi.org/10.17660/ActaHortic.2012.937.37

Barakat MN, El-Sammak H (2011) In vitro mutagenesis, plant regeneration and characterization of mutants via RAPD analysis in baby's breath *Gypsophila paniculata* L. Am J Crop Sci 5:214–222

Barakat MN, Abdel Fattah RS, Badr M, El-Torky MG (2010) In vitro mutagenesis and identification of new variants via RAPD markers for improving *Chrysanthemum morifolium*. Afr J Agric Res 5:748–757

Bolon YT, Haun WJ, Xu WW, Grant D, Stacey MG et al (2011) Phenotypic and genomic analyses of a fast neutron mutant population resource in soybean. Plant Physiol 156:240–253

Bolon YT, Stec AO, Michno JM, Roessler J, Bhaskar PB, Ries L et al (2014) Genome resilience and prevalence of segmental duplications following fast neutron irradiation of soybean. Genetics 198:967–981. https://doi.org/10.1534/genetics.114.170340

Bowen HJM (1965) Mutations in horticultural chrysanthemum. Radiat Bot Suppl. 5:695–700

Bowen HJM, Cawse PA, Dick MJ (1962) The induction of sports in chrysanthemum by gamma irradiation. Radiat Bot 1:297–303

Brock RD (1966) Second cycle of mutation and selection for quantitative variation in Trifolium subterraneum. Radiat Bot 6:357–369

Broertjes C (1966a) Mutation breeding of *Chrysanthemum*. Euphytica 15:156–162

Broertjes C (1966b) Mutation breeding in Chrysanthemum. Induced mutations and their utilization. Gatersleben 357

Broertjes C (1971) Dose-fractionation studies and radiation-induced phenomena in African violet. In: Benson DW, Sparrow AH (eds) Survival of food crops and livestock in the mission. Oak Ridge, TN, pp 325–342

Broertjes C (1972) Use in plant breeding of acute, chronic or fractionated doses of X-rays or fast neutrons as illustrated with leaves of *Saintpaulia*. Thesis, Centr. Agric. Pubbl. Doc., Wageningen, Agric. Res. Rep., 776:74

Broertjes C (1976) Mutation breeding of auto-tetraploid *Achimenes* cultivars. Euphytica 25:297–304

Broertjes C, Jong DE (1984) Radiation-induced male-sterility in daisy type of *Chrysanthemum morifolium* Ram. Euphytica 33(1):433–434

Broertjes C, Van Harten AM (1988) Applied mutation breeding for vegetatively propagated crops. Elsevier, Amsterdam

Broertjes C, Koene P, van der Veen JWH (1980) A mutant of a mutant of a mutant of a … Irradiation of progressive radiation-induced mutants in a mutation-breeding programme with *Chrysanthemum morifolium* Ram. Euphytica 29:525–530

Broertjes C, Pouwer A, Evers JFM (1983) Kleurenkeus in overvloed. Alleen geel ontbreekt nog bij Achimenes. Vakbl. Bloemisterij 38(35):34–35

Buiatti M, Tesi R (1968) Gladiolus improvement through radiation induced somatic mutation. Publication No. 1. Laboratorio di Mutagenesi e differeziamento del C.N.R., Pisa, pp 1–6

Bush SR, Earle ED, Langhans RW (1976) Plantlets from petal segments, petal epidermis, and shoot tips of the periclinal chimera Chrysanthemum morifolium 'Indianapolis'. Am J Bot 63(6):729–737

Caroline VJE, Mallaiah B (2011) In-vitro mutagenesis in an endangered medicinal cucurbit *Bryonopsis laciniosa* (L.) *Naud.* Int J Genet Eng Biotechnol 2(1):67–75. ISSN 0974-3073

Cassels AC, Walsh C, Periappuram C (1993) Diplontic selection as a positive factor in determining the fitness of mutants of *Dianthus* 'Mystere' derived from X-irradiation of nodes in in vitro culture. Euphytica 70:167–174

Castillo-Martínez CR, de la Cruz-Torrez E, Carrillo-Castañeda G, Avendaño-Arrazate CH (2015) Inducci6n de mutaciones en crisantemo (*Dendranthema grandiflora*) usando radiaci6n gamma y etil metano sulfonato. Agroproductividad 8(2):60–64. http://132.248.9.34/hevila/Agroproductividad/2015/vol8/no2/9.pdf

Cawse PA (1965) Production of Chrysanthemum sports by gamma radiation. Radioisot. Rev., Sheet. Wantage Res. Lab. (A.E.R.E.), Wantage

Chakrabarty D, Mandal AKA, Datta SK (1999) Management of chimera through direct shoot regeneration from florets of chrysanthemum (*Chrysanthemum morifolium* Ramat). J Hortic Sci Biotechnol 74(3):293–296

Chakrabarty D, Mandal AKA, Datta SK (2000) Retrieval of new coloured chrysanthemum through organogenesis from sectorial chimeras. Curr Sci 78(9):1060–1061

Chaleff RS (1983) Isolation of agronomically useful mutants from plant cell cultures. Science 214: 676–682

Chen F, Chen S, Guo W, Ji S (2003) Salt tolerance identification of three species of Chrysanthemums. Acta Hortic 618:299–305. https://doi.org/10.17660/ActaHortic.2003.618.34
Dalsou V, Short KC (1987) Selection for sodium chloride tolerance in chrysanthemum. Acta Hortic 212:737–740
Das PK, Ghosh P, Dube S, Dhua SP (1974) Induction of somatic mutations in some vegetatively propagated ornamentals by gamma irradiation. Technology (Coimbatore, India) 11(2,3): 185–188
Datta SK (1985a) Gamma ray induced mutant of a mutant chrysanthemum. J Nucl Agric Biol 14(4): 131–133
Datta SK (1985b) Colchi-mutation (C-mutation). Everyman's Sci XIX(3):70–72
Datta SK (1987) 'Colchi Bahar'—a new chrysanthemum cultivar evolved by colchi mutation. The Chrysanthemum 43(1):4
Datta SK (1990) Induction and analysis of somatic mutations in garden chrysanthemum. Adv Hortic For 1:241–254. Article No. 31
Datta SK (1991) Evaluation of recurrent irradiation on vegetatively propagated ornamental: *Chrysanthemum*. J Nucl Agric Biol 20(2):81–86
Datta SK (1992) Radiosensitivity of garden chrysanthemum. J Ind Bot Soc 71(I–IV):283–284
Datta SK (1994) Induction and analysis of somatic mutations in vegetatively propagated ornamental. D. Sc. Thesis, Kanpur University, Kanpur, India
Datta SK (1997) Ornamental plants—role of mutation. Daya Publishing House, Delhi, p 219
Datta SK (2000) Mutation studies on garden Chrysanthemum—a review. In: Singh SP (ed) Scientific horticulture, vol 7. Scientific Publisher, Jodhpur, pp 159–199
Datta SK (2005a) Technological advances for development of new ornamental varieties. In: Trivedi PC (ed) Advances in biotechnology. Agrobios (India), Jodhpur, pp 285–304
Datta SK (2005b) In: Datta SK (ed) Role of classical mutation breeding in crop improvement. Daya Publishing House, Delhi, pp 1–314
Datta SK (2012) Success story on induced mutagenesis for development of new ornamental varieties. Bioremediat Biodivers Bioavailab 6(Special issue 1):1–26. Global Science Book
Datta SK (2015) Indian floriculture; role of CSIR. Regency Publications, A Division of Astral International(P) Ltd., New Delhi
Datta SK (2019) Floriculture work at CSIR-National Botanical Research Institute, Lucknow. Sci Cult 85(7–8):274–283
Datta SK (2023) Induced mutagenesis in chrysanthemum. In: Penna S, Jain SM (eds) Mutation breeding for sustainable food production and climate resilience. Springer, pp 565–598
Datta SK, Banerji BK (1991) Analysis of induced somatic mutations in chrysanthemum. J Ind Bot Soc 71(I–IV):59–62
Datta SK, Banerji BK (1993) Gamma ray induced somatic mutation in chrysanthemum cv. 'Kalyani Mauve'. J Nucl Agric Biol 22(1):19–27
Datta SK, Banerji BK (1995a) Chromosome counting in original and gamma ray induced mutants of chrysanthemum. J Nucl Agric Biol 24(1):58–61
Datta SK, Banerji BK (1995b) Improvement of garden chrysanthemum through induced mutation. Flora Fauna 1(1):1–4
Datta SK, Teixeira da Silva JAT (2006) Role of induced mutagenesis for development of new flower colour and type in ornamentals. In: Book chapter: floriculture, ornamental and plant biotechnology. Global Science Book Ltd., pp 640–645. ISBN: 490331300X
Datta SK, Gupta MN (1980) Effects of gamma irradiation on rooted cuttings of small flower chrysanthemum. New Bot VII:73–85
Datta SK, Gupta MN (1981a) Effects of gamma irradiation on rooted cuttings of Korean type Chrysanthemum cv. 'Nirmod'. Bangladesh J Bot 10(2):124–131
Datta SK, Gupta MN (1981b) Studies on Chrysanthemum cultivar 'Otome Zakura' and its mutants. Bot Prog 4:88–92
Datta SK, Gupta MN (1983) Somatic flower colour mutation in Chrysanthemum cv. 'D-5'. J Nucl Agric Biol 12(1):22

Datta SK, Gupta MN (1984) Effects of colchicines on rooted cuttings of *Chrysanthemum*. The Chrysanthemum 40(5):191–194
Datta SK, Gupta MN (1987) Colchicine induced somatic flower color mutation in *Chrysanthemum*. Bangladesh J Bot 16(1):107–108
Datta SK, Mandal AKA (2005) In vitro mutagenesis—a quick method for establishment of solid mutant in chrysanthemum. Curr Sci 88(1):155–158
Datta SK, Banerji BK, Gupta MN (1985) 'Tulika'—a new chrysanthemum cultivar evolved by gamma irradiation. J Nucl Agric Biol 14(4):160
Datta SK, Chakraborty D, Mandal AKA (2001) Gamma ray-induced genetic manipulations in flower colour and shape in Dendranthema grandiflorum and their management through tissue culture. Plant Breed 120:91–92
De Jong J (1984) Genetic analysis in *Chrysanthemum morifolium*. I. Flowering time and flower number at low and optimum temperature. Euphytica 33:455–463
De Jong J, Custers JBM (1986) Induced changes in growth and flowering of chrysanthemum after irradiation and in vitro culture of pedicels and petal epidermis. Euphytica 35:137–148
Dermen H (1947) Periclinal cytochimeras and histogenesis in cranberry. Am J Bot 34:32–43
Dermen H (1953) Periclinal cytochimeras and origin of tissues in stem and leaf of peach. Am J Bot 40:154–168
Dilta BS, Sharma YD, Gupta YC, Bhalla R, Sharma BP (2003) Effect of gamma-rays on vegetative and flowering parameters of chrysanthemum. J Ornam Hortic 6(4):328–334
Din A, Qadri ZA, Rather ZA, Iqbal S, Malik SA, Rafiq S (2020) Congenial in vitro γ-ray induced mutagenesis underlying the diverse array of petal colours in chrysanthemum (*Dendranthemum grandiflorum kitam*) cv. "Candid". Published: 01 December 2020 by MDPI in The 1st International Electronic Conference on Plant Science session Plant Cell and Developmental Biology. https://doi.org/10.3390/IECPS2020-08780
Din A, Qadri ZA, Wani MA, Rather ZA, Iqbal S, Malik SA, Rafiq S, Nazki IT (2021) Congenial in vitro γ-ray induced mutagenesis underlying the varied array of petal colours in chrysanthemum (*Dendranthemum Grandiflorum* Kitam). 'Candid'. Research Square. https://doi.org/10.21203/rs.3.rs-617238/v1
Dommergues P (1962) Mutagenese experimentale. Ann Amel Plant 12:67–78
Dowrick GJ, El-Bayoumi A (1966a) The origin of new forms of the garden chrysanthemum. Euphytica 15:32–38
Dowrick GJ, El-Bayoumi A (1966b) The induction of mutations in chrysanthemum using X- and gamma radiation. Euphytica 15:204–210
Du Y, Luo S, Li X, Yang J et al (2017) Identification of substitutions and small insertion-deletions induced by carbon-ion beam irradiation in *Arabidopsis thaliana*. Front Plant Sci Sect Plant Genet Genomics 8. https://doi.org/10.3389/fpls.2017.01851
Dwimahyani I, Widiarsih S (2010) The effects of gamma irradiation on the growth and propagation of in vitro chrysanthemum shoot explants (cv. Yellow Puma). Atom Indonesia 36(2):45–49
Dwivedi AK, Banerji BK, Chakraborty D, Mandal AKA, Datta SK (2000) Gamma ray induced new flower colour chimera and its management through tissue culture. Indian J Agric Sci 70(12):853–855
Flick CE (1983) Isolation of mutants from cell culture. In: Evans DA, Sharp WR, Ammirato PV, Yamada Y (eds) Handbook of plant culture, vol 1. Macmillan, New York, pp 393–441
Fujii T (1962a) Mutations induced by radiation in vegetatively propagated plants with special reference to flower colour. Gamma Field Symp 1:51–59
Fujii T (1962b) A comparison of biological effects of acute and chronic irradiation. Rec Adv Breed 4:209–218
Fujii T, Mabuchi T (1961) Irradiation experiments with chrysanthemum. Seiken Hiho 12:40–44
Fujii T, Matsumura S (1967) Somatic mutations induced by chronic gamma irradiation. Seiken Jiho 19:43–46
Fukutaku Y, Yamada Y (1984) Sources of proline nitrogen in water stressed soyabean (Glycine max) II, fate of ^{15}N-labled protein. Physiol Plant 61:622–628

Furutani N, Matsumura A, Hase Y, Yoshihara R, Narumi I (2008) Dose response and mutation induction by ion beam irradiation in chrysanthemum. JAEA-Review 2008-055, p 69

Gautam AS, Soodand KC, Richarria AK (1992) Mutagenic effectiveness and efficiency of gamma rays, ethylmethane sulphonate and their synergistic effects in black gram (*Vigna mungo* L.). Cytologia 57:85–89

Gautam AS, Soodand KC, Mittal RK (1998) Mutagenic effectiveness and efficiency of gamma rays and ethylmethane sulphonate in rajmash (*Phaseolus vulgaris* L.). Legume Res 21(3&4): 217–220

Ghormade GN, Tambe TB, Patil UH, Nilima G (2020) Yield and quality of chrysanthemum varieties as influenced by chemical mutagens in VM1 generation. J Pharmacogn Phytochem 9(4):3100–3104

Girija M, Dhanavel D (2009) Mutagenic effectiveness and efficiency of gamma rays, ethyl methane sulphonate and their combined treatments in cowpea (*Vigna unguiculata* L. Walp). Global J Mol Sci 4(2):68–75

Greene EA, Codomo CA, Taylor NE et al (2003) Spectrum of chemically induced mutations from a large-scale reverse-genetic screen in *Arabidopsis*. Genetics 164(2):731–740

Grunewald J (1983) In vitro mutagenesis of *Saintpaulia* and *Pelargonium* cultivars (genetic variability, vegetative propagation). Acta Hortic 131:339–343

Gupta MN (1966) Induction of somatic mutations in some ornamental plants. In: Proc. All India Symp. Hortic, pp 107–114

Gupta MN (1979) Mutation breeding of some vegetatively propagated ornamentals. In: Progress in plant research, Silver Jubilee Publication National Botanical Research Institute, vol 2, pp 75–92

Gupta MN, Datta SK (1978) Effects of gamma rays on growth, flowering behavior and induction of somatic mutations in chrysanthemum. SABRAO J 10(2):149–161

Gupta MN, Jugran HM (1978) Mutation breeding of chrysanthemum II. Detection of gamma ray induced somatic mutations in vM_2. J Nucl Agric Biol 7(2):50–59

Gupta MN, Shukla R (1971) Mutation breeding of chrysanthemum. I. Production of new cultivars by gamma ray induced somatic mutations in vM1. In: Intnl. Symp. on use of isotopes and radiation in agric. and Animal Husbandry Research, New Delhi, Dec. 1971, pp 164–174

Hamatani M, Iitsuka Y, Abe T, Miyoshi K, Yamamoto M, Yoshida S (2001) Mutant flowers of dahlia (Dahlia pinnata Cav.) induced by heavy-ion beams. RIKEN Accel Prog Rep. 34:169

Hara Y, Abe T, Sakamoto K, Miyazawa Y, Yoshida S (2003) Effects of heavy-ion irradiation in rose (Rosa hybrida 'Bridal Fantasy') (II). RIKEN Accelerator Prog Rpt 36:135

Haspalot G, Kunter B, Kantoglu Y (2022) Determination of mutagenic-sensitivity and induced variability in the mutant populations of 'Bacardi' chrysanthemum cultivar. Genetika 54(1): 147–160. https://doi.org/10.2298/GENSR2201147H

Haspolat G (2022a) Induction of mutagenesis on *Chrysanthemums*. Ornam Hortic 28(4). https://doi.org/10.1590/2447-536X.v28i4.2523

Haspolat G (2022b) Changes on mutant pot chrysanthemums (Dendranthema ×grandiflora grandiflora Tzelev). Not Bot Horti Agrobo 50(4):13002

Haspolat G, Senel U, Taner Kantoglu Y, Kunter B, Guncag N (2019) In vitro mutation on chrysanthemums. Acta Hortic 1263:261–266. https://doi.org/10.17660/ActaHortic.2019.1263.34

Heslot H (1964) L'induction experiment de mutations chezles plantes florales. 17 CR Acad Agri Fr:1281–1308

Heywood VH, Humphries CJ (1977) Anthmidene systematic review. In: Heywood VH, Hartome JB, Turner BL (eds) The biology and chemistry of compositae. Academic, New York, pp 851–898

Hossain Z (2005) In vitro development of NaCl tolerant line in *Chrysanthemum morifolium* Ramat. using antioxidant enzymes as biochemical marker. Ph.D. Thesis, Kalyani University, Kalyani, West Bengal, India

Hossain Z, Mandal AKA, Shukla R, Datta SK (2004) NaCl stress—its chromotoxic effects and antioxidant behavior in roots of *Chrysanthemum morifolium* Ramat. Plant Sci 166:215–220

Hossain Z, Mandal AKA, Datta SK, Biswas AK (2006a) Isolation of a NaCl-tolerant mutant of *Chrysanthemum morifolium* by gamma radiation: in vitro mutagenesis and selection by salt stress. Funct Plant Biol 33:91–101

Hossain Z, Mandal AKA, Datta SK, Biswas AK (2006b) Development of NaCl-tolerant strain in *Chrysanthemum morifolium* Ramat. through in vitro mutagenesis. Plant Biol 8(4):450–461

Hossain Z, Mandal AKA, Datta SK, Biswas AK (2007) Development of NaCl-tolerant strain in *Chrysanthemum morifolium* Ramat. through shoot organogenesis of selected callus line. J Biotechnol 129:658–667

Huitema JBM, Cussenhoven GC, Dons JJM, Broertjes C (1986) Induction and selection of low temperature tolerant mutants of *Chrysanthemum* Ramat. Acta Hortic 197:89–96

Huitema JBM, Preil W, Gussenhoven GC, Schineidereit M (1989) Methods or the selection of low-temperature tolerant mutants of *Chrysanthemum morifolium* Ramat. by using irradiated cell suspension cultures. I. Selection of regenerations in vivo under suboptimal temperature conditions. Plant Breed 102:140–147

Ichikawa S, Yamakawa K, Sekiguchi F, Tatsuno T (1970) Variation in somatic chromosome number found in radiation-induced mutants of *Chrysanthemum morifolium* Hemsl. cv. Yellow Delaware and Delaware. Rad Bot 10:557–562

Iizuka M, Yoshihara R, Hase Y (2008) Development of commercial variety of osteospermum by a stepwise mutagenesis by ion beam irradiation. JAEA Takasaki Ann Rep 2007:65

Ikegami H et al (2005) Mutation induction through ion beam irradiations in protoplasts of chrysanthemum. Bull Fukuoka Agric Res Cent 24:5–9

Ikegami H, Suyama T, Kunitake T, Kuroyanagi N, Matsuno T, Hirashima K, Tanigawa T, Sakai Y, Hase Y, Tanaka A et al (2006) Mutation induction through ion beam irradiations in chrysanthemum cultivars, 'Jinba', and breeding a new line 'J CH1029' by this method. Bull Fukuoka Agric Res Cent 25:47–52

Jain HK, Bose AK, Sathpathy D, Sur SC (1961) Mutation studies in annual chrysanthemum. I. Radiation-induced variation in flower form. Indian J Genet Plant Breed 21:68–74

Jank H (1957a) Experimental production of mutations in *Chrysanthemum indicum* by X-rays. (Translation of Jank 1957a). U.K. At. Energy Established, Harwell. 36pp

Jank H (1957b) Experimentelle Mutationsauslosung dursh Rontgenstrahlen bei *Chrysanthemum indicum*. Zuchter 27:223–231

Jayakumar S, Selvaraj R (2003) Mutagenic effectiveness and efficiency of gamma rays and ethylmethane sulphonate in sunflower (*Helianthus annuus* L.). Madras Agric J 90(7–9): 574–576

Jerzy M (1990) In vitro induction of mutation in *Chrysanthemum* using x-and gamma radiation. Mutat Breed Newsl 35:10

Jerzy M, Lubomski M (1991) Adventitious shoot formation on ex in vitro derived leaf explants of *Gerbera jamesonii*. Sci Hortic 47:115–124

Jerzy M, Zalewska M (1992) In vitro adventitious bud techniques for mutation breeding of *Gerbera jamesonii*. Acta Hortic 314:269–274

Jerzy M, Zalewska M (1996) Polish cultivars of *Dendranthema grandiflora* Tzvelev and *Gerbera jamesonii* Bolus bred in vitro by induced mutations. Mutat Breed Newsl 42:19

Johnson RT (1980) Gamma irradiation and in vitro induced separation of chimeral genotypes in carnation. HortScience 15(5):605–606

Jong DE (1978) Selection for wide temperature adaptation in *Chrysanthemum morifolium* Ram. Hemsel. Nath J Agric Sci 26:110–118

Jung-Heliger H, Horn W (1980) Variation nach mutagener Behandlung von Stecklingen and in vitro-Kulturen bei Chrysanthemum. Z Pflanzenzlichg 85:185–199

Kanaya T, Saito H, Hayashi Y, Fukunishi N, Ryuto K, Miyazaki K, Kusumi T, Abe T, Suzuki K (2008) Heavy-ion beam-induced sterile mutants of verbena (*Verbena hybrida*) with an improved flowering habit. Plant Biotechnol 25:91–96

Kapadiya DB (2014) Induction of variability through mutagenesis in Chrysanthemum (*Chrysanthemum morifolium* Ramat) varieties Jaya and Maghi through mutagenesis. Thesis M.Sc. (Horti.), Navsari Agricultural University, Navsari, Gujarat, India. p 180

Kapadiya DB, Chawala SL, Patel AI, Ahlawat TR (2014) Exploitation of variability through mutagenesis (*Chrysanthemum morifolium* Ramat.) var. Maghi. The Bioscan 9(4):1799–1804

Kapadiya DB, Chawla SL, Patel AI, Bhatt D (2016) Induction of variability through in vivo mutagenesis in chrysanthemum (*Chrysanthemum morifolium* Ramat.) var. Jaya. Indian J Hort 73(1):141–144

Kaul MLH, Bhan AK (1977) Mutagenic effectiveness and efficiency of EMS, DES and gamma-rays in rice. Theor Appl Genet 50(5):241–246

Kaul A, Kumar S, Ghani M (2011) In vitro mutagenesis and detection of variability among radiomutants of chrysanthemum using RAPD. Adv Hort Sci 25(2):106–111

Kim YS, Sang YS, Jo JD, Lee HJ, Sang HK (2016) Effects of gamma ray dose rate and sucrose treatment on mutation induction in chrysanthemum. Eur J Hortic Sci 81(4):212–218. https://doi.org/10.17660/eJHS.2016/81.4.4

Kitamura S (1978) *Dendranthema* at Nippoanthemum. Acta Phytotaxonomica et. Geobotanica 29: 165–170

Kodym A, Afza R (2003) Physical and chemical mutagenesis. In: Grotewold E (ed) Plant functional genomics, vol 236. Humana Press, Inc., Totowa, NJ, pp 189–203. https://doi.org/10.1385/1-59259-413-1:189

Kodym A, Afza R, Forster BP et al (2012) Methodology for physical and chemical mutagenic treatments. In: Shu QY, Forster BP, Nakagawa H (eds) Plant mutation breeding and biotechnology. Joint FAO/IAEA Division of nuclear Techniques in Food and Agriculture, IAEA, Vienna, pp 169–180

Konzak CF, Nilan RA, Wagnerand J, Foster RJ (1965) Efficient chemical mutagenesis. The use of induced mutations in plant breeding Rept. FAO/IAEA Tech. Meet. Rome

Koornneet M, Dellaert LW, Van Der Veen JH (1982) EME-and radiation-induced mutation frequencies at individual loci in *Arabidopsis thaliana* (L). Heynh. Mutat Res 93:109–123

Kovalchuk O, Arkhipov A, Barylyak I et al (2000) Plants experiencing chronic internal exposure to ionizing radiation exhibit higher frequency of homologous recombination than acutely irradiated plants. Mutat Res 449(1):47–56

Kumari K, Dhatt K, Kapoor M (2013) Induced mutagenesis in Chrysanthemum morifolium variety 'Otome Pink' through gamma irradiation. Bioscan 8:1489–1492

Laneri U, Franconi R, Altavista P (1990) Somatic mutagenesis of *Gerbera jamesonii* hybr.: irradiation and in vitro culture. Acta Hortic 280:395–402

Latado RR, Tulmann Neto A, Mendes BMJ (1996) In vitro mutation breeding of Chrysanthemum (*Dendrathema grandifolia* Tzvelev.) cv. Pink repin (in Portuguese). Pesqui Agropecu Bras 31: 489–496

Latado RR, Adames AH, Neto AT (2004) In vitro mutation of chrysanthemum (*Dendranthema grandiflora* Tzvelev) with Ethylmethane sulphonate (EMS) in immature floral pedicels. Plant Cell Tiss Org Cult 77:103–106

Lema-Rumińska J, Sliwinska E (2009) Evaluation the stability of plants obtained from the somatic embryos in chrysanthemum (*Chrysanthemum* × *grandiflorum*/Ramat./Kitam.). Zesz Probl Post Nauk Rol 539:425–432

Lema-Rumińska J, Sliwinska E (2015) Evaluation of the genetic stability of plants obtained via somatic embryogenesis in *Chrysanthemum* × *grandiflorum* (Ramat./Kitam.). Acta Sci Pol Hort Cult 14:131–139

Lema-Rumińska J, Zalewska M (2002) Evaluation of ploidy in chrysanthemum mutants (*Dendranthema grandiflora* Tzvelev) obtained in mutagenesis induced in vitro and in vivo by ionizing radiation. Acta Sci Pol Hort Cult 1:43–48

Li W, Meng R, Liu Y, Chen S, Jiang J, Wang L, Zhao S, Wang Z, Fang W, Chen F, Guan Z (2022) Heterografted chrysanthemums enhance salt stress tolerance by integrating reactive oxygen species, soluble sugar, and proline. Hortic Res 9:uhac073. https://doi.org/10.1093/hr/uhac073

Love JE, Constantin MJ (1965) The response of some ornamental plants to fast neutrons. Tenn Farm Home Sci 56:10–12
Love JE, Malone BB (1967) Anthocyanin pigments in mutant and nonmutant Coleus plants. Rad Bot 7:549–552
Lysikov VN (1990) Experimental mutagenesis in plants. Izvestiya Akademii Nauk Moldavskoi SSR. Biologicheskie i. Khimicheskie Nauki 4:3–10
Mahna SK, Garg R (1989) Induced mutation in *Petunia nyctaginiflora* Juss. Biol Plant 31:152–155
Mahure HR, Choudhry ML, Prasad KV, Singh SK (2010) Mutation in chrysanthemum through gamma irradiation. Indian J Hortic 67:356–358
Maliga P (1984) Isolation and characterization of mutants in plant cell cultures. Annu Rev Plant Physiol 35:519–552
Maluszyuski M, Ahloowalia BS, Sigurbjornsson B (1995) Application of in vivo and in vitro mutation techniques for crop improvement. Euphytica 85:303–315
Mandal AKA, Datta SK (2005) Direct somatic embryogenesis and plant regeneration from ray florets of chrysanthemum. Biol Plant 49(1):29–33
Mandal AKA, Chakrabarty D, Datta SK (2000a) Application of in vitro techniques in mutation breeding of chrysanthemum. Plant Cell Tissue Organ Cult 60:33–38
Mandal AKA, Chakrabarty D, Datta SK (2000b) In vitro isolation of solid novel flower colour mutants from induced chimeric ray florets of chrysanthemum. Euphytica 114:9–12
Mangaiyarkarasi R, Girija M, Gnanamurthy S (2014) Mutagenic effectiveness and efficiency of gamma rays and ethyl methane sulphonate in Catharanthus roseus. Int J Curr Microbiol Appl Sci 3(5):881–889
Matsubara H (1982) Mutation breeding in ornamental plants. Techniques used for radiation induced mutants in Begonia, *Chrysanthemum*, Abelia and Winter Daphne. Gamma Field Symp 21:65–68
Matsubara H, Shigematsu K, Suda H et al (1971) The isolation of the mutated plants from sectorial chimera induced by irradiation in *Begonia* and *Chrysanthemum*. In: Japan Atomic Industrial Forum (ed) Proc. 10th Jpn. Conf. Radioisot., Tokyo Metrop. Isotop Res. Cent., Tokyo, pp 374–376
Matsumura A, Furutani N, Hase Y, Yokota Y, Tanaka A (2007) Dose response and mutation induction by ion beam irradiation in chrysanthemum. JAEA-Review 2007-060-78-3-12, p 78
Matsumura A, Nomizu T, Furutani N, Hayashi K, Minamiyama Y, Hase Y (2010) Ray florets color and shape mutants induced by 12C5+ ion beam irradiation in chrysanthemum. Sci Hortic 123: 558–561. https://doi.org/10.1016/j.scienta.2009.11.004. ISSN: 0304-4238
Mergen F, Thielges BA (1966) Effects of chronic exposures to Co60 radiation on *Pinus rigida* seedlings. Radiat Bot 6(3):203–211
Miao H, Wenlie G, Hui GY, Xue Y, Bin JH, Wei ZY (2016) Polyploidy induced by colchicine in *Dendranthema indicum* var. *aromaticum*, a scented chrysanthemum. Eur J Hortic Sci 81(4): 219–226. ISSN: 1611-4426. http://www.pubhort.org/.../index.htm
Mile N, Kulus D (2018) Microwave treatment can induce chrysanthemum phenotypic and genetic changes. Sci Hortic 227:223–233
Miler N, Kulus D, Sliwinska E (2020) Nuclear DNA content as an indicator of inflorescence colour stability of in vitro propagated solid and chimera mutants of chrysanthemum. Plant Cell Tissue Organ Cult (PCTOC) 143:421–430. https://doi.org/10.1007/s11240-020-01929-9
Mishra PN, Chandra V, Kappor LD (1973) Cultivation of plants for perfumes in India, Part I. Observations on performance of tuberose (*Polianthes tuberosa* L.) on saline-alkaline soils. Indian Perfumer 16(1):38–39
Mishra P, Datta SK, Chakrabarty D (2003) Mutation in flower colour and shape of *Chrysanthemum morifolium* induced by gamma irradiation. Biol Plant 47(1):153–156
Misra P, Datta SK (2007) Standardization of in vitro protocol in Chrysanthemum cv. Madam E. Roger for development of quality planting material and to induce genetic variability using gamma-radiation. Indian J Biotechnol 6:121–124

Misra P, Banerji BK, Kumari A (2010) Effect of gamma irradiation on chrysanthemum cultivar 'Pooja' with particular reference to induction of somatic mutation in flower color and form. Abst: national symposium on lifestyle floriculture: challenges and opportunities. Session-2, crop improvement, biotechnology, and biodiversity, March 19–21, 2010 at DR. Y.-S. Parmar University of Horticulture and Forestry, Nauni, Solan, H.P., India, Abstract No. 2.56: 41

Miyazaki K, Suzuki K, Abe T, Katsumoto Y, Yoshida S, Kusumi T (2002) Isolation of variegated mutants of Petunia hybrida using heavy-ion beam irradiation. RIKEN Accel Prog Rep 35:130

Miyazaki KK, Suzuki K, Iwaki T, Kusumi T, Abe T, Yoshida S, Fukui H (2006) Flower pigment mutations induced by heavy ion beam irradiation in an interspecific hybrid of *Torenia*. Plant Biotechnol 23:163–167

Momin KC, Gonge VS, Dalal SR, Bharad SG, Kulwal PL, Ratnawkar DV (2010) Radiation-induced variability studies in chrysanthemum under the net house. Abst: National Symposium on Lifestyle Floriculture: Challenges and Opportunities. Session-2, Crop Improvement, Biotechnology, and Biodiversity, March 19–21, 2010 at DR. Y.S. Parmar University of Horticulture and Forestry, Nauni, Solan, H.P., India, Abstract No. 2.59:43

Murashige T, Skoog FA (1962) A revised medium for rapid growth and bioassays with tobacco tissue culture. Physiol Plant 15:473–479

Nagatomi S (1991) Enlargement of induced variations by combined method of chronic irradiations with callus culture in sugarcane. Gamma Field Symp 30:87–110

Nagatomi S (1992) Effects of various irradiation methods on mutation induction of regenerants through leaf trimming and tissue culture of pineapple. Tech News 41:1–2

Nagatomi S (2002) Effectiveness of gamma ray chronic irradiation on in vitro mutagenesis in crops. SEMINARI: Agriculture & Viosciences, pp 49–58

Nagatomi S, Degi K (2009) Mutation breeding of chrysanthemum by gamma field irradiation and in vitro culture. In: Shu QY (ed) Induced plant mutations in the genomics era. Food and Agriculture Organization of the United Nations, Rome, pp 258–261

Nagatomi S, Katsumata K, Nojiri C (1993a) Induction of dwarf-type plants derived from in vitro cultures by gamma-ray irradiation in Cytisus genus. Tech News 51:1–2

Nagatomi S, Degi K, Yagaguchi M, Miyahira E, Skamoto M, Takaesu K (1993b) Six mutant varieties of different flower colour induced by floral organ culture of chronically irradiated chrysanthemum plants. Technical News of Institute of Radiation Breeding no. 43:1–2

Nagatomi S, Tanaka A, Kato A, Watanabe H, Tano S (1995) Mutation induction on chrysanthemum plants regenerated from in vitro cultured explants irradiated with 12C5+ ion beam. TIARA Annu Rep 5:50–52

Nagatomi S et al (1996a) Combined effect of gamma irradiation methods and in vitro explant sources on mutation induction of flower color in *Chrysanthemum morifolium* Ramat. Gamma Field Symp 35:51–68

Nagatomi S et al (1996b) Three mutant varieties in Eustoma grandiflorum induced through in vitro culture of chronic irradiated plants. Tech News 53:1–2

Nagatomi S et al (1996c) Mutation induction on chrysanthemum plants regenerated from in vitro cultured explants irradiated with C ion beam. TIARA Annu Rep 5:50–52

Nagatomi S, Tanaka A, Tano S, Watanabe H (1997a) Chrysanthemum mutants regenerated from in vitro explants irradiated with 12C5+ ion beam. Technical News of Institute of Radiation Breeding, No. 60

Nagatomi S, Tanaka A, Kato A, Watanabe H, Tano S (1997b) Mutation induction on chrysanthemum plants regenerated from in vitro cultured explants irradiated with 12C5 ion beam. TIARA Ann Rep 1995:50–51

Nagatomi S, Tanaka A, Kato A, Yamaguchi H, Watanabe H, Tano S (1998) Mutation induction through ion beam irradiations in rice and chrysanthemum. TIARA Ann Rep 1997:41–43

Nagatomi S, Miyahira E, Degi K (2000) Induction of flower mutation comparing with chronic and acute gamma irradiation using tissue culture techniques in *Chrysanthemum morifolium*. Ramat. Acta Hortic 508:69–73

Nagatomi S, Watanabe H, Tanaka A, Yamaguchi H, Degi K, Morishita T (2003) Six mutant varieties induced by ion beams in chrysanthemum. Institute of Radiation Breeding, Technical News 65

Nakagawa H (2009) Induced mutations in plant breeding and biological researches in Japan. In: Shu QY (ed) Induced plant mutations in the genomic era. Food and Agriculture Organization of the United Nations, Rome, pp 48–54

Nasri F, Zakizadeh H, Vafaee Y, Mozafari AK (2022) In vitro mutagenesis of *Chrysanthemum morifolium* cultivars using ethyl methanesulphonate (EMS) and mutation assessment by ISSR and IRAP markers. Plant Cell Tissue Organ Cult 149(2):657–673. https://doi.org/10.1007/s11240-021-02163-7

Natarajan AT, Maric MM (1961) The time-intensity factor in dry seed irradiation. Radiat Bot 1:1–9

Novak FJ (1991a) Mutation breeding by using tissue culture techniques. In: Gamma field symposia no. 30. Inst. of Radiation Breeding, NIAR, MAFF, Japan, pp 23–32

Novak FJ (1991b) In vitro mutation system for crop improvement. In: Plant mutation breeding for crop improvement, vol 2. IAEA, Vienna, pp 327–342

Oka-Kira E et al (2005) Klavier (klv), a novel novel-hyper nodulation mutant of Lotus japonicus affected in vascular tissue organization and floral induction. Plant J 44:505–515

Okamura M, Yasuno N, Takano M, Tanaka A, Shikazono N, Hase Y (2002) Mutation generation in chrysanthemum plants regenerated from floral organ cultures irradiated with ion beams. TIARA Ann Rep 2001:42–43

Okamura M et al (2003) Wide variety of flower-color and-shape mutants regenerated from leaf cultures irradiated with ion beams. Nucl Inst Meth Phys Res B206:574–578

Okamura M et al (2006) Advances in mutagenesis in flowers and their industrialization. Floric Ornam Plant Biotechnol 1:619–628. Global Science Books

Okamura M, Shimizu A, Watanabe S, Hase Y, Narumi I, Tanaka A (2007) Ion beam breeding of flower color variations in transgenic plants with multi-disease tolerance. JAEA-Review 2007-060-82-3-16, p 82

Okamura M, Shimizu A, Onishi N, Hase Y, Yoshihara R, Narumi I (2008) Ion beam breeding of flower color variations in transgenic plants with multi-disease tolerance. JAEA-Review 2008-055, p 62

Okamura M, Hase Y, Furusawa Y, Tanaka A (2015) Tissue dependent somaclonal mutation frequencies and spectra enhanced by ion beam irradiation in chrysanthemum. Euphytica 202: 333–343

Oladosu Y, Rafii MY, Abdullah N et al (2016) Principle and application of plant mutagenesis in crop improvement: a review. Biotechnol Biotechnol Equip 30(1):1–16

Padmadevi K (2009) In vivo and in vitro mutagenesis and molecular characterization in chrysanthemum (*Dendranthema grandiflora* Tzvelev). Ph.D. Thesis submitted to Tamil Nadu Agricultural University

Padmadevi K, Jawaharlal M, Kannan M, Ganga M (2010a) In vitro mutation studies in chrysanthemum (*Dendranthema grandiflora* Tzvelev). Abst: national symposium on lifestyle floriculture: challenges and opportunities. Session-2, crop improvement, biotechnology, and biodiversity, March 19–21, 2010 at DR. Y.S. Parmar University of Horticulture and Forestry, Nauni, Solan, H.P., India, Abstract No. 2.47:35

Padmadevi K, Jawaharlal M, Kannan M, Ganga M (2010b) A protocol for regenerating mutagen-treated explants of chrysanthemum (*Dendranthema grandiflora* Tzvelev). Abst: national symposium on lifestyle floriculture: challenges and opportunities. Session-2, crop improvement, biotechnology, and biodiversity, March 19–21, 2010 at DR. Y.S. Parmar University of Horticulture and Forestry, Nauni, Solan, H.P., India, Abstract No. 2.48:35–36

Palekar AR, Tambe TB, Ghormade GN (2022) Growth and flowering of chrysanthemum varieties influenced by induced mutagenesis in M1 generation. Pharma Innov J 11(2):737–740

Patil UH, Parhe SD, Pulate SC (2015) Induction of somatic mutation in chrysanthemum cultivar 'Local Golden'. The Bioscan 10(2):789–792

Patil UH, Karale AR, Katwate SW, Patil MS (2017) Mutation breeding in chrysanthemum (*Dendranthema grandiflora* T.). J Pharmacogn Phytochem 6(6):230–232

Predieri S (2001) Mutation induction and tissue culture in improving fruits. Plant Cell Tissue Organ Cult 64(2–3):185–210

Predieri S, Divrgilio N (2007) In vitro mutagenesis and mutant multiplication (Chapter 30). In: Jain SM, Haggman H (eds) Protocols for micropropagation of woody trees and fruits. Springer, pp 323–333

Preil W, Engelhardt M, Walther F (1983) Breeding of low temperature tolerant poinsettia (*Euphorbia pulcherrima*) and *Chrysanthemum* by means of mutation induction in in vitro cultures. Acta Hortic 131:345–351

Purente N, Chen B, Liu X, Zhou Y et al (2020) Effect of ethyl methanesulfonate on induced morphological variation in M3 generation of Chrysanthemum indicum var. aromaticum. HortScience 55(7):1099–1104

Puripunyavanich V, Piriyaphattarakit A, Chanchula N, Taychasinpitak T (2019) Mutation induction of in vitro chrysanthemum by gamma irradiation. Chiang Mai J Sci 46(3):609–617. http://epg.science.cmu.ac.th/ejournal

Rahi TS, Datta SK (2000) Performance of chrysanthemum (*Chrysanthemum morifolium*) cut flower varieties in saline water irrigated soil. Indian J Agric Sci 70(7):469–471

Rahi TS, Singh B (2011) Salinity tolerance in *Chrysanthemum morifolium*. J Appl Hortic 13(1): 30–36

Rahi TS, Shukla R, Pandey RK, Datta SK (1998) Performance of ornamental crops in salt affected soils and use of gamma rays to develop salt resistant strains. J Nucl Agric Biol 27(4):253–263

Rahi TS, Shukla R, Datta SK (2001) Effect of gamma radiation on the performance of Gladiolus vs. in normal and salt affected soil. In: Exploring the *Gladiolus* in India. Proceedings of the national conference on gladiolus, January 24–25, 2000, Lucknow, pp 73–74

Rai SK, Katiyar RS, Singh SP (2001) Prospects of gladiolus crops on sodic waste land. In: Exploring the Gladiolus in India. Proceedings of the national conference on *Gladiolus*, January 24–25, 2000, Lucknow, pp 104–104

Rana RS (1964a) Multiple allelism in radiation induced ray floret mutants of annual chrysanthemum. Curr Sci 33:592–593

Rana RS (1964b) Radiation induced chromosomal arrangements in annual chrysanthemum. Experientia 20(11):617

Rana RS (1964c) A radiation induced chimera in annual chrysanthemum. Naturwissemschaften 51: 642–643

Rana RS (1964d) Phenotypic variability of an induced mutant of annual *Chrysanthemum*. Genetica 35:236–240

Rana RS (1965a) Radiation induced variation in ray floret characteristics on annual chrysanthemum. Euphytica 14:296–300

Rana RS (1965b) A radiation induced useful mutant of annual chrysanthemum. Curr Sci 34:58

Richter A, Singleton WR (1955) The effect of chronic gamma radiation on the production of somatic mutations in carnations. Proc Natl Acad Sci U S A 41(5):295–300

Sakamoto K, Takatori Y, Chiwata R, Matsumura T, Tsukiashi K, Hayashi Y, Abe T (2016) Production of mutant line with early flowering at low temperature in spray-type chrysanthemum cultivar induced by C-ion beam irradiation. RIKEN Accel Prog Rep 49:262

Salleh S, Zaiton A, Kahar SA, Abu HA, Shuhaimi S, Yahya A (2010) Early field observations on the effect of gamma and ion beam irradiations on Chrysanthemum morifolium. Malaysia: N. p., 2010. Web

Salleh S, Ahmad Z, Hassan AA, Awang Y, Sandrang AK, Abdullah TL (2012a) Effect of ion beam irradiation on morphological and flowering characteristics of chrysanthemum. J Sains Nuklear Malaysia 24(2):59–70

Salleh S, Zaiton A, Hassan AHA, Shuhaimi S, Yahya A, Kahar SA, Lee ATB (2012b) Studies on the effectiveness of acute gamma and ion beam irradiation in generating flower colour mutation for *Chrysanthemum morifolium*. Malaysia: N. p., 2012. Web

Sarker A, Sharma B (1988) Efficiency of early generation selection for induced polygenic mutations in lentils (*Lens culinaris* Medik.). Indian J Genet 48:155–159

Sasaki K, Aida R, Niki T, Yamaguchi H, Narumi T, Nishijima T, Hayashi Y, Ryuto H, Fukunishi N, Abe T et al (2008) High efficiency improvement of transgenic torenia flowers by ion beam irradiation. Plant Biotechnol 25:81–89

Satina S, Blakeslee AF (1941) Periclinal chimeras in Datur stramonium in relation to development of leaf and flowers. Am J Bot 28:862–871

Sato T, Ohya Y, Hase Y, Tanaka A (2006) Studied on flower color and morphological mutations from chrysanthemum in vitro explants irradiated with ion beams. JAEA Takasaki Ann Rep 2004:74–75

Schum A (2003) Mutation breeding in ornamentals: an efficient breeding method. Acta Hortic 612: 6

Schum A, Preil W (1998) Induced mutations in ornamental plants. In: Jain SM, Brar S, Ahloowalia (eds) Somaclonal variation and induced mutations in crop improvement, pp 333–366

Shafiel MR, Hatamzadeh A, Azadi P, Samizadeh LH (2019) Mutation induction in chrysanthemum cut flowers using gamma irradiation method. J Ornam Plants 9(2):143–151

Shatnawi M, Fauri A, Shibli R, Al-Mazraawi M, Megdadi H, Makhadmeh I (2009) Tissue culture and salt stress in *Chrysanthemum morifolium*. Acta Hortic 829:189–196. https://doi.org/10.17660/ActaHortic.2009.829.27

Sheenan TJ, Sagawa Y (1959) The effects of gamma radiation on chrysanthemum and gladiolus. Proc Flo State Hortic Soc 72:388–391

Shikazono N, Suzuki C, Kitamura S et al (2005) Analysis of mutations induced by carbon ions in Arabidopsis thaliana. J Exp Bot 56:587–596

Shirao T, Ueno K, Minami K, Tanaka A, Imakiire S, Hase Y, Tanaka A (2007) Ion beam breeding of chrysanthemum cultivar 'Sanyo-ohgon'. JAEA-Review 2007-060 3-08, p 74

Shirao T, Ueno K, Abe T, Matsuyama T (2013) Development of DNA markers for identifying chrysanthemum cultivars generated by ion-beam irradiation. Mol Breed 31:729–735

Shukla R, Datta SK (1993) Mutation studies on early and late varieties of garden chrysanthemum. J Nucl Agric Biol 22(3–4):138–142

Simard MHN, Michaux-Ferriere, Silvy A (1992) Variants of carnation (*Dianthus caryophllus* L.) obtained by organogenesis from irradiated petals. Plant Cell Tissue Organ Cult 29:37–42

Singh LB (1970) Utilisation of saline-alkali soils without prior reclamation—*Rosa damascena*, its botany, cultivation and utilization. Econ Bot 24(2):175–179

Singh LB (1971) Utilization of saline-alkali soils for agro-industry without prior reclamation. III. Tuberose. Econ Bot 26:361–363

Singh M, Bala M (2015) Induction of mutation in chrysanthemum (*Dendranthema grandiflorum* Tzvelev.) cultivar Bindiya through gamma irradiation. Indian J Hort 72(3):376–381

Skirvin RM, Janick J (1976) Tissue culture induced variation in scented *Pelargonium* spp. J Am Soc Hortic Sci 10(3):281–290

Smith M (1985) In vitro mutagenesis. Ann Rev Genet 19:423–462

Smith AG, Noyszewski A (2018) Mutagenesis breeding for seedless varieties of popular landscape plants. Acta Hortic 1191(1191):43–52. https://doi.org/10.17660/ActaHortic.2018.1191.7

Sneepe J (1977) Selection for yield in early generation of self-fertilizing crops. Euphytica 26:27–30

Solanki IS, Sharma B (1994) Mutagenic effectiveness and efficiency of gamma rays, ethylene eimine and N-nitroso-N-ethyl urea in macrosperma lentil (Lens culinaris Medik.). Indian J Genet 54(1):72–76

Solanki IS, Sharma B (2001) Early generation selection of polygenic mutations in lentils (*Lens culinaris* Medik.). Indian J Genet 61(4):330–334

Soliman TMA, Lv S, Yang H, Hong B (2014) Isolation of flower color and shape mutations by gamma radiation of *Chrysanthemum morifolium* Ramat cv. Youka. Euphytica 199:317–324

Srivastava HP (1976) A note on the cultivation of Damasc Roses on usar soils. Indian Perfumer XVIII(part II):8–11

Srivastava HP, Sharma ML (1976) Cultivation of *Jasminum grandiflorum* L. on saline alkali soils. Indian Perfumer XVIII(part I):25–22

Stewart RN, Derman H (1970) The origin of adventitious buds in Chrysanthemum. Am J Bot 57:734–735

Su JS, Jiang JF, Zhang F, Liu Y, Ding L, Chen SM, Chen FD (2019) Current achievements and future prospects in the genetic breeding of chrysanthemum: a review. Hortic Res 6:109

Sugiyama M, Saito H, Ichida H, Hayashi Y, Ryuto H, Fukunishi N, Terakawa T, Abe T (2008a) Biological effects of heavy-ion beam irradiation on Cyclamen. Plant Biotechnol 25:101–104

Sugiyama M, Hayashi Y, Fukunishi N, Ryuto H, Terakawa T, Abe T (2008b) Development of flower color mutant of *Dianthus chinensis* var. semperflorens by heavy-ion beam irradiation. RIKEN Accel Prog Rep 41:229

Suryawati S, Aisyah I, Marwoto B, Kumiati R (2022) Comparing the response of callus of two chrysanthemums (*Dendranthema grandiflora* Tzvelev) cultivars to the chemical mutagen ethyl methane sulfonate (EMS). Acta Hortic. https://doi.org/10.17660/ActaHortic.2022.1334.10

Suzuki K, Takatsu Y, Gonai T, Nogi M, Sakamoto K, Fukunishi N, Ryuto H, Saito H, Abe T, Yoshida S et al (2005) Flower color mutation in spray-type chrysanthemum (*Dendranthema grandiflorum* (Ramat.) Kitamura) induced by heavy-ion beam irradiation. RIKEN Accel Prog Rep 38:138

Tamaki K, Yamanaka M, Hayashi Y, Abe T, Koyama Y (2017) Effect of the cultivar characteristics on the appearance of flower color mutants by C-ion irradiation in chrysanthemum. Hortic Res (Jpn) 16:117–123

Tamari M, Tanokashira Y, Nagayoshi S, Kido K, Tojima F, Hase Y, Oono Y (2017) Development of low-temperature-flowering chrysanthemum varieties 'Ryujin' and 'Touma'. QST Takasaki Ann Rep 2015:130

Tanaka A (2009) Establishment of ion beam technology for breeding. In: Shu QY (ed) Induced plant mutations in the genomics era. Food and Agriculture Organization of the United Nations, Rome, pp 216–219

Tanaka A, Shikazono N, Hase Y (2010) Studies on biological effects of ion beams on lethality, molecular nature of mutation, mutation rate, and spectrum of mutation phenotype for mutation breeding in higher plants. J Radiat Res 51:223–233

Tanokashira Y, Nagayoshi S, Hirano T, Abe T (2015) Effects of heavy-ion-beam irradiation on flower-color mutation in chrysanthemum. RIKEN Accel Prog Rep 47:297

Tanokashira Y, Nagayoshi S, Watanabe G, Hase Y (2016) Efficient breeding of yellow mutants by ion-beam irradiation in spray chrysanthemum 'Southern Chelsea'. JAEA Takasaki Ann Rep 2014:98

Telem RS, Sadhukhan R, Mandal N, Sarkar HK, Wani SH (2015) Gamma rays induced mutagenesis for identification of new variants via RAPD markers in chrysanthemum (*Chrysanthemum morifolium* Ramat.). J Cell Tissue Res 15(3):5289–5294. www.tcrjournals.com. ISSN: 0973-0028; E-ISSN: 0974-0910

Toyoda T, Watanabe H, Emoto K, Yoshimatsu S, Hase Y, Kamisoyama S (2007) Induction of new color variation by irradiation of ion beams to light yellow 'Jinba'. JAEA-Review 2007-060-81-3-15, p 81

Ueno K et al (2002) Effects of ion beam irradiation on chrysanthemum leaf discs and Sweet potato callus. JAERI-Review 35:44–46

Ueno K et al (2004) Additional improvement of chrysanthemum using ion beam re-irradiation. JAERI-Review 25:53–55

Ueno K, Shirao T, Nagayoshi S, Hase Y, Tanaka A (2005) Additional improvement of chrysanthemum using ion beam reirradiation. TIARA Ann Rep 2004:60

Ueno K, Nagayoshi S, Imakiire S, Koriyama K, Minami T, Tanaka A, Hase Y, Matsumoto T (2013) Breeding of new chrysanthemum cultivar 'Aladdin 2′ through stepwise improvements of cv. 'Jimba' using ion beam re-irradiation. Hortic Res (Jpn) 12:245–254

Ukai Y (1982) Irradiation service at the Institute of Radiation Breeding. Gamma Field Symp 21:105–112

Usenbaev EK, Imankulova K (1974) Radiation mutants of roses. Proc Int. Hort Congr. 19 109
Van Harten AM (1998) Mutation breeding: theory and practical applications. Cambridge Univ. Press, Cambridge, p 353
Velu S, Mullainathan L, Arulbalachandran D, Dhanavel D, Poonguzhali R (2007) Effectiveness and efficiency of gamma rays and EMS on cluster bean (Cyamopsis tetragonoloba (L.) Taub.). Crop Res 34(1, 2&3):249–251
Verma AK, Prasad KV, Kumar S (2010a) Isolation of yellow-colored mutant in Chrysanthemum cv. Thai Chen Queen through in vitro regeneration of ray florets. Abst: national symposium on lifestyle floriculture: challenges and opportunities. Session-2, crop improvement, biotechnology, and biodiversity, March 19–21, 2010 at DR. Y.S. Parmar University of Horticulture and Forestry, Nauni, Solan, H.P., India, Abstract No. 2.27: 24–25
Verma AK, Batra A, Misra P, Banerji BK, Dwivedi AK (2010b) Effects of ethyl methane sulphonate on chrysanthemum and induction of flower color mutation. Abst: national symposium on lifestyle floriculture: challenges and opportunities. Session-2, crop improvement, biotechnology, and biodiversity, March 19–21, 2010 at DR. Y.S. Parmar University of Horticulture and Forestry, Nauni, Solan, H.P., India, Abstract No. 2.37: 29–30
Verma AK, Batra A, Misra P, Banerji BK, Dwivedi AK (2010c) Induction of flower color mutation in Chrysanthemum. Abst: national symposium on lifestyle floriculture: challenges and opportunities. Session-2, crop improvement, biotechnology, and biodiversity, March 19–21, 2010 at DR. Y.S. Parmar University of Horticulture and Forestry, Nauni, Solan, H.P., India, Abstract No. 2.38: 30–31
Verma AK, Prasad KV, Singh K, Kumar S (2012) In vitro isolation of red coloured mutant from chimeric ray lorets of chrysanthemum induced by gamma-ray. Indian J Hort 69(4):562–567
Wakita N, Kazama Y, Hayashi Y, Ryuto H, Fukunishi N, Yamamoto K, Ijichi S, Abe T (2009) Induction of floral color mutation by C-ion irradiation in spray-type chrysanthemum. RIKEN Accel Prog Rep 41:230
Walther F, Sauer A (1985) Entwicklung von "Sortengamilien" bei *Gerbera amesonii* Dtsch. Gartenbau 39(45):2097
Walther F, Sauer A (1986a) Analysis of radiosensitivity—a basic requirement for in vitro somatic mutagenesis. II. Gerbera jamesponii. In: Nuclear techniques and in vitro culture for plant improvement. IAEA, Vienna, Aug. 1985, Vienna, International Atomic Energy Agency, STI/PUB/698, pp 155–159
Walther F, Sauer A (1986b) In vitro mutagenesis in Gerbera jamesonii. In: Horn W, Jenson CJ, Odenbach W, Schieder O (eds) Genetic manipulation in plant breeding. Proc. Symp. Eucarpia, Berlin, 1985. Walter de Gruyter Publ, Berlin, pp 555–562
Walther F, Sauer A (1989) Increase of genetic variation in 'Blue Daisy' (Brachycome multifida) by in vitro-mutagenesis and polyploidization. Mutat Breed Newsl 33:3–4
Wani AA (2009) Mutagenic effectiveness and efficiency of gamma rays, ethyl methane sulphonate and their combination treatments in chickpea (Cicer arietinum L.). Asian J Plant Sci 8(4): 318–321
Wasscher J (1956) The importance of sports in some florist flower. Euphytica 5:163–170
Watanabe H, Toyota T, Emoto K, Yoshimatsu S, Hase Y, Kamisoyama S (2008) Mutation breeding of a new chrysanthemum variety by irradiation of ion beams to 'Jinba'. JAEA Takasaki Ann Rep 2007:81
Weigle JL, Butler JK (1983) Induced dwarf mutant in *Impatiens platypetala*. J Hered 74(3):200
Wu JL, Wu C, Lei C, Baraoidan M, Bordeos A, Madamba MR, Ramos-Pamplona M, Mauleon R, Portugal A, Ulat VJ, Bruskiewich R, Wang G, Leach J, Khush G, Leung H (2005) Chemical- and irradiation-induced mutants of Indica rice IR64 for forward and reverse genetics. Plant Mol Biol 59(1):85–97
Yamaguchi H, Nagatomi S, Morishita T, Degi K, Tanaka A, Shikazono N, Hase Y (2003) Mutation induced with ion beam irradiation in rose. Nucl Instrum Methods Phys Res B 206:561–564

Yamaguchi H, Shimizu A, Degi K, Morishita T (2008) Effects of dose and dose rate of gamma ray irradiation on mutation induction and nuclear DNA content in chrysanthemum. Breed Science 58(3):331–335. https://doi.org/10.1270/jsbbs.58.331

Yamaguchi H, Shimizu A, Hase Y, Degi K, Tanaka A, Morishita T (2009) Mutation induction with ion beam irradiation of lateral buds of chrysanthemum and analysis of chimeric structure of induced mutants. Euphytica 165:97–103

Yamaguchi H, Shimizu A, Hase Y, Tanaka A, Shikazono N, Degi K, Morishita T (2010) Effects of ion beam irradiation on mutation induction and nuclear DNA content in chrysanthemum. Breed Sci 60:398–404. https://doi.org/10.1270/jsbbs.60.398-404

Yamakawa K, Sekiguchi F (1968) Radiation induced internal disbudding as a tool for enlarging mutation sectors. Gamma Field Symp 7:19–39

Yoder BI (1967) Evolution of the Indianapolis family. Grow Circ Newsl No 51:1–4

Yoshida S, Kusumi T (2002) Isolation of variegated mutants of Petunia hybrida using heavy-ion beam irradiation. RIKEN Accel Prog Rep 35:130

Zalewska M, Lema-Rumińska J, Miler N (2007) In vitro propagation using adventitious buds technique as a source of new variability chrysanthemum. Sci Hortic 113:70–73. https://doi.org/10.1016/j.scienta.2007.01.019

Gladiolus

9

Abstract

Gladiolus is one of the largest genera in the family Iridaceae. It is highly heterozygous blended with polyploidy. Extensive induced mutagenesis work has been done on gladiolus and a wide range of physical and chemical mutagens have been used. Mutagens that have been applied are X-rays, fast neutrons, thermal neutrons, electric shock, aluminum chloride, colchicines, diethyl sulfate, formalin, glycol, methyl methane sulfonate, dimethyl sulfate, ethyl methane sulfonate, nitroso dimethyl urea, N-nitroso-N-ethyl urethane, and N-nitroso-N-methyl urethane. Efforts have been made to represent selective references chronologically so that one can get the full essence of mutation work on gladiolus.

Keywords

Gladiolus · Bulb size · LD50 dose · Mutant

Gladiolus is one of the largest genera in the family Iridaceae. It is highly heterozygous blended with polyploidy. It is cultivated in almost countries of the world where spring and summer conditions are favorable. Gladiolus is a very important cut flower in the floriculture industry. The demand for gladiolus has increased worldwide due to its remunerative cut flowers. Its popularity is due to its impressive and glamorous spike (form and size), and beautiful and touchy colorful florets. Florets are especially appreciated for their appealing color, shape, and size. Its recognition as a cut flower is due to its sequential opening of florets for a lengthy period, excellent prolonged vase life, and easy cultivation in open field conditions. It can be propagated by seeds, corms, and cormels and through tissue culture.

An extensive amount of induced mutagenesis work has been done on gladiolus and a wide range of physical and chemical mutagens have been used. Mutagens that have been applied are X-rays, fast neutrons, thermal neutrons, electric shock,

S. Datta, *Role of Mutation Breeding in Floriculture Industry*,
https://doi.org/10.1007/978-981-99-5675-3_9

aluminum chloride, colchicines, diethyl sulfate, formalin, glycol, methyl methane sulfonate, dimethyl sulfate, ethyl methane sulfonate, nitroso dimethyl urea, N-nitroso-N-ethyl urethane, and N-nitroso-N-methyl urethane.

For mutation experiments mostly seeds, corm, cormel, anther calli, etc. were used as propagules. Work is so voluminous it will not be practicable to spotlight all references. All effort has been made to represent selective references chronologically so that one can get the full essence of mutation work on gladiolus. Before highlighting individual references, a tentative summary list of different mutagens, mutagen doses, author lists, etc. is provided as a ready reference.

Fast neutrons: Dose: 142, 284, 426, 568, 710, 852, and 994 rads; Dryagina et al. 1967; Abraham and Desai 1976.

Gamma rays: Dose: 1–10 Krad; 300 and 600 Gray; 2.5, 5.0, 7.5, 10.0, and 12.5 Krad; acute dose: 1, 3, 5, 7, and 10 Krad; 2 and 4 Krad; gamma field irradiation dose 10 Krad; 1000–20000R; prolonged irradiation; 2–12 KR; 2500, 5000, 10,000 and 15,000 rad; 10 and 15 krad; 500, 1000, and 2000 R; 0.5, 1.0, 5.0, and 10 krad; 3, 4, 5, 7, and 10 krad; Sanzonova and Syrovatka 1974; Dhaduk 1992; Dryagina 1964; Dryagina and Kazarinov 1965, 1966; Dryagina and Akhramona 1962; Grabowska and Mynett 1970, 1974; Dhara and Bhattacharya 1972; Gupta and Banerji 1977; Pandey and Gaur 1984; Banerji et al. 1981; Banerji and Datta 1987a, 2001; Iba et al. 1964, 1965; Isaev et al. 1960; Kaicker and Singh 1983; Meshitsuka et al. 1963; Moes 1966, 1969; Raghava et al. 1988; Grabowaska 1971, 1975; Rao et al. 1990; Sheehan and Lorz 1960; Mynett 1971 c.f. Grabowska and Mynett 1970; Misra 1976a, b, 1983; Misra and Bajpai 1978, 1983a, b; Uzenbaev and Nazarenko 1970.

X-rays: Dose: 400 and 6000 R; 10,000 r; 10–40 krad; 4000 r—Buiatti et al. 1969; Marck 1954; Broertjes and van Harten 1978; Drust 1973; Dryagina 1962a, b; Hubbard 1966; Jenkins 1961; Sax 1955.

Nitroso-methyl urea and nitroso ethyl urea: Dose: 0.05, 0.1% for 12 h; 0.2–0.4%; Dryagina 1975a, b; Kaicker and Singh 1983.

Methyl methane sulfonate: Dose: 0.05, 0.1, and 0.2% for 8 h; Banerji and Datta 1986; Banerji and Gupta 1982; Gupta and Banerji 1984.

Ethyl methane sulfonate: Dose: 0.2–1%; 0.5%; 0.9, 1.2 and 1.5%; Kaicker and Singh 1983; Misra and Bajpai 1978, 1983a, b; Moes 1966; Annonymous 1977, 1978, 1979.

Diethyl sulfonate: Misra and Bajpai 1978, 1983a, b; Dryagina 1975a, b.

N-nitroso-methyl urea: Misra and Bajpai 1978, 1983a, b.

Electrical field: 7.7 m wavelength 40.7 MHZ and 30 V; Sanzonova and Syrovatka 1974.

Formalin: 1, 2, and 4% for 6, 12, and 24 h; 0.1%—8, 16, 24 h; Banerji and Gupta 1985.

Ethylamine: 01%; MNU (0.2% + gamma radiation 2 krad); EI (0.1–1% for 18 h); IMU (0.2 to 0.4%); Kaicker and Singh 1983.

Determination of LD_{50} is very important for mutation work. Different authors have determined LD_{50} doses based on different parameters: environment, season, population size, etc. and therefore, a wide range of LD_{50} doses of different mutagens have been reported by different authors.

Gamma rays: Between 10 krad and 15 krad; sprouting basis 7.2 krad; survival basis 4.7 krad; between 7 and 10 krad; survival basis between 10 and 12.5 krad; 10.15 krad by Raghava et al. 1988; Abraham and Desai 1976; Gupta and Banerji 1977; Banerji et al. 1981; Banerji and Datta 1987a, b and Negi et al. 1983.

X-rays: Approx. 4 krad; 15,000 and 20,000 r by Broertjes and Van Harten 1978.

Marck (1954) exposed inbred F2 seeds of var. "White Star" to the dentist X-ray machine and observed reduction in survival, different types of morphological abnormalities (splited Y type spike, smaller spike, increased petal number, leathery texture, multi-folding petals, etc.), and changes in petal colors. The novel trait was isolated from a 24-s exposure group and suggested a maximum of 30-s exposure for success. Sax (1955) irradiated a bulb with 4000 r X-rays and noticed stimulation for early sprouting in two varieties ("Bloemfontic" and "Boulongne"). Corms were treated with colchicines and their effects on different morphological characters were studied. Reduced sprouting, dwarf growth habit, development of extra petals, variation in floret placement, etc. were observed (Mathews 1943; de Mol 1939, 1952a, b). Spencer (1955) reported earlier anthesis after treatment of corm with 5200 r X-ray and early flowering after treatment of bulbs with 4000 r X-rays. Yamamoto et al. (1958) detected flower color mutation by X-rays and gamma rays treatment. Development of leaf abnormalities and a decrease in blooming were reported by Sheehan (1959) after gamma irradiation. Sheehan and Sagawa (1959–1960) studied the effects of gamma radiation on different characters of gladiolus. Sheehan and Lorz (1960) observed no effect on genotypic bending in gladiolus after treatment of corms with 500,000 r gamma rays but presented the opening of florets. Isaev et al. (1960) studied the effects of prolonged gamma garden treatment on growth and cormel production over a 4-month growth period. The number of leaves, the height of the plants, and the number and size of newly formed cormlets were reduced, bearing the lowest at the higher dosage and greatest at the lower dosage. Jenkins (1961) determined the effects of thermal neutrons and X-rays on eight varieties. Low rates of irradiation reduced germination and plant height. Higher dosages were lethal (15,000 and 20,000 r X-ray).

Meshitsuka et al. (1963) mentioned that mutation can be easily induced by radiation. Iba et al. (1964, 1965) studied the effect of gamma irradiation on different cultivars. The plant growth, flowering, and yields of cormels were reduced by irradiation. Different abnormalities in vegetative and floral characters were observed at different growing stages. In the cultivar "Blue Diamond" the color changed from blue to white and from white to pink in "Tajimanaru." A full-color change in the whole flower from blue to white was found in "Blue Diamond." The pure yellow petal appeared in the X1 generation, but in the X3 and X4 generation, the color changed to pure yellow again. Rodriguez (1965) treated corms of the variety "Mansi" with 0.1% formalin for 8, 18, and 24 h. He noted 40% sprouting in 24 h of treatment and increased pollen sterility and somatic mutation in petals as white spots. Hubbard (1966) treated seeds by X-rays and found faster seed germination, the maximum number of blooms, floral abnormalities (raffling of petals, unopening of florets, increased floral parts), and development of larger bulbs in the treatments. Moes (1966, 1969) irradiated the dormant corms of tetraploid variety "Hawaii" and

Gladiolus gandavensis V.H. with 2500, 5000, 10,000, and 15,000 rad gamma rays and observed reduced survival and spike per plant. Morphological abnormalities in foliage included narrow leaves, asymmetrical development of leaf, leathery texture, notching sickle-shaped leaf, etc. and floral abnormalities included increased and decrease in floret organs, changed aestivation, fusion, and fission of tepals, notching, disharmony in a sequence of floret opening, delayed and reduced flowering. Chimeric flower color mutations were observed in MV1 and MV2 generations. In the second generation, entire mutant spikes appeared in 23.9% of the corms treated with 25,000 rad and 60.98% in those irradiated with 5000 rad. Moes (1966) treated corms of variety "Hawaii" with 0.9, 1.2, and 1.5% EMS and observed a reduction in survival and spike emergence and no mutation. Gamma ray induced somatic flower color mutations have been reported (Anonymous 1967). Cantor et al. (1967) tested the effect of gamma irradiation and magnetic field exposure of gladiolus corms and cormels of the cultivars: Her Majesty, Applause, and Speranţa. A significant effect was obtained at the variants which were irradiated with 1 Gy gamma radiation and 3 Gauss magnetic fields. Scarascia-Mugnozza (1968) from his mutation experiment reported that the mutated sector size in the petal of gladiolus is derived from 1.4 initial cells.

Uzenbaev and Nazarenko (1970) exposed corms to 5, 10, 15, and 25 krad gamma rays and recorded delayed sprouting and extensive morphological and physiological changes. Mynett (1971) studied the effect of gamma irradiation on different bulbous ornamentals including gladiolus. Sanzonova and Syrovatka (1974) studied the effect of an ultrahigh frequency of electrical field on the growth and formation of vegetative organs in hybrid gladiolus. Treatment in an electrical field (7.7 m wavelength 40.7 MHZ and 30 V) stimulated the growth of corms and the optimum results were obtained by treating the plants for 14 min. Seilleur (1975) studied the physiological behavior of plants after gamma irradiation of corms of the cultivar "Hawaii." He recorded fewer shorter flower spikes bearing fewer flowers, delayed flowering, a mutation in floret color, etc. Abraham and Desai (1976) treated the corms of the variety "Scarlet Double" with fast neutrons (142, 284, 426, 568, 710, 852, and 994 rad) and gamma rays (1–10 krad). They observed LD_{50} of gamma rays based on sprouting was 7.2 krad and 4.7 krad for survival. LD_{50} for fast neutrons was 852 rad for survival and radiobiological effectiveness was 5.5%. Fast neutrons were more effective than gamma rays. The role of induced mutations and the development of new varieties through mutagens treatment in gladiolus have been reported (Anonymous 1977, 1978, 1979). Awad and Harried (1985) conducted anatomical studies and kinetin, gibberellin, and etherphon concentrations on gladiolus cultivar "Eurovision" after applying r-irradiated doses. The differentiation and development of lobes, flower primordia, and the extension of the flower spike were retarded at higher concentrations of kinetin, gibberellin, and ethephon and high gamma irradiation doses. The flower spikes elongated and differentiation of individual flowers occurred as a result of all treatments at low concentrations or doses. Broertjes and Van Harten (1978) reported 4 krad of X-rays as the optimum dose for corms of Gladiolus tubergmii cv. "Fair Lady" but no mutation was detected. Buiatti et al. (1969) studied the mutation frequency induced by gamma rays in the variety "Oscar" after treating dormant corms with 400 and 6000 R. They could trace out the mutated cell linkage through the whole plant and suggested that a petal could derive from either 1, 2, or possible more initial cells. They suggested that mutations were found to cover more than one flower and to be induced before flower differentiation in the spike primordium of the dormant corms and mutation frequency was lower in top flowers than in bottom flowers (Buiatti et al. 1969, 1970). Buiatti et al. (1967) treated dormant corms of cv. "Johann Straus," "Oscar," "Sans Sauci," "Mansoer," and Aristocrate with 4000 and 6000 rad of gamma rays. Higher doses in the first year induced mutation in flower color in "Oscar" and "Sans Sauci." The lower doses in the second year induced 3 color mutations each in "Sans Souci" and "Mansoer" and 6 in Oscar. In separate experiments mutation frequency and mutation sector size were studied in VM1 plants grown from primary bulbs and in VM2 plants from secondary bulbs. Maximum mutation sector size was found in VM2 and VM3 but the mutation frequency was lower than in VM1 (Buiatti and Tesi 1968; Buiatti et al. 1965, 1967). Dhaduk (1992) treated corms of four varieties "Melody," G. psittacinus hybrid, "Pusa Suhagin," and "Rose Supreme" with acute gamma ray doses (1,3,5.7, and 10 krad) and determined LD50 doses between 7 and 10 krad. Reduction in the mitotic index was observed with the increase in dose and different types of chromosomal aberrations was observed. Five flower color mutants were induced and mutations were associated with a corresponding change in the anthocyanin content of the petals. Several good commercial varieties of gladiolus have been developed through spontaneous mutation/bud sports (Anonymous 1972). The first reports of such mutation were "GX Colvillicialbus" and the "bride" both of which are white and developed from bright scarlet "GX Colvillei." Color change through bud sport is common in gladiolus and especially in some cultivars. Variety "Picardy" has developed several white sport varieties ("Leading Lady," "Silver Wing," "Bingo," and "Wanda"). Another sport variety "Lady Luck" is light pink than "Picardy." Variety "Elizabeth" (lavender) has produced many white and dull white bud sports. Dhara and Bhattacharya (1972) treated dormant corms and studied the effect of gamma irradiation on different parameters. Root and leaf initiations were delayed at higher doses. Stunted growth, development of narrow leaves, short spike length with less number of flowers, and a wide range of chromosomal aberrations during root tip mitosis were observed in treated populations. Based on responses to different parameters examined the red cultivar was found to be more sensitive. Drust (1973, 1975) detected lighter flower color mutant after treating the seeds of gladiolus by x-rays (6000 r) and further he developed a variety having 33 florets on a spike after treating the corms of gladiolus with aqueous and alcoholic seeds extract of *Merenera sopolifera, Littonia, Orinthoglosum,* and *Gloriosa superba.* Effects of neutrons and gamma rays have been studied on the growth, development, and morphogenesis of gladioli. Higher doses inhibited germination, growth, and development. Few varieties showed stimulatory effects on germination, sprouting, and growth. Irradiation disturbed the normal course of the physiological and biochemical processes. Different varieties showed differential sensitivity to radiations. 5 krad was found most suitable to induce changes in flower color and other conspicuous features (Dryagina 1962a, b, 1964, 1968, 1970; Dryagina et al. 1967). Dryagina (1975a, b,

1977) applied different types of chemical mutagens (DMSI, N-nitroso-N-methyl urethane—NMU), nitroso dimethyl urea—NMU, nitroso-methyl urea, and nitroso ethyl urea) on corms, cormels, and the seed of different cultivars. Treatment with the solution of 0.05 or 0.1% was applied for 12 h before planting and gas treatment improved corm sprouting and the vegetative propagation coefficient and resulted in the selection of new productive forms. Stimulation in growth, flowering, and vegetative reproduction was observed along with the development of a new commercial flower color mutant ("Girl Friend"). Dryagina and Akhramona (1962, 1963, 1966a, b); Dryagina and Kazarinov (1965, 1966, 1972), and Dryagina et al. (1967) observed many changes in vegetative and floral characteristics after treatment with ionizing radiation, chemical (NMU) and chronic gamma field irradiation of bulbs and seeds. They detected some early flowering mutant forms and suggested radiation treatment is very effective for mutation breeding work of gladiolus. A series of experiments were conducted on different cultivars ("Polska," "Jesien," "Zloeista," "Barburlsa," "Mazur," and "Polska Jesien") using both X- and gamma-radiations (doses 500–20,000 R) (Grabowaska 1971, 1972, 1975, 1978; Glazurina and Chemagin 1974; Grabowska and Mynett 1970, 1974; Grabowska et al. 1971). The lethal dose for gladiolus corm was reported from 10 to 40 krad. Variants with dwarf nature, narrow leaves, changed flower color, etc. were detected but all the changes were not profitable for the cultivar investigated. Durst (1975) developed one white flower color new mutant cultivar "Triplex" having 33 florets on a spike in three columns with 12 buds open after treating corms of gladiolus with seed extracts (aqueous and alcoholic) of Merendera sobolifera, Littonia, Ornithoglossum, and Gloriosa superba. Glazurina and Chemagin (1974) studied the sensitivity of vegetative characters of certain flower crops to radiation. For induction of mutations large numbers of varieties were treated with gamma rays and chemical mutagens (ethyl methane sulfonate, diethyl sulfonate, and N-nitroso-methylene urea). A few notable varieties were treated with gamma rays 1.5, 3, 4, 5, 7, and 10 krad dosages ("Blue Lilac," "Himphabha," "Jo Wagenaar," "Picardy," "Ratna's Butterfly," "Sans Sauci," "Snow Princess," "Snow Princess," "Sylvia," "Green Finch," "Mayur," "Rose Momento," and "Wind Song," and one species *G. callianthus*) and chemical mutagens ("Blue Lilac," "Himphabha," "Jo Wagenaar," "Himbrabka," "Picardy," "Ratna's Butterfly," "Sans Sauci," "Snow Princess," and "Snow Princess and "Sylvia" and one species *G. callianthus*) (Misra 1975, 1976a, b, 1977, 1978a, b, 1983, 1996; Misra and Bajpai 1978, 1982, 1983a, b; Misra and Banerji 1978; Misra and Choudhary 1979; Misra and Mahesh 1993). Treatments showed delayed sprouting, reduced spike length and several flowers per spike, and different types of variations in vegetative and floral characters. Leaf variegation and deformities, stem branching, stunted plant growth, color changes in petals, bicoloration of petals, fusion, and fascination, formation of twin florets, increased floral organ, and notching in petals were recorded in most of the varieties. LD50 for survival was between 1.25% and 1.5% in each of EMS and DES for all the varieties except "Sans Sauci" and between 0.1% and 0.3% in MNU for most of the varieties. NMU at 0.05% advanced heading and flowering and increased the number of leaves, florets per spike, and cormel weight. Gamma rays at the 3 krad level proved very favorable for the production of more shoots per plant. Variations were observed in VM1 and VM2 generation. Stable mutations were isolated in varieties "Picardy," "Sans Souci," "Himbrabka," and "Ratna's Butterfly" and concluded that mutations can be induced by radiation. Raikov and Bulykov (1975) Studied the effects of ultraviolet light on different characteristics of gladiolus. Kaicker (1975) and Kaicker and Singh (1983) treated corms with gamma rays (2–12 KR) and chemical mutagens (ethyl methane sulfonate: 0.2–1%) and ethylamine 0.1%, nitroso-methyl urea 0.2–0.4% in combination with dimethyl sulfonate) for induction of mutation. Cultivar "Jo Wagenaar" produced in VM1 double flowers, stunted spikes, and leaf variegation (chlorophyll mutation) in 2 and 4 krad of gamma ray but in VM2 these abnormalities disappeared. The spectrum of flower color mutations was maximum in "Sylvia" at 2–4 KR treatment. Combined treatment of 2 KR of gamma rays and 0.2% MNU for 13 h has given highly attractive color mutants. Raghava (2000) and Raghava et al. (1981, 1988) reviewed mutation work on gladiolus and did experiments to develop new varieties. They applied gamma rays (10, 15 Gy) on different varieties and determined the LD_{50} dose based on sprouting. Desirable mutants were detected and isolated. One mutant was released as "Shobha."

Seilleur (1975, 1977a, b) studied quantitative evaluation of pigments in flowers of the cultivar "Hawaii" and 12 mutants using two-dimensional thin-layer chromatography. Six common anthocyanins (pelargonidin, peonidin, malvidin, cyanidin, petunidin, and delphinidin) were present in cv "Hawaii" and most mutants contained 18 anthocyanins. The complex organization of molecule coloration and modification due to mutation effects have been discussed. A reasonably good amount of work has been done at CSIR-NBRI, Lucknow which covered experimental research and a review of the literature. For experimental purposes, the cultivars included in the experiment were Gladiolus Psittacinum var. Hookeri cultivar "Red" and "Orange," "White Friendship," "Kajal," "Nilofar," etc., and mutagens applied were gamma rays (300 and 600 Gray, 2.5, 5.0, 7.5, 10.0, and 12.5 krad), MMS (0.05, 0.1, and 0.2%—8 h) and formalin (1, 2, and 4%—6, 12 and 24 h). LD50 on a survival basis was found to be between 10 and 12.5 krad of gamma rays for both the cultivars "Red" and "Orange." LD50 was found to be above 4% in formalin and 0.2% in methyl methane sulfonate for bot both cultivars. Reduction in sprouting, sprout number, plant height, number of leaves and plant spike length, leaf, and corm size, number of florets per spike, delayed flowering, pollen sterility, and survival was recorded after irradiation. Various types of morphological abnormalities in foliage, flower, and spike and chromosomal aberrations during root tip mitosis were observed. Chlorophyll variegation was observed in the leaf and spike. Flowering was completely ceased in some varieties at the highest doses. Flower color mutation was detected in a few plants as a sectorial chimera in M1V1. Different varieties were differentially sensitive to mutagens. Pink flower color mutation in "White Friendship" (gamma rays) and changed flower shape mutation (4% formalin) in cultivar "Red" were detected and isolated in VM2 (Banerji 1982; Banerji and Datta 1986, 1987a, b, 1988, 2000; Banerji and Gupta 1982, 1985; Banerji et al. 1981, 1994, 2000; Gupta and Banerji 1977, 1984). Negi et al. (1983) irradiated corms of three cultivars and determined LD50 to be 10.15 krad. Delayed sprouting was recorded in all three cultivars after 15 krad treatments. Radiation treatment caused the development of narrow leaves and leathery texture, delayed flowering, decrease in spike length, number of florets per spike, and floret size. A desirable and stable mutant with shell pink floret color observed in VM2 generation as a chimera in 1 krad treatment was isolated from the cultivar "Wild Rose" with Rosine purple floret color in VM4 generation and released as new variety "Shobha" (Raghava et al. 1988). One mutant from the red cultivar has been isolated in pure form. In the case of mutant, the arrow spot becomes significantly broadened in all three inner tepals which provide additional beauty to the florets. Kaicker and Singh (1983) treated the corms with 0.2–0.4% solution of NMU for 18 h, EI 0.1–1% for 18 h, and 0.2–0.4% IMU and 0.5% EMS in combination with 5% PMSO. They concluded that chemical treatment alone was not much effective. Pandey and Gaur (1984) reported from their experiment that a low irradiation dose (1 Krad) exhibited early sprouting, stimulation of oxygen uptake, and a rise in sugar content. Rao et al. (1990) gamma-irradiated (0.5, 1.0, 5.0, and 10 krad) young undifferentiated green calli from the anther of gladiolus cultivar "Picardy" and determined LD50 to be between 1.0 and 5.0 krad. Sheehan and Sagawa (1995) studied the effect of gamma irradiation on gladiolus. They irradiated the corms of Gladiolus var. "Spice" and "Span" with 4 h of continuous radiation from a ^{60}Co source and morphological abnormalities in foliage were recorded in the form of leathery texture, wrinkled leaves, and leaves which increased with an increase in exposure to gamma rays. Treatment at higher doses caused decreases in the production of flower spikes. Twin buds and semi-double florets on a few spikes were also observed sporadically. Kasumi et al. (1999) treated cormels of cultivar "Traveler" (pink floret) with gamma rays and detected changed pale pink floret as a sectorial chimera. The ovary culture technique was followed to isolate new floret chimeras in pure form. Thirty-four plants out of 52 regenerated plants (67.3%) from pale pink florets produced pale pink flowers but 17 plants (32.7%) reverted to the original pink color. Tanabe and Dohino (1993) irradiated cut flower of gladiolus by an electron beam with cultivar "Traveler" dose of <=0.4 Kgy delayed opening of the upper small buds. Cantor et al. (2002) studied the synergetic effect of gamma radiation and exposure to magnetic fields on Gladiolus. Corms and cormels of cultivars: "Her Majesty," "Applause," and "Speranţa" were irradiated for 72 h with 137 Cs gamma source on cylindrical exposure geometry. Variants were detected at 1 Gy gamma radiation and 3 Gauss magnetic fields. Dobanda (2004) studied the effects on the frequency and spectrum of genetic variability in gladiolus after gamma irradiation and in vitro regeneration. The in vitro regenerants differed from the initial varieties in quantitative and qualitative characteristics. The regenerants exhibited longer phenological phases of germination, germination-flowering, and initiation-cessation of flowering. The length of phenological phases increased directly proportional to the applied radiation doses. Bhajantri and Patil (2013) detected a wide range of variability in gladiolus varieties (Gladiolus hybridus Hort.) in M1V1 after treatment with EMS. Kasumi (2001) experimented to obtain a solid mutant in gladiolus by applying gamma rays and tissue culture. Various explant sources (cormel shoot apex, cormel pieces, leaf blades) were used for regenerating plants through embryogenesis. Histological observation confirmed that the induced regenerates were derived from adventitious embryos. There was a varietal difference in the ability of somatic embryogenesis. The plants derived from cormel apices were almost identical to the original variety "Traveler" in the flower characteristics. A simple and efficient culture system in gladiolus was established for obtaining regenerants by using cormel shoot apices as explants. It was mentioned that isolating solid mutants in flower color from mutation sectors in the perianth tissue culture is most suitable. Srivastava et al. (2007) studied the effects of gamma ray doses on different vegetative and floral characteristics of gladiolus. Patil and Dhaduk (2009) exposed corms of three cultivars (American Beauty, Nova Lux, and Eurovision) of gladiolus (Gladiolus hybrida L.) to gamma rays (1 kR to 7 kR) to induce genetic variability. Most of the characters were stimulated till 3 kR treatment but started to reduce at higher doses. The doses of 6 kR and 7 kR were found injurious. Color variation in florets and whole spike, increase and decrease in the number of floral organs, and their fasciations and spike bifurcation were observed from 4 kR to 7 kR treatments in all varieties. Three desirable color mutants were obtained from three varieties and one mutant with a bifurcated spike, having all similar characteristics as the original except for the doubled number of florets. Treatments of 5 kR and 6 kR were observed to be good for the induction of mutants and a total of four mutants were isolated in the vM2 generation. Patil et al. (2010a, b) treated three gladiolus cultivars ("American Brauty," "Nova Lux," and "Eurovision") with seven doses of gamma rays (1 to 7 kR) for induction of genetic variability. Most characters were stimulated till 3 kR treatment and started to reduce on higher doses but the treatment of 2 kR proved better overall treatments including control. 6 and 7 kR were found to be injurious floral characters. Different varieties were differentially sensitive to gamma rays. Different types of chromosomal abnormalities were recorded during root tip mitosis and the total aberrations increased in all the varieties with an increase in gamma ray doses. Changed flower color mutants were detected in all the varieties. Karki et al. (2010a, b) treated uniform-sized corms of 20 gladiolus varieties with gamma rays (0, 0.5, 1.5 kR) and examined different growth and flowering characters. Lower doses were effective in improving some important vegetative and floral characteristics. Different varieties were differentially sensitive to gamma rays. Tiwari et al. (2010) treated uniform (corm 30 g, 25 cm diameter) corms of gladiolus four varieties ("Peter Pears," "Advance Red," "White Prosperity," and "Nova Lux") with gamma rays (5, 10, 15 KRD) to induce variability. Reduction in survival rate, root length, number of spikes/plant and florets/spike, days to flower, shelf life, vase life, floret size, number of corms/plant, and weight of corm/plant was observed after irradiation and with an increase in exposure of gamma rays. In the M1 generation, a significant increase in plant height and spike length at 5 KRD and stunted growth of plant and spike length at 10 KRD gamma rays irradiation were observed in all cultivars. Morphological abnormalities in leaves and plants with abnormal leaves increased with an increase in exposure to gamma rays. Based on analysis of data it has been concluded that 5–10 KRD doses of gamma rays are suitable for induction of somatic mutation in various cultivars of gladiolus. The frequency and spectrum of mutations differed depending on the mutagen used and the dose applied. 20 varieties of gladiolus were used in the experiment to study the effects of gamma irradiation on various growth and flowering attributes. It was observed that lower doses, i.e. 0.5 and 1.5 Kr were effective in improving some important vegetative and floral parameters. Four mutants were developed (Karki and Srivastava 2010; Karki et al. 2010a, b). Singh and Kumar (2013) studied the effects of gamma rays on ten gladiolus cultivars ("Gulal," "Gunjan," "Her Majesty," "Jessica," "J.V. Gold," "Jyotsana," "Picoee," "Rose Supreme," "Shabnam," and "Urmil") on morphological characters. Sprouting of 50% corms, plant height at 30 days after planting, and number of leaves/plant at 30 days were studied. The earliest sprouting of 50% corms was recorded at 2.00 kr in cultivar "Jyotsana" and late sprouting was recorded at 7.00 kr in variety "Gunjan." Cultivar "J.V. Gold" showed the best performance concerning the number of leaves per plant at 2.00 kr treatment. Sudha (2014) studied gamma radiation (1, 2, 3, 4, 5, 6, and 7 kR) effects on three varieties ("American Beauty," "Nova Lux," and "Eurovision") of gladiolus. Sprouting and survival were affected significantly at 1 kR to 4 kR. LD50 was found to be beyond the 7 kR dose for both sprouting and survival. 4 kR and above doses were detrimental to vegetative and floral characters. Plants treated with 6 kR and 7 kR did not produce flower spikes in cv. "Nova Lux" and "Eurovision," whereas cv. "American Beauty" had produced few flower spikes. Plant height, leaf number, and leaf size, number, and size of florets were reduced significantly; leaves become narrow and leathery; delay in spike initiation at higher doses. Color variations in florets and whole spikes were also increased with an increase in dose rate along with an increase or decrease in the number of floral organs. Vase life and yield per plot at lower doses responded positively. Three desirable mutants with light colors were isolated from all three varieties in 5 kR, whereas one mutant had a bifurcated spike at 6 kR from cv. "American Beauty" and they were isolated for further study in the vM2 generation. Patil (2014) studied gamma radiation effects on three cvs. "American Beauty," "Nova Lux" and "Eurovision." LD50 was found to be beyond 7 kR. 4kR and above doses were detrimental. Color variations in florets and whole spikes were increased with an increase in dose rate along with an increase or decrease in the number of floral organs. Higher doses caused delays in spike initiation with a decrease in spike length, number and size of florets, vase life, and yield per plant while lower doses responded positively. Three desirable mutants with light colors were isolated from all three varieties in 5 kR whereas one mutant have a bifurcated spike at 6 kR from cv. American Beauty and they were isolated for further study in the vM2 generation. 4 cm and healthy corms of gladiolus varieties, viz. "Punjab Glance" and "Sylvia" were irradiated with different doses of gamma rays (0,50,100,125,150 Gy) and results showed that pollen fertility, size, stomata number, size, and the number of chloroplasts decreased with the increasing dose of irradiation. Pollen fertility was recorded as maximum in non-irradiated pollen in both varieties, while minimum viability was observed in both varieties treated with 150 Gy. Maximum pollen size, the number of stomata, stomata size, and several chloroplasts per guard cell were recorded at 50 Gy and minimum at 150 Gy as compared to control in both varieties (Dogra et al. 2017). Kumari and Kumar (2015) did mutation experiments with eight gladiolus varieties ("Yellow Golden," "Nathan Red," "White Friendship,"

"American Beauty," "Red Majesty," "Purple Flora," and "Algarve") treating corms with gamma rays (0, 25, 40, 55, and 70 Gy). Plants treated with higher doses (55 Gy and 70 Gy) showed deleterious effects although at the lowest dose (25 Gy) plants were not affected much. Plant height was reduced after irradiation and was recorded as a minimum at the highest doses. The number of shoots per plant was slightly increased at 25 Gy treatment in vM2. Effects of gamma radiation were very prominent on corm and cormel characters. A maximum number of corms per plant was recorded in plants treated with 25 Gy and a minimum at 70 Gy. "Nathan Red" variety had the maximum number and weight of corms per plant in vM1 as well as vM2 generation. The number of cormels per plant was also reduced with the increase in gamma rays dose. A maximum number of cormels per plant was observed in the "Algarve" variety after treatment with 25 Gy in vM2. Homeotic mutants were isolated from all the varieties at 55 and 70 Gy except the "American Beauty" and "Red Majesty" varieties. The corms of gladiolus cv. "Pink Beauty" was treated with gamma rays (0, 3, 6, 9 kR), and EMS (0.0.2, 0.4, 0.6%) both alone and in combination. Different vegetative, floral, and corm parameters were selected as parameters to study the effects of radiation and EMS. The days taken for the sprouting of corms were delayed after treatment with gamma rays and recorded maximum at 6 kR + 0.2%. Duration of sprouting was decreased by EMS and was observed to be maximum at 3 kR gamma rays. It has been concluded that 0.2–0.6% EMS doses are suitable for inducing mutation in gladiolu (Yadav et al. 2016). Kuldeep et al. (2017) treated bulbs of ten varieties ("Candyman Rose," "American Beauty," "Chandni," "Red beauty," "Punjab Morning," "White Prosperity," "Jester," "Srijana," "Psittacinous Hybrid," and "Priscilla") of gladiolus (Gladiolus hybrida L.) with gamma rays (0.5, 1.0, 1.5, 2.0, 2.5, 3.0, 3.5, 4.0, 4.5, and 5.0 kR) to induce variations and improvement. The earliest sprouting was recorded in 3.0 kr gamma treatment. Higher doses of gamma irradiation show lethal effects on various vegetative and floral traits. A maximum number of days taken to spike emergence was recorded with 3.5 kr. The earliest flowering was recorded with the interaction of control with cv. "Candyman Rose." The number of florets per spike is reduced at higher doses. Higher doses caused delays in spike initiation with a decrease in spike length, number, and size of florets. Patil et al. (2017) treated corms of gladiolus cv. "American Beauty" with gamma rays (5, 6, 7, 8, 9, and 10 kR), EMS (0.5%, 1.0%, and 1.5%), and DES (0.5%, 1.0%, and 1.5%) and studied their effects on different vegetative, floral and yield characters were taken under study. Parameters analyzed were early sprouting and flowering with the maximum number of sprouts, plant height, number of leaves, leaf area, number of florets per spike, spike length, floret size, blooming period, number of spikes per plant, and number of corms. Higher doses of all mutagens adversely affected the growth, flowering, and yield parameters of gladiolus. A series of mutation experiments were conducted with the same mission on gladiolus. All experiments aimed to study the effects of different doses of mutagens mainly gamma rays on the growth, morphological and floral characteristics of gladiolus (Singh et al. 2017; Sisodia 2015). Singh and Sisodia (2015) further reported the effects of gamma rays (0.0, 1.5, 2.5, 3.5, 4.5, and 5.5 kR) on different gladiolus cultivars ("Aldebaran," "Jyotsana," "Legend," "Praha,"

"Punjab Dawn," "Pusa Kiran," and "Tiger Flame"). Plant height and leaf area index were almost the same in the control and 1.5 kR treatment. Early floret color was induced with 5.5 kR. Maximum diameter of first floret and spike length were observed at 1.5 kR treatment. 2.5 and 4.5 kR gamma irradiation induced different morphological changes in the plants and flowers in the "Pusa Kiran," "Jyotsana," and "Praha" varieties. One yellow color lower petal mutant was observed in the variety "Tiger Flame" (original color orange) at 4.5 kR. The flowers of the mutant were arranged densely on the spike due to the short internodal length, resulting in an attractive look. Sisodia and Singh (2015) did another experiment to find out the influence of various gamma rays doses (1.5, 2.5, 3.5, 4.5, and 5.5 kr) on different varieties of gladiolus ("Aldebaran," "Jyotsna," "Legend," "Praha," "Punjab Dawn," "Pusa Kiran," and "Tiger Flame"). A higher dose (5.5 kr) resulted in minimum sprouting. Gamma ray doses 3.5 kr, 4.5 kr, and 5.5 kr were lethal in the case of three varieties, i.e. "Aldebaran," "Legend," and "Praha," and failed to produce any spike during first year. 1.5 kr induced early spike emergence. Late opening of first floret was observed at 2.5 kr. There was no flower opening at higher doses in the variety "Pusa Kiran" during second year. A maximum number of flowers/spikes was recorded with the control which was at par with a lower dose of gamma irradiation (1.5 kr). The durability of florets (2nd and 3rd) was maximum at 2.5 kr. The maximum length of the spike was noticed in control followed by 1.5 kr, 2.5 kr, 3.5 kr, 4.5 kr, and 5.5 kr. The maximum duration of flowering was noticed at 2.5 kr during both years. 4.5 kr induced abnormal plant growth, a bunched spike in some plants of "Pusa Kiran." Floret color mutations were recorded in "Jyotsna," "Tiger Flame," and "Punjab Dawn" which were stable in VM2. Sisodia and Singh (2014) examined the gamma radiation (1.5, 2.5, 3.5, 4.5, and 5.5 kR) effects of different gladiolus cultivars ("Aldebaran," "Jyotsana," "Legend," "Praha," "Punjab Dawn," "Pusa Kiran," and "Tiger Flame"). A maximum number of opened florets in the vase was recorded with control during first year, whereas during second year it was higher with 2.5 kR. Maximum opened florets were noticed on CVS. "Jyotsana" and "Tiger Flame" during first and second, respectively. The longevity of the first floret was more with 2.5 kR treatment. Maximum vase life was recorded at 5.5 and 2.5 kR during first and second years, respectively. Maximum vase life was recorded with cv. "Punjab Dawn" followed by cv. "Tiger Flame" during first and second year. Different doses created different types of morphological changes and an unstable mutant in the variety Pusa Kiran was observed at 2.5 kR. Variation in flower shape, size, and color was observed in different cultivars. A stable mutant was recorded during third year in cv. "Tiger Flame" at 3.5 kR. This mutant has orange-yellow color flowers. Patel et al. (2018) applied gamma rays (5.5, 6.0, 6.5, 7.0, 7.5, and 8.0 kR) and DES (0.1%, 0.2%, 0.3%, 0.4%, and 0.5%) for induction of mutations on corms of gladiolus variety "Psittacinus Hybrid." It was observed that there was a negative correlation between sprouting and survival percent with the different treatments and it was also noticed that the survival percent of DES-treated corms was less compared to gamma rays. DES was more effective on vegetative characters and produced larger spikes and more florets compared to gamma ravys. Tirkey and Singh (2019) treated bulbs of the variety "White Prosperity" with gamma rays (2 Kr,

4 Kr, 6 Kr, 8 Kr, 10 Kr, and 12 Kr), DEMS (0.25%, @ 0.5%, @ 0.75%, @ 1.0% and 1.25%) and SEMS (0.25%, @ 0.5%, @ 0.75%, @ 1.0%, and @ 1.25%) for mutation studies. Data were recorded on different characters in the second year. All mutagens had a significant effect on sprouting. Treated plants showed a reduction in plant height, number of leaves per plant, and number of cormlets per corm. Increased number of leaves and shoot per corm were recorded in 2 Kr gamma rays and 0.25% SEMS. A maximum number of spikes per plant was recorded in 0.25% of SEMS treatments. 0.25–0.50% dry and soaked EMS doses are suitable for inducing mutation in gladiolus. Sahariya et al. (2017) studied the effects of gamma rays (with 0.5, 1.0, 1.5, 2.0, 2.5, 3.0, 3.5, 4.0, 4.5, and 5.0 kR) on ten varieties of gladiolus (Gladiolus hybrida L.) ("Candyman Rose," "American Beauty," "Chandni," "Red beauty," "Punjab Morning," "White Prosperity," "Jester," "Srijana," "Psittacinous Hybrid," and "Priscilla"). The earliest sprouting was recorded in 3.0 kr gamma treatment. Higher doses showed a lethal effect on various vegetative and floral traits. A maximum number of days taken to spike emergence was recorded at a 3.5 kr dose. "Candyman Rose" variety showed maximum early flowering. The number of florets per spike reduced with an increase in doses and a minimum number of florets per spike was recorded at the higher dose. Higher doses caused delays in spike initiation with a decrease in spike length, number, and size of florets. Sathyanarayana et al. (2019) irradiated corms of five cultivars of gladiolus ("American Beauty," "Dull Queen," "Saffron," "Candy Man," and "Summer Sunshine") with gamma rays doses (0, 15, 25, 35, 45, and 55 Gy) and studied effects on vegetative and floral characters. Plants treated with higher doses (45 Gy and 55 Gy) showed deleterious effects although at the lowest dose (15 Gy and 25 Gy) plants were not affected much. The minimum number of days taken to sprouting, number of days taken to spike initiation, and maximum plant height observed in the control treatment and at par with the gamma rays dose at 15 Gy. The maximum number of tillers per corm and number of leaves per plant were observed at 25 Gy. The flower diameter and the number of florets were slightly increased at the 15 Gy treatment. Maximum plant height, number of florets per spike, and flower diameter were at 15 Gy and 25 Gy, respectively. Homeotic mutants were isolated from all the varieties at 35, 45, and 55 Gy. Kaur et al. (2021) review different aspects of gladiolus with special reference to induced mutagenesis. The role of different mutagens to induce mutations has been discussed. Pawadashetti et al. (2022) treated corms of gladiolus cultivars "Arka Gold," "Arka Pratham," and "Arka Amar" with gamma rays (15, 25, 35, 45, 55 Gy) to induce variability. The number of days taken for sprouting was delayed with increased gamma dosage. The results of two generations of data showed that 25 Gy gamma rays enhanced most of the quality parameters. Arka Amar showed maximum variations. 55 Gy gamma ray dose was not good for quality parameters. Belwal et al. (2023) treated standard-size corms of gladiolus var. Prince of Orange with gamma rays (4.0–5.5 Kr) and EMS (0.2–0.6% for 6 h) for inducing desirable variability. Vegetative and floral characters in each treatment were assessed and four potential mutants characterizing variations based on floret color, plant height, and number of florets have been detected and screened due to their desirable qualities. Devi et al. (2023) treated corms of three gladiolus cultivars ("Praha," "Tiger Flame," and

"Snow Princess") with gamma rays (0kR, 1.5kR, 3kR, 4.5kR, and 6kR) for induction of mutation. Stimulation in vegetative characters was noted up to 3 kR and deleterious effects were observed above 3 Kr. 6kR treated corms produced leathery and narrower leaves with shorter plant height, less leaf number, fewer sprouts/corm, and the least spikes/plant with an abnormal spike. 1.5 kR treatment was found to be the most beneficial for various growth parameters including plant height, the number of leaves, early sprouting, early flowering, longest flower spike, and more spikes per plant. "Tiger flame" and "Praha" were found to respond well to lower doses of gamma radiation. Cultivar "Prabha" developed discolor basal floret in 1.5 kR treatment chimeric flower developed on "Tiger Flame" at 3 kR treatment.

References

Abraham V, Desai BM (1976) Biological effectiveness of fast neutrons and gamma rays in some bulbous ornamentals V. Indian J Genet Plant Breed 36(2):230–237

Anonymous (1967) Somatic mutations in gladioli caused by hmm irradiation. Ann Facult Agric Univ Pisa 27:66–78

Anonymous (1972) Gladiolus breeding. Splash 47:45–47

Anonymous (1977) Annual report. IIHR, Bangalore, p 80

Anonymous (1978) Annual report. IIHR, Bangalore, p 45

Anonymous (1979) Annual report. IIHR, Bangalore, p 45

Awad ARF, Harried AA (1985) Anatomical studies on gladiolus stem apex as affected by kinetin, gibberellin, ethephon concentration, and gamma-irradiation doses. Acta Hortic 167:177–185

Banerji BK (1982) Reckoning LD50 in *Gladiolus psittacinus* var. Hookeri cultivar red in VM1 accurately by reed-munch and Liniar interpolation method. N Am Gladiol Coun Bull 149:74–75

Banerji BK, Datta SK (1986) Improvement of ornamental plants by induced mutation: *gladiolus*. Floriculture 6:3–5

Banerji BK, Datta SK (1987a) Physical mutagenesis in gladiolus: a review. N Am Gladiol Coun Bull 87(170):90–96

Banerji BK, Datta SK (1987b) Induction and analysis of somatic mutation in gladiolus L. N Am Gladiol Coun Bull 170:96–98

Banerji BK, Datta SK (1988) Improvement of gladiolus by induced mutations. In: National symposium on nuclear and allied techniques in agriculture, medicine, and environment research. IARI, New Delhi, p 11; Abst. No. 16. Organised by Indian Society for Nuclear Techniques in Agriculture and Biology, NRL, IARI, New Delhi-110012

Banerji BK, Datta SK (2000) Mutagenesis in gladiolus—a review. Paper presented at NBRI Lucknow in National Conference on gladiolus, Abst. No. 3. Improvement

Banerji BK, Datta SK (2001) Mutagenesis in Gladiolus-a review. Exploring the Gladiolus in India. In: Proceedings of the National Conference on Gladiolus, pp 24–38

Banerji BK, Gupta MN (1982) Effects of methyl methane sulphonate on Gladiolus psittacinus var. Hookeri cv. Orange. N Am Gladiol Coun Bull 150:64–66

Banerji BK, Gupta MN (1985) Induction of variability in Gladiolus psittacinus var. Hookeri cv. Orange by formalin treatment. N Am Gladiol Coun Bull 163:70–72

Banerji BK, Gupta MN, Datta SK (1981) Effects of gamma irradiation on gladiolus L. II Cytomorphological studies on Gladiolus psittacinus var. Hookeri cv. Orange. N Am Gladiol Coun Bull 148:50–57

Banerji BK, Datta SK, Sharma SC (1994) Gamma irradiation studied on gladiolus cv. 'White friendship'. J Nucl Agric Biol 23(3):127–133

Banerji BK, Dwivedi AK, Datta SK (2000) Gamma ray induced variability in Gladiolus. National Conference on Gladiolus. NBRI, Lucknow, Abst. No. 8

Belwal S, Srivastava R, Singh NK, Bhuj BD, Kumar A, Chand S, Karki K (2023) Screening of mutants of gladiolus cultivar prince of orange in M1 and M2 generations. Pharm Innov J 12(3): 5820–5822

Bhajantri A, Patil VS (2013) Studies on ethyl methane sulphonate (EMS) induced mutations for enhancing variability of gladiolus varieties (*Gladiolus hybridus* Hort.) in M1V2 generation. Karnataka J Agric Sci 26(3):403–407

Broertjes C, Van Harten AM (1978) Application of mutation breeding methods in the improvement of vegetatively propagated crops. Elsevier Scientific Publishing Company, Amsterdam

Buiatti M, Tesi R (1968) Gladiolus improvement through radiation induced somatic mutation. Publication No. 1, Laboratorio di Mutagenesi e differeziamento del C.N.R., Pisa, pp 1–6

Buiatti M, Ragazzini R, D'Amato F (1965) Somatic mutations in the carnation induced by gamma radiation. In: The use o induced mutations in plant breeding

Buiatti M, Molino M, Tesi R (1967) Prove di miglioramento genetic mediante radiazioni ionizzanti in una piñata da fiore a propogazione vegetative (gladiolo). Riv Ortoflorofruttic Ital 51(6): 616–627

Buiatti M, Tesi R, Molino M (1969) A developmental studies of induced somatic mutations in gladiolus. Radiat Bot 9:39–48

Buiatti M, Baroncelli S, Tesi R, Boscariol P (1970) The effect of environment on diplontic selection in irradiated gladiolus corms. Radiat Bot 10(6):531–538

Cantor M, Pop I, Korosfoy S (1967) Studies on concerning the effect of gamma radiation and mutagenic field exposure on gladiolus. J Central Eur Agric 3(4):279–284

Cantor M, Pop I, Korosfoy S (2002) Studies concerning the effect of gamma radiation and magnetic field exposure on gladiolus. J Cent Eur Agric 3(4):277–284

de Mol WE (1939) Experiment with colchicine-V, VI, VII, VIII. Herba Topiasia 3(8):1–6; 18-20

de Mol WE (1952a) Colchicine tests with gladiolus. Biological Jaarb 19:80–96

de Mol WE (1952b) The influence of colchicine on gladiolus bulb. Biological Jaarb 19:97–122

Devi S, Jamja N, Tabing R, Tagi N (2023) Influence of gamma radiation on growth, flowers and morphological changes in *gladiolus*. Environ Conserv J. https://doi.org/10.36953/ECJ.13382384

Dhaduk BK (1992) Induction of mutations in gladiolus (*gladiolus* L.) by gamma rays. Ph.D. Thesis. I.A.R.I, New Delhi

Dhara J, Bhattacharya NK (1972) Effect f gamma radiation on *Gladiolus* sp. proc. 59th Indian Sci. Congr. Assocn., pp 1–4

Dobanda E (2004) Evaluation of variability induced by gamma radiation on quantitative and qualitative traits in gladiolus. Cercetari de Genetica Vegetala si Animala 8:149–156

Dogra N, Dhatt KK, Kaur N (2017) Palynological and stomatal characters influenced by gamma rays in gladiolus cultivars. Int J Curr Res Acad Rev 5(9):44–50

Drust J (1973) Comments on articles and reprints in '72'. SPLASH: 52–57

Drust J (1975) Glads in Hungry and my work with them. N Am Gladiol Counc Bull 122:77–79

Dryagina IV (1962a) The effects of ionizing radiation on growth, development, and morphogenesis on gladiolus. Referat Zhur Biol 14:472

Dryagina IV (1962b) The effects of ionizing radiation on growth, development and morphogenesis on gladiolus. Soveschehanic po morfgenezu Rast Trudy 1(2):173–176

Dryagina IV (1964) The use of ionizing radiation in experiments with gladioli. Vestn Mosk Univ Ser VI Biol Pochv 19(5):46–56

Dryagina IV (1968) Radiation and chemical mutations. Tsvetovodstvo 8:12

Dryagina IV (1970) Ionizing radiation and chemical mutagens as factors increasing genetic diversity in gladioli. Izv Akad Nauk 6:879–883

Dryagina IV (1975a) Forms of gladiolus produced by treatment with chemical mutagens. Moscow Univ Biol Sci Bull 30(1–2):109–112

Dryagina IV (1975b) Gladioli forms obtained as a result of treatment with chemical mutagens. Vestn Mosk Univ 6(2):57–61

Dryagina IV (1977) Stimulation of gladiolus propagation by treating the corms with chemical mutagens in the gageous phase. Referativenji Zhurnal 3(55):1067
Dryagina IV, Akhramona VF (1962) Vital activity of the vegetative progeny of gladiolus bulbs which have been exposed to chronic irradiation on a gamma field (Russian). Biol Nauk 4:98–102
Dryagina IV, Akhramona VF (1963) After effect of chronic 60Cobradiation on the morphogenesis of gladiolus. Experimental Morphogenesis, Moscow University, Moscow, pp 409–415
Dryagina IV, Akhramona VF (1966a) Use of ionizing radiation in gladiolus breeding. Trudy Moscovwkowo Obszaczestwa Ispytatieliej Prirody 23:272–279
Dryagina IV, Akhramona VF (1966b) Effects of ionizing radiation on gladiolus tubers and seeds. Nucl Sci Abst 20:352–390
Dryagina IV, Kazarinov GE (1965) Effects of ionizing radiation on corms and the seed of gladiolus (Russian). Biol Nauk 1:91–94
Dryagina IV, Kazarinov GE (1966) Effects of ionizing radiation on gladiolus tubers and seeds. NSA 20: 35290
Dryagina IV, Kazarinov GE (1972) The effect radiation and chemical mutagens on variability in gladiolus cultivars. In: Khimicheskit Mutagenezei I Sozdanic Sozdanic Selektsionnogo Materiala. USSR, Nauka, Moscow, pp 349–352
Dryagina IV, Kazarinov GE, Kalinichenko IM (1967) Mutations in fruit and ornamental plants induced by the action of neutrons and gamma rays 60Co. Symposium on mutation process. Mechanism of mutation and induction factors, G. Mendel memorial symposium Prague, p 131-135
Durst J (1975) Glads in hungry and my work with them. N Am Gladiol Coun Bull 122:77–79
Glazurina AN, Chemagin NG (1974) Radiation resistance of the vegetative organs of certain flower crops. Byulleten Gosudarstvennogo Nikitskogo Botanicheskogo Sada 1(23):51.54
Grabowaska B (1971) Proby indukowania muteejiu mieczylkow ogrodowych (gladiolus hybr. Hort.) pod wplywemdzialania promiei gamma Co60 Sympozium Muteeja radoeyjnew hodowu roslin Watgawa: 7-8. Grabowaska, B. 1972. Attempts to induce mutation in gladiolus (*gladiolus hybr.* Hort.) by Co60 gamma rays. Biul Inst Hodowli Aklim Rosl 5-6:267–268
Grabowaska B (1972) Attempts to induce mutation in gladiolus (Gladiolus hybr. Hort.) by Co60 gamma rays. Biul Inst Hodowli Aklim Rosl 5–6:267–268
Grabowaska B (1975) The effects of gamma irradiation on gladioli after treatment in different stages of organogenesis. Prace Instytutu Sadownietwa Seria B Tom 1:41–50
Grabowaska B (1978) The beginning and current state of gladiolus breeding in Poland. Gladiol Ann 1:34–37
Grabowska B, Mynett K (1970) Induction of changes in garden gladiolus (*Gladiolus hybrid* horticulture) under the influence of gamma rays 60Co. Biul Insto Hodowli Aklim Rosl 1-2: 85–88
Grabowska B, Mynett K (1974) The effects of Co60 gamma radiation on gladioli, hyacinth, lilies and tulip. Second international symposium on flower bulb, England
Grabowska B, Saniewski B, Rudricki R (1971) An analysis of anthocyanidins in certain gladiolus (gladiolus hybr. Hort), varieties and their mutants. Biul Instytutu Hodowli I Aklimatyzaeji Roslin 3/4:55–58
Gupta MN, Banerji BK (1977) Effects of gamma irradiation on gladiolus L. I. studies in sprouting and flowering in *Gladiolus psittacinus* var. Hookeri cultivar red. N Am Gladiol Coun Bull 1977: 59–61
Gupta MN, Banerji BK (1984) Chemical mutagenesis in gladiolus—a review. N Am Gladiol Coun Bull 1977:76–78
Hubbard CP (1966) Atomic irradiation of *Gladiolus* seeds. Gladiol 41:59–65
Iba S, Matsubara H, Inokoshi Y, Oka M, Meshitsuka G (1964) Effects of gamma irradiation on gladiolus. II Tokyo Metrop. Isot Res Cent Ann Rep 3:109–113
Iba S, Oka M, Meshitsuka G, Matsumoto H, Kudo T (1965) Effects of gamma irradiation on gladiolus. III Tokyo Metrop. Isot Res Cent Ann Rep 4:85–88

Isaev SI, Dryagina IV, Versikina IM (1960) The effect of prolonged irradiation on 60Co on the growth and cormel production of gladioli. Doklady Akad Nauk 135:1250–1253

Jenkins JG (1961) Research report: irradiation experiment. Gladiol 36:67–70

Kaicker US (1975) Gladiolus program and growing problem in India. N Am Gladiol Coun Bull 122:73

Kaicker US, Singh HP (1983) Somatic mutations in gladiolus. Abstract of contributed paper. XV international congress of genetics, New Delhi, Dec. 12-21, part II, session C-IV D to C VII D, abstract no. 835

Karki K, Srivastva R (2010) Effect of gamma irradiation in gladiolus (*Gladiolus grandiflorus* L.). Pantnagar J Res 8(1):55–63

Karki K, Srivastava R, Chand S (2010a) Effect of gamma irradiation in gladiolus (Gadiolus grandiflorus L.). National Symposium on lifestyle floriculture: challenges and opportunities. Session-2, crop improvement, biotechnology and biodiversity, march 1921, 2010 at DR. Y.-S. Parmar University of Horticulture and Forestry, Nauni, Solan, H.P., India, abstract no. 2.9:14

Karki K, Srivastava R, Chand S (2010b) Effect of gamma irradiation in gladiolus (*Gladiolus grandiflorus* L.). Abst: National Symposium on life style floriculture: challenges and opportunities, YSPU H&F, Nauni, Solan (H.P.).14

Kasumi M (2001) Studies on induction of flower color mutants in gladiolus (gladiolus x grandiflora Hort.) by gamma irradiation and tissue culture. Bull Plant Biotechnol 4:1–49

Kasumi M, Takatsu Y, Yomotsune H, Sakuna F, Iida S (1999) The isolation of varied flower color plants by ovary culture of sectorial chimera of gladiolus. J Jpn Soc Hortic Sci 68(1):195–197

Kaur DP, Singh J, Singh B (2021) Mutation in gladiolus. Just Agric 2(2):2582–8223

Kuldeep SRA, Khan KR, Sarolia D (2017) Influence of gamma irradiation on flowering of gladiolus (Gladiolus hybrida L.). Int J Curr Microbiol App Sci 6(11):1362–1368

Kumari K, Kumar S (2015) Effect of gamma irradiation on vegetative and propagule characters in gladiolus and induction of homeotic mutants. Int J Agric Environ Biotechnol 8(2):413–422

Marck JJ (1954) Mutations: notes on mutation induced by x-rays. Glagiol 29:138–143

Mathews LEO (1943) Colchicine mutation of gladiolus. N Engl Gladiol Soc 8:93

Meshitsuka G, Oka M, Iba S, Matsubara H, Inokashi Y (1963) Effects of gamma irradiation on gladiolus. Tokyo metropolitan isotope research centre, Ann Rep 2: 169–173

Misra RL (1975) Spontaneous mutation in gladiolus variety salmon's sensation. Gladiol Breeder's Assoc Newslett 12:169–173

Misra RL (1976a) A colour mutant in gladiolus L. variety picardy. Sci Cult 42(11):569–571

Misra RL (1976b) Ionizing radiation 60Co is a factor influencing variation in gladioli. N Am Gladiol Coun Bull 125:80–85

Misra RL (1977) Induction of mutations in gladiolus. Plant Sci 9:99–100

Misra RL (1978a) Mutation studies in gladioli (gladiolus L). Ph.D. thesis. CSA University of Agriculture and Technology, Kanpur

Misra RL (1978b) Induced mutations in gladiolus through gamma radiation. Gladiol Ann:71–78

Misra RL (1983) A review on the mutation breeding of *Gladiolus*. Haryana J Hortic Sci 12(3/4): 149–155

Misra RL (1996) 60Co radiation source for creating variability in gladiolus. In DAE-BRNS symposium on nuclear techniques in increasing crop and animal productivity. Oct. 7-9, BARC Mumbai. Organized by Indian Society for Nuclear Techniques in agriculture and biology, NRL, IARI, New Delhi

Misra RL, Bajpai PN (1978) Mutation breeding of gladioli. Indian J Ornam Hortic 9(3):3–7

Misra RL, Bajpai PN (1982) Mutational studies in gladioli (gladiolus L.) IV. Effect of mutagens on flowering, the durability of florets and seed set. Hortic J 1(1–2):48–53

Misra RL, Bajpai PN (1983a) Mutational studies in gladioli (gladiolus l.) I. effect of physical mutagens on sprouting and survival of corms. Haryana J Hortic Sci 12(1/2):1–6

Misra RL, Bajpai PN (1983b) Mutational studies on gladiolus. II. Effect of mutagens on corm multiplication and storage. Punjab Hortic J 23(3/4):233–237

Misra RL, Banerji BK (1978) Mutation studies in gladiolus—a review. Can Gladiol Ann:1–15

Misra RL, Chaudhury B (1979) Fascinating mutants in gladioli. Indian Hortic 23(4):21
Misra RL, Mahesh KS (1993) Studied on mutation induction in gladiolus through 60Co gamma rays. J Ornam Hortic 2:42–46
Moes A (1966) Mutation induiles chez le glaicul (gladiolus). Bull Rech Agron Gembloux 1:76–95
Moes A (1969) Mutation breeding in tetraploid gladiolus (*Gladiolus gandavensis* V.H.). In: Chisa GC, Haussmann G (eds) Fifth Eucarpia congress, vol 23. Genet Agrar, Milan, pp 176–179
Mynett K (1971) The effect of gamma radiation 60Co on green house carnation n, dwarf dahlias, gladiolus, tulips, hyacinth and lilies. In: Polish yugoslavia symposium on the application of radioisotopes and ionizing radiation in Agriculture, Veterinary Science and Forestry, 4–5 VI Krakow, Materialy S. 34–35
Negi SS, Raghava SPS, Sharma TVRS (1983) Induction of mutations in gladiolus by gamma rays. Abstract of contributed paper XV international congress of genetics, New Delhi, Dec. 12-21, part II. Session C-IV D to VII D: 480
Pandey RN, Gaur BK (1984) Post-irradiation physiological changes in dormant gladiolus cormels. Indian J Pl Physiol 27:87–96
Patel D, Patil S, More SJ, Dohiya TP (2018) Comparative effect of physical and chemical mutagens in inducing variability in *gladiolus* variety 'Psittacinus hybrid'. Int J Curr Microbiol App Sci 7(01):645–652. https://doi.org/10.20546/ijcmas.2018.701.078
Patil DS (2014) Induction of mutation in commercial varieties of gladiolus using physical mutagen CO-60 gamma rays. Int J Adv Res Biol Sci 1(6):15–20
Patil S, Dhaduk BK (2009) Effect of gamma radiation on vegetative and floral characters of commercial varieties of gladiolus (*Gladiolus hybrida* L.). J Ornam Hort 12(4):232–238
Patil SD, Patil HE, Dhaduk BK (2010a) Response of gamma radiation on vegetative and floral characters of commercial varieties of gladiolus (*gladiolus grandifloras* L.). Abst: National Symposium on lifestyle floriculture: challenges and opportunities. Session-2, crop improvement, biotechnology, and biodiversity, march 19–21, 2010 at DR. Y.S. Parmar University of Horticulture and Forestry, Nauni, Solan, H.P., India, Abstract No. 2.21: 21
Patil SD, Patil HE, Dhaduk BK (2010b) Effect of gamma radiations on morphological and cytological characters of gladiolus (*gladiolus hybrid* L.). Abst: National Symposium on lifestyle floriculture: challenges and opportunities. Session-2, crop improvement, biotechnology, and biodiversity, march 19-21, 2010 at DR. Y.S. Parmar University of Horticulture and Forestry, Nauni, Solan, H.P., India, Abstract No 2.22: 21–22
Patil S, Chawla SL, Chaudhary P (2017) Induction of mutation through mutagens in gladiolus (Gladiolus hybridus) cv. American beauty. Int J Chem Stud 5(5):2305–2308
Pawadashetti DV, Kumari RV, Thimmarayappa M (2022) Effect of gamma irradiation on quality parameters in gladiolus (Gladiolus hybrida L.). biological forum–an. Int J 14(4):882–888; ISSN No. (Online): 2249-3239
Raghava SPS (2000) Genetic improvement of gladiolus in India. Natl. Conf. on gladiolus, Invited lecture, NBRI, Lucknow
Raghava SPS, Negi SS, Sharma TVRS (1981) New cultivars of gladiolus. Indian Hortic 26:2–3
Raghava SPS, Negi SS, Sharma TVRS, Balakrishnan KA (1988) Gamma ray induced mutants in gladiolus. J Nucl Agric Biol 17:5–10
Raikov, Bulykov (1975) The effect of short-wave, ultraviolet light on gladiolus growth and development. Byulletin Glavnogo Botanic Heskogo Sada 98:65–68
Rao TM, Negi SS, Dore Swamy R (1990) Effect of gamma irradiation on anther calli of gladiolus. VII Int. Congr. Plant Tissue Cell Culture, Amsterdam, pp A4–A82
Rodriguez AA (1965) Mutation induced in gladiolus through formaldehyde treatments. Bot Genet 1:17–19
Sahariya K, Kaushik RA, Khan R, Sarolia D (2017) Influence of gamma irradiation on flowering of gladiolus (*Gladiolus hybrida* L.). Int J Curr Microbiol App Sci 6(11):1362–1368
Sanzonova SP, Syrovatka EE (1974) The effects of an ultra-high frequency electrical field on the growth and formation of vegetative organs in hybrid gladioli. In: Adaptatsiya Rastenni V Usloviyakh Sredy Dnepropetrovsk Ukrainian SSR, pp 43–51

Sathyanarayana E, Sharma G, Tirkey T, Das BK, Divya K (2019) Studies of gamma irradiation on vegetative and floral characters of gladiolus (*Gladiolus grandiflorus* L.). J Pharm Phytochem 8(5):227–230

Sax K (1955) The effect of ionizing radiation on plant growth. Am J Bot 42:360–364

Scarascia-Mugnozza GT (1968) Mutation breeding of plants in Itali. Gamma Field Symposia 7: 111–121

Seilleur P (1975) Effects of irradiation on gladiolus corms on the physiological behavior of plants derived from them. Bull des Res Agronomiques de Gembloux 10(3):291–306

Seilleur P (1977a) Physicochemical analysis of the anthocyanic content of gladiolus flower cv. Hawaii and related mutants. III. Quantitative evaluation of pigments separated by thin layer chromatography and analysis of the results. Bull des Res Agronomiques de Gembloux 12(1–2):121–134

Seilleur P (1977b) Physicochemical analysis of the anthocyanic content of gladiolus flower cv. Hawaii and related mutants. I. Separation and identification II. Two-dimensional thin-layer separation. Bull des Res Agronomiques de Gembloux 12(1-2):97–106; 107-120

Sheehan TJ (1959) Plant improvement with gamma radiation. A.R. Flo Agric Stat 59:130–131

Sheehan TJ, Lorz AP (1960) Plant improvement with gamma radiation. A.R. Flo Agric Exp stat 1: 126–167

Sheehan TJ, Sagawa Y (1959–60) The effect of gamma radiation on chrysanthemum and gladiolus. Proc Flo St Hort Soc 72: 388–391

Singh AK, Kumar A (2013) Studies of gamma irradiation on morphological characters in gladiolus. Asian J Hortic 8(1):299–303

Singh A, Sisodia A (2015) Effect of gamma irradiation on morphological changes, flowering, and induced mutants in gladiolus. Indian J Hortic 72:84. https://doi.org/10.5958/0974-0112.2015.00015.8; Corpus ID: 84936202

Singh AK, Sah R, Sisodia A, Pal AK (2017) Effect of gamma irradiation on growth, flowering, and gladiolus. J Ornam Hortic 10(2):135–136

Sisodia A (2015) Studies on gamma rays induced morphological changes and mutants in gladiolus varieties. Ph.D. Thesis. Institute of Agricultural Science, BHU, Varanasi

Sisodia A, Singh AK (2014) Influence of gamma irradiation on morphological changes, post-harvest life and mutagenesis in *gladiolus*. Int J Agric Environ Biotechnol 7(3):535–545

Sisodia A, Singh AK (2015) Studies on gamma-ray induced mutants in gladiolus. Indian J Agric Sci 85(1):79–86

Spencer JL (1955) The effect of X-radiation on the flowering of certain cultivated bulbs and corms. Am J Bot 42:917–920

Srivastava P, Singh RP, Tripathi VK (2007) Response of gamma radiation (60Co) on vegetative and floral characters of gladiolus. J Orna Hort 10(2):135–136

Sudha DP (2014) Induction of mutation in commercial varieties of gladiolus using physical mutagen 60Co gamma rays. Int J Adv Res Biol Sci 1(6):15–20

Tanabe K, Dohini T (1993) Effects of electron beam irradiation on cut flowers. Res Bull Pl Prot Japan 29:1–9

Tirkey P, Singh D (2019) Effect of physical and chemical induced mutation on different character of gladiolus (*Gladiolus grandiflorus* L.). Int J Curr Microbiol App Sci 8(11):1510–1516. https://doi.org/10.20546/ijcmas.2019.811.17

Tiwari AK, Srivastava RM, Kumar V, Yadav LB, Misra SK (2010) Gamma rays induced morphological changes in gladiolus. Progress Agric 10:76–82

Uzenbaev EH, Nazarenko LG (1970) Some changes in growth and development of gladiolus of gladiolus under the effects of gamma irradiation from 60Co. Trudy Bot Sadov Akad Nauk Kaz SSR II:26–30

Yadav M, Rajput V, Saharan RS, Sehrawat SK, Ravika (2016) Influence of gamma radiations and EMS on morphological characteristics of *Gladiolus* cv. Pink beauty. Asian J Hortic 11(1): 114–118. https://doi.org/10.15740/HAS/TAJH/11.1/114-118

Yamamoto M, Matsuoka K, Otani S, Kondo N (1958) The flower color mutant of gladiolus induced by x-rays and gamma rays. Jpn J Breed 8(3):201–202

Rose

10

Abstract

Roses are available in numerous attractive colors and shades. Mutation breeding has been most successful in roses in inducing novelties. The chapter covers most of the induced mutation breeding work carried out at CSIR-National Botanical Research Institute, Lucknow, India and at different other institutes. A wide range of mutagens, their working doses, creation of mutant varieties, etc. have been mentioned with full support of all references.

Keywords

Rose · Mutation · Recurrent irradiation · Flower color mutant

Rose belongs to the family Rosaceae and all the species of this flower, with minor exceptions, belong to the genus Rosa. Prolonged history, symbolism, color, fragrance, and fine exquisite form made the rose the world's most popular flower. Roses are the most ancient and highly valuable ornamentals. A wide range of variations in flower type and plant growth have been developed in the genus Rosa due to significant advancements in rose breeding technology for the last 200 years. Rose can be propagated by seeds, cuttings, layering, budding, and tissue culture.

Roses are available in numerous attractive colors and shades. All the present-day colorful roses are the result of extensive hybridization, spontaneous and induced mutations, and selections. Mutation breeding has been most successful in roses in inducing novelties. The chapter covers most of the induced mutation breeding work carried out at CSIR-National Botanical Research Institute, Lucknow, India and an attempt has also been made to review the mutation studies on roses at different other Institutes.

Before citing important works a brief description of applied mutagens, doses, and authors is mentioned:

S. Datta, *Role of Mutation Breeding in Floriculture Industry*,
https://doi.org/10.1007/978-981-99-5675-3_10

X-irradiation (7–8 Krad; 4.5 Krad) (Gelin 1965; Chan 1966; Heslot 1966, 1968; Walther and Sauer 1986; Walther 1986).

Gamma rays (8–9 Krad; 4 or 8 Krad; 4–5 Krad; 5, 7.5, and 10 Krad; 4 kR; 2,4, 6, and 8 kR; 1.5–6; 8 Krad; 0.5–3 krad; 2, 3, 4, and 6 Krad; 2.5 and 5 Krad; 3, 4, and 5 Krad; 2, 3, 4, 5, and 6 krad) (Kaicker and Swarup 1972, 1978, 1985; Dommergues 1976, 1980; Heslot 1966, 1968; Gupta 1966; Gupta and Shukla 1970, 1971a, b; Swarup et al. 1971, 1973; Lata 1973, 1975; Lata and Gupta 1971a, b, 1975; Streitberg 1964, 1966a, b, c, 1967; Nakajima 1965, 1970, 1973; Usenbaev and Imankulova 1974; Desai and Abraham 1978; James 1983; Smilansky and Zieslin 1986; Benetka 1985; Kaicker 1982, 1983, 1985, 1987, 1988, 1991, 1995; Kaicker and Dhyani 1985a, b, 1986a, b, 1987; Khalatkar 1986; Datta and Gupta 1983a, b, 1985a; Gupta and Datta 1984; Datta 1986a, 1987a, b, 1988, 1989a, b, c, d, 1990a, b, 1991a, 1992; Dohare and Mathew, 1991).

Ethyl methane sulfonate (EMS) (0.5–0.8% for 24 h at 20 °C; 8 ppm; 0.25%) (Chan 1966; Dommergues 1976, 1980; Kaicker and Dhyani 1985a, b, 1986a, b, 1987; Khalatkar 1986).

Chemicals like EMS, N-nitroso-N-methyl urethane (NMU) and ethyleneimine (EI) (Swarup et al. 1971, 1973; Kaicker and Swarup 1978).

Colchicine (0.625 and 0.125% for 4 h) (Khalatkar 1986; Gupta and Datta 1983; Datta and Gupta 1985b; Datta 1994b).

Recurrent irradiation (Datta 1986b, 1993; Kaicker and Kumar 1992).

Mode of treatment: Physical mutagen—If a gamma garden facility is available, the entire plant can be exposed at developmental stages. Otherwise experimental plant size depends upon the size of the radiation source. Budwood of 13 cm in height, containing 2–4 auxiliary buds (eye) or actively growing terminal buds, is treated with desired doses. The dose rate and time required for treatment are calculated based on the activity of the cobalt-60 radiation source. After irradiation, individual eyes are removed and budded on *Rosa indica* var. odorata (or any other suitable) rootstock. Chemical mutagen: Any of the following three processes may be adopted for use of chemical mutagens: (a) about an inch of the budwood containing 2–4 axillary buds can be dipped in the desired concentration of mutagens for a specific period; (b) auxiliary buds are removed and dipped in the desired concentration of chemical mutagen for desired time; (c) shoot apex or auxiliary buds are covered with cotton and chemical mutagen is applied periodically up to desired time and then cotton is removed.

Gelin (1965) induced mutations in rose through X-irradiation. Chan (1966) irradiated five rose cultivars (Peace, Queen Elizabeth, Better Times, Baccara, and Tropica) with 7–8 Krad of X-ray and found several mutations in flower color and growth habit. Dommergues (1976) made a comparative study of the effects of both physical (gamma rays) and chemical (EMS) mutagens among diploid ("Gloria," "Mundi," "Border King"), triploid ("Orange Trimph"), and tetraploid ("Peace" and several others) cultivars. One-year-old plants were treated with 8–9 Krad gamma ray and 0.5–0.8% EMS for 24 h at 20 °C. Differential sensitivity to both the mutagens was recorded among the cultivars and radiation was found to be more effective in inducing mutation, especially in polyploids. Dommergues (1976, 1980) also studied in vitro-induced autogenesis in roses. Heslot (1966, 1968) attempted to induce mutations through gamma irradiation (4 or 8 Krad) and EMS (8 ppm) in diploid, triploid, and tetraploid cultivars and detected a wide range of mutations in flower color. High mutation frequency was noted in cultivars "Orange Triumph," "Gloria," and "Mundi." Dormant buds of three cultivars ("Christian Dior," "Kiss of Fire," and "Queen Elizabeth") were treated with gamma rays (5, 7.5, and 10 Krad), chemicals like EMS, N-nitroso-N-methyl urethane (NMU), and ethyleneimine (EI). Mutations were detected in both treatment groups but gamma rays were found to be more effective. Several mutants ("Madhosh," "Abhisarika," "Pusa Christiana," "Striped Christian Dior," etc.) were established. The solid color mutant was established in cultivar "Folklore" after radiation and chemical treatments followed by in vitro propagation (Kaicker 1982, 1983, 1985, 1987, 1988, 1990, 1995; Kaicker and Dhyani 1985a, b, 1986a, b, 1987; Kaicker and Kaicker 1988; Kaicker and Swarup 1972, 1978; Kaicker et al. 1991; Swarup et al. 1971, 1973). These experiments covered the mode of treatment, use of rootstock, and basic process of mutations. Dohare and Mathew (1991) studied the radiobiological effect of gamma rays (2, 3, 4, 5, and 6 krad) on four rose cultivars ("Raktagandha," "Raja Surendra Singh of Nalagarh," "Jantar Mantar," and "Neelambari") and detected six types of flower color mutants. Streitburg (1964, 1966a, b, c 1967) did large-scale mutation experiments on 42 cultivars using gamma radiations. She determined three optimum doses based on LD_{50} (1.5–6, 1–3, and 0.5–1 Krad) and reported differential sensitivity among the cultivars and mutations in flower type, color, and form. Nakajima (1965, 1970, 1973) tested a wide range of gamma ray doses and found the highest mutation frequency after 8 Krad treatments in the cultivar "Peace." Usenbaev and Imankulova (1974) applied 0.5–3 krad of gamma rays on the buds of four cultivars and detected approximately 20 mutations with changed flower color and petal number after 5 years. James (1983) used radioactive isotopes (personal communication, American Rose *Magazine*, Dec. 1961) and Cobalt-60 for irradiation of terminal buds of different cultivars and developed two mutations. One mutant ("Paula") was developed from cv. "Queen Elizabeth" and the other mutant ("Pink Hat") were developed from an unnamed floribunda. Desai and Abraham (1978) used gamma radiation for the improvement of rose. Smilansky and Zieslin (1986) treated stem cuttings of cv. "Mercedes" with gamma rays and EMS and found gamma rays to be more effective in inducing mutations than EMS. Benetka (1985) observed 4–5 Krad as the optimum dose for single cutting after exposure to 0, 2, 3, 4, and 6 Krad gamma rays. (Walther and Sauer 1986; Walther 1986) find out 4.5 Krad X-rays as the optimum dose for radiosensitivity of in vitro micro-propagated material based on inhibition of shoot development. But a higher dose was required for flower color mutations. Khalatkar (1986) from his experimental results suggested the treatment of axillary buds with colchicines or chemical mutagens (EMS, sodium azide) for inducing new flower color mutations. Huang and Chen (1986) detected several mutants with improved novelty from a starting variety. Murugesan et al. (1993) attempted to induce mutations in rose cv. Stg. Boniface used a tuber extract of *Gloriosa superba* and detected mutations with changed stalk length, flower diameter, petal size, and number. Hara et al. (2003) studied the effects of heavy ion irradiation in rose (Rosa hybrida "Bridal Fantasy"). Yamaguchi et al. (2003) studied the effects of carbon and helium ions on the axillary buds of different varieties of rose. They detected solid mutants after repeated cutting back. Mutants were detected in both the varieties ("Orange Rosamini" and "Red Minimo") after treatment with helium ion and carbon ion. Both ion beams induced mutants in the number of petals, flower size, flower shape, and in flower color in each cultivar. Canli and Skirvin (2008) reported that chimera thornless roses can be separated into pure types using high levels of BA. They experimented on Fairmount 1 thorny ("FM1 thorny") (a *Rosa multiflora* Thunb ex. J. Murr.) and a thornless sport of "FM1 thorny" ("Fairmount 1" ("FM1")). "FM1 thorny" yielded only thorny plants. "FM1" segregated in vitro into its constituent genotypes and yielded thorny and thornless plantlets, suggesting that "FM1" is chimeral. Madhu Bala and Singh (2010) treated axillary buds of the Hybrid Tea Rose cultivar "Pusa Mohit" with gamma rays (0, 5, 10, 15, 25, 40, 55, 65, 70, and 80 Gy) and allowed them to grow in vitro. All in vitro steps were followed properly to develop shoot buds. Two types of flower color variants (one producing large, compact flowers with dark pink petals induced by 25 Gy, and the second with medium-sized, compact, and attractive flowers with pink petals along with white streaks induced by 40 Gy as compared to red color in original cv. "Pusa Mohit"). Ryu et al. (2019) developed 12 flower color mutants by treating rooted cuttings of six commercial rose varieties with 70 Gy gamma radiation. Bhagat et al. (2022) treated developing buds of the Rose cultivar "Gladiator" with EMS and MMS ((0.20%, 0.25%, 0.30%, 0.35%), respectively, for 8 h. Chemical mutagens significantly influenced plant height, plant spread, leaf area, number of thorns on a flower stalk, days required for first flower bud initiation, the average weight of the flower, length of the peduncle, number of petals per flower, vase life, and petal area. The results on days required for bud sprouting and the color of the flower were not found significant. 0.20% EMS was found to be promising for plant spread, leaf area, the average weight of the flower, number of petals per flower, length of the peduncle, vase life, and petal area.

CSIR-NBRI, Lucknow, India did a series of experiments with different cultivars to induce mutations. About 60 varieties belonging to different categories like Hybrid Tea, Floribunda, Grandiflora, Miniature, Scented cultivars, Essential oil bearing, etc. were selected for the experiment. Some important varieties included in the experiments were "Contempo," "Montezuma," "Super Star," "Eiffel Tower," "First Prize," "Pink Parfait," "Queen Elizabeth," "Mrinalini," "Neelambari," "Sindoor," "Arjuna," "Golden Times," "Michelle meilland," "Mrinalini," "Raja Surendra Singh of Nalagarh," "Raktagandha," "Rose Gaujard," "Salmon Beauty," "Sonia Meilland," "Magic Carrousel," "Windy City," "Grenada," "John F. Kennedy," "Blue Moon," "Criterion," "Garden Party," "Pasadena," "Pristine," "Zorina", *Rosa damascene.* Budwood of all the cultivars was treated with 2, 3, 4, 5, 6, and 8 Krad gamma rays. Cultivar "Contempo" was used for testing the effects of colchicine and recurrent irradiation experiments. Treated budding eyes were budded on *Rosa indica* var. odorata rootstock.

Data were recorded on different parameters like sprouting, survival, plant height, morphological abnormalities, number of flowers per plant, petal structure, pollen fertility, and induction of somatic mutations in flower color and form. Radiosensitivity was determined mainly based on sprouting and survival. Chromosomal behavior was studied in control and treated populations. Reduction in sprouting, survival, plant height, and the number of leaflets was recorded in treated plants. Various types of leaf abnormalities were recorded. The oil content of scented cultivars was estimated. Differential radiosensitivity was observed among the experimental cultivars. Some cultivars were found to be the most sensitive and some were most resistant to gamma rays. Different cultivars of different groups were differently sensitive to gamma rays indicating that radiosensitivity was genotypic dependent. LD_{50} dose varied from cultivar to cultivar. Within the same group, there was a wide range of variation in LD_{50} dose.

Somatic mutations in flower color were detected as follows: white flower color in cultivar Queen Elizabeth (pink); yellow flower color, tangerine flower color, and stripe mutation in Contempo (orange with the yellow eye); light pink (Blossom pink) in cv. Mrinalini (Phlox pink); pink flower color in cultivar Zorina (Grenadine Red); creamish yellow mutation in Salmon Beauty (Salmon color). Gamma rays produced a bunch of flowers at the center of each flower in cv. "Imperator." Miniature cv. "Magic Carrousel" (white with red edges in petals) developed mutant without red color at the edges of petals. "Windy City" (deep pink) produced a very light pink mutant. Mutation frequency and spectrum varied with cultivar and doses of gamma rays (Lata 1973, 1975, 1980, 1981; Lata and Gupta 1971a, b, 1975; Datta 1986a, 1987a, b, 1988, 1989a, b, c, d, 1991a, 1992; Datta and Gupta 1983a, b, 1985a; Gupta and Datta 1983, 1984; Gupta 1966; Gupta and Shukla 1970, 1971a, b).

Colchicine Treatment: Budwood of rose cv. "Contempo" (orange with the yellow eye) was treated with 0.625 and 0.125% colchicines for 4 h. Delay in sprouting and reduction in sprouting, survival, sprout height, and branch and flower number were recorded after colchicines treatment. Various types of leaf abnormalities were recorded. Mutations in flower color were induced and the mutant colors were Tangerine Orange and Empire Yellow. This is, perhaps, the first report of the induction of flower color mutation in roses by colchicine (Datta 1994a).

Recurrent irradiation: Recurrent gamma irradiation showed cumulative effects on sprouting, survival, and plant height. The percentage of somatic mutation and spectrum of mutation was higher after recurrent irradiation in comparison to single irradiation in rose cv. Contempo (Datta 1986b, 1991b, 1993). Reduction in sprouting survival and plant height after exposure to gamma rays has been attributed to non-heritable physiological disturbances of growth substances. Recurrent gamma irradiation made drastic changes in the plant height of some induced mutants of Folklore. Comparative study of the two dwarf mutants, namely "Dwarf Striped Pink" and "Dwarf Bright-Yellow" revealed that there were marked differences in terms of morphological, internode length, peduncle length, number of petals, petal area, and flower colors between dwarf mutant and the original rose cv. Folklore (Kaicker 1982; Kaicker and Kumar 1992).

Stem cuttings of rose cv. "Grussanteplitz" (Sir Joseph Paxton x Fellemberg x Papa Gortier x Gloire des Rosomanes) were treated with different doses of gamma rays (2,4,6, and 8 krad) to induce variability in floral character. Radiation treatment showed a reduction in survival, plant height, number of branches, leaf and flower, and size of leaf and flower. 8 krad was found to be lethal and there was no survival of cuttings after 5 months. Low doses stimulated sprouting and growth of cuttings. Vegetative and floral organs showed different types of abnormalities and the frequency of abnormal plants and leaves increased with an increase in the dose of gamma rays. The leaf abnormalities included changes in shape, size, margin, apex and fission, and fusion of leaves. There were no dose-specific abnormalities. All the flowers produced during the blooming period were normal and the flower color of Grussanteplitz was red purple group (Horticulture color chart 73D Fan 2) except one plant after 2 krad treatment showed a mutation in flower color (c.f. Datta 2015, 2019a, b).

Essential oil bearing roses: A mutation experiment was also carried out with essential oil roses. Lata et al. (1977) studied the preliminary effects of gamma rays on Rosa damascena. Arinshtein and Krapivenko (1980) injected shoots with N-nitroso-N-ethyl urea (NEU) repeatedly and dechimerized the shoots by cutting them back twice. They detected mutations that were not commercial ones. Guo et al. (1983a, b) induced flower color mutations in "Grussan Bulin" and "Johan straurs." Klimenko et al. (1977) also induced flower color mutations in "Chrysler Imperial" and "Montezuma grandiflora." Raev (1984) used both radiation and chemicals and induced mutants with higher flower yield, higher oil, and better oil quality. Gupta et al. (1982) irradiated cuttings of Rosa Damascena with 0,1, and 2 Krad of gamma rays. Reduction in sprouting, sprout number, and plant height was recorded after irradiation. Somatic flower color mutation induced after irradiation was isolated in pure form.

References

Arinsthtein AI, Krapivenko EF (1980) Developing a method of producing mutations of essential oil rose (Russian). Plant Breed Abstr 51:10835

Madhu Bala, Singh KP (2010) Effects of gamma-ray irradiation on in vitro mutation induction and evolution of novel flower color variants in rose cv. Pusa Mohit. Abst: National Symposium on lifestyle floriculture: challenges and opportunities. Session-2, crop improvement, biotechnology, and biodiversity, march 19-21, 2010 at DR. Y.S. Parmar University of Horticulture and Forestry, Nauni, Solan, H.P., India, Abstract No 2.52: 38–39

Benetka V (1985) Some experience of methodology with isolation of mutations in rose cultivar 'Sonia'. Acta Pruhon 50:9–11

Bhagat SR, Sable PB, Vaidya SB (2022) Mutation breeding in rose (*Rosa indica* L.). Int J Curr Microbiol App Sci 11(2):166–170. https://doi.org/10.20546/ijcmas.2022.1102.019; ISSN: 2319-7706. http://www.ijcmas.com

Canli FA, Skirvin RM (2008) In vitro separation of a rose chimera. Plant Cell Tissue Organ Cult 95: 353–361. https://doi.org/10.1007/s11240-008-9449-y

Chan AP (1966) *Chrysanthemum* and rose mutations induced by X-rays. Proc Amer Soc Hort Sci 88:613–620

Datta SK (1986a) Improvement of miniature roses by gamma irradiation. Indian Rose Ann V:36–40

Datta SK (1986b) Effect of recurrent gamma irradiation on rose cv. 'Contempo'. J Nucl Agric Biol 15:125–127

Datta SK (1987a) Sensitivity of garden roses to gamma irradiation. Indian Rose Ann VI:121–126

Datta SK (1987b) Induction and analysis of somatic mutation in rose. In: Proc 6th all Indian Congr cytology genetics, radiation genetics selection, p 46
Datta SK (1988) Sensitivity of seven more H.T. roses to gamma irradiation. Indian Rose Ann VII:101–106
Datta SK (1989a) Mutation breeding in ornamental plants; the VI International Congr of the Sco. For the advancement of breeding research in Asia and Oceania (SABRAO), Tsukuba, p 130
Datta SK (1989b) Induction and analysis of somatic mutation in roses. In: Manna GK, Sinha U (eds) Perspectives in cytology and genetics, vol 16, pp 797–799
Datta SK (1989c) Effects of gamma rays on hybrid tea roses. Proc. 76th Ind. Sci. Congr., part III, section botany, p 191–192
Datta SK (1989d) Gamma ray induced somatic flower colour mutation in Mrinalini. Am Rose Ann 74:119–120
Datta SK (1990a) Role of mutation breeding in floriculture. In: Intl Symp on contribution of plant mutation breeding to crop improvement. IAEA, Vienna, pp 46–47
Datta SK (1990b) Role of mutation breeding in floriculture. In: Proc Symp. Pl. Mut. Breed. For crop improvement, I. IAEA, Vienna, pp 273–281
Datta SK (1991a) Improvement of garden roses by induced mutation. In: Golden Jubilee symp on genetic research and education: current trends and the next fifty years, New Delhi, p 767
Datta SK (1991b) Evaluation of recurrent irradiation on vegetatively propagated ornamental: chrysanthemum. J Nucl Agric Biol 20(2):81–86
Datta SK (1992) 'Mrinalini Stripe'—a new rose cv. Evolved by gamma irradiation. Indian Rose Ann X:44–47
Datta SK (1993) Recurrent irradiation effects on vegetatively propagated ornamentals. XV International Botanical Congr. Japan Abst. No. 8008
Datta SK (1994a) Rose breeding through colchicines treatment. Rose Ann 1:29–30
Datta SK (1994b) Induction and analysis of somatic mutaion in Rose cultivar 'Salmon Beauty'. Indian Rose Ann XII:79–85
Datta SK (2015) Indian floriculture: role of CSIR. Regency Publications, A division of Astral International (P) Ltd, New Delhi, p 110002
Datta SK (2019a) Floriculture work at CSIR-National Botanical Research Institute, Lucknow. Sci Cult 85(7–8):274–283
Datta SK (2019b) Present status of research on floriculture in India. LS An Int J Life Sci 8(2):71–93
Datta SK, Gupta MN (1983a) Effects of gamma rays on pigment synthesis of rose cv. 'Contempo'. J Nucl Agric Biol 12:79–80
Datta SK, Gupta MN (1983b) Mutation breeding in rose, XV International Cong of genetics held at New Delhi, p 286
Datta SK, Gupta MN (1985a) Mutation breeding on garden roses. Am Rose Ann 1:119–123
Datta SK, Gupta MN (1985b) Treatment of budding eyes with colchicine for induction of somatic mutations in rose. Indian Rose Ann IV:80–83
Desai BM, Abraham V (1978) Use of radiation in rose improvement, first all India rose convention, Bombay rose society, p 42–44
Dohare SR, Mathew V (1991) Mutation studies in garden roses. Golden Jubilee symposium on genetic research and education: current trends and the next fifty years, New Delhi, pp 767–768
Dommergues P (1976) Induced mutagenesis in roses. Hortic Fran 14:135
Dommergues P (1980) Rapid propagation of roses *in vitro* induced mutagenesis in roses. Sci Hort 13:383
Gelin O (1965) Studies on the X-ray mutation Stral pea. Agr Hort Genet 13:183–193
Guo J, Liu Y, Tang J, Liu C (1983a) Mutations in roses. J Nanjing Tech Coll Forest Prod 2:68–74
Guo J, Tang L, Liu G (1983b) Mutations in roses induced by X-rays jour. Manjini Tech Coll of Forest Hort, Calcutta, pp 107–114
Gupta MN (1966) Induction of somatic mutation in some ornamental plants. In: Proc all India symp hortic, pp 107–114

Gupta MN, Datta SK (1983) Effects of colchicines on rose cv. 'Contempo'. Indian Rose Ann III:51–54
Gupta MN, Datta SK (1984) Induction of somatic mutation in rose cv. 'Contempo' proc. 71st Indian Sci Cngr Part III section botany, p 82
Gupta MN, Shukla R (1970) Mutation breeding of garden roses: studies on induction of somatic mutations after gamma irradiation of bud wood. Prog Hort 2:69–83
Gupta MN, Shukla R (1971a) Mutation breeding of garden roses: recovery of somatic mutations after three years of gamma irradiation of bud wood. Recent Adv Hortic Kanpur 1:81–86
Gupta MN, Shukla R (1971b) Mutation breeding of garden roses. Jpn J Breed 21:129–136
Gupta MN, Nath P, Datta SK (1982) Effects of gamma irradiation on Rosa damascena mill. Indian Rose Ann II:55–61
Hara Y, Abe T, Sakamoto K, Miyazawa Y, Yoshida S (2003) Effects of heavy-ion irradiation in rose (Rosa hybrida 'bridal fantasy') (II). RIKEN Accel Prog Rep 36:135
Heslot H (1966) Induction de mutation par des agents mutagens physiques et chimiques. In: Mutations in plant breeding. IAEA, Vienna, pp 139–149
Heslot H (1968) Mutation research done in 1967 on barley, roses and marigolds, a progress report. In: Mutation in plant breeding. IAEA, Vienna, pp 153–159
Huang S, Chen Y (1986) New roses by irradiation: an update. Am Rose Ann 1:99–101
James J (1983) New roses by irradiation: an update. Am Rose Ann 1:99–101
Kaicker US (1982) Mutation breeding in roses. Indian Rose Ann II:35–42
Kaicker US (1983) A silver Jubilee of rose breeding in India. Chron Hortic 20:40
Kaicker US (1985) The creation of new roses at the Indian Agricultural Research Institute—a silver Jubilee. Indian Rose Ann IV:50–58
Kaicker US (1987) Vegetative propagation of mutated sectors of rose. Indian Rose Ann VI:112–119
Kaicker US (1988) Stable and unstable induced mutants of rose cv. 'Folklore', Abstr. XVI Intnl. Genetics Cong Canada, Genome, p 335
Kaicker US (1990) Anthocyanins in cv. Folklore and its mutants. Indian Rose Ann VIII:63–69
Kaicker US (1995) Rose breeding. In: Chadha KL, Bhattacharjee SK (eds) Advances in horticulture Vol. 12 ornamental plants. Malhotra Publishing House, New Delhi, pp 335–349
Kaicker US, Dhyani D (1985a) Induced mutations in roses. Indian J Genet Plant Breed 32:257–265
Kaicker US, Dhyani D (1985b) Induced mutations in roses. Mutat Breed Newsl 25:6–7
Kaicker US, Dhyani D (1986a) Effect of physical and chemical mutagens in the rose. Abstr Internl Hort Cong No 624 Hort Sci 21:245
Kaicker US, Dhyani D (1986b) Induced colour mutations in roses through gamma irradiation. Indian J Hort 43:287–290
Kaicker US, Dhyani D (1987) Induced color mutations in two cultivars of roses through gamma irradiation and EMS the rose Publ. By R.N.R.S., U.K, p 207
Kaicker US, Kaicker US (1988) Stable and unstable induced mutants of rose cv. 'Folklore', Abstr. XVI Intnl. Genetics Cong. Canada, Genome, p 335
Kaicker US, Kumar S (1992) Recurrent gamma ray induced dwarf mutants of rose cv 'Folklore'. Indian Rose Ann X:125–130
Kaicker US, Swarup V (1972) Induced mutations in roses. Indian J Genet Plant Breed 32(2): 257–265
Kaicker US, Swarup V (1978) Induced mutations in the rose cv. Gulzar and effects of chemical and physical mutagens on plant growth. Acta Agronomica Acad Sci Hung 27:43–48
Kaicker US, Swarup V (1985) Induced mutations in rose cv. Gulzar and effects of chemical and physical mutations in plant growth. Acta Agronomica Acad Sci Hung 27:43–48
Kaicker US, Kumar S, Singh SP (1991) Cytogenetics and rose breeding in India. Golden Jubilee Symp on genetic research and education: current trends and the next fifty years, New Delhi, p 765–766
Khalatkar AS (1986) Create a rose. Indian Rose Ann V:25–26

Klimenko ZK, Zykov KI, Shanin EV (1977) Effects of exposure of garden rose cuttings to gamma radiation on development and morphological variability. Radiobiology 17:152–155

Lata P (1973) Effects of ionizing radiation on roses with special reference to induction of mutations Ph.D. Thesis. Kanpur University, Kanpur, p 245

Lata P (1975) Hybridization in modern roses iv. Hybrids between control and mutants of cultivar Montezuma. Sabrao J 7:103–108

Lata P (1980) Effect of ionizing radiation on roses: induction of somatic mutation. Environ Exp Bot 20:325–333

Lata P (1981) Irradiation of mutants of rose. Stud Biophys 83:57–61

Lata P, Gupta MN (1971a) Mutation breeding of garden roses. Effect of gamma rays on essential oil content in some scented roses. Parfum Kosmetik 52:267–270

Lata P, Gupta MN (1971b) Effects of gamma rays on stem cuttings of essential oil breeding Rosa spp. Flavour Ind 2:421–425

Lata P, Gupta MN (1975) Effects of ionizing radiation on roses meiotic studies on control and gamma ray induced mutants of rose cultivar Montezuma. Cytologia 40:213–219

Lata P, Gupta MN, Murty SA (1977) Effects of ionizing radiation on Rosa damascena mill. New Bot IV:23–27

Murugesan S, Thamburaj S, Rajamani TK (1993) Preliminary studies on the induction of mutations in rose by tuber extract of *Gloriosa superba*. Indian Rose Ann XI:129–132

Nakajima K (1965) Induction of spots in roses by gamma ray irradiation. Gamma Field Symp 4:55–67

Nakajima K (1970) Gamma ray induced sports from a rose variety (Peace) Tech News 4, Inst of Radiat Breed, p 2

Nakajima K (1973) Induction of useful mutations of mulberry and roses by gamma rays. In: Induced mutations in vegetatively propagated plants. IAEA, Vienna, pp 105–117

Raev RT (1984) Mutation breeding of Kaznlik essential oil bearing rose (*Rosa damascena* mill). Rastenievud Nauki 21:92–99

Ryu J, Nam BN, Kim BR, Kim SH, Jo YD, Ahn JW, Kim JB, Jin CH, Han AR (2019) Comparative analysis of phytochemical composition of gamma-irradiated mutant cultivars of *Chrysanthemum morifolium*. Molecules 24:3003

Smilansky Z, Zieslin N (1986) Mutagens in roses (cv. Mercedes). Environ Exp Bot 26:279–283

Streitberg H (1964) Abschlusssbericht zur Forschungsarbeit wirtschaftlich wertvoller Mutationen bei Zierpflanzen (Rhododendron and *Rosa Arten*) durch Behandlung mit Rontgenstrahlen Teil b Mutationen bei Rosen Dtsch Akad landwirtsch. Wiss Berlin

Streitberg H (1966a) Neue Rosen und Sorten mit Hilfe der Rontgenbestrahlung. Dtsch Gartenbau 13:267–268

Streitberg H (1966b) Rosenzuchtung mit Hilfe der Rontgenbestrahlung. Arch Gartenbau 14:81–88

Streitberg H (1966c) Schaffung wirtschaftlich wertvoller Sprossvarinten bei Rosen durch Behandlung mitg Rontegenstrahlenl. Z Pfanzenzucht 55:165–182

Streitberg H (1967) Production of economically valuable variation in roses and azaleas by means of X irradiation. In: Stubbe H (ed) Induced mutations and their utilization. Gatersleben, Akademie, Verlag, Berlin, pp 359–362

Swarup V, Kaicker US, Gill HS (1971) Induced mutations in French beans and roses. Intnl Symp on use of isotopes and radiations In: Agric and animal husbandry res, p 13–25

Swarup V, Kaicker US, Malik RS, Singh AP (1973) Rose breeding in India. Indian J Genet Plant Breed 33(1):43–51

Usenbaev EK, Imankulova KI (1974) Radiation mutants of roses. Proc Int Hort Congr 19:109

Walther F (1986) Analysis of radiosensitivity a basis requirement for in vitro somatic mutagens III. Rose cultivars. Gartenbauwissenschaft 51:40–43

Walther F, Sauer A (1986) In vitro mutagens in roses, in proc. First Int symp on the research and cultivation of roses, Israel. Acta Hortic 189:37–46

Yamaguchi H, Nagatomi S, Morishita T, Degi K, Tanaka A, Shikazono N, Hase Y (2003) Mutation induced with ion beam irradiation in rose. Nucl Instrum Methods Phys Res B 206:561–564

Tuberose 11

Abstract

Tuberose (*Polianthes tuberosa* Linn.) is a monotypic monocot genus, perennial, bulbous plant belonging to the family Amaryllidaceae comprising about a dozen of species. The application of mutation technology for creating genetic variability in tuberose is limited. Almost all mutagenesis experiments were conducted with white flowers. Sincere efforts were made by different scientists to induce new flower color mutations. Different mutagens, their application doses and their effects on different vegetative and floral characters have been presented chronologically. The classical mutation technique could not establish any mutant with changed flower color. Mutations in growth habit, flower type, and leaf chlorophyll variegations have been recorded.

Keywords

Tuberose · Variegated mutant · Morphological variants

Tuberose (*Polianthes tuberosa* Linn.) is a monotypic monocot genus, perennial, bulbous plant belonging to the family Amaryllidaceae comprising about a dozen of species. All the species are wild except *P. tuberosa.* These species range in color from white, orange red, and red to striped. *P. tuberosa* is grown all over the world for cut flower production for the floriculture trade and as a source of oil in Egypt, China, France, and Morocco. Based on available flower types in India, tuberose has been classified into three types, i.e. single flowered, semi-double flowered, and double flowered. In addition to these, two more mutant varieties ("Rajat Rekha" and "Swarna Rekha"), four hybrid varieties ("Shringar," "Suvasini," "Prajwal," and "Vaibhav"), and one Sikkim variety are available. White, tubular flowers are loosely arranged on spikes that can reach 3–4 ft. in height. Foliage is long, slender, and

S. Datta, *Role of Mutation Breeding in Floriculture Industry*,
https://doi.org/10.1007/978-981-99-5675-3_11

grass-like. In the "double" cultivar some bulbs develop white flowers but the lower portion of the flower tube and leaf base are tinged with pinkish-red cells.

The application of mutation technology for creating genetic variability in tuberose is limited. Only a few reports are available. All mutagenesis experiments were conducted with available existing materials with white flowers. Younis and Borham (1975) applied 500–3000 rad gamma rays to bulbs and determined 2.5 krad as the LD_{50} dose. They studied and reported the genetic and morphological variability induced by gamma irradiation. Abraham and Desai (1976a, b) applied gamma rays and fast neutrons separately and in combination to create genetic variations in both single and double-type cultivars. Radiation effects were studied on vegetative and floral characters and found double flower type cultivar was more sensitive than the single type. Leaf color mutations were detected in both cultivars. They observed that bulb size is not very critical in determining lethality in both varieties. Patil et al. (1975) reported large flower size mutation after treatment with 0.5 Krad gamma rays. Gupta et al. (1974) successfully developed two chlorophyll-variegated mutants using 2.5 Krad gamma rays in both single and double-type cultivars. Both chlorophyll-variegated mutants have been commercialized in the name of "Rajat Rekha" (Single type, white-flowered, silvery white streaks along the middle of the blade, the mutant was detected M1V2) and "Swarna Rekha" (Double type, white-flowered, golden yellow streaks along the margins of the blade, detected M1V2). Krasaechai (1976, 1992) after irradiation of bulblets with 0, 5, 10, 15, 20, 25, and 30 Gy gamma rays observed a reduction in survival at 10 Gy or above. Treated plants had leaf chimeras but no change in flower color. Sambandamurthi (1983) studied sensitivity in tuberose and the LD_{50} values optimized for tuberose bulbs for gamma rays when @ 2 kR and the LC_{50} for EMS @ 60–70 mM and DES@ 25 mM, respectively. Datta and Shukla (1996) selected three chlorophyll-variegated mutants ("Rajat Rekha," "Svarna Rekha," and "Sikkim" selection) and treated with 500 and 1000 rads of gamma rays for testing their radiosensitivity and to determine further effects of gamma rays on their variegation pattern. "Sikkim" selection was found to be more sensitive to gamma rays as measured based on the severity of chromosomal damage. "Rajat Rekha" was found to be less sensitive. Gamma radiation changed the variegation pattern in all the mutant cultivars. Ali (2002) recommended low radiation doses for superior results in growth, sprouting, plant height, and number of leaves per plant compared to higher doses. Bulbs of four tuberose cultivars (Single, Double, Shringar, and Suvavini) were exposed to 5.0–25 Gy gamma radiations, and four mutants (dwarf mutant, high tiller mutant, nonflowering mutant, and compact inflorescence mutant) were isolated (Krishnan et al. 2003). Srivastava et al. (2007) reported the effects of different doses of gamma rays on various vegetative and floral characteristics of tuberose varieties. Guo et al. (2009) applied 20 and 25 Gy gamma rays and observed single tuberose was more tolerant than double type and recommended 20 Gy as the most suitable dose based on analysis of isoenzyme and agronomic traits. Estrada-Basaldua et al. (2011) exposed tubers to ^{60}Co gamma radiations (5, 10, 15, 20, 25, and 30 Gy) to induce genetic variability. Lateral bulbs were separated and regenerated in vitro. LD_{50} for in vitro regenerated shoots was determined at 9.09 Gy and 25.91 Gy for in vivo grown plants. 30 Gy was found to be lethal for in vivo established plants. Higher radiation doses reduced the growth of plants in both cases. Irradiated plants showed variations in leaf size and developed abnormal leaves and rosette-shaped buds. No mutation was detected. Mutagenesis experiments were conducted with both physical (gamma rays) and chemical (EMS) mutagens on bulbs of tuberose cultivars ("Calcutta Double," "Prajwal," and "Shringar"). Treated populations showed changes in both qualitative and quantitative characteristics like anatomical features of the scape, stomatal area, distribution pattern, number and arrangement of vascular bundles and mechanical tissue of the scape, changes in the number of tepals, etc. (Singh et al. 2011, 2013, 2015, 2016). Navabi et al. (2016) treated mature bulbs of various double-type tuberose with 10, 50, 100, 200, and 400 Gy gamma rays and observed a delay in sprouting, reduction in sprouting percentage at a lower dose, and changed plant morphology. 200 and 400 Gy-treated plants did not survive. Floret's length and weight increased at a lower dose. Interestingly one plant from the 10 Gy treatment developed quadruple florets. Inflorescence length decreased with an increase in doses. No change in flower color was detected. 10 Gy has been recommended as the most suitable dose for induction of mutation. Bulbs of four varieties, viz. "Kalyani Single," "Kalyani Double," "Suvasini," and "Prajwal" were treated with gamma rays (0.5 Kr, 1.5 Kr) and EMS (0.1, 0.2%) for induction of mutations. Mutagens doses showed their effects on different morphological and floral characters like sprouting number, sprouting time, leaf length, spike and floret characters, flowering duration, vase life, etc. Variants with changed plant height, increased petals per floret, fusion-type florets, etc. were isolated but without any change in floret color (Kainthura and Srivastava 2015). Panigrahi and Saiyad (2013) standardized large-scale clonal multiplication of Polianthes tuberosa L. var. "Calcutta Double" using rhizomes in solid MS medium containing different concentrations of IAA and BAP. The best response of multiple shoot production was found in MS medium containing 0.5 mg/L IAA and 3 mg/L BAP as compared to the control. The protocol may be used in in vitro mutagenesis. Pohare et al. (2012) studied the morphological effects of UV radiations on in vitro raised P. tuberose. Pohare et al. (2013) noticed a decrease in plant height and leaf length, increased leaf number, and changes in leaf color after treatment with 30 Gy gamma rays. Among the alkylating agents, three compounds, viz. ethyl methane sulfonate (EMS), 1-methyl-1-nitrosourea, and 1-ethyl-1-nitrosourea, are commonly used for inducing the variability in crops, including tuberose, which accounts for 64% of the total varieties (Wani et al. 2014). Kaintura et al. (2015, 2016) treated healthy bulbs of four cultivars (Kalyani Single, Kalyani Double, Suvasini, and Prajwa) of P. tuberosa L. with gamma rays (5,15 Gy), X-rays (6, 12 Gy) and EMS (0.1, 0.2%) and evaluated mutagenic effects of various vegetative and floral characters. Lower mutagen doses had a significant simulative effect on vegetative parameters, viz. sprouting percentage, days to sprouting, survival rate, leaf length, number of spikes per plant, florets per spike, flowering duration, and vase life. Higher doses of all mutagens had detrimental effects. Mutants exhibited variation from the parent plant concerning vegetative characteristics. Navabi et al. (2016) gamma-irradiated (dose 10, 50, 100, 200, and 400 Gy) bulbs to create new variations in Tuberose. Observations were recorded on survival and sprouting rate, flowering time, leaf number and size, size, and weight of the florets, number of petals, internode distance, plant height, and stem diameter. Experimental results showed that at the dose treatment of 10 Gy, all plants sprouted with a time delay, and at doses of 50 and 100 Gy, 57% and 29% of plants sprouted and grew, respectively. Though, at dosages of 200 and 400 Gy, none of the plants survived. Changes were observed in plant morphology but no change in flower color was observed. Initially, single, semi-double, and double flower type three varieties were available in Indonesia. Crossbreeding resulted in the development of 29 genotypes that showed differences in several characteristics, especially in the arrangements of flowers and flower stalks' straightness and rigidness. Through mutation technique, gamma rays application on tubers caused morphological damage on *Polyanthes*, reduced bulb growth by less than 30%, and reduced the plant height by more than 400% than control plants. Polyploidy-induced mutations by using colchicine led to changes in the length and diameter of the flowers and harvesting time. Until now, there is no result of the crossbreeding and mutation techniques that released a new variety. Despite this, based on the positive selection breeding on local cultivars, there have been two superior genotypes released as new varieties, namely Dian Arum (double flower type; origin of Cianjur, West Java) and Roro (semi-double flower type; origin of Pasuruan, East Java) (Sihombing 2017). Jyothi and Singh (2017) treated two cultivars ("Phule Rajani" and "Prajwal") of tuberose to induce genetic variability. Flower and bulb characters were selected to determine the effects of radiations in the M1 generation. Reduction in the number of spikes, flower number, number of bulbs, weight of bulb, and bulb diameter was observed with increasing doses. Both stimulations at lower doses and inhibition at higher doses were observed in morphological characters. Variants with uniform flowering, reduced a number of bulbs, and spike number were detected in the M1 population. But these characters were not stable in VM2. Mutants with changed flower shapes were also detected from the VM1 generation. Freshly harvest bulbs of cvs. "Prajwal" and "Phuel Rajani" responded more to gamma irradiation. 7.5 Gy and 10.0 Gy were found optimal for mutation induction in these two tuberose cultivars. Kaintura et al. (2018) studied the mutagenic effect of EMS on four cultivars Kalyani Single, Kalyani Double, Suvasini, and Prajwal after treating bulbs with 2 doses of EMS (0.1%, 0.2%). Both doses of EMS created morphological variation related to a floral character in the cultivar of Prajwal and Suvasini. The lower doses of EMS (0.1%) increased the floral size of a few florets within a spike.

Kayalvizhi et al. (2016a, b, 2017a, b) did a series of experiments on different cultivars of tuberose using both physical and chemical mutagens. Different treatments covered different doses of gamma rays and different concentrations of EMS and DES. They treated bulbs of the cultivar "Prajwal" with gamma rays (0.5, 1.0, 1.5, 2.0, 2.5, 3.0 kR), EMS (15, 30, 45, 60, 75, and 90 mM) and DES (10, 15, 20, 25, 30 and 35 mM) and find out their effects on sprouting, survival, plant height, and several leaves to determine LD50 and LC50. Based on the probit curve, the LD 50 value for gamma ray (2.0 kR), the LC 50 value for EMS (62.92 mM), and DES (21.65 mM) were determined. By increasing the dose/concentration of physical and chemical mutagens beyond LD 50 and LC 50 levels, decreased sprouting percentage, survival percentage, plant height, and number of leaves per plant were observed. To study the mutagenic effectiveness on various vegetative and floral characters four cultivars (Kalyani Single, Kalyani Double, Suvasini, and Prajwal) of P. tuberosa treated with gamma rays (5,15 Gy), X-rays (6, 12 Gy) and EMS (0.1, 0.2%). The mutagens at lower doses had a significant simulative effect on vegetative parameters viz., sprouting percentage, days to sprouting, etc. Higher doses of all mutagens had detrimental effects on the vegetative and floral characters. Mutants exhibited variation from the parent plant concerning the vegetative characteristics (Pooja et al. 2016).

Kayalvizhi et al. (2017) treated bulbs of tuberose var. Prajwal with gamma rays (0.5, 1.0, 1.5, 2.0, and 2.5 kR), DES (15, 20, 25, and 30 mM), and EMS (30, 45, 60, and 75 mM) and observed changes in morphological and floral characters. Sah et al. (2017) treated bulbs of 14 varieties with gamma rays and observed changes in different morphological characters but no desirable improvement. Yadav et al. (2018) treated bulbs of the cultivar "Prajwal" with EMS (0, 0.25, 0.5, 0.75, and 1.0%) for 5 min, 4 h, and 8 h. Lower doses initiated early sprouting and spike emergence, increased plant height, leaf numbers, decrease in leaf size, etc. but these initial parameters were delayed at higher doses. They have suggested 0.25% EMS and a treatment period of 8 h for improving growth parameters (early sprouting, increased plant height, more leaf number, early spike emergence, etc.). Abhangrao et al. (2020) exposed bulbs of cultivar "Phule Rajani" to gamma rays (0.5, 1.0, 1.5, 2.0, 2.5) for inducing variability. Dose-specific variability was recorded on different vegetative and floral characters. Stable desirable mutants, i.e. early mutant, tall mutant, dwarf mutant, and late mutant have been isolated (Abhangrao 2019; Abhangrao et al. 2019). Bulbs (1.5–2.0 cm in diameter) of double petal variety ("Calcutta Double") were exposed to 10, 20, and 30 Gy gamma rays, and seeds of variety "Arka Nirantara" were treated with different concentrations (0.25%, 0.5%, and 1%) of EMS for induction of genetic variability. Several mutants have been isolated from treated populations. One tall and branched spike mutant and another very tall mutant were isolated from the 10 Kr gamma ray treatment population of "Calcutta Double." Another unique chlorophyll-variegated leaf mutant ("Pranta Rekha") has been isolated from 10 Kr gamma rays. 1% EMS treatment produced two mutants: one dwarf and another with altered floral characteristics in "Arka Nirantra." The branched spike mutant and chlorophyll-variegated mutant have commercial importance. They have recommended gamma rays as having more mutagenic efficiency than EMS (Singh and Sadhukhan 2019). Jyothi et al. (2019) treated bulbs at three harvest stages of two cultivars with gamma rays. They detected desirable but unstable variants (tall flowering, dwarf, flower color, double spike head). Freshly harvested bulbs have been recommended as suitable for mutation experiments. Sharavani et al. (2019) detected variegated and albino leaves after treatment of bulbs with gamma rays (5–20 Gy). Kutty et al. (2020) reported that a lower dose (10 Gy) enhanced growth rate, emerged earlier flowering, and higher volatile content compared to control plants in both MV1 and MV2 generation while, higher doses, i.e. 20 Gy and 25 Gy took a longer time to flower initiation. Plants obtained from 10 Gy and 15 Gy rays displayed seven tepals against control flowers where the number is usually six. A dose of 10 Gy induced a slightly higher number of new tubers, while 15 Gy, 20 Gy, and 25 Gy decreased tuber formation. The flowers obtained from the 10 Gy irradiation dose had significantly higher total benzenoids and total terpenoids in MV1 generation. The amount of methyl 2-amino benzoic acid, 1, *H*-indole was higher in the floral emitted profile from both plants irradiated with 10 Gy and 15 Gy. It was further noted that flowers of MV2 generation developed from 10 Gy irradiation dose showed an enhancement in volatile emission.

Kumar et al. (2021) made an elaborative review covering many aspects of induced mutations in the improvement of tuberose. The review highlights the works carried out on tuberose by various workers worldwide have witnessed conventional and mutation approaches made in the genetic improvement of tuberose. The review also highlights the role of selection in the identification of desirable traits. Factors that affected mutation efficiency have also been presented and discussed. Hajizadeh et al. (2022) treated tubers with eight different gamma ray doses (0, 20, 30, 40, 50, 60, 70, and 80 Gy) via a source of cobalt-60 and studied the effect of gamma irradiation on morpho-physiological characteristics and molecular-induced variations in Polianthes tuberosa L. Results indicated that the effect of different levels of γ-rays on some morphological and physiological traits was significant as the gamma ray level was increased up to 50 Gy. The 50 Gy gamma irradiation reduced germination by 70.59%, germination rate by 66.36%, dry weight by 88.15%, fresh weight by 87.41%, flowering stem height (cm) by 69.22%, leaf area (cm^2) by 57.35%, leaf number by 34.41%, chlorophyll content (mg g − 1 FW) by 44.79%, number of florets by 92.57%, spike height (cm) by 27.80%, bulblet number by 32.57%, and bulblet diameter (mm) by 30.21%. Above 50 Gy there was no growth. RAPD analysis indicated that among 10 primers tested, nine primers showed clear bands as the highest number of amplified fragments (90) was related to the OPM13 primer and the lowest number (40) to the OPM10 primer. However, the DNA polymorphism was dose-dependent. Regar et al. (2022a, b) studied the effects of gamma rays on different vegetative characters of tuberose (Polianthes tuberosa L.) cv. Pearl Double. The results revealed that the lowest decline trends were recorded on days to 50% sprouting, sprouting percentage at 30 DAP, plant survival percentage, plant height at 60 and 90 DAP, leaves/plant at 60 and 90 DAP, leaf length at 60 and 90 DAP, and leaf width at 60 and 90 DAP at 2.50 Gy and highest declined trends at 20.0 Gy gamma irradiation in V1M1 and V1M2 generation, respectively, as compared to control. Regar et al. (2022a, b) experimented with tuberose cv. "Pearl Double" to study the response of gamma irradiation on vegetative parameters. They observed the lowest decline trends on days to 50% sprouting, sprouting percentage at 30 DAP, plant survival percentage, plant height at 60 and 90 DAP, leaves/plant at 60 and 90 DAP, leaf length at 60 and 90 DAP, and leaf width at (cm) 60 and 90 DAP at 2.50 Gy and highest declined trends at 20.0 Gy gamma irradiation in V1M1 and V1M2 generation, respectively, as compared to control. Sravani et al. (2022) treated bulbs with different concentrations of colchicine (0, 100, 200, 300, 400, and 500 ppm) for 3, 6, and 9 h and studied the performance of treated materials. The study revealed significant differences among colchicine concentrations, dipping times, and their interactions for different growth and yield characteristics. Bulbs treated with colchicine (500 ppm) for a dipping time of 9 h performed best concerning plant height, number of leaves per plant, leaf length, leaf width, leaf chlorophyll content, and number of bulbs per plant. They have mentioned that the bulbs treated with colchicine @ 500 ppm for 9 h of dipping time exhibited superior performance concerning most of the growth and yield characteristics in tuberose. Hajizadeh et al. (2023) treated bulbs of tuberose with gamma rays (0, 20, 30, 40, 50, 60, 70, and 80 Gy) and investigated the effect of gamma irradiation on morpho-physiological characteristics and molecular-induced variations. The effect of different levels of γ-rays on some morphological and physiological traits was significant as the gamma ray level was increased up to 50 Gy. The 50 Gy gamma irradiation reduced germination, germination rate, dry weight, fresh weight, flowering stem height (cm), leaf area, leaf number, chlorophyll content, number of florets, spike height, bulblet number, and bulblet diameter. Doses higher than 50 Gy were found harmful. The bulbs treated with 20 Gy gamma ray had the highest germination percentage and dry weight. The RAPD-PCR molecular marker technique was further used to identify the mutants of phenotypic variation flowers. Overall results showed that although the plant morphology was changed with gamma ray level, no changes occurred in tuberose color.

The classical mutation technique could not establish any mutant with changed flower color except two leaf chlorophyll-variegated mutants ("Rajat Rekha" and "Swarna Rekha"). It is suggested that before starting the mutation experiment we should select strains with maximum pink red cells as starting material. The chances of induction of pink color mutation will be more. There is a need to intensify in vitro mutagenesis work using the lower part of the flower tube and leaf base with maximum pink cells as explants to induce further genetic variability. Recurrent irradiation experiments may be tested in tuberose. There are possibilities to recover pink and white flower color mutants.

References

Abhangrao AK (2019) Studies on mutation breeding in tuberose (*Polianthes tuberosa* Linn.). Doctoral dissertation. Vasantrao Naik Marathwada Krishi Vidyapeeth, Parbhani

Abhangrao AK, Yadlod SS, Manisha D (2020) Effect of physical mutagens on reproductive and flowering characters of tuberose (*Polianthes tuberosa* L.). Int J Curr Microbiol App Sci 11: 1284–1291

Abraham V, Desai BM (1976a) Biological effectiveness of fast neutrons and gamma rays in some bulbous ornamentals. Indian J Genet Plant Breed 36(2):230–237

Abraham V, Desai BM (1976b) Radiation induced mutants in tuberose. Indian J Genet Plant Breed 36(3):328–331

Ali J (2002) Effect of gamma irradiation on vegetative and floral characteristics of tuberose corms (*Polianthes tuberosa* L.) var. single. UAF, Faisalabad, p 90

Datta SK, Shukla R (1996) Effect of gamma irradiation on mutants of tuberose (*Polianthes tuberose* L). Herberita 59:139–141

Estrada-Basaldua JA, Pedraza-Santos ME, Cruz-Torres EL, Martínez-Palacios A, Sáenz-Romero C, Morales-García JL (2011) Effect of 60Co gamma rays in tuberose (*Polianthes tuberosa* L.). Rev Mexicana Cienc Agríc 2:445–458

Guo W, Wen-Wen W, Yan X (2009) Mutagenic effects of co-gamma rays on *Polianthes tuberosa* L. J Nucl Agric Sci 23:799–804

Gupta MN, Sumiran R, Shukla R (1974) Mutation breeding of tuberose (*Polianthes tuberosa* L.). In: Symposium, use of radiations and radioisotopes in studies of plant productivity, Pantnagar, pp 169–179

Hajizadeh HS, Mortazavi SN, Ganjinajad M, Okatan V, Kahramanoglu I (2022) Evaluation of the optimum threshold of gamma-ray for inducing mutation on *Polianthes tuberosa* cv. Double and analysis of genetic variation with RAPD marker. Int J Radiat Biol 99:1204. https://doi.org/10.1080/09553002.2023.2159566

Hajizadeh HS, Mortazavi SN, Ganjinajad M, Okatan V, Kahramanoglu I (2023) Evaluation of the optimum threshold of gamma-ray for inducing mutation on *Polianthes tuberosa* cv. Double and analysis of genetic variation with RAPD marker. Int J Radiat Bot 4:1–13. https://doi.org/10.1080/09553002.2023.2159566

Jyothi R, Singh KP (2017) Effect of acute gamma irradiation on flower, bulb characters and stability of mutants in tuberose (*Polianthes tuberosa*). Indian J Agric Sci 87(7):968–974

Jyothi R, Singh KP, Mohapatra T, Kumar NK (2019) Induction of novel mutants and their stability using gamma irradiation in tuberose (*Polianthes tuberosa* L). Int J Curr Microbiol App Sci 8(8): 1815–1824

Kainthura P, Srivastava R (2015) Induction of genetic variability and isolation of mutants in tuberose (*Polianthes tuberosa* L). Trop Agric Res 26(4):721–732

Kaintura P, Srivastava R, Kapoor M (2016) Effect of physical and chemical mutagens on different cultivars of tuberose (*Polianthes tuberosa* Linn.) with particular reference to induction of genetic variability. Int J Agric Sci 8(15):1257–1260

Kayalvizhi K, Kannan M, Ganga M (2016a) Effects of gamma irradiation and chemical mutagens in tuberose *Polianthes tuberosa* L. Res Environ Life Sci 9(8):1030–1032

Kayalvizhi K, Kannan M, Ganga M (2016b) Mutagenic effects of chemical mutagens on tuberose (*Polianthes tuberosa* L.) var. Prajwal J Innov Agri 3(2):11–13

Kayalvizhi K, Kannan M, Ganga M (2017a) Effect of physical and chemical mutagens on morphological characters in M1V2 generation of tuberose (*Polianthes tuberosa* L.). Int J Curr Microbiol Appl Sci 6(4):2492–2499

Kayalvizhi K, Kannan M, Ganga M (2017b) Effect of mutagens on vegetative and floral characters in M1V2 generation of tuberose (*Polianthes tuberosa* L.). Bull Env Pharmacol Life Sci 6(1): 422–429

Krasaechai A (1976) Effect of gamma radiation on tuberose (*Polianthes tuberosa*). Kasetsart J Nat Sci 9(2):15–19

Krasaechai A (1992) Effect of gamma radiation on tuberose (*Polianthes tuberosa*). Kasetsart J Nat Sci 26(1):6–11

Krishnan AG, Geetha CK, Rajeevan PK, Valsalakumari PK, Saifudeen N (2003) Induced mutation in tuberose (*Polianthes tuberosa* Linn.) by gamma rays. In: Proceedings of the National symposium on recent advances in Indian floriculture. Kerala Agricultural University, Trichur, pp 255–261

Kumar M, Chaudhary V, Kumar M, Sirohi U, Yadav MK (2021) Application of conventional and mutation approaches in genetic improvement of tuberose (*Polianthes Tuberosa* L.): a review on recent development and future perspectives. Int J Agric Env Biotechnol 14(03):277–297. https://doi.org/10.30954/0974-1712.03.2021.1

Kutty NN, Ghissing U, Kumar M, Maiti MK, Mitra A (2020) Intense floral scent emission in *Polianthes tuberose* L. variants sprouted from γ-irradiated tubers. J Plant Growth Regul 39(1): 112–121

Navabi Y, Norouzi M, Arab M, Daylami SD (2016) Mutagenesis via exposure to gamma-rays in tuberose (*Polianthes tuberosa*). Electron J Biol 12(2):168–172

Panigrahi J, Saiyad MSL (2013) In vitro propagation of *Polianthes tuberosa* L. cultivars (Calcutta single). Int J Plant Animal Environ Sci 3:76–79

Patil AV, Kale PN, Kaulgud SN (1975) Some observations of gamma irradiation on growth and flowering of tuberose bulbs. Curr Sci 44:406–407

Pohare M, Rathod HP, Shahakar SB, Kelatkar SK, Suryawanshi P (2012) Effects of UV radiations on morphological characters in vitro regenerated *Polianthes tuberosa*. Res J Agric Sci 3:1307–1308

Pohare M, Batule B, Bhor S, Shahakar SB, Kelatkar SK, Varandani S (2013) Effect of gamma radiations on the morphological characters in in vitro regenerated *Polianthes tuberosa*. Indian Hortic J 3:95–97

Pooja K, Srivastava R, Kapoor M (2016) Effect of physical and chemical mutagens on different cultivars of tuberose (*Polianthes tuberosa* Linn.) with particular reference to induction of genetic variability. Int J Agric Sci 8(15):1257–1260; ISSN: 0975-3710&E-ISSN: 0975-9107

Regar AL, Mahawer LN, Bairwa HL, Rathore RS, Sharma H, Kumar S, Jat L, Husain S (2022a) Response of gamma irradiation on vegetative parameters of tuberose (*Polianthes tuberosa* L.) cv. Pearl double. Pharma Innov J 11(6):753–756; © 2022 TPI www.thepharmajournal.com

Regar AL, Mahawer LN, Bairwa HL, Rathore RS, Sharma H, Kumar S, Jat L, Husain S (2022b) Response of gamma irradiation on vegetative parameters of tuberose (*Polianthes tuberosa* L.) cv. Pearl double. Pharma Innov J 11(6):753–756

Sah R, Singh AK, Sisodia A, Padhi M (2017) Influence of gamma dose on growth, flower and bulb parameters in tuberose varieties. Int J Curr Microbiol App Sci 6(8):2038–2043

Sharavani CSR, Kode SL, Priya T, Bharti TU, Sekhar MR, Ruth C, Ramakrishna M (2019) Studies on effect of gamma irradiation on survival and growth of tuberose (*Polianthes tuberosa* L.). Adv Biores 10(1):109–113

Sihombing D (2017) *Polianthes tuberose* breeding in Indonesia. KnE Life Sci 2(6):587. https://doi.org/10.18502/kls.v2i6.1081; In ICSAFS conference proceedings

Singh PK, Sadhukhan R (2019) Identification of variants induced by physical and chemical mutagens in tuberose (*Polianthes tuberosa* L.). J Crop Weed 15(2):40–45

Singh R, Goyal RK, Godra NR (2011) Effect of gamma irradiation on growth, flowering, and bulb production in tuberose (*Polianthes tuberosa* L.) cv. Double. Haryana J Hortic Sci 40(3 & 4): 128–130

Singh PK, Sadhukhan R, Sarkar HK, Roy K (2013) Effect of EMS on morpho-anatomical changes in tuberose (*Polianthes tuberose* L.). Floric Ornam Biotechnol 7:103–105

Singh PK, Sadhukhan R, Dudhane AS, Kumar V, Sarkar HK (2015) Preliminary study on the mutagenic effect of EMS on tuberose (*Polianthes tuberosa* L.). Environ Ecol 33(3A):1386–1390

Singh PK, Sadhukhan R, Roy K, Sarkar HK (2016) Effect of EMS on morpho- anatomical changes in tuberose (*Polianthes tuberosa* L.). Floric Ornam Biotechnol 7:103–105; ©2013 global science books

Sravani K, Dorajee Rao AVD, Salomi Suneetha DR (2022) Studies on effect of colchicine on growth and yield of tuberose (Agave amica) B.L. biological forum. An Int J 14(3):1428–1433; ISSN No. (Print): 0975-1130 ISSN No. (Online): 2249-3239

Srivastava P, Singh RP, Tripathi VR (2007) Response of gamma radiation on vegetative and floral characters of postharvest characters in tuberose varieties. Int J Curr Microbiol App Sci 6(8): 1985–1991

Yadav G, Beniwal BS, Sheoran V, Kumar S, Vijay S, Kaur M, Kumar S (2018) Effect of mutagen ethyl methane sulfonate on growth characters of tuberose (*Polianthes tuberosa* L.) cv. Prajwal. Int J Chem Stud 6(4):412–416

Younis SEA, Borham JH (1975) The effects of gamma-irradiation on *Polianthes tuberosa*. Egypt J Bot 18(1,3):205–217

Part III

Other Ornamentals

Mutation Work on Other Ornamental Plants 12

Abstract

The chapter covers mutation work (mutagens, working dose, mutants) carried out throughout the world on approximately 120 ornamental crops.

Keywords

Ornamental plants · Working dose · Mutant

12.1 *Abelia grandiflora* Rehd.

Family Caprifoliaceae; hybrid species; deciduous semi-evergreen shrub; fragrant flowers; flowers white/pink; propagation by cuttings, bare root.

The only report available on *A. grandiflora* by Matsubara et al. (1976). They have developed one commercial mutant by treating stem cuttings with 30 Gy gamma rays.

12.2 Abutilon

Family Malvaceae; known as Indian mallow, Chinese Lantern; woody; annual shrub; propagation by seed, stem cuttings.

Different aspects related to radiosensitivity and how radiations affect different histogenic layers have been analyzed in *A. hybridum* hort (Potsch 1966a, b, 1967).

12.3 Acalypha

Family Euphorbiaceae; annual/perennial, herb/shrub/small tree; propagation by stem cuttings.

S. Datta, *Role of Mutation Breeding in Floriculture Industry*,
https://doi.org/10.1007/978-981-99-5675-3_12

Das et al. (1977) treated growing plants of Acalypha var. Macafiana with chronic gamma irradiation and detected mutations in leaf color and form. Ten centimetres long rooted cuttings were treated with different doses (0, 10, 20, 30, 40, 50, 60, and 70 Gy) of gamma rays. Treated cuttings showed a significant decrease in sprouting, maximum longest leaf, number of sprouts, and diameter of longest sprouts. 3 Gy treatment showed maximum plant height, leaf length, and leaf color. 50 Gy showed maximum leaf breadth (Kumar et al. 2008). To evaluate the influence of gamma rays rooted cuttings of *Acalypha hispida* were treated with gamma rays (0, 20, 25, 30, 35, and 40 Gy), and data were recorded on morphological changes, and survival, plant height, number of leaves and length of inflorescences. No significant differences were observed in survival and length of inflorescences after treatment. Plant height and number of leaves showed significant differences. Plants treated with 30 Gy and above showed color changes (pink and white) in inflorescences. 40 Gy has been recommended as the optimum dose (Rifnas et al. 2022).

12.4 Acer (*Acer negundo*)

Family Sapindaceae, known as boxelder, boxelder maple, Manitoba maple, or ash-leaved maple, fast-growing and fairly short-lived, yellow-green flowers, propagated by seed, used as ornamental shade and shelter belt tree.

Seeds of *A. negundo* were treated with gamma rays/X-rays and auxin and studied radiation effects on germination, survival, somatic mutation, etc. Details of experimental results are not available (Privalov 1965, 1967, 1968a, b, 1974). Agafonova et al. (1974) analyzed esterase isoenzymes from different tissue of mutants of *A. negundo* (c.f. Broertjes and Van Harten 1988).

12.5 Acer (*Acer ginnala*)

Family Sapindaceae, known as Amur maple, deciduous shrub or small tree, flowers yellow-green, grown as an ornamental plant/landscape, often planted as a shrub along borders, propagated by seed, stem cuttings.

A mutation technique was applied to develop a seedless variety. Seeds and unrooted cuttings were treated with gamma rays (0.1–5) and neutrons (2–14 krad) and detected radiosensitivity and seedlessness (Smith and Noyszewski 2018).

12.6 Achimenes

Family Gesneriaceae, also known as Magic flowers or Orchid Pansy; rhizomatous perennial evergreen herbs; hairy leaves and tubular Petunia-like flowers in a wide range of colors like pinks, blues, purples, lavenders, reds, yellows, mix color and white; propagation by stem, leaf, and scaly rhizome rooting, division and seed; leaves color green light to dark, in oval to elliptic shape, margin can be serrated.

Mutation work started to induce variability in Achimenes with radiations (gamma rays and X-rays) and colchicine. Radiation doses applied were 10–40 Gy and the explant used for treatment was leaves and adventitious rhizomes. Mutation work has enriched the literature on the development of non-chimeric solid mutants through adventitious shoots. Several commercial mutants have been developed from early experiments (Broertjes 1971, 1972a, b, c, 1973, 1974, 1977; Broertjes et al. 1983). Freshly detached leaves of colchicine-induced autotetraploid of 3 *Achimenes* cultivars were irradiated with X-rays or fast neutrons and different parameters like production (the number of rhizomes) and mutant frequency (color, form, size, etc.) were determined to find out the optimum dose. The mutation frequency of autotetraploid was 20–40 fold higher as compared to the corresponding diploid (Broertjes 1972a, b, c, 1976). Freshly cut leaves of *Achimenes* cv. Paul Arnold was irradiated with X-rays or fast neutrons, or treated with colchicine and then allowed to root and form adventitious buds. A wide mutation spectrum was observed. In addition to flower color early flowering mutant and compact form mutant were detected. The mutants were mostly slid, non-chiral (Broertjes 1971, 1972a, b, c, 1974, 1976).

12.7 *Aechmea fasciata*

Family Bromeliaceae; known as a silver vase or urn plant; houseplant; evergreen perennial; propagation by seeds, division.

Seedlings and seeds of *A. fasciata* and *A. bromelifolia* were exposed to X-rays and EMS (0.5 and 1%) to induce variability. The optimum dose was determined around 3000–4000 Gy. A high percentage of striped chlorophyll-deficient seedlings were detected in M1 (c.f Broertjes and Van Harten 1988). Callus and decapitated plantlet explants were treated with sodium azide and gamma rays to induce variability. The survival rate of decapitated plantlet explants treated with 0.5 mM sodium azide for 60 min was 51.3%, about a half-lethal dose. All explants treated with 2.0 mM sodium azide showed browning and/or dead. The survival rates of decapitated plantlet explants of *Aechmea fasciata* irradiated by γ-ray showed 74.2–100%. Mutant plantlets showed a great deal of chimeras in leaf (Huang 2021).

12.8 Alnus (*A. glutinosa*)

Family Betulaceae; medium-sized, short-lived tree; monoecious; male inflorescence and cone-like flowers are used in floriculture for decoration; propagation by seed, cuttings.

Ohba and Murai (1966) did a mutation experiment with *Alnus* to generate information on the breeding of broad-leaved trees under chronic gamma irradiation in a gamma field during the period 1963–1965. One-year-old saplings of *Alnus* species were planted in the experimental field. A hundred percent lethal dose rate was estimated as 60–160 r/day in the second year and 60–80 r/day in the third year. Tree height growth was severely inhibited at the high dose rate and the dose rate that reduced 50% of the tree height growth in 1964 and 1965 was about 60–80 r/day. Leaf form narrowed with increasing dose rate. Somatic mutations with variegation of leaves and narrow leaves were detected. All mutations were detected in the dose range of 50–70 r/day and this dose range has been recommended to be the optimum one for the creation of somatic mutations in broad-leaved trees under chronic gamma irradiation.

12.9 Alstromeria

Family Alstroemeriaceae, known as Peruvian lily or lily of the Incas, majority perennial, the color range including white, yellow, orange, apricot, pink, red, purple, and lavender, long vase life, propagation by seed, rhizomes, bulbs/bulb divisions, tuberous roots,

Broerties and Verboom (1974) treated young rhizomes of different *Alstroemeria* cultivars with X-rays and standardized 400 rad as the optimum dose for diploid cultivars and 500–600 rad for triploid ones. Mutations detected were mostly solid and a few mutants (Canaria Stagula, Yellow Tiger Stavero, White Wings Staretto, Harmony Stabroza, and Rosita Stareza) have been released to the trade. Przybyla (1992) treated growing rhizomes of young plants of *Alstroemeria* with gamma rays (^{60}Co) and fast neutrons (cyclotron U-120) were used in the range of 2.5–100 Gy to induce mutations. Mutations commonly recorded were a change in flower color and a change of design on the flower petals. Less frequent mutations were noticed in flower petals' shape, changes in flower size, mutations of growth vigor, productivity, and season of bloom. Most of the mutants were stable. No differences were observed in the spectrum of mutations induced by gamma rays or fast neutrons. Three to 7 Gy was the optimum dose for gamma rays and 50 Gy was lethal. Several new mutant cultivars (Catalina, Ines, and Juanita) with attractive flower color, high yielding, year-round blooming, and also miniature have been established. Anastassopoulos and Keil (1996) applied RAPD markers to study genetic variations among the phenotypically visible mutations developed either through X-irradiation or somatic variations. Polymorphisms were detected in some mutants and suggested that the RAPD technique is an inappropriate tool for the rapid screening of *Alstroemeria* for induced variation. Przybyla (2000) treated actively growing rhizomes of *Alstroemeria* sterile hybrids and sterile mutants were irradiated with gamma rays (2.5–100 Gy) and fast neutrons (3–6 Gy) to induce mutations. Dose 10 Gy caused maximum (40%) retardation of plant development and mortality and dose 25 Gy caused 70–90% mortality. A dose of 50 Gy of gamma rays was lethal. Desirable mutations were produced by 3–6 Gy gamma rays doses. New promising mutants line (Catalina, Ines, Juanita, Carlota, Isabel, Paula, Matilde, Azucena, Erendira) have been selected for trade. Tomo et al. (2012) treated in vitro cultured rhizomes of *A. aurea* with different doses of gamma rays and recorded radiation effects on different characters. Rhizome sprouting was affected at higher doses but sprouted early. Treated rhizomes showed bleaching and necrosis at higher doses. LD_{50} was established at about 40 Gy and the optimum dosage to induce mutation was suggested between 2.5 and 5 Gy.

12.10 Amaryllis/Hippeastrum

Family Amaryllidaceae; known as belladonna lily, Jersey lily, naked lady, amarillo, Easter lily, etc.; flowering bulb; funnel-shaped flowers, flowers are large and showy with various colors and forms; multiplied through separation of bulblets, seeds, in vitro, and a plant produces 2–3 bulblets in a year of growth.

Kaicker and Singh (1979) treated bulbs with 5 and 10 Gy γ-rays and detected few mutations in flower color in the cv. Dutch Red. Few private breeders conducted mutation experiments with irradiated bulbs and were successful to develop new flower color mutants (c.f. Broertjes and Van Harten 1988). New amaryllis var. "Orange Biogamma" has been developed through a combination of biotechnology and nuclear technology. The new variety with its appealing characteristics has great potential to be used as a landscaping plant as well as a potted plant (Affrida et al. 2009). Ariffin et al. (2016) studied radiation effects on *Amaryllis* "Orange Biogamma" after treating bulbs with gamma rays (0, 10, 15, 20, 25, 30, 40, and 50 Gy). Reduction in shoot length and inhibition of root production were recorded with increasing doses.

12.11 Anemone

Family Rannunculaceae; known as windflower; perennial; propagated by seeds and tubers.

Horovitz and Zohary (1966) studied the genetic background of perianth color, which developed spontaneously, in wild *Anemone coronaria*. Alkema (1974a, b) treated seeds and small tubers of *A. coronaria* with X-rays to determine the radiosensitivity. The optimum dose for seed was reported to be 100 Gy and small tubers were resistant even up to 150 Gy and radiosensitivity was cultivar dependent (c.f. Broertjes and Van Harten 1988).

12.12 Antirrhinum (Snapdragon)

Family Araceae; herbaceous; known as dragon flowers, snapdragons, and dog flower; flowers are white, yellow, orange, red, pink, or lavender or mixed colors; perennial but usually cultivated as an annual plant; flowers are produced on a tall spike; propagation by seeds, cuttings.

Association Euratom–ITAL, Netherlands initiated mutation breeding work and detected several mutants (dwarf growth habit, changed leaf shape and flower size, flower color, and several flowers per plant) and determined 7.5 Gy as the optimum dose for in vitro irradiation (c.f. Broertjes and Van Harten 1988). Realizing the importance of in vitro mutagenesis, in vitro propagation techniques were standardized for the induction of non-chimeral mutants and their faster multiplication (Pierik 1975, 1976; Pierik and Steeginans 1978; Pierik et al. 1979). The pilot experiment was conducted by treating seeds with 0, 0.10, 0.25, 0.5, 0.75, and 1.0%

EMS for 8, 10, and 12n h. O the basis of these results large-scale experiment was conducted by treating seed with 0.5, 0.75, and 1.0% EMS for 10-h. Variants like dwarfism, chlorophyll deficiency, and leaf abnormality were detected in the M1 population (Heffron et al. 2006). Actively growing plants were treated for 10 days with 50–6000 R/20 h-day with ^{60}Co gamma rays. Radiosensitivity, morphogenesis of apical meristems, and regeneration of lateral adventitious buds were studied along with the detection of somatic mutation in petals (Sekiguchi et al. 1971). Dry (20–320 kR) and soaked (10–80 kR) seeds of pink and white varieties were exposed to gamma rays (^{137}Cs) and the exposure rate was about 650 R/min, with a distance of 2 cm between the source and the material. Dry seeds were also treated with thermal neutrons with exposures ranging from 4.32×10^{13} *nth*/cm^2 to 1.08×10^{14} *nth*/cm^2. The objective of the experiments was to study the effects of water content and oxygen concentration to modify the radiation effects for large-scale irradiation experiments to induce beneficial mutations (Gupta and Matsuo 1969). Dyer et al. (2007) studied snapdragon flowers including mutants (mixta and nivea) to understand the effects on pollinator behavior of single mutations in plant genes controlling flower appearance using psychophysical tests with bumblebees. Eltayeb and Ei-Metainy (2012) reported the effects of gamma rays on three snapdragon varieties for the induction of mutations. More reports are also available on mutagenesis work on *A. majus* which covered a wide range of basic aspects like acute and chronic irradiation, radiation damage of shoot apical meristem, somatic mutations, etc. (Cuany et al. 1985a, b; Sekiguchi et al. 1971; Stubbe and Doring 1938a, b; Sheela and Sheena 2014; Doring and Stubbe 1938). Heffron and Korban (2022) treated seeds of a proprietary snapdragon line OAK564 with different doses (0, 0.10, 0.25, 0.5, 0.75, and 1.0%) of EMS for 8, 10, and 12 h to induce genetic variations and identify commercially useful variants of snapdragon. A large-scale experiment was conducted based on the findings of the pilot experiment. Several useful alterations were observed in M_1 populations for plant architecture (dwarfing, multiple branching, and secondary branching), plant color (chlorophyll chimeras, yellowish-green tissues, and dark green tissues), leaf morphology (small leaf size, leaf curling, wavy leaf margins, and thickened striations), and floral morphology (rolled petals, elongated petals, and curtailed petals). They concluded that the EMS treatments induced various mutations for various phenotypic traits, thereby increasing the genetic diversity in these M_1 snapdragon populations.

12.13 Anthurium

Family Araceae; known as tailflower, flamingo flower, laceleaf; herbs; perennial; inflorescence bears small perfect flowers; flowers are in spadix; flowers mostly red, pink, white; grown for its brightly colored flower spathes and ornamental leaves; mostly propagated by seeds, cuttings, in vitro methods.

An early mutation breeding work was initiated (Association Euratom–ITAL) which resulted in the development of mutants with changed plant stature like dwarf habit, changed leaf types, flower size, new flower colors, number of flowers per plant, etc. The mutants were mostly solid and the optimum dose was reported 7.5 Gy (c.f. Broertjes and Van Harten 1988). Puchooa and Sookun (2003) did experiments to standardize tissue culture technique for three varieties ("Nitta," "Osaki," and "Anouchka") of *A. ondreanum* and including variety "Nitta" to study the phenotypic and genotypic effects of gamma irradiation for mutation. Media composition for callus initiation, regeneration, rooting, and plantation have been determined. Explants were treated with gamma rays (5, 10, and 15 Grays/Gy), and observed that 5 Gy gave the best response in terms of callus formation and regeneration while the 15 Gy doses were lethal to the *Anthurium* tissues. Boosting effects of phenotypic characters were observed after the treatment of leaf tissues and seeds with 5 Gy. Variability was observed at the cellular level of the tissues but these changes could not be detected through the DNA fingerprints. Puchooa (2005) standardized in vitro culture of three varieties of *A. andreanum* in terms of media composition, time for callus induction, shoot formation, rooting, etc. After irradiation of explants (leaves, seeds, in vitro plantlets) 5 Gy showed the best response in terms of callus formation and regeneration and 15 Gy was lethal. Phenotypic variations could not be properly distinguished. EMS has been successfully used to induce flower color mutation in *Anthurium* cv. "Valentine" (Te-Chato and Susanon 2005). Characteristics of chimeras, development of leaf color chimera, and mutants have been studied in detail by Bin et al. (2006) and Chen (2009). Sheela and Sheena (2014) through their review proposed combining transgenic techniques with induced mutagenesis will greatly facilitate crop improvement in ornamentals including anthuriums. Suraninpong and Wuthisuthimethavee (2015) irradiated multiple shoots of 3-month-old "Angle" with acute gamma rays (0, 10, 20, 30, 40, and 50 Gy) to induce new variety through mutation. The percentage of surviving shoots decreases with increasing doses. 10 Gy-treated shoots showed higher quality or quantity in different morphological characters (shoot induction, number of shoots per explant, plant height, number of leaves, leaf length, leaf width, etc.). 10 Gy-treated shoots produced the highest percentage of the mosaic leaf. Genetic variations of mutation plants were investigated by the AFLP technique. Mosaic leaf mutants have been multiplied as a new cultivar. Yang et al. (2015) analyzed the micro- and ultrastructural and physiological features of leaves on wild-type and leaf color mutants (dark green, rubescent, etiolated, albino) in *A. andraeanum* "Sonate" using next-generation sequencing technology followed by *de novo* assembly and mentioned that the leaf color formation is greatly affected by the activity of chloroplast development and pigment biosynthesis. López-Martínez et al. (2022) exposed explants of leaves, nodes, and roots obtained from in vitro plants to 0.1% colchicine for 0, 2, 3, and 4 h to induce mutations. Colchicine effects were evaluated based on lethal dose (LD_{50}), survival, number of explants that generated callus, number of explants that formed shoots, and the number of shoots per explant. Leaves were found to be most sensitive to colchicine. The chromosome number was $2n = 29$ in explants exposed for 2 and 3 h to colchicine and $2n = 31$ in explants treated with colchicine for 4 h.

12.14 Asclepias Species

Family Apocynaceae; perennial ornamental; herbaceous; known as milkweeds; propagation by seeds, cuttings.

Kobayashi et al. (2011) treated seeds with ion beams to induce flower and plant form mutation. Seed germination decreased over 100 Gy and no seed germination at 150 Gy. Survival decreased below 50% at 50 Gy. Irradiation of 50–200 Gy was recommended as the optimum dose for seed irradiation.

12.15 Begonia

Family Bigoniaceae; one of the largest genera of flowering plants; rhizomatous/ tuberous; monoecious; grown for their foliage and attractive flowers; propagated by vegetative means like rooting of shoots, rooting of leaves, adventitious plantlets, and in vitro.

Begonia is considered to be a very appropriate material for the application of mutation technology due to its easy propagation (in vivo and in vitro), excessive heterozygosity, variability in chromosome number, etc. Mutation studies yielded interesting data, especially on adventitious plantlets. Leaf cuttings of four clones of *Begonia* × hiemalis were treated with different doses (0, 1500, 2000, or 2500 rad) of X-rays. Irradiation reduced the number and development of adventitious shoots but the mutation frequency and spectrum (flower color, shape, and size; growth habit, leaf characteristics, increased significantly, and mostly the mutants were non-chimeric (solid mutant) (Doorenbos 1973a, b; Doorenbos and Karper 1975). Shigematsu and Matsubara (1972) highlighted the technical procedure of how in vitro technique can be utilized for the isolation of radiation-induced sectorial chimeric mutant tissue. Mature leaves of Begonia cv. "Winter Queen" were cut into pieces (3 mm × 3 mm in size) and cultured in vitro. 0.5 mm length adventitious buds developed from cut fragments were exposed to gamma rays (dose rate of 0.2 krad/h) and again cultured in vitro. Two plants developed chimeric variegations on their leaves (sectors of green leaf color with silver-white spots). The original leaf color of "Winter Queen" is covered with silver white. The chimeric mutant sectors were regenerated successfully and developed a new variety with leaf color as that of the sector. Matsubara et al. (1974) irradiated *Begonia* cv. "Winter Queen" and standardized cultural practices (pruning techniques) for the development of solid mutants from a small mutated sector. They developed a new mutant whose leaves were green with maximum small silver-white spots by repeated cutting back, layering, and divisions. Molnar (1976) developed a single flowered semi-double year-round profusely blooming mutant (*Begonia* x hiemalis) by X-irradiation of Renaissance. Similarly, a good number of interesting radiation-induced mutant varieties have been developed and commercialized (Linderman 1968; IAEA 1972; Mikkelson 1976; Mikkelson et al. 1975; Anonymous 1977a, b; Lin and Molnar 1983; Benetka 1987; Soedjono 1988). An in vitro propagation of two genotypes of *Begonia* x hiemalis was achieved through adventitious shoot formation on subcultured leaf-disc explants and subsequent transplantation to the soil of explant parts with adventitious shoots. After irradiation of detached leaves with different doses of X-rays and two cycles of adventitious shoot formation on in vitro cultured leaf-disc explants, plantlets were produced. About 30% of these plants were mutated concerning, e.g. the color, size, and form of the leaves and flowers. The great majority of the mutants (98.5%) proved to be solid (non-chimeric) (Roest et al. 1981). A good number of experiments were conducted to explore the possibilities of utilization of the adventitious bud technique to develop solid mutants including ozone-sensitive mutants after mutagen treatments (Broertjes 1982; Brown and Harney 1974; Matsubara 1982; Matsubara et al. 1971, 1975). Two genotypes of B. x hiemalis were developed through adventitious shoot formation using leaf-disc explants and detached leaves were exposed to different doses of X-ray. Maximum (98.5%) solid (non-chimeric) mutants were recovered from mutated plants (30%) in flower color, size; and leaves and flowers form (Roest et al. 1981). Ito et al. (2007) studied the relative biological effectiveness of ion beam irradiation on Begonia and developed a few useful mutants. Several types of ion beams and doses were applied and evaluated based on survival and growth. The optimum dose to wet seeds of begonia was estimated at 5–10 Gy of Ne and N ion beam. The optimum dose to dry seeds was estimated at 10–20 Gy on the Ne ion beam. The optimum dose of C ion beam irradiation to the cutting of begonia new variety "Coral Fantasy" was estimated at 5–10 Gy. Flower color (yellow) mutant was selected from those of C ion beam irradiated at 10 Gy. EMS-induced *Begonia* x *hiemalis* Fotsch. "Schwabenland Red" mutants were evaluated for severity of stem rot disease caused by *Rhizoctonia solani* Kuhn and morphological changes. Three highly disease-resistant mutants were recovered. The mutants had changed leaf color and shape, and shape and number of flowers (Chen et al. 2014).

12.16 *Berberis thunbergii*

Family Berberidaceae; known as Japanese barberry, Thunberg's barberry, or red barberry; small deciduous perennial shrubs; leaves green turning to red in the autumn; used for landscape; propagated by seeds, cuttings.

The objective of mutation was to develop a seedless variety. Seeds and unrooted cuttings were treated with gamma rays (0.1–5) and neutrons (2–14 krad) and detected radiosensitivity and seedlessness (Smith and Noyszewski 2018).

12.17 Betula

Family Betulaceae; deciduous hardwood ornamental tree; propagation by stem/branch cuttings, root cuttings, grafting, and layering.

One-year-old seedlings of *Betula ermanli* were planted in the radiation field. A 100% lethal dose rate was estimated as 60–160 r/day in the second year and 60–80 r/day in the third year. Height growth was severely inhibited at the high dose rate. The dose rate that reduced 50% of height growth was about 60–80 r/day. Percentages of plants having blind buds in the main stem and abnormal branching increased sharply at 50–200 r/day. Variegation of leaves developed at 50–70 r/day (Ohba and Murai 1966). Seeds of *B.verrucosa* and *B. pubescens* were treated with different doses of X-rays and detected sterile chlorophyll mutants in M1 (Scholz 1957). Vaarama (1970) published a report which covers seed germination and variation of germination of some *Betula* spp. due to effects of light, temperature, and storage conditions; photosynthesis and growth; biology and origin of *Betula tortuosa* Ledeb; description of a few spontaneous Betula mutants (dwarf and variegated), etc. Ocokoljic and Milosevic (2004) detected a rare type of spontaneous somatic mutations in *B. verrucosa*. Two different phenotypes were formed by the spontaneous mutation from the callus tissue produced at the place of cutting. The somatic mutation had a high degree of pleiotropism, and consequently, it was considered within the context of the whole changed genotype.

12.18 Blue Daisy (*Brachycome multifida*)

Family Asteraceae; semi-shrubby; leaves bright green deeply divided; flowers mauve, pink or white ray florets with yellow disk florets; propagated by seeds, softwood cuttings.

B. multifida has no genetic variation and does not produce seeds under European environmental conditions. To increase genetic variability in "Blue Daisy" (Brachycome multifida) Walther and Stauer (1989a) applied in vitro mutagenesis through polyploidization (0.1% colchicine) and X-ray treatments (10–50 Gy) to freshly cut nodal segments. In vitro raised axillary buds developed from nodal segments were used as explants. Two polyploid morpho-types were obtained: one more erect and one more hanging phenotype, both having enlarged leaves and flowers. Variants detected in X-ray treated population were temperature-dependent chlorophyll-defect (20 Gy); almost white flowers (40 Gy); petals 50% shorter than control (40 Gy); early, rich, and extended flowering (semperflorens), graceful leaves, dwarf type (30 Gy).

12.19 Bouvardia

Family Rubiaceae; flowering plant; evergreen herbs/shrubs; perennial; propagation by seed, root/stem cuttings.

Association-ITAL, the Netherlands treated rooted cuttings with X-ray/gamma rays and standardized 20 Gy as the optimum dose. 10 Gy was the optimum dose for in vitro raised plants. Few mutations were detected (c.f. Broertjes and Van Harten 1988).

12.20 *Buddleja davidii*

Family Scrophulariaceae; known as summer lilac, butterfly bush, or orange eye; shrub; flowers perfect hermaphrodite; propagated by softwood or greenwood cuttings, and semi-ripe cuttings.

Seeds of *B. davidii* (butterfly bush) were treated with EMS (dose range 0.05–5%) and isolated one periclinal chlorophyll-variegated (along the edges of the leaves) mutant ("Summer Skies") (Smith and Brand 2012).

12.21 Cactus and Succuents

Induced mutagenesis work on this group of plants is limited. Few reports are available on variegation in succulents due to mutation. Jack (2009) started an experiment by irradiating very small seedlings with a UV-C sterilizing lamp to induce mutations. He also tested different chemicals (nitrous acid, sodium nitrite, colchicine) for the induction of mutations. Milica Filipovic (2019) reported genetic variation due to mutation is responsible for variegation in succulents. Variegated strains are multiplied and utilized as beautiful outdoor and indoor plants. Sometimes the presence of pigments in variegated plant masks the chlorophyll pigments and produces darker colors, like purples and reds. Variegations have been found in two distinct forms: one is medio-variegation where the variegation streak down the middle of the leaves and another is margin-variegation where variegation appears on the sides of the leaves. *Trichocereus* are very popular cacti for their horticultural and medicinal potential. Oxyzalin was used and a few polyploidies were induced. Hybrid seeds and seedlings (HB02 × HB01 seeds and open pollinated of HB07) and cuttings of T. peruvianus "Huarimayo monstrosa" were treated with 100, 150, and 200 Gy gamma rays. Seeds were found to be very susceptible to radiation. The next treatment dose applied was 20, 40, and 60 Gy. No change in seedlings was observed. Further experiments with *Trichocereus* seeds were conducted and used 250, 350, 450, 550, and up to 1000 Gy gamma rays. The *Trichocereus* "HB07 OP" seedlings were a better success story. Very few seedlings survived some treatments. The Huarimayo developed a patch of variegation after treatment (Anonymous 2022).

12.22 *Calendula officinalis*

Family Asteraceae; known as pot marigold; herbaceous perennial; ligulate and tubular floret; propagation by seeds.

Kaur et al. (2017) studied the effects of different doses of gamma rays (20, 40, 60, 80, 100, 120, and 140 Gy) on the induction of mutations, on cytomorphological and phytochemical characters after seed treatment of *Calendula officinalis* var. "Fiesta Gitana Yellow." Promising mutants (total phenolic content and antioxidant activity, flower color and plant height, etc.) were screened and tagged at 20 and 40 Gy for release as new varieties after assessment. Lethal doses induced desirable novelties. Three promising mutants were finally selected and screened for cytomorphological studies, phytochemical attributes, and possible use of these novel traits (flower color and plant height) as ornamental novelties. Pre-soaked seeds were treated with 2.5, 5, and 7.5 KRD of gamma rays. Stimulation in growth was observed at 2.5 KRD and stunted growth at 5 KRD. Abnormal leaves increased with an increase in dose. 2.5 KRD has been recommended as a suitable dose (Tiwari and Kumar 2011). El-Nashar (2012) treated seeds of *Calendula officinalis* (L.) c.v. "Calypso Yellow" with different concentrations of SA and DES (1000, 2000, 3000, 4000, and 5000 ppm) and studied their effects on germination percentage, plant height, number of leaves, number of branches, fresh weight, dry weight, number of inflorescences, inflorescences diameter, flowering date, leaf total chlorophyll, and root length in M_1 and M_2 generations. Both mutagens had significant effects in M1-generation concerning dry weight, inflorescences diameter, flowering date, and total leaf chlorophyll. A significant increase in dry weight and total leaf chlorophyll was recorded in DES treatment. Both mutagens showed significant effects on different characters in M_2. Samatadze et al. (2019) studied the morphological, meiotic, and chromosomal variability in two cultivars of *Calendula officinalis* L. and their mutant lines obtained through chemical mutagenesis using diethyl sulfate (DES) (0.04%, 0.08%) and dimethyl sulfate (DMS) (0.025%, 0.05%). Cultivars were differentially sensitive to DMS and DES. *C. officinalis* cv. "Zolotoe more" showed morphological changes than in cv. "Rajskij sad." DMS and DES at low concentrations had positive effects on main agro-metrical traits in both cultivars including plant height, inflorescence diameter, and number of inflorescences per plant. The mutant line also showed a dose-dependent increase in several various meiotic abnormalities. Comparative karyotype analysis and FISH-based visualization of 45S and 5S rDNA indicated a high level of karyotype stability in M1 and M2 plants. Studies indicated that DMS and DES at low concentrations have great potential in calendula mutation breeding. Elateek et al. (2020) evaluated the effects of three mutagens to select the proper mutagen for the improvement of the morphology of the plant. Seeds were treated with three doses of gamma ray (2500, 5000, 7000 Gy) and two chemical mutagens (colchicine and EMS at concentrations of 1000, 3000, and 10,000 ppm). Different parameters were studied like plant morphology, its active compounds, and protein profiling before and after treatments for comparative assessment of physical and chemical mutagens. A gamma dose of 7000 Gy and an EMS concentration of 10,000 ppm showed the highest plant height. The lowest dose of colchicine (1000 ppm) made the most impact on the plant height. All other characteristics like the number of leaves per plant, number of flowers per plant, and flower diameter responded differentially to different doses of different mutagens. Both types of mutagens showed noticeable changes in flower shape. The biochemical parameters and carotenoid contents increased at 5000 Gy, while flavonoids increased at 2500 Gy. The protein-profile analysis showed fluctuation in the gene expression. The highest performance appeared at 7000 Gy of gamma treatment, and 1000–3000 ppm of colchicine, while the same result was obtained by EMS at 3000 ppm.

12.23 Calathea

Family Marantaceae; known as calatheas or prayer plants; herbaceous perennial; potted ornamental; leaves colorful; flowers asymmetrical, yellow, purple, and white; bracts often more attractive; propagation by seed, root division.

Healthy plants or cuttings of *C. crocata* (dark purple-green foliage with orange inflorescences) were exposed to an acute dose of X-rays or gamma rays to increase variability in different desirable characters. The optimum dose was determined 11 Gy and the plants were treated repeatedly over the years. One mutant ("Esther"—large umbrella-shaped flowers with orange petals and green tips) was detected from a 1982 experiment and released in 1987 (Arkesteijn, J.A. and Hoorn, D.) (c.f. Broertjes and Van Harten 1988; Anonymous 1988). Rhizomes of *C. louisae* cv. "Albertii" were treated with different doses (0, 20, 30, 40, 50 Gy) of gamma rays and determined the optimum dose for irradiation. The next experiment was conducted using in vivo methodology (cutting back method) and 25 and 35 Gy were found to be most suitable. After six multiplications one interesting yellow variegated leaf mutant was selected and multiplied and named "Yellow Cais" for commercialization (Latado and Tulman Neto 1996; Neto et al. 2001).

12.24 *Camellia japonica*

Family Theaceae; known as common camellia, or Japanese camellia; flowering tree or shrub; flowers red to pink to white; flowers single, semi-double, or double; propagation by seeds.

Zhou et al. (1990) treated shoot cuttings of 1-year seedlings and 3–4 year plants of *C. japonica* with gamma rays and observed differential radiosensitivity among the cultivars and different plant parts. 1–3 krad was found most suitable dose for shoot cuttings. Doses below 2 krad developed high mutation frequency and the frequencies of mutation in flower color and type were 2.5–12.0%.

12.25 *Canna generalis* B., *C. indica*

Family Cannaceae; herbaceous perennials; rhizomatous rootstock; flowers red, orange, or yellow, or any combination of those colors; propagation by seeds, rhizomes.

Dwarf varieties were treated with X-rays to find out the radiosensitivity of rhizomes and determined the optimum dose of 5–10 Gy (c.f. Broertjes and Van Harten 1988). Nakronthap (1965) treated 22 cultivars comprising diploid and triploid cultivars with different doses of gamma rays (10, 15, and 27 Gy) and detected many variabilities in flower color, form, and size. Gupta (1966), and Desai and Abraham (1974) treated different varieties with a wide range of gamma rays (5, 10, 20, 30 Gy) and determined differential radiosensitivity among the cultivars of different ploidy levels. Mutations were detected mostly as small patches on plants and flowers as chimeras which could not be isolated. Few stable mutants were isolated. Mukherjee and Khoshoo (1970) irradiated rhizomes of diploid and triploid cultivars with different doses (10, 20, 30 Gy) of gamma rays to create new variability in plant and bract characters. They made a very interesting study on the morphological and cytological parameters of the irradiated materials and compared them with bud sport varieties. They observed that 3 kR gamma rays were (almost) lethal to the diploids but did not affect the triploids. They suggested starting with red flower color if a yellow flower color mutation is required. Chuantang and Li Yazhi (1989) treated dormant rhizomes of *Canna* ("Crimson Canna") with gamma rays and fast neutrons and observed high flower color variations. The optimum dose ranged from 9.2 to 36.8 Gy. Gamma rays were more effective than fast neutrons. Nine flower color mutants were detected of which three mutants with special floral features were selected for commercial exploitation. Canna generalis B. seed and rhizome were irradiated by ^{60}Co-γ ray with five irradiative dosages and the mutation efficiency was analyzed. The seed germination, seedling growth, and the rate of rhizome survival of Canna generalis B. was decreased and the mutation rate was raised with the increment of irradiative dosage. The mutation rate did not vary and the plant was heavily damaged without ornamental value over 60 Gy dosage. There was a low frequency of mutation and little variation of color and pattern of the flower in the irradiation of gamma ray to the rhizome. Results indicated that the suitable radiation dosage for Canna generalis B. seed was 40–60 Gy and for rhizome, 10–20 Gy (Fucui et al. 2011). Roy et al. (2010) applied 1, 2, and 3 kRad gamma rays to rhizomes of Canna generalis cv. "Cleopatra" and studied radiation effects on vegetative and floral characters. Reduction in survival, plant height, and growth, number of leaves, leaf size, delay in spike emergence, first color shown and full bloom, number of flowers per spike, floret number and size, and increase in morphological abnormalities in foliage and flowers were observed after gamma irradiation. Different types of stripes were noticed on leaves. Variations in flower color were noticed. Finally, yellow flower color and bronze leaf color mutants were isolated in pure form. Fucui et al. (2012) treated the seed and rhizome of *C. generalis* B. with gamma rays and observed decreased rate of seed germination, rate of seedling growth, and rate of rhizome survival and increased rate of mutation with the increasing dose of irradiation. Plants were severely damaged over 60 Gy and no change in mutation rate. Treated rhizomes showed a low frequency of mutation and limited variation of color and pattern of the flower. The suitable radiation for seed was 40–60 Gy and for rhizome 10–20 Gy. Several reports on mutation work are available on canna where the main objectives were to determine the radiosensitivity, determination of optimum dose, and creation of genetic variability in plant, leaf, bract, and flora characters (Nakronthap 1965; Chemarin et al. 1973; Deng and Liu 1987, 1990; Fucui et al. 2011; Khalaburdin 1991; Niu and Li 1988; Siranut et al. 2000).

12.26 *Catharanthus roseus*

Family apocyanaceae; formerly known as *Vinca rosea*; known as bright eyes, Cape periwinkle, Madagascar periwinkle; perennial and self-pollinated tropical and subtropical ornamental plant; propagation by seeds.

Evaluation of ethyl methane sulfonate-induced dwarf mutant of *Catharanthus roseus* (L.) G. Don revealed that the mutant exhibited marked variation in morphometric parameters. The *in vitro* antibacterial activity of the aqueous and alcoholic leaf extracts of the mutant and control plants was investigated against medically important bacteria. The mutant leaf extracts showed enhanced antibacterial activity against all the tested bacteria except Bacillus subtilis. Verma et al. (2013) reviewed mutation work on *C. roseus* and recommended the use of mutation for its further improvement. A series of articles have been published to induce desirable variables in different cultivars ("Nirmal," "Dhawal") *of C. roseus* using gamma rays, EMS, nitroso-methyl urea, etc. Experiments resulted in the development of mutants with changed characters like dwarf, semi-dwarf with high alkaloid content in the roots, wavy leaf margin, high alkaloid content in leaves, indehiscent anther, smaller anther, with lesser number of pollen grains, the embryo had only one cotyledon, salt resistant with mono-cotyledonary and hsf mutant with herkogamous flower, high alkaloid, morphological mutants (very tall, dwarf with obvate leaves, medium tall with small leaf area, nontrichomous, upright oriented elliptical leaf, and spoon-shaped leaf), leafless inflorescence architecture with increased flower frequency, small pollen, pollen fewer anthers, another with "Pin" flower, enhanced antibacterial activity, macro-mutants with enhanced total alkaloid contents of root and leaf, etc. Keeping in view the results it has been proposed that induced mutagenesis is an efficient approach for the development of improved genotypes (Venkateswarlu et al. 1983, 1988; Bhattacharjee 1998; Chatterjee 1978; Kulkarni and Baskaran 2003, 2008; Kulkarni and Suresh 1999; Kulkarni et al. 1999; Mishra and Kumar 2003; Rai and Kumar 2000; Anjalika et al. 2005; Kumar et al. 2007; Baskaran et al. 2013; Verma and Singh 2010). Mangaiyarkarasi et al. (2014) studied the effects of different doses of gamma rays (30, 40, 50, 60, and 70 KR) and EMS (30, 40, 50, 60, and 70 mM) and determined LD_{50} doses. EMS was found to be more effective and efficient in causing mutations as compared to gamma rays treatment. El-Mokadem (2014) attempted to study the effects of gamma rays and in vitro culture to develop new flower colors in C. roseus. Several plants were regenerated from irradiated internode explant. Two cultivars from the previous study Rosea and Alba, and four doses of gamma irradiation (0.0, 0.5, 0.75, and 1.0 Gy) were applied. Results indicated that both 0.75 and 1.0 Gy irradiation doses were the most effective in inducing mutation in flower shape and color, while the dose of 0.5 Gy was effective in inducing mutation in leaf form. Hassan et al. (2017) did experiments to find out the response of different parameters like morphological characteristics, proline, and alkaloids content, total carbohydrates content in the leaves, variations, mutations, and peroxidase isozyme after treating seed with gamma rays (0, 5, 10, 15, and 20 kr) and irrigation with saline water (0,100, and 150 mM). Following different variations were detected from treatment combinations: morphological characters (growth habit; leaf size, form, and color; stem color; flower structure and color), increased proline content, significant increases in the leaf vindolen and catharanthine content, variations in branching, flowering, and salt tolerance, etc. Zahra et al. (2020) irradiated seeds of C. roseus with gamma rays (30, 60 Gy), and the seeds were cultured in vitro. Morphological variability was observed in the surface area of leaves, stem length, root length, and fresh weight of regenerated shoots. Morphological results showed a significant change in leaf surface and twisted leaf in 60 Gy irradiated plantlets. The molecular analysis revealed significant genetic differences among treatments and control. They have concluded that gamma irradiation produced morphological and genetic changes in C. roseus regenerated in vitro plantlets providing to be a suitable method for further mutation breeding in periwinkle. Mistry et al. (2022) applied a mutation technique to enhance the production of MIAs in a plant system (*Catharanthus roseus*). Chemical and physical mutagenesis with sodium azide, ethyl methanesulfonate, and X-rays were adopted on C. roseus to determine their possible effects on the transcriptional modulation of MIA biosynthetic pathways in plants. The chemical mutagenesis resulted in delayed seed pod development in mutated *C. roseus* plants, with distinct leaf morphology and flower color. X-ray mutagenesis resulted in pollen-less sterile flowers. These data and HPLC analysis of catharanthine, vindoline, and vinblastine content indicated that the induced mutagenesis of *C. roseus* provides a prospective strategy to modulate plant transcriptomes and enhance the biosynthesis of pharmaceutically important antineoplastic vinblastine in the plant.

12.27 *Celosia cristata* L

Family Amaranthaceae; known as cockscomb; distinctive inflorescence in the form of the wavy crest; annual herbaceous; propagation by seeds.

Qiang et al. (2005) irradiated seeds of *C. cristata* with gamma rays and studied their effect on seed germination, seed growth, and different morphological parameters. Jala Hayati and Krisantini (2016) treated 8-week-old plantlets with 0, 25, 50, and 75 Gy gamma rays. Based on the evaluation of the decrease in growth, abnormal growth, internode, and leaf characters LD_{50}, LD_{30}, and LD_{20} were determined 68.73, 46.68, and 35.65 Gy, respectively. Muhallilin et al. (2019) irradiated plantlets with 0, 25, 50, and 75 Gy gamma ray doses to induce diversity in morphology and chemical contents in *C. cristata*. Promising morphological (color and shape of the stem, leaves, and flowers) and chemical (appearance of triterpenoid compounds) changes were observed in one clone irradiated with 25 Gy. Mostafa et al. (2014) treated plants of *C. argentea* var. "Spicata" with DMS (0, 1000, 2000, 3000, and 4000 ppm) as a soil drench to study the effects on plant growth and phytochemical composition. The concentration of 3000 ppm increased significantly the number of leaves, the concentration of anthocyanin in the leaves, inflorescences in both generations, and the concentration of phenols in the M_2 generation. All treatments decreased significantly stem diameter, and length of the inflorescence, delayed flowering in both generations and decreased plant height in the M_1

generation and chlorophyll content in the M2 generation. Different treatments induced changes in many characteristics like changes in leaf and inflorescence shape, dwarf nature of the plant, bigger growth, green vegetative growth, yellow nodes, lack of anthocyanin pigments, etc. The genetic variations were measured using ISSR. Celosia cristata is an edible ornamental plant and it has a potential role in floriculture. *Celosia cristata* L. is a multifunctional crop that is beneficial as an ornamental and medicinal crop. An attempt was made to improve the ornamental value of the plant through induced mutagenesis. Ayudhaya et al. (2022) treated seeds of C. cristata with EMS (0, 0.5, 0.6, 0.7, 0.8, 0.9, 1, 2, 3, and 4%) for 4 h. LC_{50} values in *C. cristata* were 2.78%. Variability was recorded in the plant height and flower width, changes in color, and shape of stems, leaves, and flowers. Aisyah et al. (2022) applied a mutation technique to improve agro-morphological characters and polyphenol content. EMS-treated plants were evaluated in M2 generation. There are six subpopulations with the highest diversity of quantitative characters in the M2 population in quantitative characters. Changes in the shape and color of leaves stems, and flowers of C. cristata were observed in the M2 population. They have detected potential plants with the highest polyphenol content and antioxidant capacity in the M2 population. They concluded that EMS-induced mutations can enhance the agro-morphological diversity, and polyphenols content along with the antioxidant activities of C. cristata, and recommended a successful mutation breeding program.

12.28 *Cheiranthus cheiri*

Family Brassicaceae; Erysimum cheiri, syn. Cheiranthus cheiri; known as a wallflower; is an herbaceous perennial/biennial; with narrow and pointed leaves; flower with purplish-green sepals; flowers are fragrant; propagation by seeds.

Plants of single flowered *C. cheiri* were grown and treated in the gamma field (*ca.* 4Kr) for about 3 months. Slightly double or semi-double flowered plants were detected and the degree of their doubling was variable, both between plant and plant and between flowerlets on the same inflorescence. These double flowers were possibly induced by a recessive gene and consisted of a few small spoon-like petallets mixed with pistils and stamens. In appearance, however, their inflorescences seemed to be inferior to those of the commercial cultivars, such as Double Early Wonder, due to somewhat tiny, poor flower stalks and lowered seed fertility (Saito 1977).

12.29 China Aster (*Callistephus chinensis* Nees)

Family Asteraceae; known as Annual Aster; annual; flowers light and dark purple, light and dark pink, red, yellow, and white; flower heads with tubular disc florets and ray florets; propagation by seeds.

Wosinska (1980a) studied the effects of different doses of gamma rays upon four varieties of China aster in M_1 and M_2 generations on the following features: the height of plants, maximal length of the ramification of the Ist order, the diameter of inflorescence, and several set inflorescences. All these characters showed negative responses to radiation after seed treatment except two varieties ("Agnieszka" and "Goplana") showed positive effects at 3 kr. Higher doses were harmful to all the varieties. Gamma radiation showed a beneficial effect on all the features in M_2. All varieties showed a differential response. In a separate experiment, Wosinska (1980b) studied radiation effects on pollen grains of four varieties ("Agnieszka," "Goplana," "Madonna," and "Plomien") in M_1 and M_2 and found deleterious effects of radiation in M1. Wosinska (1986) developed a bicolored ligulate flower mutant through gamma irradiation. Wosinska (1982, 2013) observed the differential response of five varieties to gamma irradiation (3–12 krad) in M1 and M2. Variations in leaf color and size and the shape of lingual flowers were noticed in M1. Callistephus chinensis has beautiful flowers with high sale prices and usually grows at highlands/ plateaus. An experiment was carried out to evaluate the effect of induced mutation by gamma rays on the performance of third-generation (M1V3) Callistephus chinensis at the lowland. Tillers were treated with gamma rays (0, 5, 10, 15, and 20 Gy) and desirable variants were selected in M2 for M3. A total of 12 genotypes were selected and planted at the lowland (125 m above sea level) in Tembalang-Semarang. The results showed that mutants showed increased height, number of leaves, number of flowers, and flower stalk length at M1V3 at lowland. The number of the tiller and flower color was not significantly different between the induced mutation plant and the control. Most flower color was purple except a few were pink. It has been concluded that C. chinensis can be adapted to lowlands due to induced mutation by gamma rays (Anwar et al. 2019). Anwar et al. (2020) evaluated gamma radiation effects on *Callistephus chinensis* at lowland. Tiller was treated with gamma rays (5, 10, 15, 20 Gy), and surviving plants were screened at M1V3. Twelve selected genotypes showed increased height, number of leaves, number of flowers, and flower stalk length. Several tillers were not significantly different between mutant plants and control, except for selections Y1-U4-11 and Y3-U3-10. The color of the flower was not significantly different between the induced mutation plant and the control, except for some of the Y3-U3-10 genotypes. Most flower colors were purple, while some of the Y1-U4-9 genotypes (Y1-U4-9-5) were pink. They concluded that *C. chinensis* can adapt in lowlands due to induced mutation by gamma rays. Shivaswamy et al. (2022) treated seedlings of China aster var. "Arka Kamini" with Diethyl sulfonate and EMS (0.05, 0.10, 0.15, 0.20, and 0.25%) and analyzed effects on different vegetative, floral, and yield characters. Maximum survival percentage of plants (94.17) and plant height (32.25 cm) was observed in 0.15% EMS and 0.15% DES, respectively. 0.25% EMS recorded maximum plant spread in E–W direction (24.65 cm) while undesirable results were obtained in 0.25% DES.

12.30 Chionodoxa

Family Asparagaceae; known as glory-of-the-snow; bulbous perennial; flowers blue, white, or pink; flowers with tepals; propagated by seeds and bulb offsets.

A preliminary experiment was conducted with *C. luciliae* to determine the radiosensitivity and to induce genetic variability in flower color by treating bulbs with different doses of X-rays. Treated bulbs survived up to 150 Gy X-rays and died above 150 Gy onwards (Alkema 1974a, b; c.f. Broertjes and Van Harten 1988).

12.31 *Chlorophytum tuberosum*

Family Asparagaceae; known as musli; rhizomatous herb; flowers white; propagated by seeds and by division of rhizomes.

Chowta and Dnyansagar (1974) treated seeds with gamma rays and EMS and observed different morphological changes in vegetative and floral characters and delay in flowering.

12.32 *Cosmos sulphureus*

Family Asteraceae; known as sulfur cosmos and yellow cosmos; annual; flowers shades of yellow, orange, and red; propagation by seeds and shoot tip cuttings.

Syakudo et al. (1964) treated dry seeds of yellow *C. sulphureus* with different doses (5000, 10,000, and 20,000 r) of X-ray and detected new large flower type and double flower mutants at 20,000 r treatment. Gupta and Samata (1967) treated plants (two heterozygous clones) of *C. bipinnatus* with acute X-ray (2 kR) and chronic gamma irradiation (3 and 2 kR). Red, light pink, and white color mutant sectors appeared on ray florets. Mutation frequency varied among the clones and stages of flower development. The mutation rate was very low after chronic γ-irradiation as compared with the acute X-irradiation. Ayudhaya et al. (2022) treated seeds of *C. sulphureus* Ca. with acute gamma rays (0, 100, 200, and 300 Gy) and studied radiation effects on survival and morphological changes periodically. No change in germination and survival was observed at 30 days at 0, 100, and 200 Gy of gamma rays. The lowest survival (25%) was recorded at 300 Gy and the LD_{50} was 253 Gy. 100 Gy induced an extra smaller petal which is an interesting feature. Results indicated that suitable gamma radiation for the cosmos was between 100 and 200 Gy.

12.33 Codiaeum (Croton)

Family Euphorbiaceae; ornamental shrub or tree; houseplant; striking foliage; evergreen perennials; more than 700 species; propagation by stem cuttings.

Radiation-induced mutation breeding work was initiated for the improvement of leaf characters (form and size) and growth habits of *Codiaeum* cultivars "Petra" and "Conte." The optimum acute dose was determined as 5–10 Gy X- or gamma rays. No mutation was reported (c.f. Broertjes and Van Harten 1988). Alabi et al. (2010) tested mutagenic effects, root regeneration, and proliferation in a plant–plant interactive hydroponic assay of cigarette tobacco extract on two reciprocal mutant cultivars (Guinea broad leaf and Guinea narrow leaf) of garden croton. The results showed that $1 \times 10^{-(4–8)}$ logarithm dilution of a standard teacup (270 mL) cigarette tobacco infusion enhanced root regeneration, elongation, and proliferation above the control level. Higher concentrations of extract inhibited root elongation and proliferation. Undiluted tobacco infusion extract (100%) delayed root initiation until 4 weeks. There was no sign of induced mutation in the plant cultivars by the extract. Saggoo et al. (2011) studied the microsporogenesis of *Croton bonplandianum* Bailland reported chromosome number, cytoplasmic connections, cytomixis, laggards, micronuclei, heterogeneous pollen size, and variation in pollen fertility. This basic information may help in designing mutation experiments.

12.34 Crocus

Family Irridaceae; seasonal flowering plant; herbaceous perennial; flowers white, yellow, orange, or purple; propagation by seeds, daughter cormels.

The dried stigma of *Crocus sativus* is the main source of saffron. Mutation breeding is one of the methods that is mostly used for ornamental plant breeding. The most commonly seen features after applying this method among new cultivars of ornamental plants are changes in color, texture, and shape of flowers and leaves, the endurance of plants, plant growth type, and the difference in flowering time, corm, and bulb production. Although Crocus has medicinal importance, it is a seasonal flowering plant. All mutation breeding work reports mainly affect flowers and other growth-related characters.

Akhund-Zade and Muzaferova (1975) induced variability in different morphological characters after the mutagen treatment of saffron. Khan (1997) observed an increase in plant height in lower doses, a decrease in higher doses, delayed sprouting, induction of flowering in VM_1 plants, and reduced fertility in higher doses after treating corms with gamma rays (0.5, 1.0, 1.5, 2.0, and 2.5 krad). They detected mutants with serrated petals and three-petal flowers. Serrated petal mutant flowered in VM_2 and VM_3 generations but it failed to flower in VM_4 generation. Khan (2004) treated corms (4–5 cm in diameter) of *C. sativus* with gamma rays (0.5, 1.0, 1.5, 2.0, 2.5, and 3.0 krad) to develop desirable mutants of commercial use. A wide range of variability was recorded on sprouting time, plant height, induction of flowering, number of petals, and shape of petals. Selected corms showed delayed sprouting and slow growth in higher doses (2.0, 2.5, and 3.0 krad) increase in plant height in lower doses (0.5 and 1.0 krad), decrease in higher doses (2.0, 2.5, and 3.0 krad), and induction of flowering in middle doses (1.0, 1.5, and 2.0 krad) in the next generation. Serrated petal mutants were found to be superior to the parental strain and other mutants concerning days of flowering, size of flower, length of stigma, and weight of 100 fresh stigmas. Delayed sprouting, slow growth in higher doses, and increase and decrease of plant height in lower and higher doses, respectively, due to the amount of auxin synthesized during sprouting. Khan (2007) treated corms of *C. sativus* L. with gamma rays (0.5, 1.0, 1.5, 2.0, and 2.5 krad) and observed variability in respect of sprouting time, plant height, induction of flowering, the shape of petals and several filaments in a stigma. An outstanding mutant with five branched stigmas (instead of three) was isolated. Khan et al. (2011) experimented on saffron to increase corm production through gamma radiations (0.1, 0.2, 0.3, 0.4, and 0.5 Kr) and chemical mutagens viz., ethyl methane sulfonate (0.1 and 0.2%), ethidium bromide (0.1 and 0.2%), and colchicine (0.05%). They applied various combinations of treatments for mutagens at different growth stages. In the M3 generation selected plants produced the highest number of daughter corms per mother corm. This was a good approach to increase saffron corms for large-scale production. Nehvi et al. (2010) induced variability by treating corms at different growth stages with colchicine, EMS, radiation, and ethidium bromide. 0.05% colchicine enhanced pistil length, leaf number, and leaf length. 0.2 Kr radiation dose harmed survival and 0.2% ethidium bromide had a significant positive effect on survival, increased number of heavier flowers, and pistil weight. Variants showed an increased number of stomata, thicker and broader leaves. The resting bud stage has been identified as an appropriate stage for inducing variation through physical and chemical mutagens. Haspolat et al. (2016) treated corms of *C. flavus* "Golden Yellow" with gamma rays (0, 5, 10, 15, 20, 25, and 30 Gy) to create a variation for improvement of traits. The effective mutation dose (EMD) was determined 43.10 Gy. Variations in flower (changes in stigma formation and tepal differentiation, formation of a new small tepal in the middle of the flower, and changes in the number of tepals) and leaves (changes in the number of white-surfaced grooves on the upper side and sticked leaves) were recorded in M1V3. Salwee and Nehvi Firdos (2017) evaluated the effects of EMS (0.1, 0.15, and 0.20%) on floral, foliar, corm, and quality attributes in saffron. Performance studies identified 18 extraordinary lines which showed significant superiority over the base population for floral, foliar, corm, and quality attributes and thus can be harnessed for commercial exploitation for enhancement in factor productivity per unit area. Magd el Din (2019) studied the effects of gamma rays (0.0, 20, 40, 60, 80, and 100 Gray) after treating rhizomes of *Cyperus alternifolius* L. Highest sprouting was observed after treatment with 20 Gy and the lowest survival was in 100 Gy. 100 Gy showed a maximum decrease in all vegetative characters. Dwarf and small plant variants were selected. RFLP analysis showed an increase in genetic polymorphism. Mahmoud A. Shraf-Eldin et al. (2018) analyzed the extent of DNA polymorphisms between *C. sativus* genotypes [(K1 & K2 = *C. sativus* var. *cashmerianus*, C1 = *C. sativus* (nonmutant), T1 = mutant (T0-2B), T2 = mutant (T1-2B), T3 = mutant (T4-2A)]. Analysis showed that two genotypes (T1-2B and T4-2A) were closely related to each other and the strain CM-cashmerianus, while the T0 of the *C. sativus* genotype showed divergence. Moghaddam et al. (2019) studied gamma irradiation effects on the *in vitro* cormogenesis and the corm size and then acclimatization of the corms of saffron.

Three ecotypes were exposed to gamma rays (2.5, 5, 7.5, and 10 Gray) and cultured in vitro. Each ecotype showed a differential response in bud formation. Their study showed that the lower dose of Gamma ray can increase the number of sprouts and thereby the number of corm production. Samadi et al. (2022) attempted to induce new variations in *C. sativus* by treating corms with EMS (0.1% and 0.2%) and colchicine (0.05% and 0.025%). The lowest survival rate of corms was related to EMS treatments. The relative expression of *ALDH*, *BGL*, and *CCD2* genes under 0.025% colchicine treatment for 12 h revealed a two-fold increase. The flow cytometric measurements of the nuclear DNA content of the colchicine-treated plants did not reveal any significant changes in 2C DNA content. Mutagens were found to be most suitable to create new variations.

Crossandra (*Crossandra infundibuliformis*): Family Acanthaceae; known as "firecracker flower"; annual or a perennial; used as a potted flowering or landscape; propagated by seeds and cuttings.

Rooted or unrooted cuttings were treated with an acute dose of X- or gamma rays and reported approximately 20 Gy as the optimum dose. No mutation was detected (c.f. Broertjes and Van Harten 1988). Dhivya (2015) reported different aspects of the induction of mutations after mutagen treatments on *C. infundibuliformis*. Thirty days old rooted cuttings of three varieties ("Arka Ambara," "Arka Shreeya," and "Arka Shravya") were treated with different doses (0, 2, 4, 6, and 8–50 Gy) of gamma rays to create further variations. Higher dosage induced morphological and flowering parameters (Tarannum et al. 2016). One private nursery (Shri Vengadapathi Reddiar, Padma Shri awardee and a proficient personality in the field of horticulture, India) developed 100 new attractive varieties of *Crossandra* (attractive color, tolerance to diseases, higher yield with good quality, shelf life, and adaptability in different agro-climatic conditions) by gamma irradiation. The nursery is using gamma irradiation and subsequent selection as an innovative approach for the improvement of other crops (Anonymous 2017). Hewawasam et al. (2004) treated in vitro shoots of *C. infundibuliformis* var. "Danica" with different doses of gamma rays and colchicine. Shoot length was reduced in both the mutagens with increasing doses. Shoot number increased with increasing dose of colchicine. Growth was stunted at higher doses (>0.03%) of colchicine. Different types of leaf abnormalities like changes in leaf size, shape, margin, and apex were recorded. One solid mutant ("Savindi") with altered leaf shape and flower color was isolated from the 3 krad treatment (Hewawasam 2003). Colchicine-induced tetraploidy has been induced in *C. infundibuliformis* Nees and in vitro techniques have been standardized which may be very helpful in future in vitro mutagenesis work (Pandey et al. 1987; Girija et al. 1999; Almeida et al. 2008). Dhivya et al. (2015) treated seeds of *C. infundibuliformis* with EMS (20, 30, 40, 50, and 60 mM) and determined its effects on different parameters like seed germination, seedling survival, and growth. Less than 50% germination and survival were recorded in all concentrations. Higher doses were detrimental. Higher internodal length, number of leaves, length, and breadth of leaf, number of branches, and length of branch were observed in 30 nM treatment. Vinodh and Kannan (2020a) exposed seeds of *C. infundibuliformis* to EMS (20, 30, 40, 50 and 60 mM) and examined its effect on days to first flowering, days for spike to complete flowering, number of spikes per plant, number of flowers per spike, length of spike, corolla tube length, diameter of flower, number of flowers per plant, yield of flowers per plant, 100 flower weight and quality. Different doses had differential effects on growth. 30 mM showed more number of spikes and flowers, flower diameter, and yield of flowers per plant than the control. Effects of different doses of gamma rays (100, 200, 300, 400, and 500 Gy) were studied on different characters like plant height, number of leaves, internodal length, number of branches, days for first flowering, number of spikes, number of flowers per plant, 100 flower weight and flower yield after seed irradiation of *C. infundibuliformis* var. "Nilakottai Local." 200 Gy treatment showed better performance in plant height, number of leaves, length, and breadth of leaf, internodal length, number of branches, length of the branch, plant spread east–west and plant spread north-south, number of spikes per plant, number of flowers per spike, number of flowers per plant, the weight of 100 flowers and yield of flowers per plant over the control (Vinodh and Kannan 2020b).

12.35 *Cryptotneria japonica*

Family Cupressaceae; monotypic genus; known as Japanese cedar, sugi; perennial; ornamental tree; evergreen; propagation by seeds.

The survey demonstrates few references on mutation-related work on *C. japonica* (sugi). Spontaneous mutations have been detected and isolated. Radiation experiments reported 0.8 Gy/day as the optimum dose for chlorophyll mutation frequency. A lower dose rate was found better to induce waxes and morphological variations (Kukimura et al. 1975, 1976; Ohba 1971a, b; Ohba and Maeta 1973). Re-irradiation experiments on mutants indicated that more mutants can be induced. Several mutants (with fat, stout and thick stems, short needles, higher rooting ability, etc.) have been induced from waxless mutants (Maeta et al. (1982). Acute and chronic irradiation induced 17 morphological mutant clones such as dwarf, slender branch, pendulus branch, abnormal needle shape, and waxless needles. The background of mutant clones was studied based on meiotic and mitotic cell division, pollen fertility, gene mutation, etc. (Kukimura et al. 1976).

12.36 Cupressus

Family Cupressaceae; common name Cypress; evergreen tree/shrubs; ornamental tree; propagation by softwood stem cuttings.

It has the potential to use as a pot plant. One early mutation work report is available on the variety "Goldcrest." Rooted cuttings were treated with X- or gamma rays and determined 5–6 Gy as the optimum dose. The mutant was selected with improved growth habits (c.f. Broertjes and Van Harten 1988). Lev-Yadun et al. (2004) studied the variegation nature of a variegated mutant tree of *Cupressus sempervirens*. The plant was propagated both by vegetative means and by seed.

Expanding megacone scales and leaders of branches of the vegetatively propagated plants and some of the seedlings exhibit variegated yellow and green coloration instead of all green. Almost all megacones showed variegation. In the variegated cones, the tips of the umbo and the connecting zone between megacone scales always remain green. No variegation was found in the branches that carried the variegated cones, but in vegetatively propagated plants and some of the seedlings, variegation was found in fast-growing leaders and side branches. Studies indicated that the variegation seems to be the outcome of a retarded expression of greening in fast-growing organs rather than from a chimera. Melati et al. (2004) studied stress-induced cytological and chemical adaptations in *Cupressus* pollens and needles from the area of polluted and control sites of urban and suburban areas. The main aim of the study was to qualitatively examine the link between air pollution and stress-induced structural damage in *Cupressus*. The chemical composition of needles and pollen was determined as an indicator of the contamination of the trees, which were also examined anatomically. Concentrations of lead were significantly higher at the polluted sites.

Such a study will be helpful to utilize pollen and needle characters to assess the effects of mutagens.

12.37 Cyclamen

Family Primulaceae, perennial flowering plant, propagated both by seeds and vegetative (tubers) means.

Breider (1959) treated young tubers with X-rays and determined very high radiosensitivity (above 100 Gy) and radiosensitivity of X-irradiated seeds of diploid cultivar was around 90–100 Gy) (c.f. Broertjes and Van Harten 1988). Wellensiek (1960) and Wellensiek and van Bren (1973) reported very important basic information that a dominant gene prevents the vegetive mutation from white to violet flower color, while in recessive condition, this gene enables the mutation to take place. They have also informed us that the mutation may involve both sides of a petal but sometimes occurs on the upper side, sometimes on the lower side only. Sugiyama et al. (2008a) studied the effects of heavy ion beam irradiation ($^{12}C^{6+}$ ion beam at doses of 10, 20, 40, 60, and 80 Gy) on cultured tissues (callus, somatic embryo, plantlet) but no mutation could be detected. In a separate experiment tubers (8–15 mm diameter) were irradiated (8–16 Gy) and detected male-sterile mutants and mutants with changed petal color and form. Kondo et al. (2008a, b) applied both tissue culture and ion beams irradiation on fragrance cyclamen to induce diversity in flower color. Petals from immature flower buds of amphidiploid were collected and two types of the etiolated petiole, one from amphidiploid seedling culture and the other from the haploid plantlets derived from anther culture were irradiated with 220 and 320 MeV carbon ion beams at doses of 0–50 Gy and cultured. Mutants with desirable traits were obtained from 0.1, 0.2, 0.5, 1, and 2 Gy. Kondo et al. (2009a, b) irradiated etiolated petioles of the cultivar “Kaori-no-mai” with a 320-MeV carbon ion beam at 0–16 Gy to induce flower color variation through mutation. Mutants with new flower colors with original novel characteristics (flower shape, flower size, and leaf color) were detected. One of the mutants bloomed novel red purple flowers, the major anthocyanin of which was delphinidin 3,5-glucoside. Because the major anthocyanins in flowers of *Cyclamen* spp. were previously restricted to malvidin, peonidin, and cyanidin types, the generation of cyclamen containing mostly the delphinidin-type anthocyanin is an important breakthrough in cyclamen breeding. The mutant is expected to become not only a commercial cultivar itself but also a valuable genetic resource for cyclamen breeding. A series of papers have been published on interspecific breeding and irradiation of etiolated petioles with a 320-MeV carbon ion beam at 0–16 Gy to increase flower color variation by mutation. Several new flower color plants recovered from experiments (Kondo et al. 2009a, b, 2010, 2011). Hase et al. (2012) reviewed induced mutation work and mentioned a more efficient mutagenesis technique using ion beam irradiation combined with sucrose pretreatment or subsequent re-irradiation. To shorten the time required for breeding new cultivars of cyclamen, they identified anthocyanin biosynthesis genes and examined the effectiveness of PCR screening of irradiated deletion-mutant candidates at early growth stages. They suggested that this research is a step toward more efficient and controlled mutation breeding using ion beams. Kameari et al. (2011, 2012) selected ion beam derived mutants ("Tennyo-no-Mai," "Miyabi-no-Mai," and a pale yellow mutant of GBCP) and a dihaploid yellow mutant of fragrant cyclamen and examined very critically biosynthesis and the role of biochemical changes of different pigments for their evolution. Chromosome doubling and conventional breeding have developed a wide range of cyclamen varieties: fragrant ("Uruwashi-no-kaori"—UR, "Kaori-no-mai"—KM, "Kokou-no-kaori"—KO) and new flower color varieties. Different parental varieties included in these methods were Cyclamen persicum "Strauss," "Pure White"; fragrant wild species of *C. purpurascens*; *C. persicum* "Golden Boy," etc. Carbon ion beam irradiation was applied to generate novel desired mutants in amphidiploids of UR, KM, and KO; dihaploids of UR, KM, and GBCP. Immature petals of fragrant amphidiploids and dihaploids cyclamen were irradiated with a carbon ion beam (0–5 krad, 220/320 MeV) to develop novel flower color. The mutants were isolated and evaluated for breeding new cultivars of fragrant cyclamens. Compositions of flower pigments (anthocyanins and flavonol glycosides or chalcone glucoside) and fragrances (monoterpenes, sesquiterpenes, phenylpropanoids, or aliphatics) were determined. Mutation breeding using ion beam irradiation combined with plant tissue culture has resulted in fragrant cyclamens with novel flower colors and pigments (Ishizaka et al. 2012; Ishizaka 2018). Nakayama et al. (2012) made a comprehensive analysis of anthocyanin and its biosynthetically related compounds in ion beam induced flower color mutants of cyclamen (*Cyclamen* spp.) and carnation (*Dianthus caryophyllus*) and discussed the mechanisms for flower color mutation.

12.38 *Cynodon dactylon* Pers. (Bermuda Grass)

Family Poaceae, tropical grass distributed in all tropical and subtropical areas, stoloniferous and rhizomatous grass.

Lu et al. (2009) did mutation breeding work on *C. dactylon* to develop a dwarf strain to reduce management costs by applying gamma radiations. Stolon treatment resulted in the detection of three dwarf-type mutant lines (7-9, 10-5, and 10-12) with lower canopy height, shorter internodes, and shorter leaves. Comparative assessment of all the mutants with control indicated that dwarf mutants exhibited dwarf characteristics and improved drought resistance.

12.39 Cyperus

Family Cyperaceae; annual/perennial; grass-like leaves; propagation by seeds, division, cuttings.

Different species of *Cyperus* are commercially important in the floriculture trade. It is mostly used in water gardens and at the margins of pools for decoration, used as a pot plant, and sold as cut foliage in the wholesale flower markets. *C. alternifolius* and *C. papyrus* are used as pot plants. Denisova and Shurshikova (1976) applied chemical mutagen in *C. esculentus* to improve growth habits and leaf characteristics and detected a few useful mutants. Honda et al. (1980) developed a commercial mutant in *C. malaccensis* through chronic gamma irradiation. Magd el Din (2019) treated rhizomes of *Cyperus alternifolius* L. with different doses (0.0, 20, 40, 60, 80, and 100 Gy) of gamma rays and radiation effects on different characters. The highest sprouting was noticed in 20 Gy and the lowest survival was in 100 Gy. 100 Gy treatment showed a significant decrease in all vegetative characters. Different doses induced dwarf plants and small plants. RFLP analysis showed a maximum increase in genetic polymorphism due to radiation treatment.

12.40 Cytisus

Family Fabaceae; shrub; flowering plant; known as brooms; propagation by seed.

Nagatomi et al. (1993) developed dwarf plant stature in the *Cytisus* genus through in vitro culture after gamma irradiation.

12.41 Dahlia

Family Asteraceae; herbaceous perennial; tuberous; considered as perennials and also annuals; wide variety of types; propagation by seed, tubers, cuttings.

Broertjes and Ballego (1967, 1968, 1969) treated tubers of several dahlia cultivars with X-rays and reported 2–3 krad as the optimum dose based on percent rooting, time of rooting, development of the young plant, and mutation frequency. Mutations in flower color and shape were detected and isolated in pure form. The propagation method was found to be very suitable for the detection of mutations. It was difficult to predict the induction of mutation in octoploid dahlia due to their complicated genetics of flower color. A good number of mutagenesis experimental reports on dormant tubers are available mostly with gamma rays. Experiments generated basic knowledge regarding the type of experimental material, dose, treatments, etc. (Das et al. 1974, 1975, 1977, 1978; Grabowska and Mynett 1964; Lantin and Decourtye 1970; Broertjes and Ballego 1967, 1968, 1969). Few reports on the development of new varieties through spontaneous mutation, plastid mutation, in vitro culture, etc. are available (Asahira et al. 1975; Khan et al. 1978; Lawrence 1931, 1942; Singh et al. 1970; Thakur and Bhagchandani 1978). Asahira et al. (1974) treated tuberous roots of two cultivars, “Kosei” and “Sunlight1” with 1000–2000 R of X-rays and detected chlorophyll and flower color mutations in both cultivars but the frequency of mutation varied. It was difficult to isolate the mutant tissues by conventional methods and therefore in vitro method was applied. Different explants like receptacles and leaves were used. It was observed that there was a serious problem in the tissue culture of dahlia that a large number of explants were endogenously contaminated with bacteria. Tissue culture was not successful to isolate mutant tissues. Dube et al. (1980) exposed 14 varieties of dahlia to gamma rays (1–8 krad) and observed a decrease in growth with increasing radiation dose. LD_{50} was determined between 3–4 and 2–3 krad as the optimum dose. Mutation frequency varied with dose as well as variety and the maximum number of mutants were found at 2 krad dose. Xicun et al. (2007) irradiated young shoots of *Dahlia pinnata* Cav (Daxueqing) cultivar by 80 MeV/u $^{12}C^{6+}$ ions and detected flower color mutant. The mutant and wild types were confirmed by the RAPD method. Hamatani et al. (2001) developed a new flower color mutant in *D. pinnata* induced by heavy ion beams. Uyama et al. (2011) treated shoots of *Dahlia* spp. with carbon ions ($^{12}C^{6+}$, 320 MeV) in the range of doses from 1 to 5 Gy to develop horticulturally important new mutants. Treated shoots were cultured properly and planted in the field for assessment. A large number of changed flower color and shape mutations were detected. Uyama et al. (2013) induced flower color mutations in Dahlia by applying ion beam irradiation. Rooted cuttings of three dahlia cultivars (“Jyotsana,” “Agni” and “Glory of India”) were exposed to acute (^{60}Co) gamma rays (1.0, 1.5, 2.0, 2.5, and 3.0 kR) to determine the lethal dose. Maximum mortality was observed at 3.0 kR while the minimum was observed at 1.0 kR. The LD_{50} values based on the survival percentage of cuttings were determined at 2.07, 1.91, and 1.87 kR for cvs. “Jyotsana,” “Agni” and “Glory of India,” respectively (Pal et al. 2017).

A lot of work has been done on Dahlia’s genetics and cytology. Noteworthy diversity in flower form and color has developed through breeding. Flower pigment composition and genetics are a little complex due to outcrossing. It is a good material to induce variability through mutation for high levels of polyploidy and heterozygosity in flower color genes. But the selection of starting material is a little difficult due to intricate genetics. For such type of highly heterozygous material mostly the selection criteria are based on the comprehensive quality of the variety. From the results of mutation experiments and the development of mutants, it has been recommended that radiation treatment should be given immediately after harvest of tubers. Early-stage irradiation gives a better chance for mutated cells to divide more freely due to less completion of diplontic selection.

12.42 *Datura innoxia* Mill.

Family Solanaceae; it is also known by many names pricklyburr, moonflower, nacazcul, toloatzin, toloaxihuitl, tolguache or toloache; perennial or annual; tuberous-rooted subshrub/herbaceous; flowers white trumpet shape; propagation by seed, tuberous roots.

Cartledge et al. (1935) from their experiments mentioned that aged seeds showed an increased mutation rate, as indicated by pollen abortion, together with a parallel increase in the rate of visible recessive gene mutations. Mutations that occurred in the mature pollen grains may be of particular interest because the simplicity, accessibility, and durability of this material seemed to make it especially useful for studies on the nature and incidence of mutation. The considerable number of mutations found in the experiments supported that the chromosomal mutations induced by aging seeds-that the metabolic conditions within the cell rather than external agents may be the cause of such mutations. Cartledge et al. (1936) studied the effects of different environmental factors (heat and moisture content) on the mutation rate of *Datura*. Pollen-abortion mutation rates increased when seeds were stored up to 10 years in the laboratory, but had not been found in seeds buried 22 years in the soil. Moisture content and temperature did not affect mutation. *Datura* seeds with 2–15% moisture content were treated from 45 to 80 °C for 2–5 h to 5 days. Maximum seeds were treated at 5% moisture and 75 or 80 °C for 2–48 h. Prolonged treatments killed the seeds, especially when these were high in moisture content, but moderate treatments increased the seedling yield over that of the controls. Germination was found to be good when seeds were treated with 5% moisture for 24 h at 80 °C, and for 36 h at 75 °C. Planting interval, moisture content, and temperature combinations were also studied. Seeds with moderate and severe heat treatments produced high percentages of plants with abnormal growth, as reflected in their types of branching. Mutations of the pollen-abortion gene type were found in 56 plants, while pollen abortion of the type caused by chromosomal mutations occurred in 37 plants. Most of the mutations involved a sector of half or less than half of the plant. In general, the mutation rate increased with increased temperature, with increased moisture content, and with increased duration of treatment. Krumbiegel (1979) treated haploid and diploid protoplasts of *D. innoxia* Mill. and *Petunia Hybrida* L. with X-rays (1.3–2.6) and MNNG. Survival of haploid protoplasts decreased exponentially with increased doses of X-rays and increased concentrations of MNNG. Diploid protoplasts were more resistant than haploids at higher doses/concentrations. Four mutants with changed pigment patterns were isolated from MNNG-treated haploid protoplasts. Three Algerian varieties of Datura seeds namely: *Datura stramonium*, *Datura inoxia*, and Tatula were treated with 5–80 Gy gamma rays. Treated and scarified seeds were germinated in vitro on an MS medium in a controlled growth chamber. Results indicated that the *Datura inoxia* variety was more sensitive compared to the two other varieties. This variety is expected to be used in an induced mutation program for the sake of obtaining mutant lines that will exhibit increased tropane alkaloid concentrations (Benslimani et al. 2011).

12.43 *Delphinium malabaricum* (Huth) Munz.

Family Rananculaceae; perennial herb; leaves palmate shaped; flower purple and blue, to red, yellow, or white; propagation by seeds.

Chinone et al. (2008) applied ion beam irradiation to induce mutations in *Delphinium*. The survival rate of leaf blades on *Delphinium* irradiated with 320 MeV carbon ions decreased at 0.5. The suitable doses for mutation induction were estimated to be around 0.5–1.0 Gy for *Delphinium*. Firdose et al. (2011) treated dry seeds of *D. malabaricum* with different doses of gamma rays (5, 10, 15, 20, and 25 kR), 12 h soaked seeds were treated with EMS and SA (0.01, 0.05, 0.10, 0.15, 0.20, 0.25, and 0.30%) for 6 h. Mutation frequency and spectrum of mutations varied with the mutagens and concentration of mutagens. Eleven different types of chlorophyll mutants (Albina, Albina-green, xantha, aurea, chlorina, viridis, yellow viridis, tigrina, striata, maculata, and variegated type) were detected in the treated populations. EMS was more effective than SA and gamma rays in inducing chlorophyll mutations. Kolar et al. (2011) studied the phenotypic response of Delphinium malabaricum to EMS, SA, and gamma rays. Different mutation frequencies and widths of mutation spectra were induced under the action of different concentrations of the applied mutagens. Eleven different types of chlorophyll mutants (Albina, Albina-green, xantha, Aurea, chlorina, Viridis, yellow viridis, tigrina, striata, maculata, and variegated type) were detected in the treated populations, and chlorophyll mutation frequency was calculated on plant population basis. The frequency of Viridis mutants was highest followed by xantha and other types in all the treatments. EMS was more efficient than SA and gamma rays in inducing chlorophyll mutations. The highest frequency of chlorophyll mutations was reported in the 0.25% EMS. They mentioned that such induced genetic variability for frequency and spectrum of chlorophyll mutations is the first report in D. malabaricum. Kolar et al. (2013) studied the effects of EMS, SA, and gamma rays on the meiotic stages of D. malabaricum. Meiotic cell division is controlled by a large number of genes and therefore attempt was made to understand the mutations in these genes may cause abnormalities that impair plant fertility. The results demonstrated that the mutagens cause various types of cytological aberrations, such as univalents, chromatin bridges, laggards, fragments, stickiness, and multinucleated cells, and the maximum aberrations were recorded at higher doses/concentrations of the mutagens. Maximum pollen mother cell abnormalities were recorded in EMS followed by gamma rays and SA. Kolar et al. (2015) attempted to induce mutations in Delphinium malabaricum using EMS, sodium azide (SA), and gamma ray. Variants were selected in M2. A variety of chlorophyll-deficient mutants and a high percentage of the flower color and morphological mutants were recorded. The maximum frequency of chlorophyll and flower color and morphological mutations were recorded in EMS-treated plants when compared to the other two mutagens. The highest percentage of color mutants arose after treatments with 0.25% of EMS and the lowest at 20 kR of gamma rays. The mutants were quite distinct, as compared to the control, and often had more attractive ornamental features. The major commercial benefit of the application of this technology has so far been obtaining novel flower mutants that can be used as an initial material for further breeding of new cultivars. Patil (2015) attempted to induce mutations in D. malabaricum using EMS, SA, and gamma rays. Mutagens effects were observed and mutants were selected in M2. A variety of chlorophyll-deficient mutants and a high percentage of the flower color and morphological mutants were recorded. The maximum frequency of chlorophyll and flower color and morphological mutations were recorded in EMS-treated plants when compared to the other two mutagens. The frequency values for the individual mutant types were varied and randomly distributed at different mutagenic treatments. Kolar et al. (2020) studied the pigment of EMS, SA, and gamma ray induced mutant of *D. malabaricum* to understand flower color variations in the mutants. The flower color mutants which ranged in color from blue to pale pink revealed a marked difference in the delphinidin content compared with the parent cultivar.

12.44 *Dianthus caryophyllus* L.

Family Caryophyllaceae; known as **Carnation**; herbaceous perennial/annual plant; most important cut flower; flowers bright pinkish-purple, including red, white, yellow, blue and green, striped variations, etc.; fragrant; hermaphrodite; propagation by seeds, stem cuttings.

Richter and Singleton (1955) evaluated carnations grown under conditions of chronic gamma irradiation and detected somatic mutations which were propagated vegetatively. Mutations were mostly in floral characters like white to red flowers, red to variegated, and standard double to single flower types. Sagawa and Mehlquist (1956, 1959) developed flower color mutants by treating vegetative propagation with different doses of gamma rays. Gamma rays were applied on varieties "Cardinal Sim" and "Elia Rosso" for induction of mutation. A decrease in the number of flowers per branch and the number of branches per plant was observed after treatment. Four types of color changes were recorded in "Elia Rosso" and one in "Cardinal Sim." The frequency of color changes was higher in the three branches which emerged first than in their laterals; it was also higher in the first-emerged branch than in the second and third ones (Buiatti and Ragazzini 1965). Chimeric mutations were established by vegetative propagation. The control flower had pelargonidin and the purple flower mutant had cyanidol. Rooted cuttings of mutant plants were exposed to 4 and 10 kR X-rays and their effects on the growth of the shoot apex were observed (Dommergues et al. 1966). In vitro, petal cultures were subjected to gamma irradiation to induce variants. The surface origin of buds and the different steps in meristem formation were revealed by histological studies. A dose of 40 Gy administered on the fourth day of culture produced variants of horticultural interest in cultivar "Niky." This period corresponded to the dedifferentiation of cells that subsequently developed into bubs (Simard et al. 1992). In vitro nodes of *the Dianthus* cultivar "Mystère" were treated with X-rays and found the production of flower color variants. The breakdown of chimeras under diplontic selection and putative selection for fitness in the mutant homohistonts was facilitated by in vitro subculture. The resultant mutants were vigorous and most were stable although pleiotropy concerning *Alternaria* disease was observed (Cassells et al. 1993). Hemlatha (1998) treated cuttings of two carnation genotypes ("Sterlite Dop" and "H-13") with gamma rays and one genotype ("Sterlite Dop") with EMS and studied their effects on morphological and floral characters, histological and histochemical changes. Sprouting and survival of cuttings decreased with increasing doses. Interesting leaf variegations and changes in floral characters (petal variegation) were recorded at 1.5 and 2 kR gamma rays. "Sterlite Dop" showed two flower color mutations at 1.5 kR and one miniature mutant at 2 kR. Mutation induction was limited in EMS treatment, only a few plants treated with 3% EMS showed creamy yellow streaks on the leaves of cv. "Sterlite dop." Histological and histochemical studies of the shoot apex revealed the stepwise development of cellular damages and considerable variation was observed in the anthocyanin, chlorophyll, and carotenoid content among the seven carnation genotypes studied. Singh et al. (2002) developed flower color variation in carnation through in vitro mutagenesis using gamma rays, nitroso-methyl urea, and ethyl methane sulfonate. They have estimated the genetic variability due to mutation at the genetic level by electrophoretic analysis of soluble proteins. Seeds were treated with EMS (0.0125, 0.250, and 0.500%) for 3 and 6 h and studied their effects on germination, plant height, etc. in M_1 (Bhattacharya 2003). Okamura et al. (2003) did in vitro experiment by treating leaf segments of carnation with the 220 MeV carbon ions and compared the mutagenic efficiency with those of gamma rays and X-rays. Carbon ion showed the maximum effect on regeneration frequency, RBE value, higher mutation frequency, and spectrum concerning gamma rays. Three new mutant varieties were established from carbon ion treatment. Dogra et al. (2010) selected two standard carnation cultivars ("Tempo" and "Raggio-di-Sole") and induced callus in vitro. Multiplied calli were exposed to gamma rays (10–50 Gy) to investigate the induced variability. The lowest survival of calli was recorded in both the cultivars at 50 Gy. In vitro, the selection of mutated calli with culture filtrate (20%) of Fusarium oxysporum f. sp. Dianthi resulted in insensitive mutants at the end of third selection cycle. Selected calli were regenerated, rooted, and inoculated with mycelia and spore suspension of F. oxysporium. Resistant and moderately resistant mutants in cv. "Tempo" and "Raggio-di-Sole" were obtained at 50 Gy. Flower color variation was obtained at higher doses (30, 40, and 50 Gy) in cv. "Tempo," whereas, in cv. "Raggio-di-Sole," a mutant with changed flower color was obtained at 50 Gy. They concluded that selected calli irradiated with 50 Gy gamma rays and 20% culture filtrate was found best for producing mutants resistant to F. oxysporium. Okamura et al. (2012) standardized a systematic and directed method by combining the advantages of ion beam breeding and genomic information and developed two highly novel varieties in *C. caryophyllus*. Experimental details have been highlighted. Okamura et al. (2013) reported admirable technical methods to diversify pigment composition in *D. caryophyllus* based on pigment analysis and other molecular bases of pigment biosynthesis. They developed a bluish-purple line displaying a highly novel metallic appearance by crossbreeding. This variety was further treated with an ion beam and developed metallic reddish purple, metallic crimson, and metallic red lines. Sterilized petals were irradiated with 320 MeV carbon ions (LET 76 keV/μm) from an AVF cyclotron and subsequently cultured in vitro and successfully isolated new colored carnations from regenerated plants. In vitro, shootlets of carnation (IIHRS-1) were exposed to gamma rays (20, 40, 60, and 80 Gy) and observed a decrease in survival with increased dosage. Explant enlargement was maximum in the case of leaf tip explant but this enlargement did not indicate survival percentage. Leaf expansion, formation of calli, and regeneration were found to be dependent on the explant used, media composition, and their interaction. Maximum regeneration was noticed from the leaf base. Regeneration was found to be dependent on irradiation dosage. Irradiation dosage had significant interaction with media and explant on survival, leaf expansion, calli formation, and on regeneration (Paramesh and Chowdhury 2005). Seeds were treated with colchicines (0.1, 0.4, and 0.7%), EMS, and SA to study the efficiency and effectiveness and to determine the effective dose based on seed germination, lethality, pollen sterility, and mutagenic effectiveness. It was noted that the increase in the dose of EMS and SA, germination percentage, and survivability were decreased; whereas colchicine doses were proportional to increase germination percentage at the seedling stage, but they were not survived till maturity. The mutagenic effectiveness was maximum (86.42%) at 0.1% EMS and minimum (13.824%) at 0.7% Col. The highest mutagenic efficiency (6.977) was recorded at 0.4% Col and lowest (0.995) in 0.7% SA based on survivability. The effectiveness of the three chemicals on Dianthus is ranked as EMS > Col > SA (Roychowdhury and Tah 2011a, b). Roychowdhury et al. (2012) studied the effects of different concentrations (0.1%, 0.4%, and 0.7%) of EMS, SA, and colchicines on seed germination, survival, and pollen sterility in M1 and M2. SA and EMS were more effective as compared to colchicines. Heavy ion beam irradiation developed mutations in flower color in *D. chinensis* var. "Semperflorens" (Sugiyama et al. 2008b). Tanaka et al. (2010) developed flower color and flower form mutants in carnation by applying ion beams which are hardly produced by gamma rays or X-rays. Mutants were UV-B resistant, had serrated petals and sepals, and anthocyaninless, etc. PCR and sequencing analyses showed that half of all mutants induced by ion beams possessed large DNA alterations, while the rest had point-like mutations. Both mutations induced by ion beams had a common feature that deletion of several bases was predominantly induced. It is plausible that ion beams induce a limited amount of large and irreparable DNA damage, resulting in the production of a null mutation that shows a new mutant phenotype. Morimoto et al. (2019) made a thorough study of how the flower color variations occurred in bud mutation carnation cultivars of the "MINAMI series," which has a wide range of petal colors. Bud-mutation carnation cultivars of the "MINAMI series" have a diversity of flower colors in which the directions of bud sports were recorded. "Poly Minami," which is the origin of the "MINAMI series," produced the eight cultivars with various petal colors through continuous bud mutations. They determined the major flavonoid pigments in the petals of each cultivar of the "MINAMI series." They also investigated the flavonoid amounts and the expression levels of the major flavonoid biosynthesis-related genes isolated from cultivars of the "MINAMI series" during flower bud development. Based on these results, they identified candidate genes which thought to be involved in the cause of each bud mutation in the "MINAMI series." The study revealed that the flower color variations in the "MINAMI series" are caused by genetic and metabolic changes associated with flavonoid biosynthesis and identified five candidate genes for flower color changes in the "MINAMI series." Patil et al. (2019) treated seedlings of Dianthus cv. "Pink Donna" with EMS (0.25, 0.50, and 0.75%) and MMS (0.1, 0.3, and 0.5%) to induce variability. Differential responses were observed in different vegetative and floral characters after treatment with both the mutagens. Carnation is a noteworthy ornamental where a good number of mutation works started and generated basic knowledge and interest for mutation work for inducing genetic variability. Early publications reported effects of different dose levels, mutation frequency and spectrum, radiation effects on shoot apex, chimeric nature of mutations, chlorophyll deficiency, developmental morphology and its ontogeny using plastogene mutants, etc. (Richter and Singleton 1955; Heslot 1964; Mehlquist et al. 1954; Sagawa 1957; Sagawa and Mehlquist 1956, 1957, 1959; Mehlquist and Sagawa 1959, 1964; Johnson 1980; Samata et al. 1979; Bugnon et al. 1965; Dulieu 1968, 1969; Gaufillier 1965; Farestveit and Klougart 1966; Farestveit 1969). Experiments on potted plants using electrons and semi-acute gamma irradiation reported mutation probability and mutation in main shoots and lateral shoots, etc. (Badr and Etman 1977; Buiatti and Ragazzini 1965; Buiatti et al. 1965; D'Amato et al. 1964; Pereau-Leroy 1969, 1970, 1974a, b, 1975; c.f. Broertjes and Van Harten 1988). Interesting results on sports analysis and radiation-induced chimeras and development of in vivo and in vitro adventitious buds and their significance in mutation were published (Dommergues and Gillot 1965, 1973; Broertjes 1982; Custers 1978; Silvy 1979). Twenty-five to 50 Gy X-rays were determined to induce variability in yield and higher doses for mutations in flower color after treating unrooted cuttings. Data were also collected on experimental design, stage of selection of mutants, target material/mutants for irradiation to recover more mutations, etc. (Sparnaaij 1974, 1978; Sparnaaij and Demmink 1970, 1971; Sparnaaij et al. 1974a, b; Carrier 1983; Silvy 1979; Silvy and Mitteau 1986; Custers et al. 1977). Despite appreciable experiments, the number of new mutant varieties in carnation is minimal. For radiation experiments mostly rooted cuttings, unrooted cuttings, and leaf axils were used as explants, and X-rays and gamma rays as radiation sources. Hentrich and Glawe (1982) developed mutants by treating axillary buds with EMS.

12.45 *Digitalis obscura*

Family Plantagince; it is also known as willow-leaved foxglove or dusty foxglove or Spanish rusty foxglove; woody perennial; propagated by seeds or cuttings.

Gavidia and Perez-Bermudez (1999) treated shoot tips of *D. obscura* genotype T4 with gamma frays (20-100 Gy) and cultured them in vitro on several selective amitrole-containing media. LD_{50} was determined 60 Gy. Variability was detected in chromosome number and cardenolide production.

12.46 *Echeveria*

Family Crassulaceae; evergreen/deciduous succulent; polycarpic; rosette shape leaves; propagation by seeds, leaf cuttings, stem cuttings, offsets.

For induction of colchicine-induced mutations, Cabahug et al. (2021) evaluated the phenotype and ploidy level of Echeveria "Peerless" after colchicine treatment. Leaf cuttings of Echeveria "Peerless" were treated with different concentrations of colchicine (0.2, 0.4, 0.6, 0.8, and 1.0%) for different durations (3, 6, 9, and 12 h). The survival and mutation rates following various treatments were determined, and the phenotypic characteristics were evaluated 12 months after the treatment. Cabhug et al. (2022) first determined the LD_{50} dose of colchicine based on the survival of *Echeveria* "Peerless." Mutants detected in MV1 were carried to MV2 generation to determine the retention of mutant characters. Mutants in MV2 exhibited shortened leaves, increased leaf width and thickness, and fewer leaves, which significantly differed from the control, indicating compactness, wider leaf apex, and varying leaf color. Stomatal characters, chromosome number, and flow cytometry analysis were conducted to confirm the induction of polyploidization. They have suggested that colchicine is suitable for developing new cultivars with novel phenotypic and cytogenetic characters for ornamental succulents.

12.47 Endymion

Family Liliaceae; synonym *Hyacinthoides hispanica* (Mill.) Rothm.; bulbous perennial; bell-shaped blue flowers; propagated by division.

Broertjes and Alkema (1970) and Alkema (1974a, b) treated matured healthy leaves of *E. hispenicus* cv. "Excelsior" and subsequently many cultivars with X-rays and colchicine to study adventitious bud formation and to induce mutations. The optimum working dose for X-ray and colchicine was reported 4 Gy and 1% for 7 h, respectively. No mutation and polyploidy could be detected.

12.48 Eryngium

Family Apiaceae; known as sea holly; more than 250 species; flowering plant; cut flower crop; annual/perennial herbs; propagation by seeds, root cuttings, divisions.

No published radiation-induced mutation work report is available. A preliminary experiment was conducted (Association Euratom–ITAL and private plant breeders, c.f. Broertjes and Van Harten 1988) by treating roots of *E. planum* and *E. zebellia* with X- or gamma rays. The optimum working dose was reported 10–15 Gy. Several plants were isolated from the colchicine-treated (root cuttings) population of *E. planum* "Blauer Zwerg" where leaves, inflorescences, and pollen grains were bigger (Pohlheim 1983).

12.49 *Etlingera elatior*

Family Zingiberaceae, herbaceous perennial; known as torch ginger, ginger flower, red ginger lily, torch lily, wild ginger, etc.; flowers yellow/pink; propagation by seeds.

Yunus et al. (2013) developed a protocol for shoot regeneration and applied a mutation technique to induce mutation in *Etlingera elatior* using 10, 20, 40, 60, 80, 100, 120, and 140 Gy gamma rays and used RAPD for early screening of mutants. They detected LD_{50} to be 10 Gy based on survival.

12.50 *Euphorbia pulcherrima*

Family Euphorbiaceae; known as poinsettia; shrub or small tree; foliage red and green; mostly propagated by stem cuttings, seed.

Love (1966, 1972) exposed rooted cuttings to fast neutron irradiation (600 rads) for induction of mutations. Mutations were detected in bract color, bract size, earliness of bloom, size and color of leaves, branching characteristics, and plant height (dwarfism). A wide range of diversity in bract color (dark red to clear white) developed, whereas the original bract color was pale red. Adventitious buds showed higher mutation frequency and the mutation sector was larger. The size of mutant sectors was larger in adventitious buds than those found in terminal or lateral buds. For induction of mutation suspension culture of *Poinsettia* cv. "Preduza" were exposed to X-rays and cultured in vitro under stress temperature (12 °C). Among a huge number of regenerated plants finally, three clones proved to be slightly better adapted to low greenhouse temperatures compared to the original cultivar (Preil and Engelhardt 1982; Preil et al. 1983). Chimeric plants were studied and through root-cutting experiments on *E. pulcherrima* it has been shown that the pink form is a chimeral sport of red and white is a mutation from red. The stem and bracts showed distinct color demarcations. The flower had 24 bracts, of which 13 were pink, 10 were red, and one divided with red on one side and pink on the other side of the midrib (Robinson and Darrow 1929; Robinson 1931). The variegated pinsettias,

Euphorbia pulcherrima Willd. ex Klotzsch. was detected in commercial cultivars which were being propagated asexually by stem cuttings and in the seedlings produced in a breeding program. Analysis showed that individual cells of variegated mutants showed the presence of both mutant and wild-type plastids. Accurate evidence indicated series of independently occurring chlorophyll mutants in *Euphorbia* were due to changes in genetic material carried in the plastids themselves (palstogenes). The mutant plastogene has been mentioned as a powerful tool for the study of developmental morphology and the ontogeny of plant tissues (Stewart 1965; Stewart and Arisumi 1966). Kleffel et al. (1986) induced a high percentage of mutants with white bract through acute and fractionated X-irradiation of embryogenic suspensions. Irradiation with 30, 40, and 60 Gy produced 4.2%, 7.7%, and 8.9% of white mutants, respectively. Koo and Cuevas-Ruiz (1974) treated rooted cuttings of *Euphorbia splendens* "Bojeri" with 20 Gy gamma rays and produced mutants with changed leaf form or colors after manifold prunings. Canul-Ku et al. (2012) treated seeds of *E. pulcherrima* Willd. Ex Klotzsch with gamma rays (50–275 Gy) to study the effects on different morphological characteristics. Significant differences were observed in germination, hypocotyl height, plant height, cyathium diameter; length and width of bract, petiole length of bract, inflorescence peduncle length; length, width, and weight of 100 seeds. The plant height was reduced to 150 Gy. Cyathium diameter, peduncle length of inflorescence, length, and width of seed was higher with 250 Gy. The mutation technique is being used routine way to develop interesting phenotypes in poinsettia. A wide range of colors covering all pink and white varieties have been developed using gamma and X-ray mutagenesis. Vilperte et al. (2011) targeted analysis of anthocyanin biosynthesis to characterize the potential mutagenesis target gene as the main responsible for the "white paradox" in poinsettia. Cultivation of *E. pulcherrima* at an optimum temperature of 20 °C in the greenhouse is expensive due to energy consumption. Walther (1981) tried to develop low-temperature tolerant ideotypes through mutation by treating suspension culture (cell groups or small calli parts) with X-rays and growing under temperature-stress conditions. Results showed an increased number of dividing cells as measured by a large calli size and an increased number of developing plants compared with the non-irradiated control. Mutation work on *Euphorbia* is more related to pigment biosynthesis to understand the biochemical changes for the development of new color combinations. Normally cyanidine derivatives develop intense red coloration in *Euphorbia*. Orange-red bract color is less common which is caused by pelargonidin-based anthocyanins by different mechanisms. Nitarska et al. (2018) attempted to select enzymes and genes of the anthocyanin pathway in four cultivars displaying four different red hues. They mentioned that F3′H is a critical step in the establishment of the orange-red poinsettia color. Although poinsettia DFR shows a low substrate specificity for dihydrokaempferol, a sufficient precursor for pelargonidin formation is available *in planta,* in the absence of F3′H activity. Pan et al. (2019) treated colchicine to existing novel cultivars to develop fertile varieties for interspecific hybridization. Colchicine concentration varied with the cultivar. Results indicated that a one-time 1 day was the optimal duration of colchicine treatment which gave higher rates of both polyploidy and morphological mutant production. Poinsettia cultivars "Dulce Rosa" (5 mg/g colchicine with lanolin; 10 mg/mL colchicine with cotton) and "Princettia-Hot Pink" (15 mg/g colchicine with lanolin; 10 mg/mL colchicine with cotton) yielded relatively high polyploidy production efficiency and morphological mutation rate. A total of three polyploidy mutants of "Dulce Rosa" and 19 polyploidy mutants of "Princettia-Hot Pink" were obtained. Both cultivars had mutants with recovered fertility, with a pollen germination rate of up to 27.5%. Moreover, unexpected non-polyploidy mutants with various morphological trait variations were also obtained.

12.51 *Eustoma grandiflorum*

Family Gentianaceae; known as Lisianthus; moderately cold-tolerant annual or biennial plant; propagation by seeds or cutting.

Nagatomi et al. (1996) developed three mutants in *E. grandiflorum* through in vitro chronic irradiation. Ohki et al. (2013) irradiated seeds of *E. grandiflorum* with a 50-MeV $4He^{2+}$ beam 3 and 10 days after imbibition. Seeds irradiated 3 days after imbibition were not affected by the irradiation. Less than 5% of the seeds irradiated 10 days after imbibition survived after 4 weeks in culture and found that these seeds already had elongated radicles. DNA polymorphism was detected in the plants from ion beam-irradiated seeds. Abou-Dahab et al. (2017a) studied the effects of gamma irradiation (0, 2, 5, 10, 20, 40, 60, 80, 100, and 120 KR) on various morphological, flowering, and anatomical characters. A wide range of leaf and flower color and form changes were observed after radiation with different doses of gamma rays. Low doses (10 KR) showed good results on shooting and rooting. Abou Dahab et al. (2017b) studied the effects of different cytokines (BA, 2ip, and Kim at 0.4 mg/L) on in vitro shootlet proliferation of *Eustoma grandiflorum* and added different concentrations of colchicine (30, 60, 120 and 240 mg/L) and sodium azide (5, 10, 15 and 20 mg/L) to free-hormone MS medium. In vitro, shootlets were cultured for 7, 14, and 28 days. Most of the morphological parameters and anthocyanin pigment contents in the flower declined by most treatments of sodium azide at the acclimatization stage. All colchicine treatments reached the morphological stage and formed bud initiation and death before flowering. The highest survival of acclimatized plants and highest values of the number of branches, branches length, leaf area, most floral parameters, photosynthetic pigments, carotenoids, and anatomical structure were obtained from 5, 10 mg/L sodium azide for 28, 7 days, and 60, 30 mg colchicine for 28 days. Abou-Dahab et al. (2019) selected *E. grandiflorum* and studied the effects of laser irradiation on in vitro growth, anatomy, flowering, chemical composition, and gene mutagenesis. Laser treatment showed an increase in most of the morphological, floral parameters, total chlorophyll, carotenoids, and anthocyanin pigment contents in the flower. Twenty minutes of green laser-treated plantlets showed the highest survival percentage of acclimatized plants and the highest values of the number of branches and branches length. The highest floral parameters, anthocyanin pigment contents in flower, and anatomical structural parameters were recorded increasing using 20 min of blue laser and 20, 25 min of green and red laser, respectively. Twenty minutes of green laser showed the lowest values of photosynthetic pigments and carotenoids. Results concluded that laser irradiation has a remarkable effect on plant morphology, flowering, chemical constituents, and gene mutagenesis. Mendoza-Gómez et al. (2020) treated in vitro four genotypes of lisianthus (C40, C43, C74, C83) with four concentrations of EMS (0.25, 0.5, 0.75, and 1 M) to induce genetic variability. They evaluated two times of application (1 and 2 h). Results showed that EMS becomes toxic at 1 h and the survival percentage of plants decreased at 1 M. Likewise the survival percentage decrease in some plants was inversely proportional to the increase in the concentration of EMS. The median lethal concentration was determined as 0.5 M.

12.52 Ferns

Adiantum capillus-veneris L: Family Pteridaceae; common name maidenhair fern, true maidenhair, southern maidenhair, venus hair fern, ladies hair, etc.; perennial deciduous; popular garden fern and houseplant; propagation by division.

Adiantum scutum: Family Pteridaceae; perennials; evergreen; houseplants/ landscapes; propagation by spores, rhizomes.

Adiantum lunulatum Burm: Family Polypodiaceae/Pteridaceae; evergreen perennial; propagation by spores, divisions.

Asplenium nidus: Family Aspleniaceae; synonym of *Oceaniopteris gibba;* epiphytic species; well-known as Bird's-nest fern; dwarf tree fern; evergreen perennial; a medium to large fern with erect, stout, unbranched rhizomes; houseplant; propagation via spores.

Blechnum gibum: family Blechnaceae; beautiful symmetry and unique palm-like appearance; propagation through spores.

Boston fern (*Nephrolepis exaltata*): Family Lomaripsidaceae; also known as sword fern; evergreen; ornamental houseplant; propagation by spores.

Hemionitis arifolia: Family Pteridaceae; known as Heart Leaf Fern; houseplant; propagation by spores, divisions.

Matteuceia struthiopteris: Family Onocleaceae; known as ostrich fern, fiddlehead fern, or shuttlecock fern; deciduous green; popular ornamental plant; propagation by spores, divisions.

Osmunda regalis: Family Osmundaceae; common name royal fern; deciduous; nonflowering but sometimes known as flowering fern; propagation by spores, division.

Platycerium alcicorne: Family Polypodiaceae; known as Common Staghorn Fern, Elkhom Fern, Platycerium Alcicome, Staghom Fern; rhizomatous; fronds are both fertile and infertile; houseplant; perennial; propagation by divisions.

***Pteris cretica* albo lineata**: Family Pteridaceae; known as Silver Ribbon Fern; evergreen; grown in gardens, potted plants, houseplants; propagation by spores, divisions.

Pteris vittata L: Family Pteridaceae; known as Chinese brake, Chinese ladder brake, ladder brake; herbaceous; ornamental foliage; propagation by spores.

Mutation work on ferns was confined at the preliminary experimental stage on Adiantum, Asplenium, Platycerium, and Pteris. To determine the radiosensitivity, spores were exposed to different doses of X-rays and the optimum working dose was noted as 200 Gy for *Adiantum scutum*, 100 Gy for *Asplenium nidus*, over 500 Gy for *Platycerium alcicorne*, and 300 Gy for *Pteris cretica* albo lineata (c.f. Broertjes and Van Harten 1988). Physical mutagens (X-rays and gamma rays) were applied to induce variability in phenotypes and aneuploids in *Osmunda regalis*, *Hemionitis arifolia* (Burm.), *Adiantum lunulatum* Burm., *Pteris vittata* L, *Adiantum capillus-veneris* L. (Partanen 1958; Nelson 1961; Carlson 1969; Haigh and Howard 1973; Howard and Haigh 1968; Khare and Kaur 1980; Kaur and Khare 1982; Palta and Mehra 1973; Khare 1994). The Housatonic River in Massachusetts is heavily contaminated with polychlorinated biphenyls. Somatic mutation frequency of ostrich fern (*Matteuceia struthiopteris*) growing heavily near the river bank, was determined (Klekowski 1976, 1984, 2011; Klekowski Jr. and Klekowski 1982; Klekowski and Kazarinova-Fukshansky 1984). Mohamad et al. (2002) irradiated plantlets of *B. gibum* with different doses of gamma rays (0, 10, 20, 40, 60, 80, 100, and 150 Gy) and assessed radiosensitivity based on survival and frond production after in vitro culture. Regenerated plantlets from treated plantlets showed stunted and slow recovery growth, especially for higher doses as compared to control plantlets. Norazlina et al. (2003) irradiated spore cultures of *Asplenium nidus* with different doses of gamma rays (0, 10, 20, 30, 90, and 180 Gy) and detected 25 Gy as the most suitable dose for spore culture induction. Pei et al. (2018) treated green globular bodies of Boston fern (*Nephrolepis exaltata* "Bostoniensis Murano") with different doses of gamma rays and determined 128 Gy as the LD_{50} dose. Multiplication and differentiation of Boston fern GGB were depressed with increasing doses. Morphological changes included darkening color, increased density, and the depression of multiplication and differentiation, and cellular damage included reduction of embryonic cells and initiation of cellular micronucleus. At higher doses the height of Boston fern plants differentiated from GGB became dwarf, and the leaf mutation rate increased. Moreover, leaf mutant *Negl* was obtained at 150 Gy.

12.53 *Festuca pratensis*

Family Poaceae; known as Meadow Fescue; perennial bunchgrass; used as an ornamental grass in the garden; propagation by seeds, tillers.

Land et al. (1971) selected a chlorotic homozygous recessive mutant plant of Festuca pratenns and analysis showed that the chlorophyll content of the leaf tissue was affected by variation in the light and temperature regimes to which the plants were exposed. Chlorophyll content was depressed when the plant grew rapidly, but was similar to that of normal green plants when grown slowly under low-light and low-temperature regimes. A reduction in dry weight, percentage of dry matter, and plastid number was recorded in chlorotic plants. The ratio of chlorophyll a: chlorophyll b increased compared to normal plants. The chloroplasts of the rapidly grown chlorotic plants showed a reduced lamellar structure. Thomas et al. (1985) selected a non-yellowing mutant genotype (Bf 993) of *Festuca pratensis* Huds. and a normal cultivar (cv. Rossa) and studied the linolenic acid-dependent oxidative chlorophyll bleaching (CHLOX) by thylakoid membranes from senescing leaf tissue of both materials. The half-life of CHLOX from the mutant was three times greater than that of the normal genotype. It was concluded that the non-yellowing mutation was not expressed through a lack of CHLOX activity. Thomas and Stoddart (1975) made a comparative study on chlorophyll levels of normal (Y) Festuca pratensis L. and a mutant genotype (NY). Chlorophyll levels in 1-cm sections of the youngest fully expanded leaves of normal (Y) Festuca pratensis L. declined almost to zero over 6 days after excision. Chlorophyll in a mutant genotype (NY) remained near the initial level for the whole of this period. Abscisic acid promoted pigment loss in Y but had no significant effect on chlorophyll in NY. Kinetin retarded pigment loss in Y but was ineffective in NY. Abscisic acid accelerated protein breakdown, whereas kinetin inhibited the loss of protein in both genotypes. The mutation thus appeared to be expressed as a highly specific lesion in pigment metabolism. They concluded that pigment breakdown may not be an inevitable part of the aging process. One cytoplasmic temperature-dependent chlorophyll-deficient mutant was isolated from the local population of meadow fescue (Festuca pratensis Huds.). The frequency of mutants varied considerably. Plants highest in vegetative mass production had more mutants in their progeny than low-productive ones. The mutant expressed the character maximum at 30–35 °C. A temperature of 20–25 °C provided a complete or partial reversal of the character within 1 week. An attempt was made to increase the frequency of temperature-dependent chlorophyll-deficient phenotypes using X-irradiation. The number of mutants per M_2 progeny decreased as the irradiation dose increased from 1 to 5 kR. These studies will be very informative for mutagenesis work. Olimpienko and Titov (1979) developed chlorophyll deficient mutants in Festuca pratensis which were cytoplasmic temperature dependent.

12.54 Ficus

Family Moraceae; wood trees/shrubs/vines/epiphytes; propagation by seeds, cuttings.

De Loose (1981) treated rooted cuttings of *Ficus benjamina* and *F. diversifolia* with different doses of X- or gamma-rays and detected several variegated mutants in *F. benjamina*. Two variegated mutants have been commercialized in the name "Golden Princess" and "Golden King."

Takahashi et al. (2012) attempted to develop desirable mutants with a high ability to mitigate pollutants (atmospheric nitrogen dioxide) through induced mutations. Shoot explants of *Ficus pumila* L. were treated with an ion beam ($^{12}C^{5+}$, $^{12}C^{6+}$, or $^{4}He^{2+}$) and selected variants. The variants were analyzed for NO_2 uptake by fumigation with 1 ppm ^{15}N-labeled NO_2 for 8 h in light, followed by mass spectrometric analysis. Based on the analysis a total of 15 lines were selected and finally, two of the 15 lines showed a mean NO_2 uptake 1.7- to 1.8-fold greater than that of the wild type. RAPD analysis demonstrated DNA variation between the progeny plants and the wild type, suggesting that the progeny were true mutants. These mutants of *F. pumila* may prove useful in mitigating atmospheric NO_2.

12.55 Forsythia

Family Oleaceae; hardy flowering deciduous shrub; tolerant to drought and pests; propagation by seed, cuttings;

Dai and Magnusson (2012) exposed shoot tips of *Forsythia* × "Meadowlark" to gamma rays (0–100 Gy) to induce desirable genetic variations. Treated shoots were grown in vitro. An average of 80.3% of shoots were rooted/recovered from the 25 Gy gamma ray treatment followed by 36.5, 5.4, and 2.1% from the 37, 50, and 70 Gy irradiation treatments, respectively. No shoots survived when exposed to 100 Gy gamma rays. Treated plants showed a certain degree of reduced growth in the greenhouse. Variations including characteristics of leaves (shape, size, hairs), stems (shape, internode length, branching), and plant stature were observed. They suggested that in vitro mutation induction using gamma ray irradiation could be a useful protocol to develop new cultivars or genetic materials for further breeding of Forsythia or other related species.

12.56 Freesia

Family Iridaceae; herbaceous perennial flowering plant; propagation by seeds and vegetative means (callus and adventitious buds).

Mutation work is limited. The early report tells that thousands of tubers were treated with X-rays, but no mutation was detected. The optimum dose was reported to be 20–50 Gy (c.f. Broertjes and Van Harten 1988; Qin et al. 1988). Ling et al. (2019) exposed bulbs of two varieties (red and purple) of *Freesia refracto* to gamma rays (25, 50, 75, and 100 Gy) to induce variability. Radiation effects were studied on different parameters like germination rate, semilethal dose, survival rate, plant height, leaf number, leaf area, root length, flowering rate, pollen viability, and pollen ornamentation, etc., and were found to decrease with increasing doses. The optimum dose of 54.04 Gy was recorded for the red variety and 58.82 Gy for the purple variety. Variations in pollen characters are good indications for large-scale mutation work on freesia. Electron beam to target turning X-ray (EBTTX) is an emerging irradiation technology that can potentially accelerate the breeding process of plants. Li et al. (2021a, b) studied the effects of EBTTX irradiation on the two freesia cultivars (the red freesia and the purple freesia) in a mutation breeding program. The germination rate, survival rate, plant height, leaf number and area, root number, and length of the two freesia cultivars decreased following different irradiation doses (25, 50, 75, and 100-Gy). Plant growth and survival rate were inhibited following the

100-Gy irradiation treatment in both varieties. The LD_{50} dose based on survival rates was 54.28-Gy for the red freesia and 60.11-Gy for the purple freesia. Irradiation significantly decreased the flowering rate, flower number, and pollen vigor. The flowering rate, flower number, and pollen viability of the two varieties reached the minimum at 75 Gy, exhibiting strong inhibitory effects. 75-Gy irradiation significantly decreased the chlorophyll content and increased the malondialdehyde (MDA) content of the two freesia cultivars. SEM analysis showed changes in the micromorphology of the leaf epidermis and pollen with the increasing irradiation doses. Results provide useful information for the mutation breeding of different freesia cultivars and other flowering plants.

12.57 Fuchsia

Family Onagraceae; small bush; showy drooping flowers; propagation by seeds, by shoot cuttings.

Mutation work on this plant is meager. One small-scale experiment was conducted with the cultivar "Red Ribbons" with gamma rays. The optimum dose was reported about 25 Gy. One purple color mutant was detected. Attempts were made to develop new forms through somaclonal variation, mutagenic treatments, and in vitro selection. Explants (stems, leaves, and flower parts) and calli were treated with *N*-methyl-*N′*-nitro-*N*-nitrosoguanidine, ethyl methane sulfonate, NaN_3, and N_2H_4 to increase the mutation frequency. Several variants (white flowers, new floral type, and cold tolerance) were isolated from callus-derived regenerated plants from the leaf segment. Few variants had increased ploidy levels (c.f. Broertjes and Van Harten 1988; Bouharmont and Dabin 1986a, b; Dabin and Bouharmont 1984).

12.58 Gentiana

Family Gentinaceae; annual or perennial; flowers deep blue but can be white, cream, yellow, or red; propagation by seeds.

Japanese cultivated gentian plants have naturally blue flowers, but some white-flowered cultivars were bred through spontaneous mutations. Nakatsuka et al. (2005) made a critical biochemical and molecular analysis to determine the molecular basis of white coloration in gentian flowers by comparing two white-flowered cultivars "Homoi" and "Polano White" to a blue-flowered cultivar "Maciry." HPLC, northern blot analysis, and Southern blot analysis showed role of anthocyanins, anthocyanidin synthase, chalcone synthase, flavone 3-hydroxylase, flavonoid 3′,5′--hydroxylase (F3′,5′H), dihydroflavonol 4-reductase (DFR), UDP-glucose: flavonoid 3-glucosyltransferase (3GT) and anthocyanin 5-aromatic acyltransferase (5AT) for development of white color. Analysis of stress-induced flower pigmentation suggested that rather than mutations in multiple structural genes being the cause, a defect in one or more regulatory factors controlling the later steps of flavonoid biosynthesis is responsible for white coloration in cv. "Polano White." Tsuji et al.

(2007) irradiated shoots of *Gentiana trifida* var. japonica with carbon ion beams using the JAEA AVF cyclotron to induce mutations. Gentiana was found to be susceptible to ion beam irradiation. The suitable doses for induction of mutations were estimated to be 0.5–1 Gy for Gentiana.

12.59 *Gerbera jamesonii* Hook.

Family Asteraceae; herbaceous perennial; flower orange, pink, red, white, yellow; propagation by seeds, in vitro.

Shoots of in vitro regenerated seeds of gerbera were treated with gamma rays and colchicine to induce variability. Flowers survived at acute doses of 1–4 krad (gamma rays) but growth was decreased as compared to the control. At dose 1 krad the survival of flowers was 100% and the treatments at doses 2, 3, 4, and 5–10 krad caused 25, 46, 50, and 100% death, respectively. 1 krad of gamma ray treatment for 1–3 times developed abnormalities like stunted shoots, divided leaves at the apex, rough and thick leaves, or narrow and thin leaves. A few flowers developed red strips on the yellow ray florets. The size of flowers was also increased but not different in chromosome number. Shoots were soaked in colchicine solution for 24 h at concentrations of 0.025 and 0.05%. All flowers died at 0.05% treatment. Some flowers survived in 0.025% treatment. From cytological studies, it was found that one plant was tetraploid ($4x = 100$), two flowers were mixoploid and the rests were diploid ($2x = 50$) Visessuwan (1985). In vitro raised plantlets of cv. "Rebecca" was treated with 20 Gy gamma rays (9.8 Gy/h and micro-propagated for two cycles. Variants with changed flower morphology (number, length, and width of ligulae) and color were detected in the form of both solid and chimeric mutants. They have recommended that more than two micropropagation cycles are necessary to increase the percentage of solid mutants (Laneri et al. 1990). Walther and Sauer (1990) investigated the effect of split versus acute X-ray doses on *in vitro*-derived micro shoots. They observed that splitting of doses between 10 Gy and 50 Gy into 2 or 3 fractions separated by periods of fourth each for recovery led to considerably rising shoot production and recommended that the application of very high X-ray doses only are usable using multiple dose fraction procedures. In vitro raised capitulum explants were exposed to EMS (0.1, 0.2, 0.5, 0.8, or 1.0% (v/v) and gamma rays (1.5, 2.0, 2.5, 5.0, or 10.0 Gy) and studied physiological effects. Activities of SOD, APX, GR, CAT, and PPO, and phenolic compounds and enzyme activities increased significantly and chlorophyll concentration reduced compared to untreated control shoots (Ghani et al. 2014). Rosna et al. (2012) studied the effects of gamma rays on in vitro growth of explants, callus, and the formation of shoots and plantlets of G. jamesonii. A gradual decline was observed in the number of shoots regenerated from irradiated petiole explants compared to the control. The number of shoots regenerated from 20 Gy irradiated petiole explant cultured on MS medium was reduced. A similar observation was reported on the effects of gamma irradiation on in vitro propagated plantlets. The gradual decline was observed based on plant height as the dose of gamma irradiation increased. A significant decline was observed in the fresh weight of the irradiated callus compared to the control. The growth responses of the callus were strongly influenced by the radiation dose. The fresh weight of the callus was reduced when callus tissues were exposed to 20 Gy. Ghani et al. (2013) treated in vitro raised shoot culture from petiole explants of gerbera cultivar "Harley" with gamma rays (1.5, 2.0, 2.5, 5.0, 10.0, 15.0, 20.0, or 30.0 Gy) and EMS (0.1, 0.2, 0.5, 0.8, or 1.0% (v/v) to generate variations in flower color. The LD_{50} values calculated for shoot survival and the induction of mutations were approx. 6.5 Gy for-rays and 0.65% (v/v) EMS for 10 min, or ≤0.1% (v/v) EMS for 20 min. Studies revealed a negative correlation between mutagen dose and plant survival, both in vitro and after acclimatization. Morphological variants showing changes in leaf shape, leaf size, scape length, flower diameter, and flower color were obtained. Early flowering was induced in all mutated plants compared to non-mutated plants. In a separate experiment, Ghani and Sharma (2019) observed the same physiological changes and detected three plants (3.33%) irradiated with 5 Gy to be tolerant to powdery mildew as these plants showed slight and delayed development of fungal colonies on the leaves. The random amplified polymorphic DNA characterization showed that the irradiated plants had DNA patterns that were different from the control and mother plants. A series of experiments were conducted with *G. jamesonii* to explore different aspects related to the role of mutation to induce variability, radiosensitivity, determination of optimum dose of radiation (X-rays), shoot production, somatic mutagenesis, chromosomal alterations, in vitro mutagenesis, the role of acute and fractionated doses, etc. Experiments generated meaningful data for induced mutagenesis in general and *Gerbera* in particular (Walther and Sauer 1986a, b, 1989a, b, 1990, 1991, 1992). Jerzy and Zalewska (1992) treated leaf explants of 16 cultivars of *G. jamesonii* with gamma rays (5–25 Gy) and cultured them in vitro and studied the development of adventitious shoots. The intensity and frequency of shoot formation were dependent on doses and cultivars. Regeneration ability was drastically restricted at 20 and 25 Gy, but there was no reduction in the development of adventitious shoots even at 25 Gy. Singh et al. (2011) treated seeds of white gerbera with different doses of gamma rays (1–5 kR) to induce variability in different characters. Induced floral fasciations were observed in M1. Ring-fasciation and linear fasciation were observed in mutant types which lead to deformed and asymmetric flower heads which were mostly male and female sterile. An increase in the number and arrangement of vascular bundles in the stalks of fasciated capitula suggests alterations in the shoot apical meristem of the mutated plants. Fasciated gerberas can be propagated vegetatively to perpetuate their unusual forms and designated as cultivars of the species. Hazbullah et al. (2012) treated petiole explants of *Garbera jamesomnii* with gamma rays (0–60 Gy), cultured them in vitro, and studied the effects on in vitro growth of explants, callus, and the formation of shoots and plantlets. The number of shoot regeneration reduced to 20 Gy. Plant height and fresh weight of irradiated callus gradually declined. Growth responses of callus were strongly influenced by the radiation dose. Singh (2014) exposed suckers of nine gerbera varieties (RCGH-12, RCGH-22, RCG-12, RCG-18, RCG-7, RCG-19, RCGH-117, RCGH-38, and RCG-10) to 1.5 Kr gamma rays and determined their radiosensitivity based on various vegetative and floral characteristics. 0.5 Kr had a detrimental effect on plant height, number of leaves, flower stalk length, flower duration, and number of flowers per plant per year in all varieties studied. Hosoguchi et al. (2021) exposed in vitro shoots of gerbera to three types of ion beams with different line energy transfers (LETs) to determine the effective and absorbed doses for mutagenesis. Higher LETs showed a lower LD_{50} based on survival. Mutations were detected in flower color and shape and male sterility. The highest mutation rates were recorded at 10 Gy in C irradiation and 5 Gy in Ar irradiation and Fe irradiation. The highest trait/morphological mutation rate was 14.1% with Ar irradiation at 5 Gy, which was one of the criteria for ion beam irradiation of gerbera in vitro shoots. Findings are very informative for more efficient mutagenesis work on gerbera. In vitro raised shoots of *G. jamesonii* cv. "Sunway" were irradiated with a single dose of 1.0 Gy gamma rays, and irradiated plantlets were micro-propagated. Radiation treatment affected the rate of propagation and induced color variations (Purshottam et al. 2021).

12.60 Gingers (Ornamental Gingers)

Family Zingiberaceae; ginger family; important flowering ornamental plants; aromatic perennial herbs; creeping rhizomes; flowers hermaphrodite; propagation by rhizomes.

Several species of shellflower (*Alpinia*) are cultivated as ornamentals. Gingers have high ornamental demand and it has high potential in cut flower arrangements as well as in landscaping. Ginger lily (*Hedychium*) produces beautiful flowers that are used for the preparation of garlands and other decorations. Three wild potential ornamental gingers (Boesenbergia siphonantha (Baker) M. Sabu & al., Curcuma inodora Blatter J., Larsenianthus careyanus (Benth.) W. J. Kress & Mood) were selected for improvement through induced mutagenesis. One percent EMS solutions in distilled water were prepared and immersed the budding portion of the materials in the solution for 12 h. Different concentration of colchicine solution was prepared like 100, 250, and 500 ppm and the materials were dipped in solution for 24 h. The materials were treated with 0.5%, 1%, and 2% acridine solutions for 10 min. All treated plants were thoroughly washed after treatment and planted. Mutagenic treatments produced variability in plants. Boesenbergia treated with colchicine and acridine showed a decrease in the overall size of leaves. Curcuma inodora treated with cochine become shorter in size but its size increased when treated with acridine. Larsenianthus careyanus treated with both acridine and colchicine showed an increase in several leaves and size in vegetative and floral attributes. One percent acridine-treated plants in L. careyanus produced beautiful white variegation on leaves. An improved variety of L. careyanus has been produced through chemical mutagenesis. Acridine-treated plants showed better changes than others. The result revealed that the efficient concentration of acridine or colchicine for inducing mutation in Boesenbergia, Curcuma inodora, and Larsenianthus careyanus is 1% (Prabhukumar et al. 2015).

12.61 Gloxinia (*Sinningia speciosa*)

Family Gesneriaceae; tuberous flowering plant; perennial/annual; propagation by division, leaf cuttings, or seed.

In vitro raised leaf explants of gloxinia were treated with UV-C (dose of 0–9 kJ/m^2) and EMS (0–1%) for 60 and 90 min and cultured in vitro. The LD_{50} of UV-C was 4.52 kJ/m^2 and multiple shoot formation in each treatment was significant. In vitro flowering was initiated from petiole explants treated with 5.4 kJ/m^2, however, this resulted in the formation of incomplete flowers, while leaf explants produced normal multiple shoots. The LD_{50} of EMS was 0.88% for 60 min and 0.73% for 90 min. Plants irradiated with UV-C showed significant differences in the flowering characters: canopy width, leaf width, leaf length, flower width, and flower length and no significant difference in stem height. Gloxinia treated with UV-C 5.4 kJ/m^2 developed the most altered flower characteristics such as alterations in color, petal thickness, a floral section with only two corollas, etc. EMS-treated plants showed significant differences in canopy width and stem height, but no significant differences were found in leaf width, leaf length, flower width, and flower length. The EST enzyme was successfully used to detect mutants obtained from mutagens treatment (Sirisom 2008). Sirisom and Te-chato (2008) treated in vitro raised leaf explants of gloxinia (*Sinningia speciosa*) with UV-C irradiation at a dose of 0–9 kJ/m^2 and cultured in vitro. After culture for 28 days, the LD_{50} of UVC irradiation was determined 4.52 kJ/m^2 and multiple shoots formation in each treatment was significant. In vitro flowering was found in the petiole of leaf explants treated with 5.4 kJ/m^2 of UV-C but those flowers were malformed. The leaf laminar produced normal shoots. Leaf explant irradiated plants showed significant differences in canopy width, leaf width, leaf length, flower diameter, and flower length while stem height was not significantly different. For floral morphologies, flower characteristics were altered in terms of petal color and thickness. Gamma radiation (10–60 Gy) was applied to the callus of *Sinningia speciosa* and studied the growth of callus tissue. Callus growth was reduced with increasing radiation dose. It was found that increasing doses of gamma radiation (10–60 Gy) had an inhibition effect on callus fresh weight in the 4-week established cultures. Gamma rays produced different effects compared to non-irradiated callus controls in morphological aspects of the callus such as textures and color of callus cultures (Daud and Taha 2011). Dong et al. (2018) studied the molecular mechanism in the domestication process of Gloxinia (*Sinningia speciosa*) by employing multiple experimental approaches including genetic analysis, genotype–phenotype associations, gene expression analysis, and functional interrogations. They mentioned that a single gene encoding a TCP protein, *SsCYC*, controls both floral orientation and zygomorphy in gloxinia. They observed that a causal mutation responsible for the development of *peloric* gloxinia lies in a 10-bp deletion in the coding sequence of *SsCYC*. They have traced the putative ancestor and reconstructed the domestication path of the *peloric* gloxinia. The results presented here suggest that a simple genetic change in a pleiotropic gene can promote the elaboration of floral organs under intensive selection pressure. Domestication is a complex process due to interspecific hybridization, chromosomal introgressions, and polyploidization events which complicate the systematic study of the sources of phenotypic variation. Hasing et al. (2019) studied the genetics of 115 individuals of *S. speciosa* representing including different species, wild representatives, and cultivated accessions. Analysis revealed that all of the domesticated varieties were derived from a single founder population. They did not detect any major hybridization or polyploidization events that could have contributed to the rapid increase in phenotypic diversity. Dong et al. (2023) reported that a single gene encoding a TCP protein, SsCYC, controls both floral orientation and zygomorphy in gloxinia based on multiple experimental approaches including genetic analysis, genotype–phenotype associations, gene expression analysis, and functional interrogations. Studies indicated that a causal mutation responsible for the development of peloric gloxinia liesin a 10-bp deletion in the coding sequence of SsCYC. They have traced the putative ancestor and reconstructed the domestication path of the peloric gloxinia, in which a 10-bp deletion in SsCYC under selection triggered its evolution from the wild progenitor. Their observations suggested that a simple genetic change in a pleiotropic gene can promote the elaboration of floral organs under intensive selection pressure.

12.62 *Gloriosa superba* L.

Family Colchicaceae; known as the flame lily, climbing lily, creeping lily, glory lily, gloriosa lily, tiger claw, agnishikha, and fire lily; herbaceous perennial vine; contains high levels of colchicine; propagated by seeds, vegetatively by dividing rhizomes.

Patwibul et al. (2001) treated tubers with gamma rays (0–10 Gy) and detected changed plant growth and morphological characteristics at 2.5 Gy and above. Radiation treatment also showed an increase in the number of shoots arising from the terminal buds of tubers, shorter stems, changes in leaf shape, flower appearance, and petal color as well as a decrease in pollen viability and germination percentage. Gamma rays induced abnormal chromosome pairing as ring multivalent, lagging chromosome and chromosome bridge in meiotic cell divisions. Anandhi et al. (2013a) detected mutants in *G. superba* after treatment with EMS, DES, and gamma ray. The mutants had morphological changes in plant height, number of leaves and flowers as well yield characters. Colchicine was estimated using HPLC of five high-yielding mutants (T8P2, T10P1, T10P4, T7P3, T9P4). These promising mutants seemed to possess high colchicine content, than the control. The maximum colchicine content of 0.707% was observed in 2.00% EMS (T7P3), followed by 0.702% in 1.00% DES (T8P2). The obtained results implicate that physical and chemical mutagens contribute to variations in alkaloid content. Anandhi et al. (2013b) studied the correlation and path coefficients of morphologically distinct mutants of *G. superba* for 17 characters in vM_2. The highest and most positive correlation for dry seed yield/plant was observed with the number of seeds/pod, number of leaves/plant, dry pod weight, fresh seed weight/pod, fresh pod weight, plant height, and number of secondary branches/plant. The path analysis of component characters on dry seed yield/plant of *G. superba* in the VM_2 generation exerted a direct positive effect through these characters: number of leaves/plant, dry pod weight, number of seeds/pod, fresh seed weight/pod. Selvarasu and Handhasamy (2013) analyzed correlation and path coefficients for 17 morphological characters of *Gloriosa superba* in the second generation. They observed maximum GCV, and PCV for several leaves/plants and several secondary branches/plants. The maximum heritability and genetic advance were recorded for stem girth. The highest and most positive correlation for dry seed yield/plant was observed with the number of seeds/pod, number of leaves/plant, dry pod weight, fresh seed weight/pod, fresh pod weight, plant height, and number of secondary branches/plant. The path analysis of component characters on dry seed yield/plant of *G. superba* in the VM_2 generation exerted a direct positive effect through these characters: number of leaves/plant, dry pod weight, number of seeds/pod, fresh seed weight/pod. Mutagenesis experiments were conducted on *G. superba* using gamma rays, EMS and DES to induce genetic variability in economic characters. The phenotypic coefficient of induced variations was in general slightly higher than the genotypic coefficient of variation for the selected traits in the VM_2 generation. High heritability and genetic advance as a percent of the mean were recorded for most of the characters under study indicating better scope for further selection. Differential patterns resulted in ISSR analysis indicating the polymorphism created by induced mutagenesis, creating scope for the selection of desirable mutants (Selvarasu and Kandhasamy 2017). Padmapriya and Rajamani (2017) treated sprouted tubers (*G. superba*, Andhra Wild) with chemical mutagens (EMS-1, 2, 3% and DES-1, 1.25%, 1.5%) to raise VM_1 and VM_2 generations for induction and detection of short-statured plants and higher seed yield. Experimental Glory lily was highly responsive to mutagenic treatments in exhibiting reduced plant height, branching height, earliness, and significant improvement in seed production. In VM_2, DES 1.5% was effective in producing short-statured plants and EMS 2% enhanced the dry seed yield per plant. Padmapriya and Rajamani (2021) treated seeds of *G. superba* with gamma rays (10, 20, 30, and 40 Gy), EMS (10, 20, 30, and 40 mM), and DES (15, 20, 25, and 30 mM) and studied chlorophyll variants. Four types of chlorophyll mutants (albina, xantha, chlorina, and viridis) were detected. Results indicated that the mutagenic efficiency was the highest, at lower and intermediate concentrations of mutagens. Among different treatments, gamma rays were more effective than the other two chemical mutagens. All parts of the Gloriosa superba, especially the tubers, and seeds, contain alkaloids namely colchicine and gloriosine. Colchicine as a mutagen is used to develop the superior properties of some plants. Polyploid plants have several advantages including larger cells, higher plants, wider leaves, larger fruits, higher production, and plants becoming more resistant to disease. Considering the role of colchicine Ernawiati et al. (2022) studied G. superba and recommended further studies for the prospect of Gloriosa superba as a source of biomutagens (natural colchicine).

12.63 Grasses

Several species of grasses are widely used turf in floriculture/horticulture, especially for golf courses, football fields, lawns, and athletic fields. Some features like coarse-leaf texture and long internodes are undesirable for good-quality golf fields. Prostrate turf varieties are desirable because of their increased low mowing tolerance, heat resistance, traffic resistance, and ground coverage compared with upright varieties. The following are the important species used for these purposes:

St. Augustine grass (*Stenotaphrum secundatum*): Family Poaceae; known as buffalo turf/buffalo grass/carpetgrass; perennial; ground cover; warm season lawn grass; popular pasture grass, lawn grass; dark green; propagation by seed, plugs, sections of rhizomes/stolons.

Bermuda grass (*Cynodon dactylon*): Family Poaceae; perennial; tropical grass; blades gray-green; annual/perennial; herbaceous; cultivated in warm climates; known in many common names (Bahamas grass, dhoub, kiri-hire, devil's grass, African couch, Indian couch, star grass, etc.); used as valuable fodder grass/lawn and turf grass; propagation by seed, stolons, rhizomes.

Centipede grass (*Eremochloa ophiuroids*): Family Poaceae; perennial; warm season lawn grass; requires infrequent mowing; widely used in transitional and warm climatic regions; propagation by seeds, vegetative means.

Kentucky bluegrass (*Poa pratensis* L.): Family Poaceae; herbaceous perennial; blades dark green, narrow; creeping rhizomes; used as lawn grass and pasture grass; propagation by seeds, rhizomes, axillary buds.

Dallis grass (*Paspalum dilatatum* var. pauciciliatum): Family Poaceae; known as Water grass, Caterpillar grass, Hairy flowered paspalum; herbaceous perennial bunchgrass/weed; occasionally cultivated as a pasture grass; propagation by seeds, vegetative means;

Perennial ryegrass (*Lolium perenne* L.).: Family Poaceae; perennial/annuals evergreen; propagation by seeds, vegetative means.

Zoysia matrella: Family Poaceae; known as Manila grass; rhizomatous herbaceous perennial lawngrass; ornamental grass; propagation by seeds, vegetative means.

A series of experiments were conducted to develop desirable mutants through induced mutagenesis. The main objectives of mutation work were to study cytogenetic effects; breaking apomixes; increase genetic variability; overcome the difficulty of facultative apomixes; resistance to root-knot nematode; tolerance to temperate winters; winterhardiness; resistance to root-knot and other nematodes; frost tolerant and with better spring growth; to improve undesirable coarse-leaf texture and long internodes, etc. Mutagens mostly used were X-ray, gamma rays, beta rays, EMS, NEU, NMU, 1,4-*bis*-diazo-acetyle butane (1,4 DAB), DMS, EI, etc. (Bashaw and Hoff 1962; Burton 1972, 1974, 1975, 1976, 1979, 1981, 1985; Burton and Hanna 1977; Burton and Jackson 1962; Burton et al. 1980, 1982; Busey 1980; De Loose 1964; Hansen and Juska 1959, 1962; Julen 1954, 1958, 1961; Kuleshov et al. 1974; Powell 1974, 1976; Powell and Murray 1978; Powell et al. 1974; Powell and Toler 1980; Lu et al. 2009).

A series of publications are available on induced mutagenesis for the improvement of different species of grasses. Basic information on cytogenetic aspects and reproduction was reported by Gustafson and Gadd (1965). This information was very informative for mutation work. Hansen and Juska (1962) reported radiation-induced breakdown of apomixes and 150–200 Gy as the optimum dose of X-rays for dry seeds. De loose (1964) used Poa pratensis and applied X-rays, gamma rays, beta rays, and EMS and detected mutants with agronomic potential. Interesting desirable mutants were reported in brome grass by treating seeds with nitroso-*N*-ethyl urethane (NEU), *N*-nitroso-*N*-methyl urethane (NMU), 1,4-*bis*-diazo-acetyl butane (1,4 DAB), dimethyl sulfate (DMS) and ethyleneimine (EI). Thermal neutrons initially induced stem rust resistant line in Merion Kentucky bluegrass (Hansen and Juska (1959, 1962). Powell and Murray (1978) developed several desirable mutants and concluded that the vegetative adventitious bud technique is very suitable in Poa pratensic cv. "Belturf." A similar concept of application of induced mutation for improvement in apomicts was reported in the Dichantium annulatum complex (Singh and Mehra 1971). Early experiments developed root-knot nematode resistance variety by treating rhizomes of four varieties ("Tifgreen," "Tifway," "Tifdwarf," and "Tifcote") of turf Burmuda grass with 70–120 Gy gamma rays (Burton 1972, 1974, 1975, 1976, 1979; Burton et al. 1982; Powell 1974, 1976; Powell et al. 1974). A good number of desirable mutants (winterhardiness, weed-free turf, resistant to root not and nematodes, frost tolerant, etc.) were recovered from Coastcross-1 Bermuda grass by treating stems with 70 Gy gamma rays (Burton 1981, 1985; Burton et al. 1980). Mutants have been induced in Dallis grass (Paspalum dilatatum Poir.) and Augustine grass (*Stenotaphrum secundatum*) after treating stolon with gamma rays (45–70 Gy) (Bashaw and Hoff 1962; Burton and Jackson 1962; Reinert et al. 1981; Busey 1980; Powell 1976; Powell and Toler 1980; Toler and Grisham 1983; Tolar et al. 1985; c.f. Broertjes and Van Harten 1988). Pederson and Dickens (1985) reported one mutant ("AU Centennial") after gamma irradiation (300–400 Gy) in centipede grass (*Eremochloa ophiuroides*).

Rhizomes of two turf bermuda grass cultivars were treated with gamma rays to evaluate the mutation breeding technique for these vegetatively propagated grasses. Dosages of 9000 and 11,300 rads produced mutations (dwarfism and changes in color hue of the stolons) (Powell et al. 1974). Busey (1980) treated stolon pieces of different accessions of St. Augustine grass [*Stenotaphrum secundatum* (Walt.) Kuntze] with gamma rays (4500 rads) and detected 40% survival. "Bitterblue" and another accession were entirely killed at 4000 rads. At 4500 rads, up to 7% recognizable mutants of accession, FA-243 were obtained. In addition to morphological variants, a chimeral anthocyanin change was noticed. From this chimera arose a stable genotype with green stolons and white stigmas, whereas the source genotype (FA-243) had red stolons and purple stigmas. Mutation breeding is effective in improving St. Augustineg rass when easily recognizable variants are needed. Powell and Toler (1980) treated stolons of "Floratam" St. Augustine grass with different doses of gamma rays (4.5–7.0 Krads) and selected mutant lines from the treated population. Mutants ranged from dwarfs to very fine leaf types and change in color. Genes already present in "Floratam" for resistance to the St. Augustine Decline Strain of Panicum Mosaic Virus were not modified in the mutants. Results suggested that mutation breeding procedures could be a major step forward in breeding this species. Burton et al. (1982) treated short dormant rhizome sections of sterile triploid hybrids (Tifgreen, Tifway, and Tifdwarf) of *Cynodon dactylon* and *C. transvaalensis* with EMS and gamma rays. Gamma radiation-induced mutants appeared to be better than the parents in one or more characteristics. Chlorophyll-deficient mutants have appeared and one mutant slightly, but significantly, more winter hardy than Coastcross-1 has been obtained. Lu et al. (2009) applied gamma rays on bermuda grass for induction of dwarf mutants from a native bermuda grass (*Cynodon dactylon*) germplasm. They isolated three dwarf-type mutant lines (7-9, 10-5, and 10-12) with lower canopy height, shorter internodes, and shorter leaves from 3000 irradiated stolons. "Raleigh" cultivar is superior cold tolerance than St. Augustine grass but, its coarse-leaf texture and long internodes are undesirable. Node cuttings and calli were exposed to gamma rays and dosages of 48.5 and 72.6 Gy were determined as LD_{50} and LD_{20} for the cuttings, respectively. Regeneration of calli reduced at higher dosages. Thirteen morphological mutants (semi-dwarf type with reduced internode length and leaf blade length) were identified. One mutant had much fewer and shorter stolons and displayed an upright and tufty growth pattern. The altered morphological traits were stable as shown by their growth performance in various locations and conditions (Li et al. 2010). Chen et al. (2011) treated 6-year maintained calli of manila grass (*Zoysia matrella* [L.] Merr.) with different doses (0, 5, 10, 20, 40, 80, 100, 150, 200, 250, and 300 Gy) of $_{60}$Co γ-rays for development of salt-tolerant variants. The regeneration rate and regeneration capacity of the calli were highest after treatment with 20 Gy ^{60}Co γ-rays, 27.08 and 91.67%, respectively. 10.42% of the calli developed shoots in 100 Gy, but at 150 Gy, both regeneration capacity and regeneration rate declined significantly, and no shoot was observed after 6 weeks of regeneration. 100–150 Gy was found to be the most appropriate irradiation span for inducing somaclonal variation. Five NaCl-tolerant variant lines, Ze1, Fv1, Te1, Tw1, Fr1, were selected on subculture medium. Mutlu et al. (2015) applied gamma rays to induce desirable variation in *Cynodon dactylon* (L.) Pers. Single node stolons were treated with gamma rays (70, 90, or 110 Gy). Dosages of 85 and 57 Gy were determined as LD_{50} and LD_{20} for the cuttings, respectively. Mutants with dwarf growth habit, higher shoot density, and finer leaf texture than the parental genotype were detected and isolated. Chen et al. (2016) applied a mutation technique using gamma radiation to develop desirable prostrate turf varieties in perennial ryegrass (*Lolium perenne* L.). They screened gamma ray mediated dominant dwarf mutants and found that about 10% of dwarf mutant lines displayed a prostrate phenotype at mature stages. One prostrate line, "Lowboy I" was isolated which had a shorter canopy, leaf blade, and internode lengths, and greater tolerance to low mowing stress than the wild type. Characterization confirmed that the dwarf and prostrate phenotypes were both due to GA deficiency. Mohd et al. (2016) selected *Cynodon dactylon* for the improvement of its undesirable morphological characteristics (coarse-leaf texture and long internodes) through gamma ray induced mutations. Single node stolons were treated with gamma rays (0, 20, 40, 60, 80, 100, 120, and 140 Gy) and determined 90 Gy as the LD_{50} dose. Higher doses significantly reduced survival. Twenty-two mutants (semi-dwarf type with reduced internode length and leaf blade length) were identified and evaluated. Kim et al. (2019a, b) studied the comparative effects of gamma rays, proton ions, and carbon ions on seed germination, plant growth parameters, and DNA fragmentation of creeping bentgrass seeds. The LD_{50} doses were 115.9 Gy (γ-rays), 225.1 Gy (proton ions), and 57.7 Gy (carbon ions). Decreasing head DNA percentages were observed for γ-rays and proton ions. Carbon ions induced the lowest frequency of DNA fragmentation. The biological effects of the ionizing radiation types on creeping bentgrass were summarizable as follows: germination, carbon ions (C) > γ-rays (G) > proton ions (P); survival, $C > P = G$; growth, $C \geq P > G$; DNA fragmentation, $G \geq P > C$. These results indicate that proton ions are useful as a physical mutagen in plant mutation breeding. Gamma radiation developed dwarf mutant with shorter stems, wider leaves, lower canopy height, and a darker green color in *Zoysia matrella* [L.] Merr. Lin et al. (2020) conducted morphological tests and physiological, biochemical, and transcriptional analyses to reveal the dwarfing mechanism in the mutant.

12.64 *Glebionis segetum* (Corn Marigold)

Family Asteraceae; formerly Chrysanthemum segetum; herbaceous perennial/annual plant; flowers produced in capitula; ray florets orange, yellow; propagation by seeds.

Seeds of *Glebionis segetum* (corn marigold) were exposed to 20, 40, 60, 80, and 100 Gy of gamma rays, and three mutants were detected at 20 Gy and two mutants at 20 Gy. Mutants showed significant changes in morphological and biochemical characters in addition to flower color. RAPD markers differentiated mutants from their parent (Kapoor et al. 2014).

12.65 *Gypsophila paniculata*

Family Caryophyllaceae; commonly known as baby's breath, herbaceous perennial; used for cut and dry floral arrangements, propagation by seed, and vegetative means.

Tsuji et al. (2007) irradiated shoots of *Gypsophila paniculata* with carbon ion beams using the JAEA AVF cyclotron to induce mutations. *Gypsophila* was found to be resistant to ion beam irradiation. The suitable dose for induction of mutations was estimated to be 2–4 Gy for Gypsophila. Barakat and El-Sammak (2011a) standardized in vitro protocol of Gypsophila paniculata from shoot tips and lateral buds. In a separate experiment, Barakat and El-Sammak (2011b) performed in vitro mutagenesis by treating shoot tips and lateral buds with gamma rays (0.25, 0.5, 0.75, and 1 Gy) for in vitro mutagenesis. Gamma rays significantly influenced and affected callus induction, number of shoots per explant, and shoot length. Not much effect was observed on shot formation. The lateral bud explants gave a significantly higher number of shoots compared to shoot tip explants. RAPD technique helped to differentiate mutants from their parents. Li et al. (2020) treated seedlings of *G. paniculata* cultivar "Cloudstar 4" (semi-double flower) using gamma radiation (3, 4, and 5 kR) to develop a double flower type. They were successful to develop a mutant (named "Huixing 1") from 4-kR gamma radiation. The mutant has bigger flowers and more flower petals.

12.66 *Guzmania peacockii*

Family Bromeliaceae; evergreen epiphytic perennial; propagation by seeds, offsets.

Self-fertilized seeds of *G. peacockii* were treated with 30 Gy of gamma rays and detected a high percentage of chlorophyll-deficient striped (white or yellow) plants in M1. One mutation was solid which was isolated as commercialized in the name "Edith" (De Loose 1966, 1969a, 1973a).

Callus and decapitated plantlet explants of Guzmania "Hilda" were treated with sodium azide and gamma rays for the induction of mutations. The survival rate was 0% when calli were treated by sodium. azide. The survival rate of decapitated plantlets treated with 0.5 mM sodium azide for 60 min was 51.3%. All explants treated with 2.0 mM sodium azide showed browning and/or dead. Survival of G. "Hilda," G. "Cherry," G. "Luna," and G. "Focus" was 74.2–100% after gamma irradiation. Survival of G. "Focus" was 45.0% when treated with 15 Gy gamma rays. Mutant plantlets showed a lot of chimeras in the leaf (Huang 2021).

12.67 Hebe

Family Plantaginaceae; species in the section are also called shrubby veronicas or hebes; flowers perfect; flower arranged in the spiked inflorescence; evergreen shrubs; flowers shades of blue, purple, pink, or white; propagation by seeds, cuttings.

Nodal explants of micro-propagated two varieties ("Oratia Beauty" and "Wiri Mist") of Hebe were treated with gamma rays (20, 30, 40, 45, 50, 60 Gy) and determined their radiosensitivity. The cultivar "Oratia Beauty" was more sensitive. The mutation was higher in "Oratia Beauty" (7.1%) than in "Wiri Mist" (2.2%) (Gallone et al. 2012).

12.68 *Hedera helix*

Family Araliaceae; known as common ivy, European ivy, and many more; evergreen climbing vine; propagation by rooting procedure, tip cuttings.

Knuth (1962) treated many cultivars of *H. helix* with different doses of X-rays and determined 40 Gy as the optimum working dose. One mutant with excessive variegation was established. Yena (2022) described a new cultivar of Hedera helix

"Peregreenus" developed due to vegetative mutations. Studies on this vegetative mutation helped to trace out the line of its six ancestors. Evidence on certain patterns in the development of variegated ivy bud mutations has been discussed. A manifestation of the law of homologous series in hereditary variability was revealed, an instance of three ways of bud mutation emergence has been described, and also examples of different rates of bud mutations' development in ivy have been shown. The sequence of occurrence of vegetative mutations in the linear genealogical chain of cultivars can serve as a model for the unfolding of anagenetic events in circumstances opposite to normal evolution.

12.69 *Hibiscus rosa-sinensis, Hibiscus moscheutos,* and *Hibiscus syriacus*

Family Malvaceae; known as Chinese hibiscus, China rose, Hawaiian hibiscus, rose mallow and shoeblack plant; annual and perennial herbaceous/woody shrubs/small trees; flowers white to pink, red, blue, orange, peach, yellow or purple; propagation by seeds, cuttings.

Semi-acute gamma rays were applied on stem cuttings and detected several flower color and form mutations. The optimum dose is reported around 100–200 Gy. Few mutants have been established in pure form (Das et al. 1974, 1977). Early mutation work reported 30 Gy X-ray as the optimum acute dose for rooted cuttings for *H. rosa-sinensis* (c.f. Broertjes and Van Harten 1988). Stem cuttings of *H. rosa-sinensis* cultivar "Alipur Beauty" (light carmine red double flower type) were treated with 1, 2, and 4 Krad of gamma rays and recorded morphological and floral changes. Two plants from the 4 krad treatment developed a five-petalled single flower type mutant as a chimera. The mutant was isolated and established as a new variety ("Anjali") for commercial utilization (Banerji and Datta 1986, 1988). One small flower mutant has been developed in *H. rosa-sinensis* after gamma irradiation (Srivastava and Mishra 2005; Anonymous 1989). Warner and Erwin (1998) exposed six Hibiscus species to total supplemental irradiance (high-pressure sodium lamps) levels of 0, 2, 4, and 8 mol/day photosynthetically active radiation (PAR), at the apical meristem, for plants grown under a 9-h photoperiod, and 2, 4, 8, and 16 mol/day PAR for plants grown under a 16-h photoperiod. Data collected was recorded on anthesis date, number of leaves below the first flower, lateral shoot and flower bud number, plant height, and flower diameter. Increasing daily light integral reduced the number of leaves below the first flower for some species in one or both photoperiods. Irradiance and photoperiod effects on the lateral shoot and flower bud number at anthesis have been discussed. Photosynthetic-response curves and light-response curves concerning flowering were constructed for each species. Naveena et al. (2020a) treated hardwood cuttings of *H. rosa-sinenis* Cultivar "Red Single" with a wide range of gamma rays (1, 1.5, 2, 2.5, 3, 3.5, 4, 4.5, and 5 kR) and EMS (0.1, 0.2, 0.3, 0.4, 0.5, 0.6, 0.7, 0.8, 0.9, 1.0% for 6 h) to induce mutations. There was a gradual reduction in survival of cuttings, shoot length, leaf length, and leaf width with an increase in dosage of gamma rays and EMS. Gamma rays were more effective than EMS. LD_{50} value for gamma radiation was determined at 2.81 kR and for EMS at 0.89%, respectively. Naveena et al. (2020b) treated Semi-hardwood cuttings of *H. rosa-sinensis* L. Cultivar "Red Single" treated with different doses of gamma rays (25, 30, and 35 Gy) and EMS (0.8, 0.9 1.0%) and studied their effects in the first generation. High frequency and a wide spectrum of chlorophyll mutations were observed in the gamma ray treated population. Mutagenic efficiency and effectiveness were higher in gamma radiation-induced plants, particularly in a lower dose of 25 Gy. *Hibiscus syriacus* (Rose of Shaaron) was included in the mutation breeding program to develop genetic variability through X-ray, gamma ray, and DES treatment. Several mutant varieties have been recovered (Hong et al. 1980; Song et al. 1999; Kang et al. 2007). Kim et al. (1997) studied the radiosensitivity of *H. syriacus* after treating cuttings with gamma rays. The rate of germination and survival and plant height increased at 4 kR. Forty-five percent reduction in plant height was recorded at 5 kR optimum dose has been recommended 10–12 kR for mutation breeding in Hibiscus. Several promising mutant varieties have been selected from the experiment. Semi-hardwood cuttings of Hibiscus rosa Sinensis L. Cultivar "Red Single" were exposed to gamma rays (25, 30, and 35 Gy) and EMS 0.8% (64.43 mM), 0.9% (72.48 mM, 1.0% (80.54 mM) separately to their effectiveness and efficiency in M1V1. Physical mutagens created a high frequency as well as a wide spectrum of chlorophyll mutants. A total of five types of chlorophyll mutant viz., albino, xantha, viridis, chlorina, and xantha-viridis were observed as a result of physical and chemical mutation. Mutagenic efficiency and effectiveness were higher in gamma radiation-induced plants, particularly in a lower dose of 25 Gy. The highest mutation rate in terms of effectiveness and efficiency was observed in gamma rays than the Ethyl Methane Sulfonate treatment (Naveena et al. 2020a). Naveena et al. (2020a) treated hardwood cuttings of Hibiscus rosa-sinensis L. with gamma rays (1–5 kR) and ten different concentrations of EMS (0.1–1.0%) to determine mutagenic sensitivity and to induce mutations. There was a gradual reduction in survival of cuttings, shoot length, leaf length, and leaf width with an increase in dosage of gamma rays and EMS. Gamma rays treated plants showed a significant gradual reduction in growth rate than the EMS-treated plants. LD_{50} dose for gamma radiation and EMS were determined as 2.81 kR and 0.89%, respectively. Priyanka and Dhanavel (2021) studied the possibilities of induction of mutations after treating healthy seeds of Hibiscus sabdariffa L.(Roselle) to Gamma rays (05, 10, 15, 20, 25, 30,35, 40, 45, and 50 KR) and EMS (05, 10, 15, 20, 25, 30, 35, 40, 45 and 50 mM). They studied seed germination and seedling survival and determined the LD_{50}. Results showed decreased seed germination and survival with increasing the dose/conc. of mutagens.

12.70 *Helianthus tuberosus* (Jerusalem Artichoke)

Family Asteraceae; known as sunroot, sunchoke, wild sunflower, topinambur, or earth apple; herbaceous perennial; used as medicinal and food (tubers) plant; flowers are very bright yellow and used as a cut bunch; propagated by seeds, tubers.

Coppola (1986) treated tubers with gamma rays (3 krad) and observed abnormal leaf shapes and sizes in some plants and white-skinned tubers instead of red in the M1 generation. Plants developed from these white tubers had unbranched stems that were thinner than those of controls. Songsri et al. (2019) treated two Jerusalem artichoke genotypes "HEL 65" and "CN 52867" with gamma rays to determine LD_{50} dose radiation effects on growth and tuber yield. The LD_{50} for the genotypes "HEL 65" and "CN 52867" were 22 and 27 Grays (Gy), respectively. Germination and plant height decreased after irradiation. The number of branches was increased with higher gamma radiation doses (20 and 25 Gy). The tuber yield increased in both genotypes at 5 Gy. The radiation treatments did not change flower characters in either genotype.

12.71 *Helianthus annus* (Sunflower)

Family Asteraceae; herbaceous annual forb; edible oily seeds; good ornamental flowers; propagation by seeds.

Omar et al. (1993) studied the effect of gamma rays and NaCl on the growth and cellular contents of soluble carbohydrates, protein, and nucleic acids in *Helianthus annuus* L. callus. Radiosensitivity of gamma rays was determined based on fresh weight changes. The LD_{50} was determined 2.8 krad. The inclusion of NaCl in the medium caused a significant reduction in callus fresh weight. The cellular contents of protein, soluble carbohydrates, and RNA were reduced, while DNA increased at a 2% NaCl level. There was a significant increase in protein, carbohydrates, and DNA, while a significant reduction in RNA content was observed. The role of such information in breeding for salt-tolerant sunflowers following physical mutagenesis in vitro has been outlined. Cecconi and Durante (2000) made a clonal analysis to study the fate of cells during the shoot meristematic growth of sunflowers. 3000 dry seeds (F1 seeds were obtained by crossing the cytoplasmic male sterile line AD-811 with 3 restorer lines carrying three different markers for epidermis pigmentation) for each cross were irradiated with 15 KR of hard X-rays. Chimerical plants were scored and classified considering the pigmentation of florets and pericarp. Mutation frequency was 8.14% for achene color, 3.6% for floret color, 2.2% for anthocyanin pigmentation of the stigma, and 0.18% for fertility restoration. The analysis of the different patterns of mutated sectors and their relative frequencies indicate that the apparent cell number contributing to the capitulum development is 2 or 4, while the analysis of the shape and the size of mutated sectors indicate the possibility of cell rearrangement during cell meristematic growth. The effect of gamma radiation on the production of new and desirable ornamental characters in sunflower (cultivars "G-101," "Hysun-354," "Mayak," and "Vidoc") plants cultured in vitro was investigated. The plant height of all somaclonal variants was substantially reduced by 81% in Hysun-354, 88% in Mayak, and 88% in Vidoc compared to their original parental material (control) (Barakat et al. 2002). The seeds of two varieties of common Sunflower (*Helianthus annuus* L.) viz: USH-430 and KL-675 were treated to estimate the mutagenicity of both gamma rays, sodium azide alone and in combination to generate morphological macro-mutations and screen and ascertain the chromosomal aberrations followed by reduction of pollen fertility with increased doses. An attempt was made to know the genetic basis of the chromosomal aberrations. Such aberrations were a source of changes in the pattern of gene regulation at the time of differentiation leading to the formation of cultivars (Ramesh Kumar and Venkat Ratnam 2009). Seeds of two varieties of sunflowers ("USH-430" and "Nidhi-999") were treated with gamma rays, sodium azide, and combined treatments. The following micro-mutants were scored *viz*: patchy albino, white margin, virescent, darker green, xantha purple, albino. The micro-mutants were increased with increased dose concentrations with all three treatments in both varieties. The micro-mutations rate was high with sodium azide followed by combined and gamma ray treatment. The macro-mutants include branched, basal stem bifurcation, rosette and compact leaf arrangement, double-headed, dwarf early-I, dwarf early-II, and mosaic leaf arrangement. The macro-mutations were scored high at gamma rays followed by combined and sodium azide treatments (Kumar and Ratnam 2010). Ramesh Kumar and Venkat Ratnam (2010) treated seeds of two varieties ("USH-430" and "SHSF-333") of *H. annus* L. with gamma rays, and sodium azide separately and in a combination of two and studied their effects on seed germination, seedling survival, pollen fertility and seed set in M2. There was a gradual decrease in seedling survival and pollen fertility with an increase in the dose of mutagen in both varieties. The percentage of lethality and achene sterility gradually increased with an increased dose. The seed set percentage increased significantly in the USH-430 variety in some treatments. Mutagenic effectiveness decreased with an increased dose or concentration of gamma rays and sodium azide in both varieties. In combined treatments, mutagenic effectiveness gradually increased with an increased dose. Mutagenic efficiency increased with an increased dose in the case of gamma-irradiated seeds in both varieties. Mutagenic efficiency decreased gradually with an increased dose in both varieties. The linear correlation coefficient was positive in the case of gamma rays and sodium azide separated, whereas in combined treatment, a negative correlation was observed. Jamdhade and Kashid (2016) treated seeds of two varieties of *H. annus* L. with EMS (0.05, 0.10, and 0.15%), SA (0.01, 0.02 and 0.03%) and gamma rays (10, 20 and 30 KR) and observed differential response of varieties in respect of seed germination. Morphological and physiological alternations result in the activation of the embryo and germination of the seed took place. Feng et al. (2016) studied the morphological, histological, and photosynthetic characteristics of the stably inherited sunflower petal-sepal mutant induced by space radiation-induced mutagenesis. The inflorescence of the petal-sepal mutant maintained the appearance and structure of the capitulum, whereas no explicit tubular flower or ligulate flower was differentiated. The petal-sepal mutant only completed the inflorescence development and the differentiation of sepal primordia and inflorescence primordia, without entering the differentiation stage of tubular flower primordia, ligulate flower primordia, stamen primordia, or pistil primordia. The photosynthetic rate, transpiration, and stomatal conductance of petal-sepal mutants were relatively weaker than the control plants. They concluded that the petal-sepal mutant of sunflower had only inflorescence differentiation, while several mutant genes were caused by radiation-induced mutation, which entered an infinitely recurrent development process rather than the floral differentiation stage. Diaz et al. (2018) treated achenes with gamma rays (0, 100, 200, 300, 400, 500, 600, 700, 800, and 900 Gy) and studied variability in plant height, root length and volume, and dry biomass. The results indicated that germination and sprouting decreased as the radiation increased. Plant height, length, root volume, and dry biomass decreased at high doses. They concluded that doses of 100 and 200 Gy have a stimulating effect on plant height and root length, being an important agent, to induce genetic variability in sunflowers.

12.72 *Heliconia psittacorum*

Family Heliconiaceae; perennial herb; known as parrot's beak, parakeet flower, parrot's flower, parrot's plantain, false bird-of-paradise; perennial; propagation by seeds.

An attempt was made to standardize methodology to induce mutations and somaclonal regeneration from irradiated explants (inflorescences buds and leaves base). To study the effects of radiation on callus multiplication, different gamma radiation doses of 10, 20, 30, 40, 50, and 75 Gy were applied. The inflorescence buds did not show a positive response to embryo induction. Embryogenic callus formation took place from the leaves base. No significant differences were observed in callus multiplication between the applied doses. Considering different aspects of the development and radiosensitivity of embryogenic calli, 40 Gy was recommended as the appropriate dose for the induction of genetic variability in *H. psittacorum* (Urrea and Ceballos 2005, 2017).

12.73 Hosta

Family Asparagaceae; herbaceous perennial; conspicuous foliage; foliage plant; propagation by seed, division.

Vaughn and Wilson (1980) examined the genetics and ultrastructure of chimera variegated Hosta mutant "Vaugn 73-2" and based on the evidence mentioned that the mutant is truly a plastome mutant. Vaughn et al. (1980) studied plastid chimeras of the genus Hosta two mutants with C4-type ultrastructure. Mesophyll plastids from the white tissue had an extensive system of peripheral reticula, characteristic of C4 plants, and macrograna or vesiculated thylakoids, characteristic of photosystem I deficient mutants. Microbodies were either absent or in low quantities in these cells. Bundle sheath plastids from the white sectors had a concentration of all organelles, with plastids of typical C4 bundle sheath ultrastructure and numerous, ornate mitochondria. Green restituted sectors of one of these mutants retained the ability to make peripheral reticulum yet had normal Hosta granal formation. Despite these changes, the plants showed a carbon fractionation like C3 plants. Possible reasons for these ultrastructural modifications have been discussed. Vaugh and Wilson

(1980) examined the genetics and ultrastructure of a chimeral Hosta mutant to understand its origin. A series of new varieties have been developed in Hosta through sports of original variety and also through sports of sports variety. Klinkhamer (2020) reported a mutant variety "Blue Ivory" which developed from Hosta "Lady in Red." The new variety Hosta "Atlantis" has been registered (AHS) in 2004 and reported in 2005. This is a mutant of Hosta "Abba Dabba Do." Interestingly, Hosta "Abba Dabba Do" was developed as a sport from Hosta "Sun Power" in 1993 and registered (AHS) in 1998 (https://www.jlbg.org/content/gallery/Hosta_(Hosta_Lilies)_-_Plant_Delights_Nursery_And_JLBG_Introductions/Hosta-Abba-Dabba-Do-(11312)/. New mutant Hosta variety "Time Traveler" has been patented and reported. It has been characterized by a round, corrugated, chartreuse-yellowish, pointed leaf having a gray-green center, and a white-edged leaf sown in the middle between the chartreuse-yellowish and the gray-green. It has developed as a result of a whole plant mutation of *Hosta* "Stitch in Time" discovered in a pot at a nursery in Raleigh, N.C. (Publication number: 20210092890, Patent Application).

Filed: Dec 18, 2019, Publication Date: Mar 25, 2021, Inventor: Tony Avent (Raleigh, NC).

Application Number: 16/602,889. https://patents.justia.com/patent/20210092890). The new variety Hosta "Gold Standard" (fortunei) was developed through mutation from Hosta "Frtumeri." The mutant had wide-oval, cordate, seersuckered, yellow leaves with irregular dark green margins. The yellow leaf color may eventually fade to white near the end of the season. Funnel-shaped, pinkish lavender flowers bloom in mid-summer on scapes rising to 3 ft tall (W. George Schmid, https://www.ballyrobertgardens.com/products/hosta-gold-standard-fortunei-v).

Lee et al. (2019) studied the genomic architecture of *Hosta* chloroplasts and examined the level of nucleotide and size variation. They sequenced four (*H. clausa*, *H. jonesii*, *H. minor*, and *H. venusta*) and analyzed six *Hosta* species (including the four, *H. capitata* and *H. yingeri*) distributed throughout South Korea. The study provides detailed information on the chloroplast genome of the *Hosta* taxa. Such work will be very helpful for developing molecular markers applicable in various research areas including induced mutagenesis. Kim et al. (2021) used EMS to develop a new form through induced mutations. From the treated plants, some plants with yellowish-green color patterns on the edge of the leaves were selected. Finally, a new variety "Hwangnarae" was developed where the edge of leaves had uniform yellowish-green color. The mutant has been recommended as a ground cover plant or pot plant.

12.74 Hoya

Family Apocyanaceae; known as Wax plant; evergreen perennial; adventitious roots; popular houseplant; propagation by stem cuttings.

Van Raalte (1980) treated rooted cuttings of *Hosta carnosa* with an acute dose of 50 Gy gamma- or X-rays and detected a large number of variability in the leaf. Mutant characters were yellow-green incurved leaves; white-edged curved leaves;

green-reddish leaves; white-green variegated leaves, etc. Some of the mutant varieties have been commercialized (c.f. Broertjes and Van Harten 1988).

12.75 *Hoya carnosa*

Family Apocyanaceae; perennial; houseplant grown for its attractive waxy foliage, and sweetly scented flowers; propagated by stem cuttings, air layering.

Mutation work on Hoya is meager. No detailed experimental methods are available. Probably rooted cuttings were treated with an acute dose (50 Gy) of X- or gamma rays and developed stable periclinal chimera through repeated cutback methods. The amount of induced variability was much and about 20 mutants have been commercialized and few have been patented. All new varieties have been developed through sports and cobalt irradiation (C.f. Broertjes and Van Harten 1988; Anonymous 1988; Van Raalte 1980).

12.76 Hyacinthus

Family Asparagaceae; bulbous perennial; propagation by seeds, offset bulbs.

Early mutation breeding works were initiated using X-rays and mutations were detected on flower color and flower form (single to double) (De Mol 1926, 1931, 1933, 1934, 1937a, b, 1940, 1953). Mutation work reports are also available from Association Euratom–ITAL (the Netherlands) on different cultivars ("Amsterdam," "Anna Maria," "Blue Bird," "Blue Jacket," "Pink Pearl," "Jan Bos," "Ostara") treating bulbs with X-rays and fast neutron (2.5, 3, 3.5–6 Gy). X-rays 2–5 Gy, fast neutrons 1–6 Gy). Radiosensitivity varied with the scooping of bulbs. Few mutations in flowers have been reported (c.f. Broertjes and Van Harten 1988; Broertjes and Alkema 1970).

12.77 Hydrangea

Family Hydrangeaceae; shrubs/small tree/lianas; deciduous/evergreen; two types of flowers are present in flowerheads: small non-showy fertile flowers are present in the center and sterile large showy flowers with bright color sepals; propagation by seeds, cuttings.

Iizuka et al. (1998) irradiated seeds of two cultivars ("Minazuki"—*Hydrangea paniculata* and "Blue sky"—*Hydrangea macrophylla*) with ion beam (220 Mev $^{12}C^{5+}$) at various doses (10–50 Gy) and cultured in vitro to induce variability. Effects on germination, survival, and morphological changes were recorded. The germination rate decreased with increasing irradiation dose. Germination time was also delayed after treatment. Dwarf-type and branch-type mutants were detected. Kodama et al. (2015) and Kudo et al. (1998) applied heavy ion beam radiation and developed mutants with both changed flower form and color in *Hydrangea macrophylla*.

12.78 *Impatiens platypetala*

Family Balsaminaceae; perennial species with a vertical growth habit and rigid stems, attractive plant with glossy ribbed leaves; propagated by seeds.

Alston and Sparrow (1962) determined the somatic mutation rate for petals of *Impatiens balsamina* heterozygous for flower color. The mutations are thought to result from marker loss of the dominant allele. The mutation rate for the *L* allele per r of daily exposure determined at 200 r/day was 1 in 0.76×10^6 cells. The dosage-response curve appears to be non-linear at the higher dose rates. Arisumi (1973) developed stable polyploids (tetraploids) by treating seedlings and cuttings of Impatiens. He also isolated octoploids and periclinal cytochimeras. Tetraploids were larger than diploids but had greater percentages of abortive pollen and less fertility than the diploids. The octoploids were sterile and ornamentally inferior to diploids or tetraploids. Arisumi (1978) detected several different types of Impatiens with more highly variegated foliage developed through somatic mutations. The new types have been commercialized through growers and breeders. Weigle and Butler (1983) treated seeds of *I. platypetala* with 0.08 M EMS and detected a dwarf-type plant in the M_2 generation and determined that this character is controlled by a single recessive gene.

12.79 Iresine

Family Amaranthaceae; herbaceous perennial; flowering plant; cultivated as an ornamental plant; leaf colorful evergreen; propagation by cuttings.

Das et al. (1977) selected Iresine along with several other vegetatively propagated ornamentals (chrysanthemum, dahlia, hibiscus, etc.) to induce mutations using acute and chronic gamma irradiation. Both types of radiation treatments reduced growth and induced various physiological abnormalities in growth and leaf characteristics. Somatic mutants were detected in Iresine and all other experimental materials as a chimera. Chimeric mutations were established by perpetuation over 3–5 vegetative generations. Changed leaf color mutant in Iresine was isolated from chronic irradiation.

12.80 Iris

Family Iridaceae; herbaceous perennial; bulb/corm/rhizome; flowers hermaphrodite; propagated by seed, rhizomes.

Konzak and Randolph (1956) mentioned the importance of induced mutation in Iris. Hekstra and Broertjes (1968) treated dormant bulbs of a sterile cultivar

("Wedgwood") of Iris with different doses of X-rays and recommended roughly 10 Gy as the optimum dose when the bulbs are just lifted. Mutations in flower color were detected. Mutation experiments were conducted on the Dutch iris (*Iris hollandica* Prof. Blaauw) for inducing genetic variability and the development of new varieties. As reported in three articles, uniform bulbs were irradiated with gamma rays (0, 600, 800, 1000, 1200, and 1400 rad; 0, 0.6, 0.8, 1.0, 1.2, 1.4 Krad; 600, 800, 1000, 1200, and 1400 rad) and studied effects on different characters like bulb sprouting, plant height and survival, leaf number and size, perianth size and bulb size, number and weight. All characters showed significant reductions with an increase in radiation dose in vM1. The LD_{50} was found to be 1200 rad. The reduction in all the characters in vM_2 was not significant. Variants with changed flower shapes were obtained (Rather and Jhon 1996, 2000; Rather et al. 2002).

12.81 *Ipomoea purpurea* (L) Roth.

Family Convolvulaceae; known as Morning glory; herbaceous annual; propagation by seeds, cuttings.

Parkhi and Khalatkar (1988) made reproductive alterations through induced mutations in *I. prupurea*. Bhate (1999) applied chemical mutagens (EMS, SA, and NMG) on two varieties ("violet blue" and "red purple") of *I. purpurea* and isolated chlorophyll, flower color, and floral morphological mutations in M_2 populations. In a separate experiment, Bhate (2001) did a mutation experiment to change the flower color and corolla morphology of two *I. purpurea* varieties. The violet blue (VB) cultivar had flowers with a typical gamopetalous corolla and the red purple (RP) cultivar had flowers with an incised corolla at one or more places at the petal union, i.e. incompletely gamopetalous and had petaloid appendages (petalomaniatic) borne on the outer side. Application of mutagens EMS, NMG, and SA induced an incompletely gamopetalous petalomaniatic mutant in cultivar VB as well as the typical gamopetalous corolla morphotype in cultivar RP. Hybridization studies were made using cv. RP and its white color mutant as one of the parents and crossed with cv. VB and studies revealed that independent loci exist for the incomplete gamopetalous nature of the corolla and its petalomaniatic tendency. The mutants obtained were corolla whorl specific and were described as meristic mutations in *I. purpurea*.

12.82 Jasmine

Family Oleaceae; leaves deciduous/evergreen; shrubs/vines; flowers white or yellow; propagation by seed, cuttings, layering, sucker, grafting, budding, and tissue culture.

Nambisan et al. (1980) developed leaf spot resistance and dwarf mutants in Jasminum. Kumar et al. (1983) irradiated softwood cuttings of *Jasminum* with gamma rays and found 30 Gy was almost lethal and the optimum dose was around 5 Gy. Ghosh et al. (2018a) treated terminal cuttings of *Jasminum grandiflorum* cv. "White Pitchi" and semi-hardwood cuttings of *J. nitidum* (clone Acc.Jn-1 selected at TNAU) and *J. multiflorum* cv. "Arka Arpan" with different concentrations of EMS (5–50 nM). There was a gradual and significant reduction in survival of cuttings, shoot length, number of leaves, leaf length, and width with an increase in the dosage of EMS. LD_{50} dosage was reported to be 37.15, 40.7, and 42.66 mM for "White Pitchi," "Arka Arpan," and ACC Jn-1, respectively. In a separate experiment, Ghosh et al. (2018b) treated semi-hardwood cuttings of three different jasmine species (*J. grandiflorum*, cv. White, *J. multiflorum* cv. Arka, Arpan and *Jaminum nitidum* pre-release culture Acc. Jn-1) with different doses of gamma rays (5–40 Gy). Based on survival and growth rate data LD_{50} dosages of gamma rays for cv. White, Arka Arpan, and ACC Jn-1 were determined 17.8, 28, and 25.1 Gy, respectively. Mutagenic (gamma rays) sensitivity has been studied and determined LD_{50} dose based on morphological and floral characters in *Jasminum* and *Jasminum sambac* (Ghosh et al. 2018c, 2019; Kannan et al. 2002). Mekala (2009) observed various vegetative and floral changes after mutagen treatments in Jasminum sambac cv. Gundumalli. Mekala et al. (2010) reported gamma radiation and EMS effects separately and in combination on semi-hardwood cuttings of Gundumalli (J. sambac). Mutations produced a higher amount of phenotypic and genotypic coefficient of variation for improvement of Gundumalli for yield and novelty by selection. Healthy terminal cuttings of *J. grandiflorum* cv. CO.1 Pitchi and semi-hardwood cuttings of *J. auriculatum* cv. CO.1 Mullai was treated with different concentrations of EMS ranging from 5 to 50 mM (5, 10, 15, 20, 25, 30, 35, 40, 45, and 50 mM). Significant reductions in the survival of cuttings, shoot length, number of leaves, leaf length, and leaf width were observed with an increase in dosage and LD_{50} dosage of EMS was determined to be 42.65 and 44.6 mM for CO.1 Pitchi and CO.1 Mullai, respectively (Ghosh and Ganga 2019). Ghosh et al. (2020) treated Jasminum grandiflorum cv. "White Pitchi" with gamma irradiation and EMS to assess the effect of the mutagens and to assess the mutants using molecular markers. Terminal cuttings were treated with gamma rays (10, 15, 20, and 25 Gy) and EMS (25, 30, 35, and 40 mM). Plant height, the number of primary branches, leaf area, flower yield, and flower bud length were reduced with an increase in the dosage of the mutagens. Lower doses of gamma rays induced earliness in flowering. Comparative biochemical analysis, scanning electron microscope (SEM) imaging, and molecular analysis were carried out to understand the nature of the mutation. Molecular analysis based on ISSR data revealed that Jaccard's similarity indicated that there was not much significant variation at the genetic level between the parent and the putative mutants. SEM analysis of one of the putative mutants revealed that the epidermal cells of exposed leaves appeared relatively deformed with sparse trichomes compared to unirradiated leaves. Nelka et al. (2021) exposed rooted cuttings of *J. officinale* to gamma rays (0, 17, 21, 25, 29, and 33 Gy) and observed a significant decrease in survival at 25–33 Gy and determined 33 Gy as LD_{50}. Different types of abnormalities were noticed at different doses like reduction in plant height and shoot length, shoot number, reduced petal width, etc. No change in floral characters was noticed. Gopitha et al. (2022) treated directly and in M1V1 semi-hardwood cuttings of *Jasminum sambac* cv. Ramanathapuram Gundumalli with acute gamma irradiation (0–40 Gy) and decapitated rooted cuttings were treated with EMS (0–90 nM) and gamma rays (rays (15, 20, 25, and 30 Gy) and EMS (45, 60, 75). LD_{50} was determined 21.37 Gy for gamma rays and 64.57 mM for EMS. Treated plants showed variable responses to mutagens for growth and flowering characteristics. Lavanya et al. (2022) treated cuttings of *Jasminum auriculatum* ecotype "Muthu Mullai" with gamma rays and EMS to determine the Lethal Dose (LD_{50}) and the growth reduction dose (GR_{50}). Parameters selected for the study were the mortality rate, survival percentage, shoot length, root length, number of leaves, number of sprouts, vigor index, leaf length, and leaf width. The LD_{50} value was 12.479 Gy for gamma rays and 13.268 mM for EMS treatment. The GR_{50} for different growth parameters ranged from 14.93 to 22.9 Gy for gamma rays and 1.05 to 19.9 mM for EMS treatment. The mutagenic efficiency and effectiveness were 214.96 and 89.36 for GI and 48.66 and 33.77 for EMS treatment, respectively. These doses can be used for generating considerable variation, which can be put to use in future crop improvement programs for Jasmine.

12.83 Kalanchoe

Family Crassulaceae; succulent; shrubs or perennial herbaceous; annual or biennial; ornamental houseplant; flowers yellow, red, orange, pink, white; propagation by offsets, stem cuttings.

Broertjes and Leffring (1972) X-irradiated freshly detached leaves of *Kalanchoë* cv. Annette and Josine and then planted to root and produce adventitious plantlets. The mutations observed were changes in flower color and size, time of flowering, type of inflorescence, leaf form, size and color, plant habit, and other directly visible characteristics. Mutations were non-chimeric which indicated that the apex of the adventitious plantlets might have originated from a single cell. Results indicated that the age of the leaf, the presence or absence of the petiole as well as the genotype proved to be important factors in the production of adventitious plantlets. X-irradiation produced mutations in leaf structure and habit in *K. laciniata* and stimulation in bulbil formations in *K. daigremontiana*. The majority of the mutants were solid and the optimum dose determined was 30 Gy (Nakronthap 1974; Sharma Rao and Singh 1976; Sharma Rao 1977). Van Dordrecht (1984) treated leaf-induced callus adventitious shoots with X-rays and observed a high percentage of development of solid mutants. The optimum dose was reported 20 Gy. Horn (1984) detected maximum solid mutants after treating leaf explants, adventitious shoots, callus, or cell suspension culture after treating with gamma rays (15 and 30 Gy). Karper and Pierik (1981) suggested in vitro propagation by adventitious shoots is a good procedure to isolate mutated sectors. The radiosensitivity of Kalanchoe has been determined at 20 Gy of X-ray or gamma rays. This has been determined based on the treatment of small plants, seeds, pollen egg cells, etc. (Johnson 1948; Schwemmle and Robbelen 1962; Stein and Sparrow 1963, 1966). Brown spot (caused by

Stemphylium lycopersici) is one of the common foliage diseases in Kalanchoe. This disease tends to infect the leaves of kalanchoe plants in hot and humid environments, reducing their aesthetic value. Li et al. (2019) attempted to develop and generate mutations resistant to brown spots in "Mary" kalanchoe through chemical mutagenesis followed by molecular marker identification. Probable variants were developed by treating embryogenic calluses with EMS at median lethal doses (LD_{50})–either a 0.8% concentration for 2 h or a 1.0% concentration for 0.5 h. Three regenerated EMS mutant lines showed no obvious brown spot lesions on their leaves after the disease resistance assay and were subjected to polymorphism identification by start codon targeted (SCoT) molecular markers. Three (SCoT40, SCoT71, and SCoT72) of 45 selected primers were chosen to identify the mutants. This work may lay the foundation for the further development of new disease-resistant cultivars of kalanchoe.

12.84 Kohleria

Family Gesneriaceae; rhizomatous; herbs or subshrubs; propagation by stem cuttings, division of rhizomes, tip cuttings.

Parliman and Stushnoff (1979) treated fresh-cut leaves of *Kohleria eriantha* and *K.* x "Longwood" with a wide range of gamma ray doses to evaluate the adventitious buds. *K. Eriantha* was not successful to propagate from leaf half cuttings. "Longwood" produced a small number of adventitious plantlets. Low and moderate doses of gamma rays increased plantlet production, several mutants per 100 surviving leaf halves, optimum production of all mutants of useful (flower color) and of dwarf mutants were obtained at 2.5 kR for noncolchicine-treated leaf halves and 1.5–3.0 kR for colchicine-treated leaf halves. Geier (1983) successfully applied chemical mutagen to adventitious shoots and developed in vitro compact growth habit mutant in Kohleria. In other experiments, Geier (1988, 1989, 1994, 2012) treated internode explants from in vitro grown shoots with NMH and isolated one mutant with enhanced flowering capacity under low light conditions. This mutant was again treated with NMH and developed another mutant that flowers under low light conditions and was coupled with significantly shorter internodes and smaller leaves. He has discussed many aspects related to chimeras.

12.85 *Lagerestroemia indica*

Family Lythraceae; shrubs/deciduous tree; woody ornamental; flowers purple, pink, red, white; ornamental flowers; propagation by seeds, cuttings.

Boddie and Whitcomb (1978) reported mutagenic effects on Lagerestroemia indica. Li et al. (2015) characterized and sequenced the leaf transcriptome of a L. indica yellow leaf mutant, named YL03. They reported that a total of 79 unigenes were involved in chlorophyll biosynthesis and degradation, photosynthesis, and chloroplast development. They have explored the possible formation of a pathway of yellow leaf mutants based on their analysis. Comprehensive transcriptome analysis discovers novel candidate genes related to leaf color in a Lagerstroemia indica yellow leaf mutant (Zhang et al. 2014). Yim et al. (2010) investigated the sensitivity, seed germination, and response of quantitative characters of 18 ornamental plants to proton beam radiation. Seeds were irradiated at the dose of 0–2 kGy of the proton beam at room temperature by 45 MeV MC-50 Cyclotron. They studied proton beam induced mutation in Lagerstroemia indica and Ligustrum obtusifolium. 750 Gy induced mutants in leaf length, leaf width, internode length, plant height, leaf color, autumn leaves, and plant width in each strain. The potential of each mutant was analyzed for registration and commercial viability. One new cultivar "Bulkkot" was selected for a patent. Xia (2020) treated seeds of *Lagerstroemia indica* with different doses of ^{60}Co-γ and an electron beam to record their germination and seedling growth and to find out the best irradiation source and dose. In both treatments, the germination rates were first promoted and then declined. Survival of seedlings decreased significantly with the irradiation dose. Plant height was inhibited after electron beam irradiation. Plant height was first inhibited and then promoted after gamma irradiation. Branch length and branch number significantly inhibited both treatments. The ^{60}Co-γ irradiation had little inhibition effect on the ground diameter of the seedlings, while the electron beam irradiation had a highly different effect on the ground diameter. They found the optimum irradiation dosage 113.06–299.63 Gy for ^{60}Co-γ and 245.5–372.24 Gy for electron beam. *L. indica* was more sensitive to ^{60}Co-γ irradiation.

12.86 *Lavandula intermedia* Emeric

Family Lamiaceae; herbaceous perennial; flowering plant; propagation by seed, cuttings, divisions.

Tsuro et al. (2008) studied the effects of different doses of $^{12}C^{6+}$ beams and gamma rays on callus growth and shoot formation of *Lavandula intermedia* Emeric. Large differences in responses of callus growth and shoot formation were observed between the two ionizing radiations. In $^{12}C^{6+}$ beam irradiation, callus proliferation was strongly restricted with an increase in irradiation dose. In gamma irradiation up to 20 Gy, callus proliferation was stimulated with an increase of irradiation dose, although, over 20 Gy, callus growth was inversely restricted with an increase of irradiation dose. Similar results were observed in shoot formation. These results suggested that a large difference in biological effect for lavandin cultured cells existed between $^{12}C^{6+}$ beams and gamma rays.

12.87 Lilium

Family Liliaceae; herbaceous flowering plant; perennials bulbous; propagated mostly vegetatively (bulb scales, bulbils, bulblets), seed and in vitro methods.

Lilium is an interesting material for mutation studies considering its nature of heterozygosity, chromosome number, and size. Radiosensitivity and mutation studies generated both basic and applied knowledge for mutation work on bulbous ornamentals. Radiosensitivity has been determined based on chromosome length, chromosomal aberrations both in meiotic and mitotic divisions, lethality, mutation, etc. (Brown and Cave 1953, 1954a, b; Brown and Zohary 1953, 1955; Bowen and Sparrow 1954; Crouse 1954; Mitra 1958; Lizuka and Ikeda 1963; Taek et al. 2005). "Faraday" is the first new flower color mutant variety developed from the variety "Fantasy" through radiation treatment (De Mol, 1926 c.f. Broertjes and Van Harten 1988). Mutants with new flower color, short flower stalk, and chlorophyll-variegated leaves have been developed through X-ray treatment and commercialized (De Mol 1949; Custers et al. 1977; Van Eijk and Eikelboom 1981a, b). Hopper and Peloquin (1968) treated styles of self-incompatible *L. longiflorum* cultivars with 24–70 kr X-rays to study radiation effects on pollen tube growth. Results indicated that X-irradiation provides a method for obtaining self-seed as well as a possible means of exploring the biochemical nature of this pollen-style interaction. Low doses of X-rays (2000–4000 r) were applied to the pollen of *L. formosanum* and studied different aspects related to seed formation, failure of ovule development, pollen tube growth, fertilization, dominant lethal, pseudogamy, etc. (Brown and Cave 1953, 1954a, b). Loh and Cooper (1966) treated pollens of *L. regale* with different doses of gamma rays (1, 2, 4, 8, and 10 kR) and studied radiation effects on germination, number of dividing generative nuclei, meiotic aberrations, chromatid breaks, isochromatic breaks, fragments, interchange, etc. and correlated their percentage with dosages. Van Groenestijn and Van Tuyl (1983) commercialized flower color and low light tolerant mutants which were developed through X-irradiation. Seeds of annual wild species of *L. grandiflorum* var. rubrum (large ruby corolla and a dark red spot at the center of the flower) were treated with EMS (0.01, 0.05, 0.1, and 0.5%) for 18 h to create variability in floral characters. High frequency of morphological mutants (chlorophyll deficiency, leaf waxy bloom, corolla color, flower shape and size, and plant habit, new var. "Aurora") were detected in M3. 0.05 and 0.1% EMS was determined as the most optimal dose to induce mutations (Largon and Lyakh 2002; Lyakh and Lagron 2005). Wang et al. (1989) treated bulb scales of *L. davidii* var. "Willmottiae" with gamma rays (500–2000 r) and cultured in vitro supplemented with colchicine (1–4 mg/l), NAA, and BA. Regenerated plantlets were further treated with gamma rays (100–500 R) and colchicine (1–4 mg/L). The proliferation and regeneration of plantlets were inhibited significantly. No mutation has been reported. A huge number of bulb scales were exposed to X-rays (2.5 Gy) for the development of new genetic variations for commercial exploitation. Many varieties developed solid promising mutants (new flower color, low light tolerant) (Cuany et al. 1958; Broertjes 1969a, 1972b; Broertjes and Alkema 1970; Anonymous 1991a, b; Chinone et al. 2008; Grassotti et al. 1987). Chiba et al. (2007) studied mutants of Asiatic hybrid lily and Lilium × formolongi Hort. induced by ion beam irradiation. Six flower color mutants were observed in the M_1 generation of the Asiatic hybrid lily, and two pollen-less mutants were observed in the M_2 generation. Carbon ion irradiation seemed to be more effective than Helium ion to induce mutants of Asiatic hybrid lily. Chinone et al. (2007) exploited possibilities of induction of mutation in Oriental hybrid Lily by treating bulb scales in vitro with 320 MeV carbon and 50 MeV helium ion. The suitable dose was estimated to be around 0.2–0.4 Gy in carbon ions and 1.5–2.0 Gy in helium ions. Leaf color mutations were observed. Kondo et al. (2007) treated sliced scales of an Oriental hybrid lily, *Lilium* cv. "Acapulco" with 0–2.0 Gy of 50 MeV 4 He^{2+} beams and cultured in vitro. The doses of 0.2–0.8 Gy suppressed the regeneration rate to 50–90%. They determined doses of 0.2–0.8 Gy of 50 MeV 4 He^{2+} to be appropriate for obtaining mutants. These results suggest that scales are more insensitive to ion beams than calluses. Kondo et al. (2008a, b) applied ion beam irradiation to *Lilium* × formolongi hort (cv. White Aga) and developed mutants where the pollen formation was affected. Akhar et al. (2016) studied the effects of EMS on in vitro mutagenesis of Lilium spp. (cv. OT Geel and Robina). They treated surface sterilized bulb scales with EMS solution and cut them into 3 mm thickness TCL explants and incubated them on a regeneration medium. 0.2% EMS was the most effective concentration in the enhancement of shoot numbers in both cultivars. Molecular changes in OT Geel and Robina mutants showed that mutation can induce high polymorphism, so in OT Geel, 11 treatments were classified into four groups with a similarity level of 0.87%, and in Robina, 11 treatments were grouped into two groups with similarity level 0.86%. In all treatments, genetic variations were observed, but higher concentrations of EMS were more effective for mutation induction. The results indicated that the efficiency of ISSR markers for the detection of genetic variants is high and leads to their early selection of them. This study has been claimed as the first report to identify EMS-treated mutants using ISSR marker in Lilium plants. Girsang et al. (2021) studied the effects of different concentrations of colchicine (0, 100, 200, and 300 ppm) and different treatment period (3, 6, 9, and 12 h) on lily. Results indicated that the concentration of colchicine (300 ppm) influenced the plant tallness in the first week until the sixth week when planting and the concentration of colchicine in 100 ppm considerably influenced the size of the bud after flowering and the span of bloom. Hajizadeh et al. (2022) treated lilium bulbs with gamma rays (0, 10, 20, 30, 40, and 50 Gy) to study the radiation effect and to induce variability. Different parameters like the number of leaves, leaf area, fresh and dry weight, stem height, leaf relative water content, ion leakage, chlorophyll content, etc. were selected to measure the radiation effects. Irradiations showed a significant difference in leaf number, ion leakage, and chlorophyll a, b, and total. The optimal dose of radiation was computed between 22.5 and 27.5 Gy. The consequence of treatment on leaf relative water content was not significant. In general high gamma irradiation, doses had harmful effects on the growth of Lilium.

12.88 *Limonium sinuatum* Mill (Statice)

Family Plumbaginaceae; short-lived herbaceous perennial/annual plant; popular cut flower for arrangement due to long keeping quality and wide range of flowers; flowers white to pink, purple, and yellow; propagation by seeds, cuttings.

Chinone et al. (2008) applied ion beam irradiation to induce mutations in *Limonium sinuatum*. The growth of the shoot on *L. sinuatum* irradiated with 320 MeV carbon ions decreased at 1.0 Gy. The suitable doses for mutation induction were estimated to be around 1.0 Gy for *L. sinuatum*. Crossbreeding has developed new varieties but no superior pink variety is available. Ogawa et al. (2014) aimed to develop a statice mutant with pink calyx by heavy ion beam irradiation. Multiple shoot cultures of statice "Kishu Fine Grape" were irradiated with a C-ion beam (5–30 Gy) and determined that 5 Gy was a suitable dose based on survival and rooting data. They further irradiated one purple cultivar ("Kishu Fine Grape") and two light purple cvs. ("Kishu Fine Lavender" and "Kishu Star") with c-ion beams at doses of 2 and 5 Gy. "Kishu Fine Lavender" developed six pale flower color mutants. "Kishu Star" developed ten mutants comprising pale, deep color, and reddish combinations. Results suggested that heavy ion beam irradiation is suitable for inducing various useful mutants in statice.

12.89 *Lonicera japonica*

Family Caprifoliaceae; shrubs or twining vine; flowers are double-tongued; propagated by seeds, cuttings, and layering.

Boyarskikh et al. (2016) studied the mutagenic effect on plants growing in active fault zones being the factors of the evolutionary transformation of plant populations. They evaluated the mutation activity of a *Lonicera caerulea* natural population in one of the active fault zones in the Altai Mountains. They studied mitotic abnormalities of meristematic cells of *Lonicera caerulea* seedlings and observed that the local geological and geophysical environment (i.e. mineralogical composition of rocks and anomalies of the magnetic field) increased the mitotic activity and the number of abnormal mitoses in the meristematic cells. The results may help to clarify the role of environmental conditions of tectonically active regions in micro-evolutionary processes. Microcuttings from five honeysuckle species were irradiated in Petri dishes at the time of subculturing. Bud survival was defined as the percentage of buds yielding shoots longer than 10 mm after 4 months of post-irradiation culture. Cultivar survival was investigated after irradiation with gamma ray doses from 10 to 60 Gy. A methodology was developed to increase the probability of obtaining a mutation induced in one cell giving rise to a mutated sector in the regenerated plant (Cambecedes et al. 1992).

12.90 Lotus (*Nelumbo nucifera* Gaertn.)

Family Nelumbonaceae; economically important aquatic plant; perennial; propagation by dividing rhizomes, fibrous root, seeds.

Arunyanart and Soontronyatara (2002) treated in vitro raised plantlets with acute gamma rays (0, 2, 3, 4, 5, or 6 krad) and/or X-rays (0, 1, 2, 3, 4, or 5 krad) to induce mutations. They have recorded differential dose-specific effects on different characters. However, the overall alterations reported were 1 and 2 krad—long secondary roots, numerous adventitious roots, good shoot growth, healthy rhizome development; 3–5 krad—vitrification, chlorosis, deformed petioles, inhibition of growth of lateral buds, secondary roots, and rhizomes; 6 krad—no survival. Aneuploid cells were detected at the doses of 3 and 4 krad and abnormal stomata at 4 krad. Wu et al. (2007) studied outer space-induced mutation of lotus seeds (No. 36 space lotus seed) and compared their proximate composition and nutritional components with the native lotus seeds. HPLC fingerprints studies confirmed that the quality of the most chemical contents of No. 36 seeds was better than that of native seeds. Lama et al. (2005) exposed seeds of *N. nucifera* to different doses of gamma rays (0, 2, 4, 6, 8, and 10 Kilorad) and embryos were cultured in vitro. Fifty percent growth reduction (GR_{50}) was recorded at 6 Kilorad. All plantlets were subcultured at regular intervals. Control plantlets had the highest shoot growth and adventitious root formation; 2-kilorad treated plants showed second highest growth; 4-kilorad treatment exhibited no adventitious root formation and the growth of all plantlets at 6–10 kilorad was inhibited. Data on adventitious shoot regeneration will be helpful for future mutagenesis work for the development of solid mutants. Arunyanart and Chaitrayagun (2005) standardized callus induction and somatic embryogenesis in *Nelumbo nucifera* Gaertn.) cv. "Satabankacha" from the culturing bud, cotyledon, and young leaf explants. Methods will be very helpful for the improvement of lotus through induced mutagenesis and genetic engineering. Vichai et al. (2011) did gamma irradiation experiments by treating rhizomes with10 Gy to induce photoperiod-response mutation in American lotus *Nelumbo lutea* and in the interspecific hybrid *N. lutea* x *N. nucifera* "Poontarik." The mutant showed a maximal day-length period of approximately 13.5 h in comparison to the control (14.5 h). The mutant was crossed and a hybrid between *N. nucifera* "Bua Luang Phrarachinee" x the mutant *N. lutea*, was selected showing large flower size and pink petals, and good adaptability character. Liu et al. (2016) standardized and induced embryogenic callus from cotyledons. Technological details for embryogenic formation will be very helpful for the future development of new varieties through transgenic transformation and in vitro mutagenesis not only in lotus but also in other ornamentals. Soontornyatara et al. (2017) attempted to induce morphological and cytological variations through gamma irradiations. Seeds of *N. nucifera* cultivar "Pathum" were exposed to gamma rays (0, 10, 20, 30, 40, 50, 60, 70, 80, 90, and 100 Krad) and determined 28 krad as LD_{50} on germination basis and 50 krad as lethal dose. Different types of morphological abnormalities (twin leaves and different types of dispersed spot leaves) were detected in the treated population and recommended optimal radiation dose to be around 20–30 krad. Liu et al. (2019) standardized and optimized an improved and efficient micropropagation method using mature embryos in *N. nucifera* "Weishan Hong." This will help future in vitro mutagenesis work in lotus.

12.91 Mesembryanthemum

Family Aizoaceae; succulents; perennial herb; propagation by seeds, cuttings.

Chaturvedi et al. (1997) treated seedlings (13 cm height) of Mesembryanthemum with gamma rays (5, 10, 20, 30, 40, and 50 Krad) and studied radiation effects on pollen features. Pollen grains in the control plants are basically 3(-4)-colpate-colporoidate. Besides, parasyncolpate grains occur as a variation. The pollen size is variable having a wide range (P x E: 17–37 x 14–35 μm). The pollen sculpture is spinulose-punctate. In the irradiated plants, the pollen size range drastically narrowed down. While the pollen grains of the plants treated with 5, 20, 30, 40, and 50 Krad dosages had the same pollen morphology as in the controls, the plants treated with 10 Krad dosage showed a complete change in the exine ornamentation exhibiting "Microoverrucate" pattern as against the spinulose pattern, characteristic of the species. Since the apertural character did not deviate from the basic 3-colpate-colporoidate condition, it has been contended that the exine surface pattern and the apertural features may be considered as separate characters controlled by different genes and that the change in the exine pattern may be attributed to the mutation of the gene/genes controlling the pattern. Seeds of *M. cristallinum* L. were soaked with colchicine (0.01, 0.05, 0.10, and 0.20%) for 24 h for polyploidy induction and mutation. Colchicine affected vegetative growth and the number of chromosomes. The colchicine concentration of 0.10% was found to affect the length, width, and thickness of the leaves and changed the chromosome number from diploid ($2n = 18$) to Octaploid ($8n = 108$) (Qalby et al. 2020).

12.92 *Moluccella laevis*

Family Lamiaceae; it is also known as Bells of Ireland or shell flower; annual; grown for spikes with small white flowers; propagated by seeds.

Dry and soaked seeds were treated with different doses (0, 2.5, 5, 7.5, 10, 12.5, 15, 17.5, and 20 Kr) of gamma rays and studied radiation effects on germination, survival, growth, and morphological changes. Maximum germination of dry seeds was recorded at 2.5 and 5 kr and maximum in wet seeds at 2.5 kr. Dry and wet seeds had a differential response to different doses as recorded based on increased plant height, increased branch number, dry weight of vegetative growth, changes in flowering date, increased flower number per branch, stimulative effect, morphological variations, etc. (Minisi et al. 2013).

12.93 Muscari (*M. armeniacum*)

Family Asparagaceae (formerly Liliaceae); perennial bulbous plant; propagated by leaf cutting and adventitious buds.

No recent mutation work has been reported on this plant. In vitro propagation methods and induced mutation breeding work have been reported long back

(Broertjes and Alkema 1970; Roest and Bokelmann 1981). Leaf pieces were exposed to different doses of X-rays and determined 0–15 Gy as the optimum working dose based on rooting and the number of bulblets per leaf. Leaf pieces were treated with 0.2–0.4% colchicine. Bulblets were selected from the treated population with different changed characteristics like flowering time, changed inflorescence color, size and form, growth habit, and leaf characters. Most of the mutants were solid and non-chimeric. Autotetraploids with changed characters (sturdy growth habit, large inflorescence, thick pedicel) were also isolated from colchicine treatment (c.f. Broertjes and Van Harten 1988). Grape hyacinth (*Muscari* spp.) is a famous bulbous blue flower. Few bicolor varieties are available in the market. Ma et al. (2023) realized that the discovery of bicolor varieties and understanding of their mechanisms are crucial to the breeding of new varieties. They have reported a significant bicolor mutant with white upper and violet lower portions, with both parts belonging to a single raceme. They made a critical biochemical analysis. Ionomics showed that pH and metal element contents were not responsible for the bicolor formation. Targeted metabolomics showed that the content of the color-related compounds was significantly lower in the upper part than that in the lower part. Further analysis revealed differentially expressed genes in which anthocyanin synthesis gene expression of the upper part was significantly lower than that of the lower part. Transcription factor differential expression analysis described the presence of a pair of *MaMYB113a/b* sequences, with low levels of expression in the upper part and high expression in the lower part. They concluded that the differential expression of *MaMYB113a/b* contributes to the formation of a bicolor mutant in *Muscari latifolium*.

12.94 Narcissus

Family Amaryllidaceae; perennial herbaceous bulbiferous; flowers hermaphrodite; propagation by seeds, offsets, stem bulblets, and division of basal sections.

De Mol (1926) and Zandbergen (1964) reported spontaneous mutations in cultivars of *Narcissus*. The first mutation breeding work was initiated using several cultivars and a huge number of bulbs of two cultivars ("Chanterelle" and "Poesy") were treated with 7–8 Gy X-rays but no mutation was detected. Similarly, two varieties ("Baby Moon" and "New Baby") of *N. jonquure* were treated with 4–5 Gy X-ray but no mutation was induced (c.f. Broertjes and Van Harten 1988). To develop the adventitious bud technique, Alkema (1974a, b) treated scales of the cultivar Golden Harvest with 4 or 5 Gy X-ray, and after three growth seasons two aberrant plants were detected. Datta et al. (2003) treated bulbs of *Narcissus tazetta* cv. "cicily white" with different doses of gamma rays (250, 500, and 1000 rad) to study the radiation effects on chromosomes and pollen grains. Early separation, bridges, and laggards were recorded in a few cells during meiotic cell division. All the gamma-radiation doses affected mitotic cell division and the percentage of dividing cells reduced with an increase in doses. A wide range of chromosomal abnormalities were detected in all treatment doses. Micronuclei formation was the major abnormality and early separation, bridges, laggards, and exclusion were the next major abnormalities recorded in all the doses and the abnormalities increased mostly with an increase in dose. Pollens of control plants of *Narcissus* are boat-shaped, monocolpate with reticulate exine where the lumina is circular and the top of muri is flat. No change in apertural character was found in the pollen of irradiated plants while significant changes in exine ornamentation pattern comprising size and shape of lumina and type of muri were observed in all plants with all irradiated dosages. The changes in exine ornamentation involve structural elements of pollen exine and thus it is pointed out that the gene/s controlling pollen exine ornamentation is more radiosensitive than the gene/s controlling aperture character of pollen. Lu et al. (2007) treated bulbs of Chinese narcissus (*Narcissus tazetta* var. "chinensis") with gamma rays ((5–100 Gy) and grown in vitro to evaluate the radiation effects on adventitious bud formation from bulb scales and the survival rate of plantlets. There was a significant decrease in survival and multiplication rate with an increased radiation dose and the optimum dose was determined approximately 10 Gy. DNA fingerprinting (RAPD and AFLP) technique was applied to find out the genetic variations among the regenerants and found a high frequency of mutants which indicates that treatment of in vitro cultures with γ-rays may be an effective way to improve narcissus cultivars.

12.95 Nautilocalyx

Family Gesneriaceae; evergreen perennial; herbaceous; leaves very glossy, attractive; propagation by seeds, stem cuttings, leaf cuttings.

No mutation work has been reported on this plant. As mentioned, the adventitious bud technique has good importance in the induction of mutations. There are few early publications on how superficial cell layers of *N. lynchiican* be grown in vitro and their organo-genetic potential can be directed by auxin, cytokinin, and sucrose (c.f. Broertjes and Van Harten 1988; Tran Thanh Van 1973; Tran Thanh Van and Drira 1971). These studies provided added information on the origin of adventitious organs as reported by others (Venverloo 1974; Venverloo et al. 1983; Choura 1938). Studies will be helpful for future mutagenesis work on *Nautocalyx*.

12.96 *Nerium oleander*

Family Apocynaceae; perennial evergreen shrub or small tree; known as oleander; flower pink; propagation by cuttings.

Saniya et al. (2023) did a mutation experiment with two varieties (pink and white) of Nerium oleander to determine the LD50 dose for EMS. Ten to 15 cm length semi-hardwood cuttings were treated with EMS (0.0, 0.1, 0.2, 0.3, and 0.4%). After incubation at room temperature for 12 h, the cuttings were thoroughly rinsed with running tap water for 10 min and planted. The results showed a steady and significant reduction in the survival of cuttings with an increase in the dosage of EMS. The probit curve analysis based on the survival percentage revealed the LD_{50} dosage of EMS to be 0.4% and 0.3% for nerium pink and white genotypes, respectively.

12.97 *Nertera granadensis*

Family Rubiaceae; known as coral bead plant, pincushion plant, coral moss, English baby tears; herbaceous perennial; covered with coral color berries; attractive foliage; suitable for pot culture; propagation by seeds, tip cuttings, divisions.

The author could not find any recent report on mutation breeding work on this plant for improvement. The only report is available in Broertjes and Van Harten's (1988) book where the objective of mutation breeding was to create variability in the size and color of berries. Plants were treated with X- or gamma rays and developed a huge number of plants through divisions of foliage. The optimum dose was reported to be 10–15 Gy and no mutation was detected. Nothing can be concluded from this preliminary experiment and considering the beauty of the plant more mutation breeding work is expected for its improvement.

12.98 *Nymphaea rubra* Roxb.

Family Nymphaeaceae; aquatic perennial rhizomatous herb; flowers bisexual; propagation by stolons, vegetative buds.

Gupta (1977) treated healthy rhizomes (4–5 cm in diameter) of *N. rubra* with gamma rays (1, 2, 3, 4, 5, and 6 krad) and studied radiation effects on morphological and cytological changes. The 5 krad and higher doses were found to be lethal. Primary unstable effects of radiation showed morphological changes in vegetative and floral parts. Stable changes were recorded in different characters like a change from toothed to the entire margin of the leaf, numerical alterations in petals and stamens, induction of narrower or twisted petals with obtuse tips, changes in the contour of androecium and intensity of the color of petals were stable. Chromosomal aberrations were recorded in treated plants and pollen sterility decreased after radiation treatment.

12.99 Orchid

Family Orchidaceae; perennial herbs; epiphytes; rhizomatous/corms/tubers; propagation by seeds, vegetative means.

Kozlowska-kalisz (1979) made comparative studies on the biological activities of endogenous growth regulators in meristematic tissue of radiation-treated and non-treated Cymbidium. The meristematic tissues after 5 weeks of culture were exposed to gamma rays (0.8, 20.0, and 70 kR) and reported 0.8 kR as a stimulating dose, and total inhibition and lethal effect were observed in 20.0 kr and 70.0 kR, respectively. Gamma radiation was found to change the activity of gibberellin, auxin, and cytokinin-like substances. Protocorms of spring orchid (Cymbidium), developed by tissue culture of aerial parts, seeds, and stems, were treated with ultraviolet (UV) radiation and colchicine. After UV irradiation, the cell nucleus became deformed. Frequencies of cell division increased with UV dose, causing phenotypic variation among regenerated plants (Fen and Chu 1997). Kuang (1999) exposed in vitro cultured protocorm-like bodies of two orchid hybrids ("Mokara Chark Kuan" and "Dendrobium Jacky") to different doses of gamma rays (0, 20, 25, 40, 50, 60, 75, 80, 100, 125 and 150 Gy) and determined optimum dose 20–40 Gy for "Mokara Chark Kuan" and 60–70 Gy for Dendrobium Jacky, respectively. The protocorm-like bodies of two novel varieties of miniature hybrid Cymbidium (bred by crossing Cymbidium hybrida with Cymbidium goeringi) were irradiated with 320 MeV $^{12}C^{6+}$ ions (doses 0–128 Gy) and cultured in vitro. Regeneration drastically decreased in both varieties at 16 Gy or more. Based on these results further experiments were conducted on three varieties and applied doses of 10–12 Gy (Yuki et al. 2007). Affrida et al. (2008) treated protocorm-like bodies (PLBs) of three orchid species (*Dendrobium crumenatum*, *Dendrobium mirbellianum*) at various doses with 320 MeV i2 C^{5+} ions and determined the optimum irradiation condition and effects on flower color and morphology, flowering habit, and insect resistance. Some morphological changes were observed in flowers of *Dendrobium crumenatum*, while one insect-resistant mutant was obtained in *Dendrobium mirbellianum*. Aurigue et al. (2008) included different species of Spathoglottis (*S. kimballiana* var. angustifolia, *S. plicata*, *S. tomentosa*, and *S. vanoverberghii*) to create genetic variability by colchicine and gamma ray treatment. No effect of colchicine was recorded. Survival of irradiated protocorms decreased with increasing dose from 0 to 50 Gy. The average height of seedlings and the length of the longest root were significantly affected by gamma radiation. 10 Gy acute gamma ray treatment at the ptotocorm stage created variations in morphological characters (purple pigmentation on the flower stalk, shorter internodes or distance between flowers, thicker substance of individual flowers and wider or stouter leaves, lack of pigments or albinism, purple pigmentation on leaves, forked leaves, split seedlings or furcation, and multiple branching). One prominent mutant was used as a female parent in the breeding of Spathoglottis "Lion of Singapore" and selected a hybrid with a different character. Micro-propagation of orchids is performed either by nodal shoot multiplication or through the induction of protocorm-like bodies (PLBs). Chen et al. (2008) standardized multiplication using PLBs. Somaclonal variation was observed in the tissue culture process and its frequency depends upon different factors related to cultural processes. Several mutant genotypes (peloric flowers) were collected from nurseries and applied molecular techniques were to compare peloric mutants with wild-type flowers. Such studies will help to understand the development of orchid plants by molecular dissection of gene functions and classical crossing and genetic analysis. Gonzales et al. (2008) studied gamma radiation radiosensitivity (lethal dose and optimum dose) of three species of ground orchids (*S. plicata*, *S. kimballiana* var. Angustifolia, and *S. tomentosa*) by treating germinated embryos (protocorms). Different characteristics like the percent mortality of seedlings, seedlings' height, number of roots, and root length were selected as parameters. Treated plants showed both increase and decrease in leaf number (at 10 and 20 Gy), albinism, pigmentation, forked leaves, furcations, multiple branching, etc. PLBs of *Dendrobium friedericksianum* Rehb. f. were treated with EMS (0, 0.25, 0.5, 0.75, and 1%) for 60 and 90 min then cultured in vitro. EMS at a concentration of 0.8% for 90 min decreased the survival rate to 50% (LD50) PLBs. EMS at a concentration of 1% for 90 min gave the highest results (12 shoots/explant). Morphological observation revealed that EMS at concentration 1% for 60 min gave 10% albino plants and 1% EMS for 90 min gave three characteristics of chimera, sectorial (Muangsorn and Te-chato 2008). Khosravi et al. (2009) observed variations in plantlets of *D. Serdang* Beauty regenerated from protocorm after colchicine treatment. RAPD technique has been standardized to detect polymorphism between the mutated regenerant DSB V, 13 species of Dendrobium, and 13 orchids across generas. Pimonrat and Suraninpong (2009) successfully induced mutations in *S. plicata* Blume by gamma irradiation. Khoddamzadeh et al. (2010) detected somaclonal variation in the protocorm culture of *Phalaenopsis bellina*. The radiation-induced mutants are mostly detected based on phenotypic changes. Ariffin et al. (2012) standardized techniques to quantify gene expression (as RNA isolation, cDNA synthesis, and primer design steps for CHS genes in different flower tissues) of functional markers for the selection of desired flower color mutants. Sequence variations of the CHS partial gene in *Dendrobium Sonia* mutants with different flower colors showed that two protein positions have been changed as compared to the control. Luan et al. (2012) standardized the maximum shoot bud proliferation medium composition for callus formation of *Paphiopedilum delenatii* and *Paphiopedilum callosum*. The LD_{50} of protocorm-like bodies, shoot buds, and in vitro, plantlets were determined for *P. delenatii* (20, 23.7, and 38 Gy) and *P. callosum* (23, 27.1 and 40.4 Gy), respectively. No variant line was detected in the gamma ray treated population. Twenty-four variant lines were screened for samples irradiated by C^{6+} ion beams (3 Gy) and analyzed the genetic relationships among six generated variant lines and wild types using RAPD techniques. Atra et al. (2014) applied gamma rays (30 to 100 Gy) to *Spthoglottis plicata* Blume to induce further genetic variability in flower color. Induced genetic variations were identified using morphological characters and ISSR markers and concluded that genetic diversity existed among orchid mutants and this variation would be of importance for further orchid breeding programs. Lee et al. (2015) studied the effects of a proton beam (45 MeV proton beam—mean linear energy transfer (LET) = 1.461 keV/μm), a 100 MeV proton beam (LET = 0.7306 keV/μm), and gamma ray (LET = 0.2 keV/μm) in the *Cymbidium* hybrid RB001 [(*C. sinensis* × *C. goeringii*) × *Cymbidium* spp.] to find out the feasibility of proton beam in mutation breeding program for *Cymbidium*. They have suggested that the proton beam is expected to be a useful tool for developing new mutant varieties of *Cymbidium*. Lee et al. (2016) investigated the effects and damage caused by γ-radiation in a *Cymbidium* hybrid, RB001. The Relative growth rate of protocorm-like bodies (PLBs) was reduced by 50% at a γ-ray dose of 40 Gy and malondialdehyde concentrations increased significantly with increasing radiation dose. Effects of radiation on several antioxidant defense enzymes indicated that gamma rays caused little DNA damage and suggest the feasibility of using physiological responses, the "comet assay," and flow cytometry to detect DNA damage after γ-irradiation. Wannajindapom et al. (2016) selected an important commercial orchid *Dendrobium* "Earsakul" for improvement through in vitro mutagenesis by treating protocorm-like bodies with sodium azide. They used morphological characters, molecular markers, and the cytological method to select and evaluate the mutants. Morphological differentiation like reduced height, higher numbers of nodes, reduced node length, shorter and thicker leaves, shorter and fewer roots, etc. was observed in some putative mutants compared with controls. Altered DNA profiles were detected in all putative mutants. There was no change in chromosome number. Results indicate that NaN_3 can be used effectively to mutagenize *Dendrobium* "Earsakul" PLBs, and ISSR is a powerful tool for the identification of mutants. One-month-old seeds of *S. plicata* were exposed to X-rays (0, 6, 12, 18, and 24 rad) and grown in vitro to induce genetic variability. Genetic variability was determined using PCR RAPD-based technology. Twelve to 18 rad doses were able to induce morphological variations in seedlings characters (leaf, root and shoot, early flowering) (Aloysius et al. 2017). Protocorm-like bodies (PLB) of Dendrobium Sonia-28, an important ornamental orchid in the flower industry, were treated with gamma rays (10–200 Gy) to induce variability. The survival rate and weight of PLBs were inversely related to the irradiation doses. LD_{50} for the PLBs was approximately 43 Gy. Plantlets infected with low doses of gamma radiation had better development of the shoot and root as well as the foliage. SEM and TEM analysis disclosed severe surface and cell organelles damage and stomatal closure in protocorm-like bodies (PLBs) infected with the high irradiation dose. Some RAPD markers discriminated between selected and non-selected variants of Dendrobium Sonia-28, showing different banding patterns for each gamma irradiation dose (Dehgahi and Joniyasa 2017). Hernández-Muñoz et al. (2017a) treated protocorms of *Laelia autumnalis* with gamma rays (0, 5, 10, 15, 20, 25, 30, 35, 40, 45, and 50 Gy) to determine the lethal dose. Radiation effects were evaluated based on different parameters like the number of leaves and roots; root length; the number of leaf and pseudo-bulbs; fresh biomass; seedling biomass, etc. Differential responses of these parameters were observed in different doses. In a separate experiment, Hernández-Muñoz et al. (2017b) exposed seeds of *L. autumnalis* to gamma rays (3–30 Gy, with 3 Gy intervals) to stimulate seed germination. Low doses showed stimulating effects whereas higher doses were negative in delaying germination and seedling development. Philodendron erubescens "Gold" is a popular climber with brilliant greenish-yellow leaves. Dayani et al. (2018, 2021) exposed rooted cuttings ($n = 200$) of P. erubescens "Gold" to gamma rays (70, 100, 150 Gy) and recovered them in a propagator. Survived and promising selections were maintained for up to 12 generations (V12) to evaluate growth and morphological variations with their genetic stability. Only one regenerated M_1 plant showed morphological variations in leaves, which were multiplied and maintained as promising lines. Several variations including characteristics of leaves (shape, size, color), stems (internodal length and branching) and plant stature were observed among M_1 lines and in subsequent vegetative generations. Leaves presented three different color patches, however, neither the color distribution pattern was uniform nor stable. M_1V_4 line shows the highest stability of color distribution in leaves; the color composition of leaves ranged as 0–10% dark bluish green, 60–90% strong yellow-green, and 10–30% brilliant greenish yellow throughout the 12 generations. They have mentioned that gamma-irradiated P. erubescens "Gold" line M_{1-4} can be a promising mutant to develop as a new Philodendron cultivar. Kim et al. (2019a, b) analyzed the effects of the total dose and irradiation duration on the growth of *Cymbidium* hybrid RB001 protocorm-like bodies (PLBs) and determined the optimum working dose (1 h, 16.1 Gy; 4 h, 23.6 Gy; 8 h, 37.9 Gy; 16 h, 37.9 Gy; and 24 h, 40.0 Gy). Results indicated that the irradiation duration affects PLB growth in response to γ-rays. Pimonrat et al. (2012) treated 3-month-old seedlings of *Spathoglottis plicata* Blume with acute gamma rays (0, 2, 4, 6, 8, and 10 Krad) to induce variability. No survival was recorded at 2 Krad higher doses and 1.43 Krad was determined as LD_{50}. Few clones were identified with significant morphological differences (dwarf stem, low tillering number, thick leaf, and short inflorescence with long spots on the sepals and petals and undulated petal margin) which finally could be promoted as a new cultivar for commercial production. The AFLP technique was applied which confirmed the genetic diversity of variants with a polymorphism rate of 1.89–29.85%. Kim et al. (2020) studied the dark/light treatments followed by γ-irradiation and noted increased chlorophyll-related leaf mutants by 1.4- to 2.0-fold compared with γ-ray treatment alone. Dark/light treatments combined with γ-irradiation increased the frequency of leaf color mutants in *Cymbidium*, which supports the wider implementation of a plant breeding methodology that increases the mutation frequency of a target trait by controlling the expression of target trait-related genes. Li et al. (2021a, b) made a review on orchid breeding and discussed the advantages of many breeding techniques to develop varieties with unique characteristics, including flower color, morphology, and resistance. They have mentioned that mutation breeding (natural and induced) is very suitable for orchids due to the high mutation rate, breaking of the relationship of traits, effective improvement of individual traits, and shortening of the breeding cycle. Above all any new changes can be easily propagated vegetatively as mentioned by others Kharkwal et al. 2004; Guo et al. 2016; Yamaguchi 2018; Toker et al. 2007). Similarly, polyploidization has been recommended as a good option for the development of new orchid varieties. Polyploidy breeding has been successfully applied in many orchid species like Cymbidium (Wang et al. 2011), Dendrobium (Li and An 2009; Zhang et al. 2011), Oncidium (Cui et al. 2010a), Phalaenopsis (Cheng 2011; Cui et al. 2010b; Yin et al. 2010; Jin et al. 2012). Sherpa et al. (2022) treated protocorm-like bodies of *Dendrobium* "Emma White" with gamma rays (10, 20, 40, 60, 80 Gy) and studied in vitro growth and induction of mutations. Both proliferation and regeneration of PLBs decreased progressively with increasing doses, except for a significantly enhanced growth response at 10 Gy. The optimum dose was reported in the range of 10–25 Gy. DNA content significantly reduced at 40 Gy. Root length, plant height, and leaf number were significantly increased, and earliness in flower development at 10 Gy. The mutation technique has been enriched with many more publications which highlighted the mutation approaches for inducing genetic variations in different orchids. These papers cover different technical aspects of induction mutagenesis using ion beam radiation, gamma rays, in vitro mutagenesis, acute irradiation, combined mutation, transgenic techniques, etc. Different species and varieties included in the experiments were *Spathoglottis plicata* Blume, *Phalaenopsis amabilis* (L.) Blume, *and Dendrobium* cv. Sonia, different species of *Dendrobium* (*D. crumenatum*, *D. mirbellianum*), etc. Proto-corm-like bodies were mostly used as explants. The optimum irradiation condition and the effect of irradiation on each species were studied, particularly on flower color and morphology, flowering habit, and insect resistance (Affrida et al. 2008; Romeida et al. 2012; Sulistianingsih 2013; Sheela and Sheena 2014; Sheela et al. 2008; Sakinah et al. 2005; Mohd Nazir and Sakinah 2001; Mohd Nazir et al. 2003; Sakinah and Mohd Nazir 2002; Hassan et al. 2010; Kurniati 2004; Khoddamzadeh et al. 2010). The black orchid (Coelogyne pandurata Lindley) is a native Indonesian orchid species that are considered difficult to be grown in cultivation and may take years to flower. Widiarsih and Dwimahyani (2023) did a mutation experiment with gamma rays (0, 5, 10, 15, 20, and 25 Gy) to investigate the mutagenic effects of gamma rays. Different parameters like survival rate, number of leaves, number of roots, number of shoots, leaf color, and plantlet type were selected for observations. Plantlets of 10 Gy have the best survival rate and the number of new shoots, but the only number of leaves in the 5 Gy dose was significantly higher than the control. Growth was almost stunted at 25 Gy treatment. Root growth was significantly reduced in plantlets of 15 Gy and above. The maximum morphological changes were found in 10 and 15 Gy, and this treatment range was recommended for future mutation breeding research of black orchids.

12.100 Ornamental Tree

It is necessary to develop the improved strains of ornamental plants with more disease-resistant and useful for landscape or phytoremediation. Gamma rays and X-rays are widely used as mutagens to induce mutations. Proton beam had higher energy than gamma-ray and worked with localized strength, so that proton beam radiation could be valuable tool to induce useful strains of ornamental plants. Proton ion beam irradiation was used to induce useful mutant in chrysanthemum and carnation in Japan. Kwon et al. (2009) reported the effects of proton beam and expected proton beam to be a new mutagen. They conducted the experiment to investigate the proton beam radiation sensitivity and seed germination rate of the various ornamental plants like as Albizia julibrissin, Ficus religiosa, Rhus chinensis, Sorbaria sorbilfolia and Spiraea chinensis and to survey the quantitative characteristics of proton beam induced strains. Seeds were irradiated at the dose of 0–2 kGy of proton beam at room temperature. Proton beam energy level was 45 MeV and was irradiated at dose of 0–2 kGy by MC-50 Cyclotron. They assessed the effects of proton beam on radiation sensitivity and morphological changes of the plants and the seed germination rate. The germination rate decreased at the higher dose. The effects of mutation induction by proton beam irradiation on seeds in Lagerstroemia indica were investigated. Irradiation with proton beam at the dose of

750 Gy induced mutants in leaf length, leaf width, internode length, plant height, leaf color, autumn leaves and plant width in each strains. Promising strain (strain 25) for commercial varieties was selected Lagerstroemia indica. It was analyzed that strain 25 showed the highest genetic dissimility from original species. The strain 25 had red leaf edge and maintained autumnal tints till late fall. They tried to promote a patent registration of the strain 25 as a new caltivar "Bulkkot."

12.101 *Ornithogalum virens* L.

Family Asparagaceae; herbaceous perennial; bulbous; propagation by seeds, bulb division, or twin-scaling.

Leaf segments of *O. thyrsoides* Jacq. were treated with 3–10 Gy doses of X-ray or 1.6–5 Gy fast neutrons. Leaf segments were also treated with 0.1–0.4% colchicine for 7 h. Optimum dose for X-ray reported 10 Gy and for fast neutron 3.3–4 Gy. Few plants were selected from all treatments with changed characters like small size, small flowers, changed flower form, white flower with a green tinge, etc. (c.f Broertjes and Van Harten 1988). Biswas and Biswas (2006) irradiated seeds of *O. virens* L. with gamma rays (20–40 krad) and developed partially sterile strains. Meiotic chromosomal behavior and pollen grain sterility were studied. They observed dicentric bridges, acentric fragments, paracentric inversion, bivalents, univalents, early movement, lagging, unequal separation, etc. in pollen mother cells. Seeds of *Galtonia candicans* (syn. *Ornithogalum candicans*) and cape hyacinth were treated with different concentrations of EMS (0, 0.2, 0.4, 0.6, 0.8, and 1%). Seed germination decreased and no seedling survived at 0.6, 0.8, or 1%. Pollen sterility was reduced to 3% in the 0.4% EMS. EMS has been suggested as a viable option to reduce plant height and decrease seed set (Contreras and Shearer 2020).

12.102 Osteospermum

Family Asteraceae; known as daisy bushes and African daisies; annual/perennial; long flowering time; propagation by seed, cuttings.

Lizuka et al. (2007) irradiated with ion beams and selected 37 flower color and morphological variants in Osterospermum "Mother Symphony." Mutant flower colors were pastel color, white and orange. The upper and lower sides of the petal had different colors. The irradiated variants included orange- and yellow-isochronous variants. No white color could be induced. Further efforts were made to induce the white variant, they irradiated the white petal variant OM7 of "Mother Symphony" with ion beams. The leaf-disc sections of re-irradiated variant OM7 exhibited higher sensitivity to ion beams than those of "Mother Symphony." The optimum irradiation dose was determined to be 0.1–5 Gy. Color variations at the upper side of the petals included purple and pale orange and color variations at the underside of the petals included pale yellow and orange. No isochroous variant was obtained. Other detected mutants were dwarf, multi-petal, and petal deformations.

The second irradiation of OM7 resulted in the acquisition of variants such as purple and multi-petals that did not appear in the first irradiation experiment, suggesting that re-irradiation induced a stepwise mutation in pigment synthesis and morphogenesis pathways. Lizuka et al. (2011) from further experiments developed a new cultivar "Viento Flamingo" through irradiation. Okada et al. (2012) irradiated Osteospermum "Mother Symphony" with ion beams, and selected the pastel color variant. Osteospermum has different colors between each side of the petals. But, they could induce the orange- and yellow-isochronous variants from irradiated "Mother Symphony." But no white could be induced in earlier experiments. They irradiated again the white petal variant OM7 of "Mother Symphony" with ion beams to obtain white-isochronous petals and developed the OM706 that has light yellow on the back side of the petal. They re-irradiated again leaf sections of OM706 with carbon ion beams (220 MeV $^{12}C^{5+}$ and 320 MeV $^{12}C^{6+}$) at a range of 0.5–5 Gy. The resultant plantlets were investigated to detect any flower color and morphological changes. OM706 which had the white petal on the upper side and the light yellow petal on the back side was derived from re-irradiated the white petal variant OM7. The third irradiation of OM706 induced a lot of mutations in petal color or plant type compared to the first and second irradiation. In a color variation, yellowish-white, whitish light orange, whitish light yellow, and pale yellow mutants and lighter yellow in the back side of the petal than OM706 were obtained.

12.103 *Paspalum notatum* Flugge (Bahiagrass)

Family Poaceae; perennial turf grass used for residential lawns and sides of roadways; propagation by stolons, and rhizomes.

Kannan et al. (2015) selected Bahiagrass for inducing genetic variability through in vitro chemical mutagenesis. The turf quality of bahiagrass is limited by its open growth habit, light green color, and prolific production of tall seedheads. The mission of induced mutation work was to develop uniform, mutant seed progeny with improved turf quality. Different concentrations of Sodium azide were applied to seeds and plants were regenerated via somatic embryogenesis through callus. Thorough screening of all selected va riants/mutants resulted in the isolation of the superior line showing higher density, finer leaves, an upright growth habit, dark green color, reduced seedhead formation, uniform seed progeny, and superior drought tolerance.

12.104 Pelargonium

Family Geraniaceae; herbaceous perennials, succulents, shrubs; known as geraniums, pelargoniums, or storksbills; propagation by seeds, cuttings.

Early experiments on induced mutagenesis were carried out on *P. zonale* cv. "Madama Salleron," "Freak of Nature," "Kleiner Liebling," "Salmon Beauty," "Larschen Muller," and. *P. graveolens* and mutagens used were colchicine, X-rays

(5–10 Gy, 10–12.5 Gy), NMU, gamma rays, etc. The objectives of early work were to discover arrangements of periclinal chimeras, change in ploidy level, and to induce morphological changes (leaf chlorophyll content, shape and color; flower color, shape, and size). Mutation frequency was low but in vivo adventitious technique and protoplasts developed non-chimeric mutants (Bergann 1967a, b; Bergan and Bergan 1959; Pohlheim 1977; Potsch 1964; Daker 1966, 1967; Pohlheim et al. 1972, 1976; Stewart et al. 1974; Craig 1963; Skirvin and Janick 1974, 1976a, b; Grunewaldt 1983; Janick et al. 1977; Kameya 1975). Lamina and peduncle pieces of Pelargonium and Saintpaulia were treated with MMU and gamma rays and regenerated in vitro. The efficiency of in vitro mutagenesis has been recommended based on increased nonchimera mutations (Grunewaldt 1983). Yu et al. (2016) exposed shoots of geranium (*Pelargonium hortorum*, Bailey) to carbon ion beam radiation (0, 10, 15, 30, and 40 Gy) and isolated one flower color mutation. They studied different characteristics like morphological, physiological, and DNA polymorphism of mutants and compared them with those of the wild type.

12.105 Peperomia

Family Piperaceae; epiphytes/lithophytes; perennial shrubs or vine; ornamental foliage; flower green, white, brown; propagated by seeds, cuttings, dividing.

No published report is available on induced mutagenesis work on peperomia although its propagation methods are suitable for generating basic information on mutation work. Adventitious buds are easily grown on detached leaf/leaf parts and in vitro (Harris and Hart 1964; Henny 1978). Adventitious buds easily develop from green and yellow portions of detached leaves of *P. obtusifolia* variegate and shoots originate from one/few cells. Bergann and Bergann (1982) from regeneration studies of three variegated Peperomia from detached leaves/leaf pieces and shoot axes mentioned that adventitious buds develop from one or more L-III cells. The optimum dose of X-ray was reported to be 20–30 Gy based on survival or plantlet production (c.f. Broertjes and Van Harten 1988).

12.106 Petunia

Family Solanaceae; technically perennial but mostly annual herb; many groups based on flower size; propagation by seeds, stem cuttings.

Moore and Haskins (1935) studied the effects of X-rays on the flower color of Petunia. They treated flower buds and seeds with X-rays and observed modifications in flower color. Colijn et al. (1979) reported that *N*-methyl-*N′*-nitro-*N*-nitrosoguanidine (MNNG) is an effective mutagen for *Petunia hybrida* suspension cells by using mercury(II)-chloride ($HgCl_2$) and Dl-6-fluoro tryptophan (6FT) as positive selection conditions. Twelve-hour soaked seeds of white diploid *Petunia nyctaginiflora* Juss were treated with MMS (0.009; 0.012; 0.018 and 0.025 M) for 3 h and *N*-methyl-*N′*-nitro-*N*-nitrosoguanidine (MNNG) (0.012; 0.025; 0.037 and

0.050 M) for 1/2 h at 28 °C. Six morphological and seven chlorophyll mutants were isolated from M_2 segregating families. The isolated induced mutants were described and genetically analyzed (Mahna and Garg 1989). Khalatkar and Kashikar (1980) did for the first time experimental evidence for the mutagenicity of aqueous sodium azide (pH 6.6) in Petunia. Dry and pre-soaked seeds of a diploid white-flowering strain of *Petunia hybrida* Hort. ($2n = 14$) were treated with aqueous SA and EMS for 18 h at 24 + 2 °C for comparative mutagenicity studies. Different parameters like germination, seedlings height, chlorophyll variegations, etc. were recorded in different treatments. Pure line white flower variety seeds were treated with EMS or gamma rays and several flower color changes were noticed in M_1 in EMS treatment and few in the gamma ray treated population. In M_2 large variability in flower color was noticed in EMS treatment. The types of flower color alterations produced by gamma radiation were different than those of EMS (Kashikar and Khalatkar 1981). Male sterility (MS) is a particularly useful tool for obtaining hybrid varieties. Male sterility prevents unwanted selfing and thus helps to produce pure hybrid seed. There are two systems of male sterility, i.e. Genie male sterility (GMS) and cytoplasmic-nuclear male sterility (CMS). They have discussed different aspects related to GMS and CMS and suggested how mutation techniques can be utilized to induce CMS (van Harten and Bal 1986). Gerats (1991) described mutations in petunia and mentioned how the petunia mutant system can be used for molecular research on floral development. The most interesting mutants were "Blind" and "Green petals" where the petals were changed. The petals in "Blind" changed toward antheroid and in "Green" toward sepals. He has also highlighted the approach to isolate developmental genes in Petunia using transposon tagging. Van Houwelingen et al. (1998) developed 50 flower pigmentation mutants in *P. hybrid* using endogenous transposable elements (TEs) as a mutagen. The majority of the mutants were unstable. The mutants showed alterations either in anthocyanin biosynthesis, in the intracellular pH of petals, or in the shape of petal cells. Molecular analysis indicated that mutants for novel loci are most likely tagged by dTph1 elements opening the way for their isolation. Berenschot et al. (2008) studied the mutagenic efficiency in terms of genomic damage by analyzing the developmental characteristics of the plantlets after treating seeds with gamma radiation (0, 20, 40, 60, 80, and 100 Gy) and EMS (0, 0.05, 0.1, 0.15, 0.2 and 0.25% (v/v). High doses showed reduced growth and seedling viability. Moderate DNA damage allowing a high saturation of mutant alleles in the genome and the generation of viable plants for reverse genetics studies was correlated to the biological parameter LD_{50}, the dose required to kill half of the tested population. Novel morphological mutants were identified. Miyazaki et al. (2002) induced variegated mutants in Petunia hybrida applying heavy-ion beam. Okamura et al. (2009) reported the production of new types of flower color variations in petunia obtained by ion beam irradiation. Buds of vigorous variety of petunia cv. "Kirin Hana-Saka Rose" (bright rose flower) were irradiated with 320 MeV carbon ions of 10 Gy and cultured in vitro. Mutations detected were a pink color, cherry color, striped white, spotted petals, etc. Hase et al. (2010) observed that petunia seedlings treated with 3% sucrose before exposure to 320-MeV carbon ions from 8 days after sowing accumulated a significant amount of pigments within 4 days compared to non-treated control seedlings. Chlorophyll mutants were obtained in both treated and non-treated groups with a similar frequency in M_2. Flower color mutants were magenta, purple, light pink, and white over the original violet color. The frequency of flower color mutants was significantly higher in the sucrose-treated group than in the non-treated group. These results suggest that sucrose pretreatment specifically increases the frequency of flower color mutation following the ion beam. Dona et al. (2013) studied exposure to different dose rates of gamma rays on *Petunia x hybrid.* They explored the dynamics of DNA damage accumulation and molecular mechanisms after exposing leaf discs to LDR (total dose 50 Gy, delivered at 0.33 Gy/min) and HDR (total doses 50 and 100 Gy, delivered at 5.15 Gy/min) γ-ray in the 0–24 h period after treatments. They observed significant fluctuations of double-strand breaks and different repair capacities were observed between treatments in the 0–4 h period following irradiation. The results have been discussed in light of the potential practical applications of LDR-based treatments in mutation breeding. Jiang et al. (2014) standardized conditions for mutagenizing petunia with EMS minimizing deleterious effects on viability and fertility. Three parameters the imbibition of seeds before EMS treatment, the EMS concentration, and EMS exposure time were selected for this study. The mutagenesis of 2000 petunia seeds for TILLING was conducted with a 12-h imbibition followed by exposure to 0.1% EMS for 12 h. They developed a mutagenized population of the doubled haploid *P. hybrida* line "Mitchell Diploid" as a resource for TILLING. Amanda et al. (2008) treated seeds with gamma rays (0, 20, 40, 60, 80, and 100 Gy) and EMS (0, 0.05, 0.1, 0.15, 0.2, and 0.25% (v/v) and recorded decreased seedling viability. They evaluated and noted moderate genomic damage allowing a high saturation of mutant alleles in the genome and the generation of viable plants for reverse genetics studies.

12.107 *Philodendron erubescens* "Gold"

Family Araneae; climber with greenish-yellow leaves; used in indoor gardening and landscaping; propagation by stem cuttings.

Karunananda et al. (2018) treated rooted cuttings of *P. erubescens* "Gold" to different doses (70, 100, 150 Gy) of gamma rays to create variations for commercial purpose. Regenerated shoots were evaluated at 12 generations based on the genetic stability of growth and morphological variations. Twelve and 1% survival were observed in 70 and 100 Gy, respectively, after 6 months. M_1 selected lines showed different morphological variations in leaves (shape, size, color), stems (internodal length and branching), and plant stature. Different color patches were detected on leaves but their distribution pattern was not uniform and stable. Stability of color distribution was observed in leaves of the M1-4 line whose color composition was dark bluish green, strong yellow-green, and brilliant greenish yellow.

12.108 Phlox

Family Polemoniaceae; perennial/annual; flowers tubular; flowers blue, purple, pink, red, white; propagated by seeds, stem cuttings.

Pathak and Raghuvanshi (1980) treated seeds of yellow and white varieties of *Phlox drummondii* Hook with different doses of gamma rays (10–50 krad) and studied radiation effects on germination, survival, pollen fertility, and seed set. A polypetalous variant was observed at 10 krad in the white variety, showing suppression of meiosis. Seeds of *P. drummondii* were irradiated with gamma rays (5, 10, 15, 20, and 25 kR) and studied effects on mitosis. Different types of chromosomal abnormalities were recorded like dicentric, tricentric, translocation, deletion, fragment, ring, minute, bridge, and micronucleus in treated root tip mitosis. Chromosomal aberration frequency increased in mitosis along with the doses of gamma rays (Verma and Raina 1980; Pillai and Verma 1992; Verma and Sharma 2000; Ramesh and Verma 2015). Seeds were treated with colchicine (0, 0.1, 0.25, and 0.5 g/l) for 24 and 36 h. Germination and survival of seedlings were decreased. Based on all experimental results they have mentioned that 0.5% colchicine is suitable for the induction of somatic mutations in *P. drummondi* (Tiwari and Kumar 2010). Tiwari and Mishra (2012) treated seeds of the red flower color variety of *P. drummondi* with colchicine (0.01, 0.025, and 0.05% solutions for two-time scales of 24 and 36 h) and seeds were sown in the field. Germination and survival were severely affected by different concentrations of colchicine. Treatments reduced plant height and number of leaves per branch, but branch number increased. Stomatal size and frequency helped identification of ploidy level.

12.109 *Plectranthus* (Coleus, *Coleus blumei*, and *Coleus amboinicus* Lour)

Family Lamiaceae; different synonyms *Coleus blumei*, *Plectranthus scutellarioides* and *Solenostemon scutellarioides;* evergreen perennial; decorative variegated leaves; propagation by seeds, cuttings. [**Coleus**: Family—Lamiaceae; herbs/shrubs, sometimes succulents; perennial (usually grown as an annual); beautiful foliage (in shape, style, and color); propagation by stem cuttings].

Love and Mullenax (1964) detected non-genetic disorder after treating cuttings of Coleus blumei with gamma radiation. Quite a large number of variegated mutants were developed after treatment of cultivar "Scarlet Red" with fast neutron (2–10 Gy). The dwarf mutant was also isolated (Love and Constantin 1965, 1966). Love and Malone (1967) analyzed pigments of mutant coleus and reported that differences among the mutants were due to variations of the anthocyanin pigments.

Cuttings were treated with gamma rays (15, 30, 45, and 60 Gy) to determine the LD_{50} dose and radiation effects. Radiation treatment significantly affected height, number of leaf nodes, and growth habits. LD_{50} for yellow/green, green/brown, and variegated green/brown of *Coleus blumei*, and *Coleus amboinicus* Lour were 48.66, 65.2, 52.81, and 37.62 Gy, respectively (Aisyah et al. 2015). Aisyah et al. (2017)

applied acute and fractionated gamma radiations on shoot cuttings of Coleus spp. to increase genetic variability. Radiation effects were recorded on different parameters in M_1. Fractionated doses (40, 45, and 50 gy) produced eight putative mutants in *C. blumei* and acute radiation (40 and 45 gy) developed three putative mutants.

12.110 Populus

Family Salicaceae; deciduous flowering plant; many are grown as ornamental trees; flowers mostly dioecious; propagation by seeds, cuttings.

Biswas et al. (2013) attempted a forward genetics approach to develop drought/salt-tolerant mutants using a combination of ion beams and in vitro regeneration system in Populus. Newly grown canes of 3-to 4-month-old plants were cultured in vitro and newly regenerated healthy rooted shoots were screened for further growth and root development in NaCl-supplemented rooting medium. Regeneration of treated shoots and roots and their survival rate in soil indicated that ion beam can be used as a potential mutagen to produce mutants with a wider mutation spectrum for drought/salt tolerance. Nishiguchi et al. (2012) induced small plant height due to short internode after gamma irradiation in Populus.

12.111 *Portulaca grandiflora* Hook.

Family Portulacaceae; succulent flowering plant; known as rose moss, 11 o'clock, Mexican rose, moss rose, the sun rose, rock rose, and moss-rose purslane; annual/perennial; flowers single and double types; propagation by seeds, cuttings.

Gupta (1966, 1970) treated both annual and perennial varieties with gamma rays and developed six mutants ("Karna Pali," "Mukta," "Ratnam," "Jhumka," "Vibhuti," and "Lalita") with changed flower color and types. 40 Gy was reported as the optimum dose. Lata and Gupta (1971) studied morphological and cytological characters of gamma ray induced mutants of perennial portulaca. Cotter (1963) exposed *P. smallii* to cobalt field (1000 r, 5000 r, 15,000 r at three dose rates, and 30,000 r) and exposed control and treated plants to two photoperiods, i.e. 14 natural which favors vegetative growth, and 10 h which helps maximum floral initiation. Both vegetative and reproductive ability was suppressed with increasing doses but reproduction was more sensitive than growth. Several mutants (flower color and type) have been developed by different scientists using gamma irradiation (Banerjee 1967; Desai 1973, 1974; Abraham and Desai 1978; Skirvin et al. 1982). The optimum dose reported was 10–25 Gy. Raghuvanshi and Singh (1979, 1980) after treating 2*n* and 4*n* cuttings of *P. grandiflora* Hook. Var. perennial with gamma rays (2, 4, and 6 krad) determined that 4*n* had a higher mutation frequency. Mutations were noted in floral and morphological characters. Flowering was suppressed in the mutant which has been correlated with the lack of florigen (Abraham and Desai 1977). Kruczkowska et al. (1997) reported the effect of EMS and NaN_3 on callus culture and plant regeneration of *Portulaca grandiflora*. Seeds of *Portulaca* *grandiflora* variety PgBmj (white) were cultured in vitro and after 1 month callus was placed on a multiplication medium supplemented with EMS at concentrations ranging from 0.033 to 0.1% (v/v) for 3 weeks. For NaN_3 treatment, the callus was immersed in the mutagen solution at concentrations ranging from 0.1 to 10 mM at pH 3 for 1 h. From the data of callus survival and shoot induction percentage the optimum dose for each mutagenic treatment was determined. A good number of morphological mutants affecting shoot length and shape were recovered. This finding has made a substantial contribution to the induction of genetic variability in *Portulaca gandiflora*. Stem cuttings of two perennial *Portulaca grandiflora* varieties, "Double Orange" and "Double Pink" were exposed to gamma radiations (0, 10, 20, and 40 grays) for induction of mutations. Three mutated vegetative characters (flower color, form, and size) were vegetatively propagated as new varieties ("Chompoo Praparat," "Pattik," and "Som Arunee"). Acute gamma irradiation application to stem cuttings induced flower color mutations (Wongpiyasatid and Roongtanakiat 1992; Wongpiyasatid and Hormchan 2000; Tangsombatvitchit et al. 2008). Stem cuttings of two varieties ("Double Orange" and "Double Pink") were exposed to gamma rays (0, 10, 20, and 40 grays) and observed mutations in flower color, form, and size. A total of three mutants ("Chompoo Praparat," "Pattik," and "Som Arunee") were isolated from treated populations of both varieties (Wongpiyasatid and Hormchan 2000). Bennani and Rossi-Hassani (2001) attempted to increase betalain synthesis through new genetic markers by seed mutagenesis. White and red seeds were treated with EMS (1.2–40% in 0.1 M phosphate buffer (pH 7.0 at 22 + 2 °C) for 4 h, or with sodium azide (NaN_3) (2.5–30 mM in 0.1 M phosphate buffer (pH 3.0 at 22 ± 2°C) for 1 h. Five morphological mutations (dwarf, late flowering, male sterile, female sterile and dwarf mutant with altered leaves) were selected from the M_2 population. Aisyah et al. (2022) aimed to evaluate the diversity of morphological characters of P. grandiflora mutants in the MV8 population resulting from the recurrent irradiation technique. The recurrent irradiation technique was applied in the P. grandiflora mutation breeding program and produced new morphological variations of P. grandiflora until the MV7 generation. Fifty-three selected mutants (MV7) of the P. grandiflora generation were multiplied by stem cuttings to develop the MV8 population. The diversity of morphological characters of P. grandiflora in the MV8 population was identified quantitatively and qualitatively. Several P. grandiflora mutants in the MV8 population have a new phenotypic appearance, chiefly in the qualitative characteristics of flowers.

12.112 *Prunus lannesiana*

Family Rosaceae; known as Oshima Cherry; deciduous tree; hermaphrodite; propagated by seeds, cuttings, and layering.

Prunus lannesiana "Gioiko" is the only Japanese ornamental cherry variety that has a double flower with 10–12 petals. The color of the flower changes from pale green to light purple during blooming time. Ishii et al. (2009) treated 150 mm length scions of *Prunus lannesiana* "Gioiko" with carbon beam irradiation ($^{12}C^{6+}$,

135 MeV/u, LET 22.6 keV/m) were 10, 15, and 20 Gy. The irradiated scions were grafted on the rootstock (*Prunus lannesiana* "Viridis"). Survival decreased with an increase in doses. No survival was recorded in 20 Gy. One changed flower color mutant was isolated from 10 Gy irradiation. The mutant flower color was pale yellow at the beginning of blooming time and then changed to yellowish pink. The color of the rims of the petals was whitish green. It was very interesting that the period from grafting to the first bloom of the mutant line was 1 year but 3 years in the original variety which is a very useful character for both breeding and commercialization of the new cultivar. The mutant cultivar was named "Nishina Zaou" and released to the market in 2008. Ishii et al. (2011) further reported the development of new flower color mutants in cherry blossoms using heavy ion irradiation.

Hayashi et al. (2019) standardized a unique technology for the induction of mutation using heavy ion beams for Prunus. Pale-yellow cherry blossom "Nishina Zao" was created by irradiating greenish "Gyoiko" (*Prunus lannesiana*) scions with carbon ions. The mutant flowered twice during a single year and did not require a cold period for flowering and produced three times the number of flowers of the original variety if the mutant was exposed to cold winter temperatures. This mutant was named "Nishina Otome" and was commercially released in 2010. They were successful in creating new cultivars using the progeny of irradiated plants that are not changed concerning either flower color or shape. Although the original cultivar, *P lannesiana* "Beni-yutaka," has a double flower with 15–20 petals, the new cultivar "Nishina Tomoka" has a large, single flower with five petals. Heavy ion beams are thus effective tools for creating new variations of *Prunus* trees through mutation breeding.

12.113 Ranunculus

Family Ranunculaceae; known as buttercups, spearworts, and water crowfoots.; more than 1800 species of flowering plants; perennial/annual/biennial; herbaceous/aquatic/terrestrial; flowers hermaphrodite; propagation by seeds, tubers.

Alkema (1974a, b) experimented with the *Ranunculus* cultivar "Barbaroux" by treating different doses of X-rays and determined the optimum working dose of approximately 100 Gy (c.f. Broertjes and Van Harten 1988). Dorion et al. (1975) isolated viable protoplasts from leaves of *Ranunculus scelerantus* L. by one-step mixed enzyme treatment. Cultured protoplasts regenerated cell walls and cell divisions were observed occasionally. Calli formed from the cell colonies differentiated into plantlets which were then grown to complete plants. This result will give feedback for in vitro mutagenesis. Papanicolaou and Kokkini (1983) reported homoeotic mutations and inheritance of hairy sepals and anthers in *Ranunculus millefoliatus*.

12.114 *Rhododendron simsii* (Chinese Azalea)

Family Ericaceae; evergreen or semi-evergreen; propagation by seeds, cuttings, grafting, and layering.

Induced mutagenesis experimental results are available in the literature on Rhododendrons and Azalea are highlighted. Both Rhododendron and Azalea are closely related and belong to the genus Rhododendron and azaleas are recognized as a subspecies of the rhododendron family. Literature highlights that all azaleas are rhododendrons but not all rhododendrons are azaleas. The majority of rhododendron varieties are evergreen and larger in height, leaf size, and flower size compared to deciduous small azalea. Different species and varieties included in mutation experiments were *R. catawbiense* ("Album" and "Boursault"), *R. impeditum*, *R. ponticum*, *R. altaclarensis* "Christoffer Wren" (under Azalea: *R. japonicum—Azalea mollis*), *R. japonicum* (*Azalea mallis*), *R. obtusum* (Japanese azalea) cv. "Vuyk's Scarlet," cv. "Silvester," *R. simsii* (syn. *Azalea indica*), etc. For treatment mostly young, actively growing plants, rooted cuttings, unrooted cuttings, pot plants, etc. were used and gamma rays and X-rays were the radiation sources. Materials were treated with acute, chronic, and recurrent doses. Radiosensitivity was determined for experimental materials and optimum doses as observed from different experiments have been reported: 40–60 Gy X-rays; 15 Gy higher doses always reduced growth and induced morphological abnormalities. Mutations could not be detected in all the experiments. Some cultivars developed several flower color mutations. Few experiments were conducted with many cultivars and a good number of mutants have been developed and commercialized (c.f. Broertjes and Van Harten 1988; De Loose 1968, 1969a, b, 1970a, b, c, d, 1971a, b, 1973a, b, 1974a, b, c, d, e, 1979; Heursel 1972, 1981; Streitberg 1965, 1966a, b, 1967a, b, Preil and Walther 1983; Akabane et al. 1973). Knowledge of pigment composition is important in mutation. A large number of different flower color hybrids, species, varieties, and bud mutations of *R. simsii* were selected and their pigment composition was studied by TLC and absorption spectroscopy. The most common pigments identified were cyanidin 3-glucoside, cyanidin 3-galactoside, cyanidin 3-arabinoside, cyanidin 3, 5-diglucoside, cyanidin 3-galactoside-5-glucoside, peonidin 3, 5-diglucoside (acylated with caffeic acid) and malvidin 3, 5-diglucoside (acylated with caffeic acid) (De Loose 1968, 1969a, 1970a, b). De Loose (1979) applied soft X-rays recurrent irradiation on a chimeric flower color sport of *Rhododendron simsii* Planch and recommended recurrent irradiation technique for further inducing mutations based on mutation frequency and efficiency. Kobayashi et al. (2007) Kobayashi et al. (2008/2007) treated seeds and cultured leaf segments of *Rhododendron kaempferi*, *R. ripense*, *R. japonicum*, and *R. matternichii* var. brevitolium with a carbon ion beam (220 MeV $^{12}C^{5+}$) at 0–50 Gy to induce flower color and shape mutation. Germination rate reduced to 30–50 Gy in different species. The optimal irradiation doses determined were 30–40 Gy for *R. kaempferi*, 20 Gy for *R. ripense*, 10–20 Gy for *R. japonicum*, and 20–30 Gy for *R. matternichii* var. brevitolium. The regeneration of callus decreased in treated materials. Shoot primordial and shoot were not able to generate. The effective doses were thought to be from 4 to 8 Gy.

Seeds of *Rhododendron ripense*, *R. japonicum*, *R. wadanum*, and *R. degronianum* var. okiense were irradiated with carbon ion beams (220 MeV $^{12}C^{5+}$) at 0–50 Gy) and seedlings were found to grow up to 20 and 30 Gy. Survival rates were higher at low-dose irradiation conditions. Development of variegated leaves was observed in some of the seedlings (Kobayashi et al. 2009). Atak et al. (2011) selected six mutant plantlets from Alfred and seven mutant plantlets from the Paars variety from M1V2 generations and applied the RAPD technique to find out the differences between Rhododendron mutants and control plants. These mutants were selected from plantlets that were irradiated with doses of 5 and 10 Gy. Results showed that Rhododendron mutants were district from controls.

12.115 Ribes

Family Grossulariaceae; deciduous shrubs; propagation by seeds, bare roots.

"Oregon Snowflake" evolved from wild-collected seeds whose growth habit was larger and whose growth performance was poor. Mutation techniques were applied to improve its plant habit and desired leaf shape. Seeds were treated (24- and 48-h durations) with different concentrations of EMS (0, 0.2, 0.4, 0.8, and 1.2% EMS in a 0.1 M sodium phosphate buffer solution—pH 36 7.0). Desired traits with highly dissected leaf morphology were selected from 300 surviving plants and multiplied by terminal softwood cutting as a new variety (Contreras and Friddle 2015).

12.116 Rudbeckia

Family Asteraceae; herbaceous; mostly perennial; hardy; flowers compact; mustard yellow; propagated by seeds, suckers.

Shukla et al. (1986) exposed rooted suckers of *Rudbeckia laciniata* cv. "Golden Glow" to 0.5, 1.0, 2.0, and 4.0 Krad of gamma rays to induce variability. Different types of morphological abnormalities were noticed in leaf and floral characters. One plant irradiated with 0.5 Krad produced one branch with lighter yellow flowers. Two plants irradiated with 1.0 Krad, one bearing flower heads with lesser and narrower flower heads (Gerbera type) and the other with more florets (Pompon dahlia type), were detected as chimeras. Oates et al. (2013) studied the effects of gamma rays (0, 5, 10, 20, or 40 Gy) on in vitro embryogenic callus of *Rudbeckia subtomentosa* cv. "Henry Eilers." The radiation effect was evaluated on different morphological characters.

12.117 *Saintpaulia ionantha*

Family Gesneriaceae; commonly known as African Violet; popular houseplant; herbaceous perennial; flowers blue, pink, purple, and white; propagation by seeds, leaf cuttings.

Broertjes (1972c) applied acute, chronic, or fractionated doses of X-rays or fast neutrons on leaves of *Saintpaulia* to create genetic variations. Warfield (1973) detected 13% of plantlets to develop mutations after 1 h treatment of petioles of leaf cuttings with 0.5 M EMS. Petioles of diploid variety were treated with X-rays (2000. and 3000 r) and developed non-chimeric mutants which persisted through repeated vegetative propagations (Sparrow and Schairer 1980). Grunewaldt (1980, 1983, 1988) studied bud sports and attempted to induce in vitro genetic variability by treating lamina and peduncle pieces with *N*-methyl-nitroso-urea or gamma rays. Cultural methods for different genotypes have been discussed to develop new variants, especially propagation of temperature tolerant of *Saintpaula*. Espino and Vazquez (1981) cultured in vitro leaf segments of *S. ionantha* supplemented with different concentrations of colchicine and caffeine and detected a higher number of polyploid plantlets in colchicine. Cytohimeric plantlets were analyzed and highlighted their possible origin. Kelly and Lineberger (1981) irradiated cuttings of *S. jonantha* "Ulery Wend" with thermal neutrons (250, 1000, and 5000 rad) and determined the radiation effects on lethality, root formation, morphology, and peroxidase isozymes. The time of root formation was increased and high lethality was observed at higher doses. No morphological variants were detected but alterations in peroxidase isozyme patterns were noticed. Radiation increased the number of isozymes and destroyed the catalytic activity or suppressed the synthesis of native isozymes. Such changes in isozymes have been suggested to be used as an effective genetic marker for radiation-induced mutants. Geier (1983) successfully applied chemical mutagen to adventitious shoots and developed in vitro low-temperature tolerance mutant in *Saintpaulia*. Warburton et al. (1984) screened anther-derived plants and commercially available cold-tolerant varieties after growing at a range of temperatures (5–25 °C) for the selection of low-temperature tolerance traits. Anther-derived plants which were designated as low-temperature tolerant were separately screened in vivo and in vitro and considered to be physiological mutants and not epigenetic variants. The mutagen EMS did not significantly increase the number of plants able to survive the screening temperature of 10 °C. Attempts to improve the tolerance of plants to low temperatures by extending their photoperiod were not successful. Lineberger and Druckenbrod (1985) made a thorough analysis of the chimeric nature of pinwheel flowering African violets by in vitro culturing of various tissues. Studies on flowering patterns supported that the cultivars "Valencia," "Dardevil," and "Desert Dawn" were periclinal chimeras. Analysis of experimental results has explained different important aspects related to the formation of adventitious shoots in African violet. Ando et al. (1986) isolated periclinal chimeric sports of *Saintpaulia* and discussed the mechanism of chimeric structures and the origin of adventitious shoots. Seneviratne and Wijesundara (2004) treated leaf petiole of four cultivars of *S. ionantha* with different concentrations (0, 0.025, 0.04, 0.05, 0.06, and 0.1%) of colchicine for a different period (18, 23.5, 27, 43, 47, and 117 h) and recorded effects on foliar and floral characters. Mutants with attractive floral and foliar variations were isolated. Adventitious shoots from in vitro leaf explants of two different *Saintpaulia ionahta* (Mauve and Indikon) cultivars were exposed to MeV carbon ion beam and 8 MeV X-ray irradiation and studied tissue increase, shoots differentiation, and morphology changes in the shoots. Differential sensitivity of both the materials was noted and several mutants have been detected. It has been concluded that the effect of mutation induction by carbon ion beam irradiation on the leaf explants of *Saintpaulia ionahta* is better than that by X-ray irradiation, and the optimal mutagenic dose varies from 20 to 25 Gy for carbon ion beam irradiation (Zhou et al. 2006a). Zhou et al. (2006b) in a separate experiment determined the effects of five different linear energy transfers (range of 31–151 keV/μm or 8 MeV of X-rays (LET = 0.2 keV/μm) of carbon ion beams on adventitious shoots from *in vitro* leaf explants of S. *ionahta* Mauve cultivar about tissue increase, shoots differentiation and morphology changes in the shoots. Seneviratne and Wijesundara (2007) used gamma rays and colchicine to induce new flower color. A decrease in inflorescence height and an increase in flower diameter were observed with increasing concentrations of colchicine. One mutant with white (gradually turned into purple) flowers with purple margins was detected from 0.06% colchicine dipped for 22.5 h. 15 Gy gamma ray treatment improved plant architecture. In vitro leaf sections of *Saintpaulia* cv. Crystobal was exposed to EMS (0, 0.2, 0.4, and 0.6%) for 30, 60, 120, and 240 min. Ten mutants (four variegated leaves and the rest flower color) were isolated from treatments and this in vitro mutagenesis has been recommended for mutation breeding (Fang and Traore 2011). Wongpiyasatid et al. (2007) treated leaf cuttings of *S. jonantha* cv. Optima Hawaii (purple flower) with gamma rays (0, 10, 20, 40, 60, 80, and 100 Gy) to create variability. LD_{50} was determined to be 56 Gy. Radiation effects were noted on different characters like the number of leaves per plant, plant canopy width, the number of inflorescences per plant, the number of flowers per inflorescence, flower size, and mutated characters. Mutations were detected in changes in leaf color, leaf margin, leaf thickness, flower size, flower form, and plant type. Seven desirable mutants were established in pure form.

There are many more reports on mutagenesis work on S*aintpaulia*. These have enhanced knowledge on many practical aspects for future mutagenesis work—optimum working dose, selection of material for an experiment at the right stage of development, etc. Several desirable mutants have been developed through radiation or colchicine treatment or chemical mutagens. The development of variegated adventitious buds needs special mention. Nitrosomethylurea treatment of leaf cuttings produced variegation due to plastome mutation (Craig and Hampson 1979; Espino and Vazquez 1981; Kelly and Lineberger 1981; Relichova 1984; Eyerdom 1981; Hentrich and Beger 1974; Jungnickel 1977; Plummer and Leopold 1957; Pohlheim 1974; 1974; Pohlheim 1977; Polheim 1980, 1981; Pohlheim and Beger 1974; Pohlheim and Pohlheim 1976).

12.118 *Sarcococca confusa*

Family Buxaceae; evergreen shrub; attractive foliage; winter fragrance; black fruit; shade and drought tolerance; high degree of apomixes; propagation by seeds, semi-hardwood cuttings.

Hoskins and Contreras (2019) treated seeds of *S. confuse* with EMS (with 0, 0.2, 0.4, 0.8, and 1.2%) for 24 and 48 h to induce variations. Data were recorded on seed germination, growth, and relative frequency of polyembryony. With increasing EMS doses germination and occurrence of polyembryony decreased. Plants with fruits decreased with increasing EMS doses. A chlorophyll mutant and several dwarf forms were detected.

12.119 Schefflera sp.

Family Araliaceae; flowering plant; known as Umbrella tree; evergreen trees or shrubs; indoor foliage plant; propagation by cuttings.

No recent report on induced mutagenesis is available on *Schefflera* sp. The preliminary mutation was initiated at Euratom-ITAI, the Netherlands by treating rooted cuttings of S. venulosa "Clementine" with X- or Gamma rays to create variability in growth habit, leaf form, leaf variegations, etc. 10 Gy was found to be the optimum dose (c.f. Broertjes and Van Harten 1988).

12.120 Silene Species

Family Caryophyllaceae; known as campion or catchfly; herbs; biennial/perennial/annual; hermaphrodite/dioecious/gynodioecious; showy flowers of various colors; propagation by seeds, cuttings.

Jiang and Dunn (2016) did mutagenesis experiments with EMS and Caffeine on four ornamental *Silene* Species (*Silene coronaria* (L.) Desr., *S.* ×*haageana* Lem. "Molten Lava," *S. chalcedonica* L., and *S. flos-cuculi* L.) to expand genetic variability and to create new valuable cultivars. An attempt was made to study the effects of EMS and Caffeine separately and in a combination of the two. The four species were chosen based on their different ornamental traits and different germination rates. The treatment combinations were included 0.6% EMS (v/v), 10% caffeine (w/v), 20% caffeine (w/v), 0.6% EMS plus 10% caffeine, 0.6% EMS plus 20% caffeine, and a control with only deionized water. Species responded differently to the mutagens. Preliminary trials showed that some species of *Silene* are EMS-sensitive. Caffeine alone was found to enhance the seed vigor of *S. flos-cuniculi*. Generally, caffeine plus EMS treatments had a greater effect on mutation rate than either treatment applied alone. Mutation effects on flower traits were observed from 0.6% EMS plus 20% caffeine treatment on both *S. chalcedonica* and *S.* ×*haageana*. Bigger flowers were observed on *S. chalcedonica* mutants, and smaller flowers were seen along with stunted plants in mutant *S.* ×*haageana* plants. There was no flower color change or variegation on these selected species except there were some bleached spots on *S.* ×*haageana* "Molten Lava" petals, which did not appear to be a beneficial trait. Another obvious change observed in flowers was that mutagenesis caused stronger pistils and weaker stamens in *S. chalcedonica* and *S.* ×*haageana*. Caffeine as a plant mutagen should be further investigated to determine the most efficient concentration as well as effects on other plant species, as several desirable mutants were obtained with leaf variegation.

12.121 *Salvia coccinea, S. splendens*

Family Lamiaceae; known as blood sage, scarlet sage, Texas sage, or tropical sage; herbaceous perennial/annual; flowers bright red; propagation by seeds, divisions.

Feiglova (1968) detected a new spontaneous mutation of the species *Salvia splendens* KER-GAWL and found different pigmentation of a part of its inflorescence. Effects of the environment especially the influence of light effect on the synthesis of anthocyanins causing coloring of the inflorescence have been discussed. Haq (1983) developed morphological (plant stature and flower color) mutants in *S. coccinea* by colchicine treatment. Wu et al. (2009) studied the dose-response of $^{12}C^{6+}$ heavy ions bombardment on mutagenesis in *Salvia splendens* Ker-Gawl. Mutants with bileaf, trileaf, and tetraleaf conglutination were detected in the treated population. The chimeric bicolor flower (dark red and fresh red flower) was isolated in the M_1 generation. RAPD analysis demonstrated that DNA variations existed among the wild-type, fresh, and dark red flower shoots of the chimera. The dark red flower shoots of the chimera were conserved and cultivated at a large scale through micropropagation. Yamaguchi et al. (2017) developed a new variety of *Salvia* through ion beam breeding. Bugallo et al. (2021) treated seeds of the selected genotype of *S. coccinea* with X-rays (100–600 Gy) to induce genetic variability for the compact nature of the plant and variegation of the leaves. LD_{50} was determined to be 312 Gy. Leaf deformations, alterations in leaf pigmentation (variegation, albinism), and stem twisting were some changed mutant phenotypes. Rebekah C.I. Maynard and John M. Ruter (2023) treated seeds of red-flowered *S. coccinia* with EMS (0, 0.4, 0.8, or 1.2% for 8, 12, or 24 h). The optimum dose was determined at 1.2% which induced desirable mutations. Mutations included differences in leaf shape and flower size in addition to albina, chlorina, virescens, and chimeral chlorophyll changes. Salvia coccinea is a valuable flowering annual but there is a limited range of petal colors and no leaf variegation. Maynard and Ruter (2023) selected red color variety and exposed seeds to EMS (0, 0.4, 0.8, or 1.2%) for 8, 12, or 24 h. The optimal treatment rate was determined to be 1.2% EMS for 8 h, which generated desirable mutations near the median lethal dose (LD_{50}). The M1 population had a 53% germination rate and was completely morphologically uniform. Mutations were detected in M2 in leaf shape and flower size, albina, chlorina, virescens, and chimeral chlorophyll changes. Six stable mutants were established.

12.122 *Sandersonia aurantiaca* Hook.

Family Colchicaceae; monotypic genus; rhizomatous; known as Christmas bells, golden lily of the valley, Chinese lantern lily, and Chinese lantern bulb; flowers orange/yellow; propagation by seed, division of the rootstock.

Seeds were treated with heavy ion beam irradiation (2–10 krad) and developed mutants (Horita et al. 2002). Morgan et al. (2002) applied several techniques like mutagenesis, polyploidy, cross-breeding, etc. to develop new forms of *Sandersonia*. New variants were selected from the treated population. Morgan et al. (2004) developed a triploid with characteristics between those of two parents (diploid and tetraploid). Davies et al. (2002) did experiments to understand the responses of the flower stem quality in response to light and temperature irradiance and recommended that the temperature and irradiance environment is required for producing high-quality flower stems.

12.123 Sansevieria (Now Under Dracaena)

Family Asparagaceae; known as the snake plant, Saint George's sword, mother-in-law's tongue, viper's bowstring hemp; evergreen perennial; many variations in foliage form; popular houseplant; propagation by cuttings, rhizomes.

Chlorophyll-variegated mutations are frequently observed after mutagen treatments. One such 20 Gay gamma ray treated stable mutant of *Dracaena sanderina* var. virescens was selected for spectral analysis of chlorophyll content. Analysis indicated a decreased ratio in chlorophylls a/b compared with the wild-type mother plant Calorimetry of isolated chloroplast preparations showed a major difference in thermotropic transition in the range of −50 °C to 100 °C at TP of 64 °C and an enthalpy of −9.68 kJ/g in the chlorophyll mutant (Lapade et al. 2001; Palamine et al. 2005). Teng, Emily Shih-wen (2007) initiated long-term breeding to create new varieties applying gamma rays and oryzalin for the foliage industry. Unrooted cuttings of four *Dracaena* varieties were exposed to Cesium-137 gamma rays at dosages ranging from 0 to 500 Grays (Gy) in the first round and 0–50 Gy in the second round to estimate the LD_{50} dosages for root and shoot formation. For root formation, the LD_{50} dosages were estimated as 14.6, 13.8, 5.7, and 17.7 Gy, and for shoot formation, the dosages were estimated as 19.4, 16.6, 22.1, and 10.9 Gy for *D. deremensis* "Santa Rosa," *D. fragrans* "Massangeana," *D.fragrans* "Victoriae," and *D. xmasseffiana*, respectively. Chlorophyll variegation mutation was most common in the treatments. Experiments were also designed to create variability by treating axillary buds and callus of *D. deremensis* "Santa Rosa" and *D. deremensis* "Lisa" with oryzalin (0, 0.05, and 0.01%) for 24 and 48 h. Mixoploid and tetraploid plants were recovered. The tetraploid has shorter internodes and shorter leaves than its diploid counterpart. Fernando Aurigue (2019) developed one chlorophyll-variegated mutant (*Dracaena* "Sun Beam"—NSIC 2014 Or 85) from *D. braunii* by treating stem cuttings with gamma rays. D. "Sun Beam" is similar to the parent material in all aspects except for the shorter leaf and its broad bar. D. "Sun Beam" is very attractive and shoots or top cuttings may be used as cut foliage in flower arrangements or grown individually or in a group as containerized plants, materials for terrariums, dish gardens, and landscaping. Taychasinpitak (2009) developed two new mutant varieties: gold-dust *Dracaena* var. Friendmanii and var. Bangkok beauty through gamma irradiations. Unrooted cuttings of four *Dracaena* varieties

(*D. deremensis* "Santa Rosa," *D. fragrans* "Massangeana," *D. fragrans* "Victoriae," and *D.* ×*masseffiana*) were exposed to Cesium-137 gamma rays (0, 5, 10, 20, 30, 40, or 50 Gy) to determine the LD_{50} dosage. The LD_{50} dosages for shoot formation were estimated as 19.4, 16.6, 22.1, and 10.9 Gy for *D. deremensis* "Santa Rosa," *D. fragrans* "Massangeana," *D. fragrans* "Victoriae," and *D.* ×*masseffiana*, respectively. Chlorophyll-variegated mutations were detected from treated plants (Teng and Leonhardt 2009a). Teng and Leonhardt (2009b) attempted to induce polyploidy in *Dracaena* using oryzalin. In one method developing axillary buds of *D. deremensis* "Santa Rosa" were treated in vivo by placing oryzalin-soaked cotton on the meristem and wrapping it in plastic. In the second method callus tissue of *D. deremensis* "Lisa" were treated in vitro by soaking the calli in oryzalin solution. Mixoploid and tetraploid plants were recovered from both treatments. The tetraploid plant exhibited shorter internodes and shorter leaves than its diploid counterpart and is under evaluation for suitability as a new variety or for use in hybridizing efforts.

12.124 *Sarcococca confusa*

Family Buxaceae; known as Sweetbox; evergreen shrub used for charming foliage; sweet-scented; flowers creamy-white; propagation by seeds, semi-hardwood cuttings.

Breeding is restricted due to high apomixis. Seeds were exposed to EMS (0, 0.2, 0.4, 0.8, and 1.2%) for 24 and 48 h and studied its effect on seed germination, growth, and relative frequency of polyembryony. The occurrence of polyembryony decreased with increasing EMS dose and chlorophyll and dwarf form mutations were detected in the treated population (Hoskins and Contreras 2019).

12.125 Schefflera

Family Araliaceae; wonderful houseplant; known as umbrella tree; leaf evergreen; trees/shrubs; propagation by cuttings.

No recent mutation work on Schefflera could be found. A very preliminary mutation work report is available after treating rooted cuttings of S. venulosa "Clementine" with X- or gamma rays. The objective of the experiment was to induce mutations with desirable characteristics like growth habits, leaf characters, etc. The optimum dose suggested ±10 Gy (c.f. Broertjes and Van Harten 1988).

From a floriculture point of view, dwarf Schefflera will be a very popular houseplant. It will be an interesting indoor bonsai specimen. Therefore, it will be a good plant for the application of mutation techniques to create variability.

12.126 Scilla

Family Asparagaceae; perennial herbaceous; bulbous; propagated by seeds and bulbs.

Mutation work on this plant is very limited. Bulbs of *S. sibirica* were treated with different doses of X-rays and the optimum dose was determined between 5 and 10 Gy. Leaf segments of *S. sibirica* cv. "Spring Beauty" were treated with X-ray and colchicine to explore the feasibility of the frequency of formation of adventitious buds and the optimum dose was found to be 3.5 Gy (Alkema 1974a, b, c.f. Broertjes and Van Harten 1988). Chakravarty and Sen (1987) standardized regenerated callus and callus tissues of *Scilla indica* were treated with different doses (2.5–20 Gy) of gamma rays and studied different aspects related to mutation. Cell division and growth rate were accelerated at low doses. Treated calli showed different types of chromosomal abnormalities and few regenerated clones showed better growth and vigor (Chakravarty and Sen 2001).

12.127 *Sophora davidii* (Franch.) Kom. ex Pavol

Family Fabaceae; spiny, multi-stemmed, deciduous shrub; landscape shrub; flowers blue, white; propagation by seed, bulb offsets.

Wang et al. (2017) studied the effects of gamma rays (20–140 Kr) on seed germination, seedling morphology, stem diameter, and branches per plant and used ISSR markers to identify the DNA polymorphism among mutants.

12.128 *Spiraea thunbergii*

Family Rosaceae; small shrub; leaves semi-deciduous; flowers white; propagation by seed, stem cuttings.

Seeds of "Pinkey" were treated with an ion beam (220 MeV $^{12}C^{5+}$) at various doses (5–140 Gy) and cultured on a modified ½ MS medium. The treated population showed decreased germination, dwarf type, acute leaf type, and flower color mutations (Iizuka et al. 2001).

12.129 *Streptocarpus*

Family Gesneriaceae; known as Cape primrose; herbaceous perennials, propagation either by seed, clumps, leaf cuttings, stem cuttings, or plantlets from the roots.

Streptocarpus is a good material that generated good knowledge on mutation technology by its easy propagation through leaf segments, leaf cuttings, adventitious plantlets, etc. These propagules help to develop non-chimeral solid mutants. For induction of mutations mostly all the experiments were conducted by treating leaves with X-rays. Few experiments were conducted with fast neutron and colchicine treatment. The optimum dose for the X-ray was determined 30 Gy. A good number of mutants and tetraploids were developed from these experiments. Mutations and mutant characters were compact plant, dwarf habit, free flowering, year-round blooming, flower dark blue, light blue, white, pink, red, etc. Chemical mutagen was not successful, although the attempt was very limited. Promising mutants have been commercialized (Anonymous 1975; Broertjes 1968, 1969b, 1970, 1973, 1974; Broertjes et al. 1969; Brown 1971, 1973, 1974; Brown and Davies 1971; Choudhary 1976; Davies 1971; Davies and Hedley 1975; Van Raatle and Van Raatle-Wichers 1974; Zeven 1972, 1973; Osiecki 1989).

12.130 *Stromantha sanguine*

Family Marantaceae; herbaceous perennial; excellent houseplant; flower white; propagation by seed, rhizome division/cutting.

The experiment was conducted by treating rhizomes of *Stromantha sanguina* cv. "Sanguinea" with different doses (0, 10, 20, 30, 40, 50, 60 Gy) of gamma rays to determine the radiosensitivity and optimum dose for large-scale irradiation. A further experiment was conducted using in vivo methodology (cutting back method) by treating 30 and 40 Gy for induction of variability. Survival and plant height decreased with increasing doses. After six multiplications one interesting gray rachis mutant was selected from "Sanguinea" with a commercial important character. The grower multiplied the mutants and named "Grey Cais" for commercialization (Tulman Neto and Latado 1996; Neto et al. 2001).

12.131 *Tagetes erecta* L. (African Marigold)

Family Asteraceae; herbaceous annual/perennial; pinnate green leaves; propagation by seed.

Gamma rays (0, 100, 200, 300, and 400 Gy) were applied to seeds of African marigold cv. "Pusa Narangi Gainda" and changes in morphological characters (plant height, number of branches, leaf number, leaf size, plant spread, stem diameter, increased foliage, number of ray florets and size, floral abnormalities, etc. and also stimulation in growth) were recorded. Exposure to 100 Gy has been recommended for better results (Singh et al. 2009). Verma et al. (2010) reported changes in different morphological and floral characters of the African marigold cultivar "Pusa Basanti Gaida" after seedling treatment with gamma rays (5, 10, 15, 20 Gy). Sarkar et al. (2016) applied in vivo and in vitro gamma ray mutagenesis to induce novelty in the marigold cultivar "Pusa Narangi Gainda." Seedlings were grown in vivo and the proliferated cultures in vitro were exposed to different doses of ^{60}Co gamma rays (5, 10, 15, 20, 25, 30, 35, 40 Gy). 20 and 15 Gy doses were identified as LD_{50} values for in vivo and in vitro conditions, respectively. 20 Gy treatment induced early flower bud compared to control under in vivo study. 15 Gy in vitro treatment changed the flower color to yellow. Several mutants were detected in M_1 in both treatment methods. The mutants which were selected from the M_1 generation were selfed to raise the M_2 generation. The study revealed six putative mutants (pm1, pm2, pm3, pm4, pm5, and pm6), which could successfully maintain their distinct traits. Among them, pm3 produced early flowering, and pm4 developed light orange-colored flowers. To induce mutation and polyploidy Sadhukhan et al. (2014) applied colchicine to T. erecta var. "Bidhan Marigold." The colchicine was applied in two ways: absorption of colchicine through roots and absorption of colchicine through buds. Stomatal size and number were used as parameters to identify polyploidy. The stomatal dimension (length and breadth) of the colchicines-treated plants appeared to be greater than the untreated control. But the number of stomata was found to decrease in all the treated plants. The colchiploid plants had larger and thick leaves with deep green pigmentation as compared to the untreated control. Majumder et al. (2018a) studied the variability and correlation of putative mutants of marigold var. "Pusa Narangi Gainda" generated through gamma irradiation (in vivo and in vitro) in M_2 generation for 11 traits. The putative mutants were assessed in M_2 for their diversity using ISSR markers and the amplified DNA fragments were compared with their respective parent genotype. The selected mutants were multiplied up to six generations to attain the desired level of homozygosity for their large-scale field evaluation. Majumder et al. (2018b) treated in vivo grown seedlings and in vitro raised proliferated cultures of marigold cultivar Pusa Narangi Gainda with gamma rays to induce variability. Mutants and variants were screened based on yield and morphological characters. The genomic DNA of putative mutants was evaluated in PCR amplification using RAPD primers and the amplified DNA fragments from mutants were compared with their respective controls. Latha and Dharmatt (2018) treated seeds of marigold cv. Double Orange with gamma rays (0, 100, 200, 300, 400, 500, and 600 Grays) to induce mutations. Germination percentage, survival percentage, plant height, and number of leaves reduced with an increase in the dose of gamma rays. The flower diameter and the number of petals per flower were significantly increased at 600 Gy. Flower yield decreased with increasing radiation doses. Higher doses (500 and 600 Gray) induced maximum chlorophyll mutants and morphological mutants. 500 and 600 Gray showed stimulation on flower characters. Seeds of *Tagetes erecta* L. were treated with gamma rays (10, 20, 30, 40, and 50 KR) and EMS (10, 20, 30, 40, and 50 mM) for inducing variability. Based on different parameters like seed germination, seedling survival, etc. LD_{50} has been determined and gamma ray was found to be a more efficient mutagen than EMS for *Tagetes erecta* L. (Aravind and Dhanavel 2021). Aravind and Dhanavel (2022) evaluated the combined effects of gamma ray and EMS ranging from [0 (Control), 10 KR + 10 mM, 20 KR + 20 mM, 30 KR + 30 mM, 40 KR + 40 mM, and 50 KR + 50 mM] to induce mutagenesis in Tagetes erecta. The study revealed that the germination percentage, germination speed index, seed vigor, and seedling survival of other combined mutagenic treatments were significantly lowered. They claim that this work is one of the pioneering approaches in T. erecta. using combined mutagenesis and providing a combined mutagenesis-based plant breeding program in *T. erecta*. Seeds of marigolds were soaked in a solution of colchicine with hydrogen peroxide as its solvent. Results indicated that the treatments caused the formation of chlorophyll mutant and an increase in the length of flower diameter. Both the control and treated plant samples were attacked by leaf miners, giant land snails, virus-like disease, and botrytis flower blight disease (Susrama and Yuliadhi 2020). Lenawaty et al. (2022) treated *T. erecta* genotype MG04 and *T. patula* genotype MG21 with acute and chronic EMS solutions to create diversity. For acute treatment, seeds were soaked in EMS for 4 h under ten different concentration levels (0, 0.2, 0.4, 0.6, 0.8, 1.0, 1.5, 2.0, and 3.0%). The chronic treatment was carried out by diluting the concentration of EMS LC50 1/10×, 1/100×, 1/1000×, and immersion time of 6, 24, and 48 h. Data were recorded on survival and quantitative and qualitative characters and found that the sensitivity of *T. erecta* to EMS treatment was quite higher than *T. patula*. The acute application method showed a relatively low mutation rate and limited diversity of flower types. The chronic application method produced various flower shapes and whiter flower colors in *T. erecta* and *T. patula* resulting in a redder flower color than the acute application method.

12.132 *Tibouchina organensis*

Family Melastomataceae; known as Glory bush; evergreen tree/shrub; hairy stems; flowers shades of purple/mauve/lilac/violet Bluish purple; propagation by softwood cuttings.

Dwarf tetraploid *Tibouchiana organensis* was developed through in vitro culture of nodal segments. Split dose gamma rays were applied as follows: first irradiation (split dose of 3× 15 Gy applied at intervals of 4 h), second irradiation (3× 15 Gy), third irradiation (3× 15 Gy), and fourth irradiation (3× 15 Gy) (c.f. Schum and Preil 1998). Unrooted cuttings of the unnamed Tibouchina selection were exposed to gamma radiation to create new Tibouchinas with different plant forms and sizes. The mutant was selected from irradiated plants. The mutant is characterized by its upright and outwardly spreading plant habit; freely branching habit; and purple-colored flowers. The mutant has been named "Grue-Tib 04" and patented (Grünewald 2002). Beaumont (2013) detected a naturally occurring branch mutation of *Tibouchina* "Carol Lyn" (not patented). The mutant is characterized by green leaves with cream-colored margins, a compact habit, and purple flowers. The mutant has been named "Blaze of Glory" and patented. Haggo (2006) selected a spontaneous mutant from *Tibouchina urvilleana* "edwardsii" (unpatented). The mutant is characterized by unique foliage variegation ranging in color from yellow-green to pink. The mutant has been named "TB01" and patented.

12.133 *Tigridia pavonia*

Family Iridaceae; summer-flowering bulbous herbaceous perennial; leaves are narrow with three petalled blooms; propagation by seeds, offsets.

Díaz-López et al. (2003) studied the effects of gamma irradiation (0, 5, 10, 15, 20, 25, and 30 Gy) on bulbs of *Tigridia pavonia* and observed effects on days to emergence, plant height, number of stems per plant, stem length, number of leaves per stem, leaf length, number of stems per bulb, number of flowers per bulb, and flower diameter. No effect of gamma rays was observed on days to flower stem apparition, number of branches per stem, and number of fruits per flower stem. Higher dose induced abnormalities in flowers. Doses between 15 and 25 Gy induced three flower color mutations.

12.134 *Tillandsis fasciculata*

Family Bromeliaceae; known as the giant air plant or cardinal air plant; epiphytic; monocarpic; gorgeous bloom spike; propagation by seeds, offsets.

Only one chlorophyll-deficient variegated *Tillandsis* (*T. cyanea* Variegata) is commercially available on a limited basis. A mutagenesis experiment was conducted to induce more desirable chlorophyll-variegated phenotypes. Seeds of *Tillandsia fasciculata* var. fasciculata were treated with gamma radiation (0, 10, 12, 15, 18, 21, 24, 27, and 29 kR.), combined gamma and thermal neutron radiation (neutron dose of 111.6 rad hÿ1 accompanied by 824.5 rad hÿ1 of gamma radiation) or by EMS (1.2% EMS3 h and 0.4% EMS5 h). Various mutagenic treatments generated transiently variegated seedlings (wild type, albino, yellow, yellowish-green, and variegated phenotypes). These variegated seedlings were either sectorial or mericlinal chimeras, consequently, the variegation of these seedlings was lost as they grew older. Gamma radiation at 21 and 27 kR produced the highest percentage of variegated seedlings and 15.8% with 1.2% EMS. Wild types had greater total chlorophyll a, b, and total chlorophyll than mutant phenotypes. This is one of the first reports on the mutagenesis of a *Tillandsia* species. Stable periclinal chlorophyll-deficient chimeras of *Tillandsia* species can likely be obtained via mutagenesis if large numbers of seeds are treated with a suitable mutagen (Koh and Davies Jr 2000, 2001).

12.135 *Torenia fournieri*

Family Lindermiaceae; herbaceous annual; flower blue, purple, white; propagation by seed.

Leaf tissues and stem internodes without lateral meristems of the *Torenia hybrid* cv. "Summer Wave Blue" (Suntory Flowers) were exposed to 14N and 20Ne ions and cultured in vitro. Irradiation treatments were conducted at a dosage range of 5, 10, 20, and 50 Gy for both ions. Two types of mutations (one with deletion of the blue gene and the other involving the deletion or duplication of a gene related to pigment production) were detected (Miyazaki et al. 2006). They increased the frequency of flower color mutations in an interspecific hybrid of *Torenia* after irradiating with beams of Nitrogen (N) or Neon (Ne) through regeneration of stem or leaf without lateral meristem. Anthocyanin assay divided mutants into two groups: one group involved the deletion of the blue gene (DFR) and the other group involved the deletion or duplication of a gene related to pigment production. Ion beams of 12C6 and 20Ne10 were applied to leaf disks from wild type and five genetically modified transgenic torenia (*Torenia fournieri* Lind. CV. "Crown Violet") lines and studied phenotypic variability. Wide variations in flower color and shape were observed in both materials. Mutation events were more in transgenic plants. A combination of genetic engineering and ion beam irradiation has been recommended for improvement within a short period. Sasaki et al. (2008) applied heavy ion beam ($^{12}C^{6+}$ and $^{20}Ne^{10+}$) irradiation to leaf disks of wild type and genetically modified (in which petal color and pattern had been modified by controlling two anthocyanin biosynthesis-related genes encoding chalcone synthase (*CHS*) and dihydroflavonol-4-reductase (*DFR*) torenia CV. "Crown Violet" plants for genetic modification. Regenerated flowering plants showed mutant phenotypes mainly in flowers with wide variation in color and shape. Mutation efficiencies for petal color and coloration pattern were higher in transgenic plants than in wild-type plants, while those for petal shape and corolla divergence were almost equivalent in the two plant groups. Among these mutants, a class B gene-deficient mutant was investigated as a model case for further study to facilitate the control of flower phenotype. From experimental results, it has been proposed that the combination of genetic engineering and ion beam irradiation greatly facilitates the improvement of agrobiological and commercial traits within a short period. Sawangmee et al. (2011) treated interspecific hybrids of *Torenia* with acute (Cs-137) or chronic (Co-60) γ-ray irradiation combined with a detached-leaf technique to induce flower color mutation. The mutation efficiencies of the acute and chronic levels of irradiation for flower color showed no differences. The mutation rate was 1% for explants. The production of mutants with yellow-colored petals suggested disruption of anthocyanin synthesis.

12.136 Tradescantia

Tradescantia is a very interesting material for radiobiological studies. Biological and genetic characteristics of tradescantia are widely used as sensitive test-object to determine the genetic effects of various mutagens.

Sparrow et al. (1972) examined the dose-response curve for pink somatic mutations in *Tradescantia* stamen hairs after exposure to neutron and X-ray irradiation with doses ranging from a fraction of rad to the region of saturation. The dose-effect relation for neutrons indicated a linear dependence from 0.01 to 8 rads and between 0.25 and 5 rads for X-rays. The observations were in good agreement with the predictions of the theory of dual radiation action and support its interpretation of the effects of radiation on higher organisms. Nauman et al. (1975) studied the rate of induction of pink mutant cells in stamen hairs of *Tradescantia* clone 02 after exposing inflorescences to 60–80 rads of X-rays at 0.3–503 rads/min. The mutation rate was calculated at different doses. Dennis (1976) studied the biological effects of

Tradescantia occidentalis after treating with 250 kVp X-rays, cobalt-60 gamma-radiation, and monoenergetic neutrons with energies between 0.08 and 15 MeV. The effect studied was that of the induction of pink sectors in the otherwise blue staminal hairs of the flowers at low doses of radiation. Limiting r.b.e. and o.e.r. values at low doses were derived. It has been concluded that the effect studied is complex and may not provide a critical test of bio-physical theories of radiation effects. Inflorescences of several clones of *Tradescantia are heterozygous* for flower color and were treated with ionizing radiation and with the gaseous form of several known or suspected chemical mutagens. Pink somatic mutations were scored in the stamen hair cells of mature flowers and dose-/exposure-response curves were constructed. Results indicated that there was no evidence of a threshold for mutation response following x or neutron irradiation. Results so far obtained for gaseous chemical mutagens were less clear but also suggested that there was no threshold for mutation response (Nauman et al. 1976). Ichikawa and Takahashi (1977) exposed young inflorescences of two different *Tradescantia* clones to acute small doses (approx. 3–50 R) of ^{60}Co gamma rays and scored increased linearly the somatic mutation frequencies from blue to pink in the stamen hairs post-irradiation with increasing dose in both clones. Schairer et al. (1978) tested the mutagenicity of pollutants SO_2, NO_2, and O_3 and vapors of mutagens such as 1,2-dibromoethane (DBE) and ethyl methanesulfonate (EMS) on stamen hair mutation assay and observed a significant number of phenotypic changes following exposures to as little as 0.14 ppm of DBE. They observed maximum sensitivity of the system with long-term or chronic exposures because the response increases linearly in proportion to the duration of exposure up to 21 days. Ichikawa and Sparrow (1978) treated inflorescences of *Tradescantia* clone 02 ($2n = 12$) to a series of γ-ray exposures at two different exposure rates (29.3 R/min and 0.026–0.52 R/min.) and studied Pink mutation-response curves, and survival curves based on reproductive integrity. Ichikawa et al. (1981) treated two triploid clones (KU 7 and KU 9) of *Tradescantia* heterozygous for flower color with 1–42.3 R of gamma rays or the scattering radiation in the gamma field and studied the occurrence of somatic pink mutations in the stamen hairs. Scattering radiation was found to have a genetical efficiency of more than two times higher than that of gamma rays. Van't Hof and Schairer (1982) reported based on experiments conducted on the hybrid diploid clone 4430 of the plant T*radescantia* both in the field and laboratory that this plant is very suitable for the detection of gaseous mutagens. Osipova and Shevchenko (1984) studied the biological and genetic characteristics of the system of stamina fiber hair of *Tradescantia* clone 2 after treatment with X-rays and various concentrations of uranium-238 salt. The mutagenic effect of uranium-238 has confirmed the high sensitivity of the SFH system. Even the low concentration like 1.7 mg/L has the mutation effect approximately equal to the effect of 0.03–0.05 Gy of X-rays. Ma et al. (1994) tested the mutagenicity of azidoglycerol (AG, 3-azido-1,2-propanediol), *N*-methyl-*N*-nitrosourea (MNU), sodium azide (NaN_3), and maleic hydrazide (MH) based on *Tradescantia* stamen hair mutation (Trad-SH) assay. Pink mutations were scored between the seventh and 14th day according to a standard protocol. The genetically effective doses of all the chemicals have been worked out and suggested that the Trad-SH assay is a reliable method for screening chemicals for their potential mutagenic effects. Sakuramoto and Ichikawa (1996) treated rooted young inflorescence of *Tradescantia* clone KU 9 (heterozygous for flower color—blue/pink) with acute (doses of 0.255–1.03 Gy) and fractionated X-ray doses and made a comparative study of induced mutation frequency. The differential effects of both irradiation processes have been explained based on DNA and/or chromosomal breaks at different time intervals. Ichikawa (1997) treated young inflorescence-bearing cuttings of *Tradescantia* clone BNL 02 with thermal neutrons for 12–120 s in the Irradiation Tube set in the Heavy Water Facility to study mutation frequency in stamen hairs. The induced somatic pink mutation frequencies in the stamen hairs increased linearly with thermal neutron flux at a rate of 1.69 pink mutant events per 10^4 hair-cell divisions. Yakovleva et al. (2011) studied the genotoxicity of benzo (a)pyrene for adaptation levels of *Tradescantia* to soil pollution. They observed plant adaptation at physiological and genetic levels and mentioned that the products of benzo(a)pyrene metabolism act as auxin on plants influencing the growth of root biomass and cell size.

Mutation research on *Tradescantia* is almost a routine exercise for testing mutagenicity and many other related problems related to mutagenicity. There are a huge number of publications and there are regular new publications. Some more literature is cited as ready reference (Evseeva and Zainullin 2000; Ichikawa et al. 1969; Nauman et al. 1977; Mericle and Mericle 1967; Underbrink et al. 1973; Evseeva and Geras'kin 2001).

12.137 *Tricyrtis hirta*

Family Liliaceae; known as "Japanese toad lily", "Hairy Toad Lily"; hardy herbaceous perennial; flowers white; propagation by seeds, stem cuttings, or divisions.

Embryogenic calluses of *T. hirta* were irradiated with $^{12}C^{6}$ ion beams (5, 10, 20, 50 Gy), and changes in different morphological characters (leaf length, leaf width, flower length, and flower diameter) were recorded in treated plants. Interesting variations were established which included dwarfism, slender and deep green leaves, and large flowers. There was no regeneration from calluses irradiated with 50 Gy. Irradiation increased the mean number of shoots per plant and decreased the mean number of nodes per shoot and shoot length. Not much radiation effect was noticed in the mean values of leaf length, leaf width, soil and plant analyzer development (SPAD) value of leaves, flower length, and flower diameter between the control and the irradiation treatments at different doses. The flower number was increased in the 20 Gy treatment. The spectrum of morphological variations increased with an increase in radiation doses. Appealing variations established in pure form were dwarfism, slender and deep green leaves, and large flowers (Nakano et al. 2010).

12.138 *Trifolium repens*

Family Fabaceae; known as white clover; herbaceous perennial; propagation by seeds, stolons.

It is a very good forage crop but four-leaf white clover has commercial and ornamental value. Davies and Wall (1960) applied acute, chronic, and fractionated gamma rays on two clonal lines of *Trifolium repens* heterozygous for the leaf marking alleles VbY and v. The objective was to induce somatic mutations at the VbY locus. Fractionated and chronic doses induced lower frequencies of mutations than acute doses. A pronounced effect of genotype on the mutagenic sensitivity of the VbY locus was demonstrated. Two elements have been separated at the VbY locus, which as far as is known is not separable by crossing-over, and mutant phenotypes produced (distinguished by the presence of a broken white V surmounted by a yellow tip) which do not occur naturally. Four-leaf white clover is not found easily in nature due to its low appearance rate. Song et al. (2009) exposed flowers of four-leaf white clover (*Trifolium repens* L.) at the pollination stage to gamma rays. The M_1 seeds produced approximately 74% germination rate, with seedling survival at 46%. The frequency of four leaflets was increased in the M_1 generation. One (Jeju Lucky-1 (JL-1) had a frequency of about 60%. M_2 generation was raised to find out that mutation was somaclonal or genetic. The same phenotype bred true in M_2. Although results demonstrated that the irradiation of fully mature flowers led to a higher frequency of 4-leaflets, they could not clearly explain the genetic mechanism involved. They mentioned that JL-1 is valuable as a new variety, without further genetic fixation, because white clover can be propagated vegetatively by stolons.

12.139 Tulip

Family Liliaceae; very appealing ornamental enriched with a massive number of cultivars developed through breeding and spontaneous mutations; herbaceous bulbiferous perennial; flowers large, showy, and brightly colored, generally red, pink, yellow, or white, different colored blotch at the base of the tepals; propagation by seeds, bulb offsets, micropropagation.

Spontaneous mutations enriched tulips adequately in the diversification of color range (De Mol 1933, 1949). Mutation breeding work, in the beginning, generated a wide range of basic information which enriched technical approaches for mutation work on bulbous ornamentals. An appreciable number of research articles and review papers have been published that cover all essential primary knowledge related to induced mutation work on the tulip. Application of high doses was recommended based on the use of soft X-rays but later advised working dose within 8 Gy; bulb size is very important and small bulbs are suitable; pink color is more susceptible to develop usually other colors; difficult to induce mutation in yellow flower color; genotype-dependent mutation; irradiation of large bulbs are advisable but freshly harvested bulbs can also be treated, etc. (De Mol 1933, 1949; Thamm

1956; Mol van Oud Loosdrecht 1956; Nybom 1961; Graboska and Mynett 1970; Matsubara et al. 1965; Matsuda 1960; Myodo 1942; Nezu 1962, 1963a, b, 1964, 1965, 1967; Nezu and Obata 1964a, b; Van Eijk and Eikelboom 1981a, b; Custers et al. 1977; Broertjes and Van Harten 1988; Anonymous 1988). It is a matter of apprehension that despite huge experiments the development of new mutant varieties was limited. One can get detailed inside stories from individual publications. Meshitsuka et al. (1962) irradiated bulbs with gamma rays (100–500 r) and reported the effects of different doses on different characters. One hundred to 500 r slightly accelerated plant growth but inhibited it at higher dosages. Lower doses induced red stripes and spots in petals and higher doses (1000 r above) reduced petal width. Ikegawa et al. (2016) irradiated flower stalk-derived callus of tulip cv. "Yumenomurasaki" using 320 MeV carbon ion (320 C) or 107 MeV helium ion (107 He) for induction of mutation. Results indicated that 0.5 Gy of 320 C or 1.5–2.0 Gy of 107 was adequate for induction of mutation. Li et al. (2022) attempted to induce variability in tulips through irradiation mutation. They found different effects of radiations on different characters at different doses: stimulation in bulb germination and improved survival in lower dose (5 Gy); higher doses (20–100 Gy) inhibited seed germination and growth, decreased flowering rate, petal number, flower stem length, and flower diameter. A decrease in total chlorophyll content and an increase in malondialdehyde content were observed above 40 Gy. Variations in stigma and flower pattern and flower color were observed. Anthocyanin and flavonoid contents decreased with increased doses of radiation (from 5 to 100 Gy). Altered micromorphology of leaf stomata and abnormal meiotic chromosomal division were recorded in 80 Gy. 80 Gy has been considered as the optimum working dose for tulip. Li et al. (2022) studied the biological effects of gamma rays on tulips and applied the ISSR-PCR molecular marker technique used to identify the mutants of phenotypic variation plants. This study showed that low irradiation doses (5 Gy) stimulated bulb germination to improve the survival rate of tulip, while high irradiation doses (20–100 Gy) significantly ($P < 0.05$) inhibited its seed germination and growth, and decreased the flowering rate, petal number, flower stem length, and flower diameter. Above 40 Gy significantly ($P < 0.05$) decreased the total chlorophyll content and increased the malondialdehyde (MDA) content in tulips. Three types of both stigma variations and flower pattern variations, and four types of flower color variations were observed. With increasing doses (from 5 to 100 Gy), the anthocyanin and flavonoid contents continuously decreased. High irradiation doses altered the micromorphology of leaf stomata. 80 Gy treatment showed root tip mitotic abnormal chromosomal division. Increasing the irradiation dose from 20 to 100 Gy enhanced the micronucleus rate. ISSR analysis showed polymorphic bands due to genetic variations in tulips. They concluded that 80 Gy may be an appropriate radiation dose to better enhance the efficiency of mutagenic breeds in tulip plants.

12.140 Verbena (*Verbena hybrid*)

Family Verbenaceae; annual/perennial herbaceous/semi-woody; diploid/polyploid; propagation by seeds, cuttings.

Kanaya et al. (2008) irradiated *in vitro* cultured nodes of fertile cultivars "Temari Sakura" (FS), "Temari Coral Pink" (FC), and "Temari White" (FW) with 1–10 Gy of the ^{14}N-ion beam (1890 MeV). Lateral shoot development of FS, FC, and FW was not affected by irradiation with up to 10 Gy. Two sterile mutants SS and SC, which were isolated from 5 Gy-irradiated FS and FC, respectively, were characterized by their flowering habits. These mutants had a larger number of inflorescences and better longevity. They mentioned based on results that heavy ion beam irradiation is an excellent tool for isolating sterile mutants without alterations in other important traits at a high frequency. Suzuki et al. (2002) induced sterile mutants in Verbena hybrid through heavy ion beam irradiation. Saito (1977) detected a pure white seedling in *V. erinoides* among several purple-red ones spontaneously and exposed the plant to the gamma field. The mutated plant was similar to normal plants except in flower color and the mutant flower size was a little smaller. Seedlings were developed from mutants in subsequent generations and all plants showed pure white flowers indicating recessive mutation.

12.141 *Vitex agnus-castus*

Family Lamiaceae; known as lilac chaste tree, chaste tree; deciduous shrub; flowers violet/blue; propagation by cuttings.

Seeds of *V. agnus-castus* were treated with different doses of gamma rays (doses (10, 20, 50, 100, 200, 300, and 400 Gray) and six antimitotic agent concentrations (0.05 and 0.1% colchicine, and 0.005 and 0.01% oryzalin and also trifluralin) to create variations and chromosome doubling. The LD_{50} was determined to be 55 Gy for seed germination and 41.3 Gy for seedling survival. Single-stemmed plant type was obtained at a 50 Gy irradiation dose. 0.05% colchicine-induced polyploidy (Ari et al. 2015).

12.142 *Weigela*

Famil Caprifoliaceae; deciduous shrubs; propagation by cuttings.

Shoot internode (0.5 cm long) of in vitro propagated Weigela plants was treated with 0.50% EMS in 3% DMSO for 90 min to promote adventitious shoot formation. Six mutations were isolated in pure form (Duron 1992).

12.143 *Zephyranthes*

Family Amryllidaceae; known as Atamasco Lily, Fairy Lily, Rainflower, Rqin Lily, Zephyr Lily; bulbous perennial; propagation by seeds, divisions.

Very few early reports are available on mutation breeding. Early experiments were carried out on different stages of bulb development with X-rays and colchicine. Variations were detected but no promising mutants have been isolated. Experimental doses were 12 and 50 Gy X- or gamma-rays. The doses were higher for this material (Banerjee 1967; Spencer 1955; Tisch 1974; c.f. Broertjes and Van Harten 1988).

12.144 *Zinnia elegans* Jacq.

Family Asteraceae; annual; flowers single, semi-double, or double; propagation by seeds, cuttings.

Swarup and Raghava (1974) treated seeds of *Zinnia elegans* Jacq. with 20 kr X-ray and isolated a mutant line resistance to leaf-curl virus disease through recurrent selection in irradiated progeny. A series of experiments were conducted by Venkatachalam and Jayabalan (1991, 1992, 1994a, b, 1997) to study the effects of gamma rays on different varieties of *Zinnia elegans* Jacq. Radiation effects were observed on morphological and flower characters (increase in plant height, branch number, flower number, and flower diameter). New flower color mutations (magenta, yellow, red, and red with white spots) and the number of ray florets were significantly increased over control. They have also noted changes in leaf proteins in gamma ray induced mutants. *Zinnia elegans* seeds were treated with 75, 100, and 125 Gy gamma rays and studied the effects on seed germination, growth, and survival of seedlings and plant height. The higher dose was lethal. Variations in plant height, number of flowers, flower diameter, and a total of eight floral variations (novel form and color) were observed in the third generation (Pallavi et al. 2017). Dry seeds of two varieties (Suttons Gaint Double Orange and Yellow) were exposed to 5, 10, and 15 kR doses of gamma rays and recorded radiation effects on different qualitative and quantitative characters (Kole and Meher 2005). Pratiwi (2010) detected diversity in different morphological characters in M_2 generation after X-irradiation on *Z. elegans* traits. Mohammed et al. (2022) treated two cultivars of zinnia (red and white) with colchicine (0, 0.05, and 0.1%) and progesterone (0.5, 10 mg/L) to induce mutations. Colchicine-induced chromosomal duplication in both varieties, and the stomata density was affected as a result of the interaction between cultivars and colchicine. The red cultivar was found superior with 0.1% colchicine significantly. The progesterone at 5 mg/L had the highest value in the stomatal area, and the triple interaction treatment between the red variety and 0.1% colchicine and 5 mg/L of progesterone produced the highest value in the stomatal density, and the highest stomatal area resulted when the triple interaction between the white variety and 0% colchicine And 5 mg/L of progesterone.

References

Abou Dahab AM, Heikal AAM, Taha LS, Gabr AMM, Metwally SA, Ali AAR (2017a) In vitro mutagenesis induction in *Eustoma grandiflorum* plant using gamma radiation. J Environ Sci Technol 10:175–185. https://doi.org/10.3923/jest.2017.175.185. https://scialert.net/abstract/?doi=jest.2017.175.185

Abou Dahab AM, Tarek AA, Amaal AMH, Taha LS, Gabr AMM, Metwally SA, Awatef IAR (2017b) Propagation and chemical mutgenic induction of *Eustoma grandiflorum* plant using tissue culture technique. Asian J Appl Sci Technol 1(9):496–511. Online ISSN: 2456-883X. www.ajast.net

Abou-Dahab A-DM, Mohammed TA, Heikel AA, Taha L, Gabr AMM, Metwally S, Ali AIR (2019) In vitro laser radiation induces mutation and growth in *Eustoma grandiflorum* plant. Bull Natl Res Centre 43:3. https://doi.org/10.1186/s42269-018-0036-z

Abraham V, Desai BM (1977) Induced tetraploidy in a *perennial Portulaca* mutant. Indian J Hortic 34(3):301–305

Abraham V, Desai BM (1978) New induced mutants in *Portulaca grandiflora* Hook. Mutant Breed Newsl 11:10

Affrida AH, Sakinah A, Zaiton A, Mohd Nazir B, Tanaka A, Narumi I, Oono Y, Hase Y (2008) Mutation induction in orchids using ion beams. JAEA Takasaki annual report 2007, p 61

Affrida AH, Zaiton A, Shakinah S, Shuhaimi S, Mohamed Najli MY, Mohd Nazir B (2009) Biotechnology-gamma generated new variety of *Amaryllis*. In: Proceedings of national conference on new crops and bio-resources. The Royale Bintang Resort & Spa Seremban, Negeri Sembilan

Agafonova OV, Korochkin LI, Privalov GF (1974) Study of the esterase isoenzymes in various tissues of mutants of *Acer negundo* L. (in Russian). Plant Breed. Abstract 47: No. 1692

Aisyah SI, Marthin Y, Damanik MRM (2015) Improvement of *Coleus* performance through mutation induction using gamma ray irradiation. J Trop Crop Sci 2:26–32. https://doi.org/10.29244/jtcs.2.1.26-32

Aisyah SI, Togatorop ER, Damanik MRM (2017) Physically induced mutation by acute and fractionated irradiation on *Coleus* spp. Acta Hortic. 1167:227–236. https://doi.org/10.17660/ActaHortic.2017.1167.35

Aisyah SI, Saraswati RAM, Yudha YS, Nurcholis W (2022) The diversity of agro morphological characters of Portulaca grandiflorain the MV8 population deriving from recurrent irradiation. Biodiversitas 23:4432–4439. https://doi.org/10.13057/biodiv/d230908

Akabane M, Takashima D, Nakaeda K, Yamanake A (1973) On new azalea varieties 'Kimbunnohana', 'Koun', 'Kobaruto' and azaleodendron 'Akanegumo'. Bull Tochigi Agric Exp Stn 17:119–122

Akhar FK, Khadem A, Sharifi A, Nemati Z, Yazdi M, Bagheri A (2016) In vitro mutation induction on TCL explants of Lilium (Lilium spp.) with ethyl methane sulfonate (EMS). J. Biol. Todays World 5(10):177–185. https://doi.org/10.15412/J.JBTW.01051002

Akhund-Zade, Muzaferova RS (1975) Studies of the effectiveness of gamma irradiation of the saffron. Radiobiologiya 15(2):319–322

Alabi OA, Akinbami OA, Ogunwenmo KO, Esan EB (2010) Effects of cigarette tobacco infusion on root regeneration and proliferation of two cultivars of garden croton (*Codiaeum variegatum*). Asian J Plant Sci 9(2):81–87. https://doi.org/10.3923/ajps.2010.81.87

Alkema HY (1974a) Mutatie veredeling bij bolgewassen door middle van radioactive bestraling en andere methoden. Rapport 24. Laborataorium voor Bloembollenonderzoek, Lisse, p 15

Alkema HY (1974b) Mutatie veredeling bij bolgewassen door middle van radioactive bestraling en andere methoden. Rapport 24. Laboratorium voor Bloembollenonderzoek, Lisse, pp 13–14

Almeida JL, Diniz JDN, Hernandez FFF (2008) Micropropagation of *Crossandra infundibuliformis* Nees cultivar Mona Wallhead. Rev Bras Hortic Ornam 2:115–122

Aloysius S, Purwantoro A, Dewi K, Semiarti E (2017) Improvement of genetic variability in seedlings of *Spathoglottis plicataorchids* through X-ray irradiation. Biodiversitas 18(1): 20–27. https://doi.org/10.13057/biodiv/d180104

Alston RE, Sparrow AH (1962) Somatic mutation rates in double and triple heterozygotes of *Impatiens balsamia* following chronic gamma irradiation. Radiat Bot 1:229–232

Amanda S, Berenschot L, Maria I et al (2008) Mutagenesis in *Petunia* x hybridaVilm. and isolation of a novel morphological mutant. Braz J Plant Physiol 20(2):95–103. https://doi.org/10.1590/S1677-04202008000200002

Anandhi S, Rajamani K, Jawaharlal M, Maheshwaran M, Gnanam R (2013a) Colchicine content in induced mutants of glory lily (*Gloriosa superba* L.). Int J Agric Innov Res 1(6). ISSN (Online) 2319-1473

Anandhi S, Rajamani K, Jawaharlal M, Maheshwaran M, Gnanam R (2013b) Correlation and path coefficients in induced mutants of Glory Lily (*Gloriosa superba* L.). Int J Agric Sci Res 3(4): 85–92. ISSN 2250-0057

Anastassopoulos E, Keil M (1996) Assessment of natural and induced genetic variation in *Alstroemeria* using random amplified polymorphic DNA (RAPD) markers. Euphytica 90:235–244

Ando T, Akiyama Y, Yokoi M (1986) Flower colour sports in *Saintpaulia* cultivars. Sci Hortic 29(1/2):191–197. https://doi.org/10.1016/0304-4238(86)90046-4. ISSN/ISBN: 0304-4238

Anjalika MR, Banerjee N, Mandal S (2005) Structural and functional parameters of mating and seed germination in mutants of *Catharanthus roseus* (Apocynaceae). J Appl Biosci 31(2):145–149

Anonymous (1975) Streptocarpus-Versuche. Gartenbauschule Wien-Schonbrunn, Versuchsergebnisse, Heft 11. Gartenbauschule, Wien-Schonbrunn, pp 113–115

Anonymous (1977a) List of mutant varieties. Mutat Breed Newsl 9(1):14–17

Anonymous (1977b) Manual on mutation breeding (2nd edn). Tech Rep Ser No. 119. IAEA, Vienna, p 288

Anonymous (1988) Mutat Breed Newsl (31):8–38

Anonymous (1991a) Plant mutation breed for crop improvement. In: Proceedings of FAO/IAEA symposium, Vienna, 1990, 2 vols. IAEA, Vienna

Anonymous (1991b) Mutat Breed Newsl (37):18–45

Anonymous (2017) Improved varieties of *Crossandra* and *Casuarina*. Laxminarayana's Crossandra Innovation Centre, Villianoor. http://nif.org.in/, http://nif.org.in/innovation/improved-varieties-of-Crossandra-spp-and-Casuarina-spp/928

Anonymous (2022) Mutation breeding of *Trichocereus cacti* using gamma irradiation. Herbalistics. https://herbalistics.com.au

Anwar S, Karno K, Kusmiyati F, Herwibawa B (2019) Induced mutation by gamma rays on performance of MV_3 *Callistephus chinensis* at lowland. IOP Conf Ser Earth Environ Sci 518: 012066. https://doi.org/10.1088/1755-1315/518/1/012066

Anwar S, Karno K, Kusmiyati F, Herwibawa B (2020) Induced mutation by gamma rays on performance of MV3 *Callistephus chinensis* at lowland. IOP Conf Ser Earth Environ Sci 518: 012066. https://doi.org/10.1088/1755-1315/518/1/012066

Aravind S, Dhanavel D (2021) Induced physical and chemical mutagenesis on Marigold (*Tagetes erecta* L.) to determine the lethality, germination and seedling survivability. Int J Bot Stud 6(3): 235–237. www.botanyjournals.com. ISSN: 2455-541X

Aravind S, Dhanavel D (2022) Germination studies on combined mutagenic treatment employing gamma rays and EMS in African Marigold (*Tagetes erecta* L.). Indian J Nat Sci 12(70). ISSN: 0976 – 0997

Ari E, Djapo H, Mutlu N et al (2015) Creation of variation through gamma irradiation and polyploidization in *Vitex agnus-castus* L. Sci Hortic 195:74–81. https://doi.org/10.1016/j.scienta.2015.08.039

Ariffin S, Mohamad A, Ratnam W (2012) Technical aspects in understanding effects of gamma irradiation on flower colour changes in *Dendrobium Sonia*. Journal Sains Nuklear Malaysia 24: 91–101

Ariffin S, Salleh S, Shamsuddin S, Musa S, Hassan AA (2016) Effects of acute gamma irradiation on *Amaryllis* 'Orange Biogamma'. In: R&D seminar 2016: research and development seminar 2016, Bangi, 8–10 Nov 2016, 6p

Arisumi T (1973) Morphology and breeding behavior of colchicine-induced polyploid Impatiens spp. L. J Am Soc Hortic Sci 98(6):599–601

Arisumi T (1978) Somatic mutations affecting leaf characteristics of new *Guinea impatiens*. HortScience 13(3, section 2):350

Arunyanart S, Chaitrayagun M (2005) Induction of somatic embryogenesis in lotus (*Nelumbo nucifera* Geartn.). Sci Hortic 105(3):411–420. https://doi.org/10.1016/j.scienta.2005.01.034

Arunyanart S, Soontronyatara S (2002) Mutation in duction by γ and X-ray irradiation in tissue cultured lotus. Plant Cell Tiss Org Cult 70:119–122

Asahira T, Yamagata H, Inagaki H, Osuga S (1974) Isolation and propagation of mutations in *Dahlia* by in vitro culture. In: Proceedings of a research co-ordination meeting held at Tokai, 30 Sept–4 Oct 1974 organized by the Joint FAO/IAEA Division of Atomic Energy in Food and Agriculture, pp 1–12

Asahira T, Yamagata H, Inagaki M, Osuga S (1975) Isolation and propagation mutations in Dahlia by in vitro culture. In: Improvement of vegetatively propagated plants through induced mutations, Tokai, 1974. IAEA, Vienna, pp 1–23

Atak C, Celik O, Acik L (2011) Genetic analysis of Rhododendron mutants using random amplified polymorphic DNA (RAPD). Pak J Bot. 43(2):1173–1182

Atra R, Sutjahjo SH, Purwita A, Sukma D, Rustikawati R, Human S (2014) Induced mutation in Orchid (*Spathoglottis plicata* Blume) originated from Bengkulu Province. In: International seminar on plant mutation breeding, Jakarta

Aurigue F (2019) DOST-PNRI mutant variety: Dracaena 'Sun Beam'. Philip J Sci 149(S1):11–14. https://doi.org/10.56899/149.S1.02

Aurigue F, Lobiano R, Gonzales M et al (2008) Mutation breeding in Philippines *Spathoglottis* orchids. In: Abst of FAO/IAEA international symposium on induced mutations in plants, Vienna, 12–15 Aug, p 115

Ayudhaya PT, Konsanuk S, Vongvanrungrueng A (2022) Effect of acute gamma radiation on survival rate and morphological change from seed of Cosmos sulphureus Cav. Wichcha J 41 (1):108–117

Badr M, Etman M (1977) Gamma-radiation induced effects on the X_1-generation in carnation (*Dianthus caryophyllus*, L.). Egypt J Genet Cytol 6:32–43

Banerjee SK (1967) Effect of gamma radiation on some perennial ornamentals. Indian J Genet Plant Breed 27(3):417–422

Banerji BK, Datta SK (1986) Induction of single flower mutant in *Hibiscus* cv. 'Alipur Beauty'. J Nucl Agric Biol 15(4):237–240

Banerji BK, Datta SK (1988) 'Anjali' - a new gamma ray iinduced single flower mutant of *Hibiscus*. J Nucl Agric Biol 17(2):113–114

Barakat MN, El-Sammak H (2011a) In vitro culture and plant regeneration from shoot tip and lateral bud explants of *Gypsophila paniculata* L. J Med Plant Res 5(15):3351–3358. http://www.academicjournals.org/JMPR. ISSN 1996-0875 ©2011

Barakat MN, El-Sammak H (2011b) *In vitro* mutagenesis, plant regeneration and characterization of mutants via RAPD analysis in Baby's breath *Gypsophila paniculata* L. Aust J Crop Sci 5(2): 214–222. Retrieved from http://www.cropj.com/barakat_5_2_2011_214_222.pdf

Barakat MN, El Shennawy OA, Raslan MR, Esmail NM (2002) In vitro mutation for sunflower (*Helianthus annuus* L.) breeding. Alexandria J Agric Res 47(3):77–84. ISSN/ISBN: 0044-7250 Accession: 003808230

Bashaw EC, Hoff BI (1962) Effects of radiation on apomictic common dallisgrass. Crop Sci 2:501–504

Baskaran K, Srinivas KVNS, Kulkarni RN (2013) Two induced macro-mutants of periwinkle with enhanced contents of leaf and root alkaloids and their inheritance. Ind Crop Prod 2013(43): 701–703

Beaumont D (2013) Tibouchina plant named 'Blaze of Glory'. US15/530,7772016-08-312017-02-27

Benetka V (1987) Induction of compact mutants in *Begonia* x hiemalis Fotch cultivar Schwabenland. Zahradnict 14:75–80. Cited in: Hort. Abstr. 1988: no. 1013.

Bennani F, Rossi-Hassani BD (2001) Seed mutagenesis in *Portulaca grandiflora* (Hook), INIS-XA-427. International Atomic Energy Agency, Vienna

Benslimani N, Khelifi-Slaoui M, Lassel A, Khiar H, Djerrad A, Mansouri B (2011) In vitro radiosensitivity study of *Datura* Sp seeds for increased alkaloid-producing mutant lines. Adv Environ Biol 381. link.gale.com/apps/doc/A253057788/AONE?u=anon~5c97b7d3&sid=googleScholar&xid=1362b24c

Berenschot AS, Zucchi MI, Tulmann-Neto A, Quecini V (2008) Mutagenesis in *Petunia hybrida* Vilm. and isolation of a novel morphological mutant. Braz J Plant Physiol 20:95–103. https://doi.org/10.1590/S1677-04202008000200002

Bergan F, Bergan L (1959) Uber experimentell ausgeloste vegetative Spaltungen und Umlagerungen an chimarischen Klonen, zugleich als Beispiele erfolgreicher Staudenauslese. I. *Pelargonium zonale*. Ait. 'Madame Salleron'. Zuchter 29(8):361–374

Bergann F (1967a) The relative instability of chimerical clones- the basis of further breeding. In: Stubbe H (ed) Induced mutations and their utilization. Abh. Dtsch. Akad. Wiss, Berlin Kl. Medizin 2, pp 287–300

Bergann F (1967b) Mutations-Chimaren: Rohmaterial zuchterischer Weiterbehandlung. Umsch Wiss Tech 67(24):791–797

Bergann F, Bergann L (1982) Zur Entwicklungeschichte des Angiospermenblatts. I. Uber Periklinalchimareen bei Peperomia und ihre experimentelle Entmischung und Umlagerung. Biol Zentralbl 101(4):485–502

Bhate RH (1999) Mutagenic response in two cultivars of morning glory *Ipomoea purpurea* (L) Roth. Indian J Genet Plant Breed 59(1):37–45

Bhate RH (2001) Chemically induced floral morphological mutations in two cultivars of *Ipomoea purpurea* (L.). Roth. Sci Hort 88:133–145. https://doi.org/10.1016/S0304-4238(00)00221-1

Bhattacharjee R (1998) Mutagenic effectiveness and efficiency of gamma rays ethyl methane sulphonate and nitroso-methyl urea in periwinkle, *Catharanthus roseus*. J Nucl Agric Biol 27(1):61–64

Bhattacharya C (2003) Effect of ethyl methanesulphonate on carnation (*Dianthus caryophyllus* L.). Environ Ecol 21(2):301–305. ISSN/ISBN: 0970-0420

Bin X, Weijie X, Guangdong W et al (2006) Characteristics of chimeras of *Anthurium andraeanum* from *in vitro* mutation. Chin Bull Bot 23:698–702

Biswas SC, Biswas AK (2006) Cytogenetic characterization of induced sterility in *Ornithogalum virens* L. Cytologia (Tokyo) 71:119–123. https://doi.org/10.1508/cytologia.71.119

Biswas KK, Mohri T, Kogawara S, Narumi I, Oono Y (2013) Developing protocol for screening of drought/salt tolerant mutants with ion beam mutagenesis in Populus sp. JAEA Takasaki Annual Report 2011. JAEA, Tokai

Boddie J, Whitcomb CE (1978) Plant modification with a mutagenic compound. Research report. Agric Exp Stn, Oklahoma State University, No. P-777:6-7. (Hortic. Abstr., 1980, 50: No. 6358. (concerning *Lagerestroemia indica* and *Maclura pomifera*)

Bouharmont J, Dabin P (1986a) Somaclonal variation in some cultivars of *Fuchsia*. In: Semal J (ed) Somaclonal variations and crop improvement. Nijhoff, Dordrecht, pp 257–260

Bouharmont J, Dabin P (1986b) Application of *in vitro* cultures for mutation breeding in Fuchsia. In: Nuclear techniques and in vitro culture for plant improvement, Vienna, 1985, pp 339–347

Bowen CC, Sparrow AH (1954) Radiosensitivity of several meiotic stages of *Lilium*. Genetics 39: 960

Boyarskikh IG, Kulikova AI, Agatova AR, Bakiyanov AI, Florinsky IV (2016) Mutation activity of *Lonicera caerulea* population in an active fault zone (the Altai Mountains). arXiv:1508.02016 [q-bio.PE]

Breider H (1959) Rontgeninduzierte Mutationen bei Vitis, Malus and *Cyclamen*. Bayer Landwirtsch Jahrb 36:396–401
Broerties C, Verboom H (1974) Mutation breeding of *Alstroemeria*. Euphytica 23:39–44
Broertjes C (1968) Mutation breeding of *Streptocarpus*. Euphytica 16(2):163–167
Broertjes C (1969a) Mutation breeding of vegetatively propagated crops. In: G.C., Chisci and G. Haussmann (Editors), Proc. 5th Eucarpia Congress, Milan, 1968. Genet Agrar 23:139–165
Broertjes C (1969b) Mutation breeding of *Streptocarpus*. Euphytica 18:333–339
Broertjes C (1970) Mutationszuchtung bei vegetative vermehrbaren Zierpflanzen. Gartenwelt 70(11):266–268
Broertjes C (1971) Achimenes. Het verkrijgen van varieteiten door bestraling. Vakbl Bloemisterij 26(34):21
Broertjes C (1972a) Mutation breeding of *Achimenes*. Euphytica 21:48–62
Broertjes C (1972b) Improvement of vegetatively propagated plants by ionizing radiation. In: Induced Mutations and Plant Improvement. IAEA, Vienna, pp 293–299
Broertjes C (1972c) Use in plant breeding of acute, chronic or fractionated doses of X-rays or fast neutrons as illustrated with leaves of *Saintpaulia*. Thesis Agricultural research report, 776, 74p
Broertjes C (1973) Achimenes en *Streptocarpus*. Nieuwe varieteien door bestraling en colchicinebehandeling. Vakbl Bloemisterij 28(28):11
Broertjes C (1974) The production of polyploids using the adventitious bud technique. In: FAO'IAEA/EUCARPIA conference on mutations and polyplois, Bari, 1972. IAEA, Vienna, pp 29–35
Broertjes C (1976) Mutation breeding of auto tetra ploid achimenes cultivars. Euphytica 25(2): 297–304. https://doi.org/10.1007/BF00041560. ISSN/ISBN: 0014-2336. Accession: 005949316
Broertjes C (1977) Nieuwe Achimenes cultivars verkregen door bestraling of door colchicines behandeling. Vakbl Bloemisterij 32(43):34–35
Broertjes C (1982) The significance of in vitro adventitious bud techniques for mutation breeding of vegetatively propagated plants II. In: Proceedings of research co-ordination meeting Coimbatore, India (1980). IAEA, Vienna, pp 1–10
Broertjes C, Alkema HY (1970) Mutation breeding in flower bulbs. In: First interantional symposium on flowerbulbs; Noordwijk/Lisse II, pp 407–411
Broertjes C, Ballego JM (1967) Mutation breeding of *Dahlia variabilis*. *Euphytica 16*(2):171–176. https://doi.org/10.1007/BF00043451
Broertjes C, Ballego JM (1968) Addendum to: Mutation Breeding of *Dahlia variabilis*. Euphytica 17:507
Broertjes C, Ballego JM (1969) Inconstances de mutation de Dahlia. Soc FR Dahlia Bull Semest 4: 6, 13–15
Broertjes C, Leffring L (1972) Mutation breeding of *Kalanchoë*. Euphytica 21:415–423. https://doi.org/10.1007/BF00039336
Broertjes C, Van Harten AM (1988) Applied mutation breeding for vegetatively propagated crops. Elsevier, Amsterdam, 316p
Broertjes C, Leffring L, Leuning B (1969) Nieuwe Streptocarpus varieteiten door bestraling. Vakbl Bloemisterij 24(22):806–807
Broertjes C, Pouwer A, Evers JFM (1983) Kleurenkeus in overvloed. Alleen geel ontbreekt nog bij Achimenes. Vakbl Bloemisterij 38(35):34–35
Brown AG (1971) X-ray treatment of F_1 hybrid plants from *Streptocarpus johannis* x *S. hybridus*. John Innes Inst Annu Rep 62:49
Brown AG (1973) Hybrid streptocarpus. J R Hortic Soc 98:201–205
Brown, A.G. 1974. *Streptocarpus* – the development of a new commercial pot plant. In: The glasshouse environment and objectives in breeding new varieties of ornamental crops. Eucarpia Meeting on ornamentals, Norwich. John Innes Institute annual report, vol 66, p 43
Brown SW, Cave MS (1953) Induced dominant lethality in *Lilium*. Proc Natl Acad Sci U S A 39: 97–102

Brown SW, Cave MS (1954a) The detection and nature of dominant lethals in *Lilium*. I. Effects of x-rays on the heritable component and functional ability of the polln grain. Am J Bot 41:455–469

Brown SW, Cave MS (1954b) The detection and nature of dominant lethals in *Lilium*. II. Cytological abnormalities in ovules after pollen irradiation. Am J Bot 41:469–483

Brown AG, Davies DR (1971) The comparative roles of artificially induced mutations and hybridization in the breeding of *Streptocarpus*. In: Eucarpia meeting, ornamentals, Wageningen. Institute of Horticultural Plant Breeding, Wageningen, pp 32–37

Brown JL, Harney PM (1974) Induction of non-chimeral mutants in *Begonia*. HortScience 9(3):276

Brown SW, Zohary D (1953) Chiasmata and crossing over in *Lilium formosanum*. Genetics 38:657

Brown SW, Zohary D (1955) The relationship of chiasmata and crossing over in *Lilium formosanum*. Genetics 40:850–873

Bugallo V, Facciuto G, Brizuela V, Prina AR, de la Torre MCP (2021) Application of X-rays to obtain ornamental mutants in Salvia coccinea. ACI Avances en Ciencias e Ingenierías 12 (22):26–35. https://doi.org/10.18272/aci.v12i3.1921

Bugnon F, Dulieu H, Gaufillier J (1965) Sur les variations topographiques, le long de la pousse, entre les deux constituants d'une chimere chlorophyllienne sectoriate ou mericline, chez l'Oeillet (*Dianthus caryophyllus* L); essaid'interpretation ontogenique. C R Acad Sci 260: 211–213

Buiatti M, Ragazzini R (1965) Gamma ray induced changes in the carnation, *Dianthus caryophyllus* L. Radiat Bot 5(2):99–105

Buiatti M, Ragazzini R, D'Amato F (1965) Somatic mutations in the carnation induced by gamma radiation. In: The use o induced mutations in plant breeding. FAO/IAEA/Eucarpia Meeting, Rome, 1964. Radiat Bot 5(suppl):719–723

Burton GW (1972) The use of mutagenic agents in grass and turf breeding. In: Constantin MJ (ed) Mutant. Breed. Workshop. University of Tennessee, Knoxville

Burton GW (1974) Radiation breeding of warm season forage and turf grasses. In: Polyploidy and induced mutations in plant breeding, Bari, 1972. IAEA, Vienna, pp 35–39

Burton GW (1975) Improving sterile turf and forage burmudagrass hybrids by gamma radiation. Improvement of vegetatively propagated plants through induced mutations. Proceedings of a research co-ordination Meeting held at Tokai, 30 September to 4 October 1974, Organized by the Joint FAO/IAEA Division of Atomic Energy in Food and Agriculture, 33. IAEA-173 Technical Document Issued by the IAEA, Vienna, 1975

Burton GW (1976) Using gamma irradiation to improve sterile turf and forage Bermuda grasses. In: Improvement of vegetatively propagated plants and tree crops through induced mutations, Wageningen, 1976. IAEA, Vienna, pp 25–32

Burton GW (1979) Induced mutations for improving millets, apomictic crop plants and veget propagated grasses. In: Induced mutations for crop improvement in Africa, Nigeria, 1978. IAEA, TWCDOC No. 222:33-40

Burton GW (1981) Tifway-2 bermudagrass. Mut Breed Newsl 18:8–10

Burton GW (1985) Registration of Tifway-2 bermudagrass. Crop Sci 25(2):364

Burton GW, Hanna WW (1977) Performance of mutants induced in sterile turf bermudagrass. Mut Breed Newsl 9(1):4

Burton GW, Jackson JE (1962) Radiation breeding of apomictic prostrate dallisgrass, *Paspalum dilatatum* var pauciciliatum. Crop Sci 2:295–297

Burton GW, Constantin MJ, Dobson JW Jr et al (1980) An induced mutant of Coastcross-1 bermudagrass with improved winterhardiness. Environ Exp Bot 20(2):115–117

Burton GW, Haanna WW, Powell JB (1982) Mutation breeding of vegetatively propagated turf and forage Bermuda grass. Joint FAO/IAEA Division of Isotope and Radiation Applications of Atomic Energy for Food and Agricultural Development, Vienna (Austria); Panel proceedings series,pp 167–174, 310p. ISBN 92-0-111182-7. IAEA; Vienna; Final research co-ordination meeting on the improvement of vegetatively propagated crops and tree crops through induced mutations; Coimbatore (India), 11–15 Feb 1980

Busey P (1980) Gamma ray dosage and mutation breeding in St. Augustinegrass. Crop Sci 20(2): 181–184. Florida Agric. Exp. Stn. Journal Series No. 1859. https://doi.org/10.2135/cropsci1980.0011183X002000020008x

Cabahug RAM, Khanh HTTM, Lim KB et al (2021) Phenotype and ploidy evaluation of colchicine-induced *Echeveria* 'Peerless'. Toxicol Environ Health Sci 13:17–24. https://doi.org/10.1007/s13530-020-00069-z

Cabhug RAM, Tran MKTH, Ahn Y-J, Hwang Y-J (2022) Retention of mutations in colchicine-induced ornamental succulent *Echeveria* 'Peerless'. Plants 11(24):3420. Published online 2022 Dec 7. https://doi.org/10.3390/plants11243420

Cambecedes J, Duron M, Decourtye L, Jalouzot R (1992) Methodology of in vitro gamma rays irradiation from *Lonicera* species; mutant description and biochemical characterization. Acta Hortic 320:119–126. ISSN/ISBN: 0567-7572, Accession: 002431821

Canul-Ku J, García-Pérez F, Campos-Bravo E, Barrios-Gómez EJ, De La Cruz-Torres E, García-Andrade JM, F de J Osuna-Canizalezy, Ramírez-Rojas S (2012) Efecto de la irradiación sobre nochebuena silvestre (*Euphorbia pulcherrima* Willd. ex Klotzsch) en Morelos (Effect of irradiation on wild poinsettia (*Euphorbia pulcherrima* Willd. ex Klotzsch) in Morelos). Revista Mexicana de ciencias agricolas, *versión impresa* ISSN 2007-0934, Rev. Mex. Cienc. Agríc vol.3 no.8 Texcoco nov./dic. http://www.redalyc.org/articulo.oa?id=263124770002

Carlson PS (1969) Production of auxotrophic mutants in ferns. Genet Res 14:337–339

Carrier LE (1983) The distribution of mutants in second flush cuttings of an irradiated carnation seedling. Acta Hortic 141:37–39

Cartledge JL, Murray MJ, Blakeslee AF (1935) Increased mutation rate from aged Datura pollen. Genetics 21. https://www.pnas.org/doi/pdf/10.1073/pnas.21.11.597

Cartledge JL, Barton LV, Blakeslee AF (1936) Heat and moisture as factors in the increased mutation rate from Datura seeds. Proc Am Philos Soc 76(5):663–685

Cassells AC, Walsh C, Periappuram C (1993) Diplontic selection as a positive factor in determining the fitness of mutants of Dianthus 'Mystere' derived from X-irradiation of nodes in in vitro culture. Euphytica 70:167–174. https://doi.org/10.1007/BF00023756

Cecconi F, Durante M (2000) Chimera induction in Sunflower (*Helianthus annus* L.). Analysis of sector patterns for different genetic markers. https://www.isasunflower.org/documents/EIT9

Chakravarty B, Sen S (1987) *In vitro* regeneration from callus cultures of *Scilla indica* (Roxb) Baker. Curr Sci 56:960–962

Chakravarty B, Sen S (2001) Enhancement of regeneration potential and variability by γ-irradiation in cultured cells of *Scilla Indica*. Biol Plant 44:189–193

Chatterjee SK (1978) The cultivation of *Catharanthus roseus* in India. In: Alfermann AW, Reinhard E (eds) Production of natural compounds by cell culture methods. Proceedings of the international symposium on plant cell culture, Munchen, pp 74–85

Chaturvedi M, Datta K, Datta SK (1997) Pollen variation in gamma irradiated *Mesembryanthemum criniflorum* L.F. (Aizoaceae). Taiwania 42(4):289–295

Chemarin NG, Glazurina AN, Zabelin JA (1973) On uses of ionizing rays in *Canna* breeding. Byull Gos Nikitsk Bot Sada 1(20):52–57

Chen XX (2009) Formation mechanism and maintenance characteristics of leaf colour chimera and mutants of *Anthurium andreaeanum* 'Sonate'. M.Sc. Thesis, Nanjing, Agricultural University of Jiangsu, China

Chen FC, Yu JY, Pei-Yin Chen PY et al (2008) Somaclonal variation in orchids and the application of biotechnology. Acta Hortic 766:315–322

Chen S, Chai M, Jia Y, Gao Z, Zhang I (2011) In vitro selection of salt tolerant variants following ^{60}Co gamma irradiation of long-term callus cultures of *Zoysia matrella* [L.] Merr. Plant Cell Tiss Org Cult 107:493–500

Chen Y, Zhang Y, Yuan S, Liu H, Zeng X, Zhang H (2014) Ethyl methane sulfonate induced disease resistance in *Begonia hiemalis* Fotsch. Hortic Environ Biotechnol 55:498–505. https://doi.org/10.1007/s13580-014-0053-2

Chen J, Thammina C, Li W et al (2016) Isolation of prostrate turfgrass mutants via screening of dwarf phenotype and characterization of a perennial ryegrass prostrate mutant. Hortic Res 3: 16003. https://doi.org/10.1038/hortres.2016.3

Cheng QQ (2011) Studies on leaf culture and induction of octoploid of Phalaenopsis cultivars. M.D. Dissertation. Shantou University, Shantou. (in Chinese)

Chiba N, Arakawa K, Nakamura S, Suzuki S, Yokota Y, Hase Y, Narumi I (2007) Mutation induction of asiatic hybrid lily and Lilium × formolongi Hort. using ion beam irradiation. JAEA-review 2007-060-77-3-11, p 77

Chinone S, Ishizawa A, Tokuhiro K, Nakatsubo K, Amano M, Hase Y, Narumi I, Tanaka A (2007) Mutation induction on oriental hybrid lily irradiated with ion beams. JAEA-review 2007-060-83-3-17, p 83

Chinone S, Tokuhiro K, Nakatsubo K, Hase K, Narumi I (2008) Mutation induction on *Delphinium* and *Limonium sinuatum* irradiated with ion beams. JAEA-review 2008-055, p 68

Choudhary DK (1976) Non-mutagenicity of sodium azide on *Streptocarpus hybridus*. Z Pflanzenzucht 76(2):167–170

Choura P (1938) Sur la nature par les hetero-auxines dans la formation provoquee raciness ou de bourgeons en n'importe quell point de boutures de feuilles. C R Acad Sci Paris 207:597–599

Chowta CD, Dnyansagar VR (1974) Abnormalities induced in flowering and floral parts by gamma rays and EMS in *Chlorophytum tuberosum*. Baker M V R Patrika 9:71–76

Chuantang N, Li Yazhi L (1989) The radiation induced mutation of canna (*Canna* L.). Acta Agric Nucleatae Sin 2(1):33–39. ISSN: 1000-8551

Colijn CM, Kool AJ, Nijkamp HJJ (1979) An effective chemical mutagenesis procedure for *Petunia hybrida* cell suspension cultures. Theor Appl Genet 55:101–106

Contreras RN, Friddle MW (2015) 'Oregon Snowflake' flowering currant. HortScience 50:320–321

Contreras RN, Shearer K (2020) Exposing seeds of *Galtonia candicans* to ethyl methanesulphonate reduced inflorescence height, lodging, and fertility. HortScience 55:621–624. https://doi.org/10.21273/HORTSCI14775-19

Coppola F (1986) Mutants obtained after gamma irradiation of *Jerusalem artichoke* cv. Violet de Rennes (*Helianthus tuberosus* L.). Mut Breed Newsl 28:9–10

Cotter DJ (1963) Effects of radiation on the photoperiodism of *Portulaca smallii*. Radiat Bot 3(3): 265–269

Craig R (1963) The inheritance of several characters in the Geranium, *Pelargonium hortorum* Bailey. Thesis, Pennsylvania State University, University Park, p 71

Craig R, Hampson SH (1979) New African violets (*Saintpaulia*) created in a mutation breeding program. Sci Agric 26(4):12

Crouse HV (1954) X-ray breakage of lily chromosomes at first meiotic metaphase. Science 119: 485–487

Cuany RL, Sparrow AH, John AA (1958) Spontaneous and radiation-induced somatic mutation rates in *Antirrhinum*, *Petunia*, *Tradescantia* and *Lilium*. Proc X Int Congr Genet 2:62–63

Cuany RL, Sparrow AH, Pond V (1985a) Genetic response of *Antirrhinum majus* to acute and chronic plant irradiation. Z Indukt Abstamm VererbLehre 89:7–13

Cuany RL, Sparrow AH, Pond V (1985b) Genetic response of *Antirrhinum majus* to acute and chronic plant irradiation. Mol Gen Genet 89:1432–1874

Cui GR, Zhang ZX, Zhang CY, Hu NB, Sui YH, Li JQ (2010a) Polyploid induction and identification of Oncidium. Acta Pratacult Sin 19:184–190. (in Chinese)

Cui GR, Zhang ZX, Hu NB, Zhang CY, Hou XL (2010b) Tetraploid of Phalaenopsis induction via colchicine treatment from protocorm-like bodies in liquid culture. J Zhejiang Univ 1:49–55. (in Chinese)

Custers JBM (1978) Plantlet formation from internode bases of carnation (*Dianthus caryophyllus* L.) in vivo – useful to mutation breeding or not? Neth J Agric Sci 26(1):31–40

Custers JBM, Van Eijk JP, Sparnaay LD (1977) New developments in mutation breeding of vegeta propa ornamental crops with special reference to quantitative characters. In: Association

Euratom –ITAL, TFAO/IAEA technical meet, Rome, 1964. Pergamon Press, Oxford, pp 302–316

D'Amato F, Moschini E, Pacini L (1964) Mutazioni somatiche nel garofano indotte dalla radiazione gamma. Caryologia 17:93–101

Dabin P, Bouharmont J (1984) Use of in vitro cultures for mutation breeding in *Fuchsia*. In: Novak J, Havel L, Dolezel J (eds) Plant tissue and cell culture, application to crop improvement. Academy of Science, Prague, pp 451–452

Dai W, Magnusson VA (2012) Morphological variations in forsythia induced by gamma ray irradiation. In: ASHS annual conference, Aug 2012, Poster Board #395

Daker MG (1966) 'Kleiner Liebling', a haploid cultivafr of *Pelargonium*. Nature 211(5048): 549–550

Daker MG (1967) Cytological studies on a haploid cultivar of *Pelargonium*, and its colchicine-induced diploids. Chromosoma 21(3):250–271

Das PK, Dube S, Ghosh P, Dhua SP (1974) Induction of somatic mutations in some vegetatively propagated ornamentals by gamma irradiation. Technology 11(2,3):185–188

Das PK, Dube S, Ghosh P, Dhua SP (1975) Mutation breeding in *Daahlia*. Indian J Ornam Hortic 6(2):3–8

Das PK, Dube S, Ghosh P, Dhua SP (1977) Improvement of some vegetatively propagated ornamentals by gamma radiation. Indian J Hortic 34(2):169–174

Das PK, Dube S, Dey AK (1978) New Dahlias through irradiation. Indian Hortic 23(1):19–21

Datta K, Shukla R, Datta SK (2003) Effects of gamma irradiation in context of palynological and cytological parameters on Narcissus tazetta cv. Cicily white. Cytologia 68(3):225–230

Daud N, Taha RM (2011) Effects of gamma radiation on growth of *Sinningia speciosa* callus tissue. Jurnal Sains dan Matematik 3(2):33–41. ISSN 1985-7918 33

Davies DR (1971) Mutation breeding. Span 14(2):01–104

Davies DR, Hedley CL (1975) The induction by mutation of all-year-round flowering in *Streptocarpus*. Euphytica 24:269–275. https://doi.org/10.1007/BF00147198

Davies D, Wall E (1960) Induced mutations at the V^{by} locus of *Trifolium repens*. 1. Effects of acute, chronic and fractionated doses of gamma radiation on induction of somatic mutations. Heredity 15:1–15. https://doi.org/10.1038/hdy.1960.52

Davies LJ, Brooking R, Catley JL, Halligan EA (2002) Effects of constant temperature and irradiance on the flower stem quality of *Sandersonia aurantiaca*. Sci Hortic 93(3-4):321–332

Dayani K, Ramya R, Wathsala A (2018) ^{60}Co gamma irradiation-induced variations in vegetatively propagated Philodendron Erubescens 'Gold'. Joint FAO/IAEA Division of Nuclear Techniques in Food and Agriculture, Vienna, p 36, 175p; FAO/IAEA International Symposium on Plant Mutation Breeding and Biotechnology; Vienna (Austria); 27–31 Aug 2018; IAEA-CN--263-42. https://www.iaea.org/sites/default/files/18/08/cn-263-abstracts.pdf

Dayani K, Ramya R, Wathsala A (2021) 60Co gamma irradiation-induced mutation in vegetatively propagated Philodendron erubescens 'Gold'. In: Sivasankar S et al (eds) Mutation breeding, genetic diversity and crop adaptation to climate change. International Atomic Energy Agency 2021, pp 386–398. https://doi.org/10.1079/9781789249095.0040

De Loose R (1964) Mutatie-onderzoek by *Poa pratensis* (Veldbeemdgras R.v.P.). Meded. Landbouwhogesch. Opzoekingsstms Staat Gent, XXIV 4:1367–1375

De Loose R (1966) Quelques enseignements preliminaries resultant de l'application des rayonnements ionisants dans la mutagenese chez Rhododendron simsii (*Azalea indica*) et Guzmania peacockii. In: Table Ronde sur l'Utilisation des Mutations Induites en Horticulture, Gembloux. Seminars in horticultural science, pp 36–40

De Loose R (1968) Kwalitatief onderzoek naar de bloemkleurpigmenten bij de Belgische hybriden van Rhododendron simsii Planch (*Azalea indica* Linn.). Meded Viaam Chem Ver 30(4):99–123

De Loose R (1969a) Erfahrungen mit radioaktiven Bestrahlung von Zierpflanzen. Gartenwelt 15: 348–350

De Loose R (1969b) The flower pigments of the Belgian hybrids of *Rhododendron simsii* and other species and varieties from *Rhododendron* subseries *obtusum*. Phytochemistry 8:253–259

De Loose R (1970a) Flower pigment composition of natural bud-variants among hybrid Chinese Azaleas, *Rhododendron Simsii* (Planch.). J Hortic Sci 45(3):265–274. https://doi.org/10.1080/00221589.1970.11514354

De Loose R (1970b) Flavonoid glycosides in the petals of some *Rhododendron* species and hybrids. Phytochemistry 0:875–879

De Loose R (1970c) Perspektieven voor het bekomen van een geelbloeiende Azalea indica. Standpunt van een scheikundige Blg Tuinbouw 51:307–310

De Loose R (1970d) Het bekomen van 'sporten' voor de bloemkeur bij de Belgische hybriden van Rhododendron simsii planch. (=Azalea indica) door middle van 60Co-gammastralen. Meded Fac LandbouwwetRijksuniv Gent 35(4):1047–1074

De Loose R (1971a) Mutatie-veredeling bij siergewassen met behulp van ioniserende stralingen. In: Grondslagen en Resultaten van de Mtatieveredling. Rijkscentrum van Landbouwkuddig Onderzoek, Gent, pp 34–56

De Loose R (1971b) Mutation research on the Belgian hybrids of *Rhododendron simsii* (Planch.). In: Eucarpia meeting on ornamentals, Wageningen. Institute for Horticultural Plant Breeding, Wageningen, pp 1–8

De Loose, R. 1973a. Bestralingen, mutatie- en isotopenonderzoek. Aktiviteitsverslag Rijksstation voor Sierplantenteelt, Melle, 1966-1971, VI, pp 90–102

De Loose R (1973b) Het bekomen van 'sporten' voor de bloemkleur bij hybriden van Rhododendron simsii Planch (Azalea indica L.) met behulp van ioniserende stralingen. Meded Fac Landbouwwet Rijksuniv Gent 38(42):2090–2109

De Loose R (1974a) Nieuwe cultivar op komst van Azalea indica L.: een bonte sport van cv. DE Waele's Favorite, bekomen door bestraling. Verbondsnieuws Belg Sierteelt 18(4):125–129

De Loose R (1974b) Mutation breeding of the hybrids of *Rhododendron simsii* Planch. (Azalea indica L.). Mutat Breed Newsl 3:15

De Loose R (1974c) Mutatie-veredelingsonderzoek op de *Azalea indica* L. Agrico ntact. Minist Landbouw Belg 34:1–5

De Loose R (1974d) Het chromatografie-patroon van de bloemkleurpigmenten bij de hybriden van Rhododendron simsii Planch. En *Rhododendron obtusum* Planch. (*Azalea indica* and *Azalea japonicum*). Meded Fac LandbouwwetRijksuniv Gent 39(1):238–254

De Loose R (1974e) Het chromatografie-patroon van de bloemkleurpigmenten bij de hybriden van *Rhoddendron simsii* Planch. En Rhododendron obtusum Planch. (*Azalea indica* L. and *Azalea japonica* L.) Deel B. Bloemkleur en bloemkleurpigmenten bij spontane en kunstmatige sporten van *Azalea indica* L. en *Azalea japonica* L. Meded Fac Landbouwwet Rijksuniv Gent 39(3): 1369–1384

De Loose R (1979) Radiation induced chimeric rearrangement in flower structure of *Rhododendron simsii* PLANCH. (*Azalea indica* L.), use of recurrent irradiation. Euphytica 28:105–113

De Loose R (1981) Mutatieyeredeling bij siergewassen. Agricontact 115:1–3

De Mol WE (1926) Heteroploidy and somatic variation in Dutch flowering bulbe. Am Nat LX:334–339

De Mol WE (1931) Somatische variation der Blumenfarbe der Hyazinthe durch Rontgenbestrahlung und andere ausseren Umstande. Z Indukt Abstamm Vererbungsl 59:280–283

De Mol WE (1933) Mutation sowohl als Modifikation durch Rontgenbestrahhung und die "Teilungshypothese". Cellule 42:149–162

De Mol WE (1934) Drei aufeinnanderfolgende Jahre der Mutation und Modifikation bei Hyazinthen, nach kaltem Aufbewahren wahrend des Zellteilungsprozess zur Blutenbildung. Cellule 43:131–146

De Mol WE (1937a) De invloed van Rontgenbestraling op bloembolgewassen (hyacinten en tulpen). Philips Tech Tijdschr 2(11):321–328

De Mol WE (1937b) Een vergelijking tussen het somatische muteren der bloemkleur bij diploid en tetraploide varieteiten van Hyacinthus orientalis. Agricultura (Leuven) 40:216–228

De Mol WE (1940) Uber Modifikationen, Mutationen und den Parallelismus dazwischen, in Zusammenhang mit kaltebehandlung von Hyazinthen. Genetica 22(1–3):231–260

De Mol WE (1949) Twenty-five years of tulip improvement by x-rays. Pap Mich Acad Sci Arts Lett 35:9–14

De Mol WE (1953) X-raying of hyacinths and tulips from the beginning, before thirty years (1922) till today (1952). Jpn J Breed 3(1):1–8

Dehgahi R, Joniyasa A (2017) Gamma irradiation-induced variation in Dendrobium Sonia-28 orchid protocorm-like bodies (PLBs). Fungal Genom Biol 7(2):4–11. In the book: Mutation Breeding, genetic diversity and crop adaptation to climate change

Deng H, Liu S (1987) The study of radiation induced breeding on *Canna indica* L. J South China Agric Univ 1987-03

Deng H, Liu S (1990) The study of radiation induced breeding in Bougainvillea. J South China Agric Univ 13(1):89–93

Denisova NP, Shurshikova GN (1976) Effect of chemical mutagens on economically useful characters in Cyperus esculentus. Fiziol J Fiz-Khim Mekhanizmy Regulyatsii Obmen. Protsessov Organizma, Dep. 2570-77:29–32 (Pl. Breed. Abst. 50: No. 525)

Dennis JA (1976) Somatic aberration induction in tradescantia occidentals by neutrons, X- and gamma radiations. II Biologigal results, r. b. e. and o. e. r. Int J Radiat Biol Stud Phys Chem Med 29(4):323

Desai BM (1973) Report on mutation experiments on ornamentals at Bhabha Atomic Research Centre, Trombay. In: 2nd Workshop on Floriculture, 29–31 Jan. Agri-Horticultural Society, Calcutta

Desai BM (1974) New cultivars of perennial Portulaca through gamma irradiation. Indian Hortic 19(2):19,23

Desai BM, Abraham V (1974) Radiation induced mutants in Canna. In: Use of radioisotopes in studies of plant productivity, Pantnagar, pp 180–186

Dhivya M (2015) Studies on induced mutagenesis in crossandra (*Crossandra infundibuliformis* (L.) Nees). Ph.D. Thesis submitted to Tamil Nadu Agricultural University, Coimbatore

Dhivya M, Balakrisnnan S, Thangaselvabai T (2015) Sensitivity and mutagenic effects of ethyl methane sulphonate on the growth of *Crossandra infundibuliformis* L. Nees. Int J Agric Sci Res 5(3):199–204. ISSN(P): 2250-0057; ISSN(E): 2321-0087

Díaz LE, García SAL, Morales RA, Báez RI, Pérez VE, Olivar HA, Vargas REJ, Hernández HP, De la Cruz TE, García AJM, Loeza CJM (2018) Effect of gamma radiation of 60Co on sunflower plants (Helianthus annuus L.) (Asteraceae), from irradiated achenes. Scientia Agropecuaria 9(3). https://doi.org/10.17268/sci.agropecu.2018.03.02

Díaz-López E, Pichardo-Riego J, de la Cruz TE et al (2003) Variabilidad inducida en *Tigridia pavonia* (L. f.) D.C. var. Sandra por irradiación de bulbos con rayos gamma de ^{60}Co. Revista Chapingo Serie Horticultura 9(2):235–241

Dogra S, Dhiman SR, Gupta YC, Sharma YD (2010) In vitro mutagenesis in carnation (*Dianthus caryophyllus* L.) for induction of resistance to Fusarium oxysporum f. sp. Dianthi. In: National symposium on lifestyle floriculture: challenges and opportunities. Session-2, Crop Improvement, Biotechnology, and Biodiversity, March 19-21, 2010 at DR. Y.S. Parmar University of Horticulture and Forestry, Nauni, Solan, Abstract No. 2.2(O):10

Dommergues P, Gillot J (1965) Variation de la reaction des boutures d'oeillet a l'irradiation gamma. In: The use of induced mutations in plant breeding. FAO/IAEA/EUCARPIA Meeting, Rome, 1964. Radiat Bot 5(suppl):713–718

Dommergues P, Gillot J (1973) Obtention de clones genetiquement homogenes dans toutes leurs couches ontogeniques a partir d'une chimere d'oeillet americain. Ann Amelior Plant 23(2): 83–93

Dommergues P, Heslot H, Gillot J et al (1966) 'L'induction de mutations chez les rosiers' Induced Mutations and their Utilization. In: Proceedings of symposium Erwin-Bauer-Gedachtnisvorlesungen IV. Gatersleben. Akademie-Verlag, Berlin, pp 319–348

Dona M, Ventura L, Macovei A, Confalonieri M, Savio M, Giovannini A, Carbonera D, Balestrazzi A (2013) Gamma irradiation with different dose rates induces different DNA damage responses in Petunia hybrida cells. J Plant Physiol 170:780–787. https://doi.org/10.1016/j.jplph.2013.01.010

Dong Y, Liu J, Li PW, Li CQ, Lu TF, Yang X, Wang YZ (2018) Evolution of Darwin's *Peloric* Gloxinia (*Sinningia speciosa*) is caused by a null mutation in a pleiotropic TCP gene. Mol Biol Evol 35(8):1901–1915. https://doi.org/10.1093/molbev/msy090

Dong Y, Liu J, Li P-W, Li C-Q, Tian-Feng L, Yang X, Wang Y-Z (2023) Evolution of Darwin's peloric gloxinia (Sinningia speciosa) is caused by a null mutation in a pleiotropic TCP gene. Mol. Biol. Evol. 35(8):1901–1915. https://doi.org/10.1093/molbev/msy090. https://www.researchgate.net/publication/324920052_Evolution_of_Darwin's_Peloric_Gloxinia_Sinningia_speciosa_Is_Caused_by_a_Null_Mutation_in_a_Pleiotropic_TCP_Gene

Doorenbos J (1973a) Breeding Elatior-Begonias. Acta Hortic 31:127–131. (reprinted in: The Begonian 40 (12) 1973: 272–277, 290–291)

Doorenbos J (1973b) 'Turo' een nieuwe Elatior-Begonia. Vakbl Bloemisterij 28(42):25

Doorenbos J, Karper JJ (1975) X-ray induced mutations in Begonia x hiemalis. Euphytica 24(1): 13–19

Doring H, Stubbe H (1938) Die Bedeutung des Ernahrungszustandes (Phosphormangel) fur die strahlen induzierte Mutabilitat bei *Antirrhinum majus*. Z Ind Abst Vererb 75:352–357

Dorion N, Chupeau Y, Bourgin JP (1975) Isolation, culture and regeneration into plants of Ranunculus sceleratus L. leaf protoplasts. Plant Sci Lett 5:325–331. https://doi.org/10.1016/0304-4211(75)90060-7

Dube S, Das PK, Dey AK, Bid NN (1980) Varietal improvement of *Dahlia* by gamma radiation. Indian J Hortic *37*(1):82–87. http://www.indianjournals.com/ijor.aspx?target=ijor:ijh&volume=37&issue=1&article=018

Dulieu H (1968) Emploi des chimeres chlorophylliennes pour l'etude de l'ontogenie foliaire. Bull Sci Bourgogne, XXV, pp 1–60

Dulieu H (1969) Mutations Somatiques Chlorophylliennes Induites et Ontogene Caulinaire. Thesis, Faculte des Sciences, Dijon, C.N.R.S. AO3016

Duron M (1992) Induced mutations through EMS treatment after adventitious bud formation on shoot internodes of Weigela cv, Bristol Ruby. Acta Hortic 320:113–118. https://doi.org/10.17660/ActaHortic.1992.320.15

Dyer A, Whitney H, Arnold S, Glover B, Chittka L (2007) Mutations perturbing petal cell shape and anthocyanin synthesis influence bumblebee perception of *Antirrhinum majus* flower colour. Arthropod Plant Interact 1(1):45–55. https://doi.org/10.1007/s11829-007-9002-7

Elateek S, Elmanabawy EA, Fahmy EM, Awad N (2020) Genetic improvement of *Calendula officinal* through mutation induction using gamma irradiation and chemical mutagens. Arab Univ J Agric Sci. https://doi.org/10.21608/ajs.2020.24242.1171

El-Mokadem HE (2014) In vitro induction of flower mutation in Catharanthus roseus using gamma irradiation. Alexandria Sci Exch J 35(1):64–68

El-Nashar YI (2012) Effect of chemical mutagens on vegetative growth and flowering in *Calendula Officinalis* L. (Cv. Calypso Yellow). Alexandria Sci Exchange J 33(2):108–115, Article 5

Eltayeb H, Ei-Metainy AY (2012) Genetical effect of ionizing radiation on *Antirrhinum majus*. LAP Lambert Academic Publishing, London. ISBN-10: 9783659137938, ISBN-13: 978-3659137938

Ernawiati E, Agustrina R, Kanedi M (2022) Kembang sungsang (*Gloriosa superba* L.): a potential plant as a source of biomutagens. Magna Sci Adv Biol Pharmacy 7(1):36–43. https://doi.org/10.30574/msabp.2022.7.1.0088

Espino FJ, Vazquez AM (1981) Chromosome numbers of *Saintpaulia ionantha* plantlets regenerated from leaves cultured *in vitro* with caffeine and colchicine. Euphytica 30:847–853

Evseeva TI, Geras'kin SA (2001) A combined effect of the radiation and nonradiation factors on *Tradescantia* (UrO RAN, Yekaterinburg. [in Russian]

Evseeva TI, Zainullin VG (2000) A study of mutagenic activity of the air and the snow cover in Syktyvkar by the test for somatic mutations in *Tradescantia* (Clone 02). Ekologiya 25(5):343

Eyerdom H (1981) Flower color sports and variations in *Saintpaulia* hybrids. Afr Violet Mag 34(4): 32–37

Fang J, Traore S (2011) In vitro mutation induction of Saintpaulia using ethyl methanesulfonate. HortScience 46:981–984. https://doi.org/10.21273/HORTSCI.46.7.981

Farestveit B (1969) Flower colour chimeras in glasshouse carnation, *Dianthus caryophyllus* L. Yearbook-Royal Veterinary and Agricultural College, Copenhegen, pp 19–33

Farestveit B, Klougart A (1966) Bestraling af nelliker med aceellererede elektroner. Horticultura 20:9–12

Feiglova B (1968) A new colour mutant in Salvia splendens KER-GAwL. Nova barevna mutace u Salvia splendens KER-GA WL. Preslia (Praha) 40:267–273

Fen L, Chu DG (1997) Studies on the inductive mutation of spring orchid. J Hum Agric Univ 23(4): 336–340

Feng H, Yang J, Li D (2016) Study on petal-sepal mutant of sunflower (*Helianthus annuus*) after space mutation. Am J Plant Sci 7:2483–2497. https://doi.org/10.4236/ajps.2016.717216

Filipovic M (2019) What causes succulent mutations? Unusual Seeds, 25 Dec. https://unusualseeds.net/rose-succulent

Firdose K, Pawar N, Dixit G (2011) Induced chlorophyll mutations in *Delphinium malabaricum* (Huth) Munz. J Appl Hortic 13(1):18–24. https://www.researchgate.net/publication/259569523_Induced_chlorophyll_mutations_in_Delphinium_malabaricum_Huth_Munz

Fucui Z, Zuoming X, Yichun H (2011) Study on irradiative mutation breeding of *Canna generalis* Bailey. J Anhui Agric Sci 39:56–58

Fucui Z, Zuoming X, Yichun H (2012) Study on irradiative mutation breeding of *Canna generalis* Bailey. J Anhui Agric Sci 39(1):56–58. ISSN: 0517-6611

Gallone A, Hunter A, Douglas GC (2012) Radiosensitivity of Hebe 'Oratia Beauty' and 'Wiri Mist' irradiated in vitro with gamma rays from 60Co. Sci Hortic 138:36–42. https://doi.org/10.1016/j.scienta.2012.02.006

Gaufillier J (1965) Recherches sur I'Utilisation des Deticiences Chlorophylliennes Induties pour I'Interpretation du Fonctionnement des Meristemes Terminaux de Tiges Feuillets: Cas de I'Oeillet (*Dianthus caryophyllus* L.). Faculte des Sciences, Dijon, 34p

Gavidia I, Perez-Bermudez P (1999) Variants of *Digitalis obscura* from irradiated shoot tips. Euphytica 110:153–159

Geier T (1983) Induction and selection of mutants in tissue cultures of Gesneriaceae. Acta Hortic. 131:329–337

Geier T (1988) *In vitro* chemical induction and propagation of *Kohleria* mutants with altered plant shape and flowering behavior. Acta Hortic 226:687–693

Geier T (1989) An improved type of '*Kohleria*' obtained through in vitro chemical mutagenesis. Mutat Breed Newsl 34:7–8

Geier T (1994) 'Orange Glow' – eine neue Kohleria-Sorte. TASPO Gartenbaumagazin ½:24–26

Geier T (2012) Chimeras: properties and dissociation in vegetatively propagated plants. In: Plant mutation breeding and biotechnology. CABI, Wallingford, pp 191–201

Gerats AGM (1991) Mutants involved in floral and plant development in Petunia. Plant Sci 80:19–25

Ghani M, Sharma SK (2019) Induction of powdery mildew resistance in gerbera (*Gerbera jamesonii*) through gamma irradiation. Physiol Mol Biol Plants 25:159–166. https://doi.org/10.1007/s12298-018-0613-5

Ghani M, Kumar S, Thakur M (2013) Induction of novel variants through physical and chemical mutagenesis in Barbeton daisy (*Gerbera jamesonii* Hook.). J Hortic Sci Biotech 88(5):585–590. https://doi.org/10.1080/14620316.2013.11513010

Ghani M, Kumar S, Thakur M (2014) Physiological and biochemical responses of gerbera (*Gerbera jamesonii* Hook.) to physical and chemical mutagenesis. J Hortic Sci Biotechnol 89:301–306. https://doi.org/10.1080/14620316.2014.11513083

Ghosh S, Ganga M (2019) Determination of lethal dose for ethyl methane sulphonate induced mutagenesis in jasmine. Chem Sci Rev Lett 8:6–10. ISSN 2278-6783.

Ghosh S, Ganga M, John Joel A (2018a) Assessment of mutagenic sensitivity in jasmine (*Jasminum* spp.) to chemical mutagen. Electron J Plant Breed 9(3):1002–1011. ISSN 0975-928X. https://doi.org/10.5958/0975-928X.2018.00125.4

Ghosh S, Ganga M, Soorianathasundaram K (2018b) Determination of radio sensitivity of jasmine (*Jasminum* spp.) to gamma rays. Electron J Plant Breed 9(3):956–965. ISSN: 0975-928X. https://doi.org/10.5958/0975-928X.2018.00119.9

Ghosh S, Ganga M, Ratna PR (2018c) Determination of mutagenic sensitivity of *Jasminum grandiflorum* to gamma irradiation. In: Proceedings of the national symposium on recent advances in floriculture and urban horticulture in global perspective, 4–5 Jan. BCKV, Kalyani, pp 68–69

Ghosh S, Ganga M, Soorianathasundaram K (2019) Studies of mutagenic effectiveness and efficiency of gamma rays and ethyl methane sulphonate in jasmine. Curr J Appl Sci Technol 38:1–8. https://doi.org/10.9734/cjast/2019/v38i630414

Ghosh S, Ganga M, Soorianathasundaram K, Kumar A, Kapoor M (2020) Induction of mutation in *Jasminum grandiflorum* with gamma rays and EMS and identification of novel mutants using molecular markers and SEM imaging. Indian J Hortic 77(4):695–703. https://doi.org/10.5958/0974-0112.2020.00101.2

Girija S, Ganapathi A, Vengadesan G (1999) Micropropagation of *Crossandra infundibuliformis* (L.) Nees. Sci Hortic 82(3/4):331–337

Girsang RMY, Kardhinata EH, Damanik RIM, Karo B (2021) The phenotypic appeal of lily (*Lilium longiflorum* Thunb.) prompted mutation by colchicine. IOP Conf Ser Earth Environ Sci 782: 042050. https://doi.org/10.1088/1755-1315/782/4/042050

Gonzales MA, Aurigue FB, Tapic RT (2008) Radiosensitivity of three species of ground orchids (*S. plicata plicata*, *S. kimballiana* var. *angustifolia* and *S. tomentosa*) to acute gamma radiation. CLSU R D J 2(1):30–53

Gopitha G, Rajamani K, Ganga M, Ravikesavan R, Gnanam R (2022) Resolving the acute gamma irradiation and ethyl methanesulphonate induced lethality for *Jasminum sambac* L. (Aiton) cv. Ramanathapuram Gundumalli. Electron J Plant Breed 13(2):341–349

Grabowska B, Mynett K (1964) The radiation of gamma Co60 rays on the dwarf Dahlia. Bull Insto Hodowli Ak Rosi 5-6:17–26. (In Polish)

Grabowska B, Mynett K (1970) Induction of changes in garden tulips (*Tulipa* hybr. Hort.) under the influence of gamma rays 60Co. Bull Inst Hodowli Aklim Rosl 1–2:89–92

Grassotti A, Mercuri A, Schiva T, Ranieri M (1987) Mutations induced by physical agents on Lilium Ann 1st Sper Floricolt 18:139–153

Grünewald TPM (2002) Tibouchina plant named 'Grue-Tib 04'. Patent number: PP13361. https://patents.justia.com/inventor/theodorus-p-m-gr-newald

Grunewaldt J (1980) Spontane und induzierte Farbschecken-von *Saintpaulia ionantha* H. Wendl. Gartenbauwissenschaft 45(3):124–128

Grunewaldt J (1983) *In vitro* mutagenesis of *Saintpaulia* and *Pelargonium* cultivars (genetic variability, vegetative propagation). Acta Hortic 131:339–343. ISSN/ISBN: 0567-7572. https://doi.org/10.17660/actahortic.1983.131.40

Grunewaldt J (1988) *In vitro* selection and propagation of temperature tolerant of *Saintpaula*. Acta Hortic 226:271–275. https://doi.org/10.17660/ActaHortic.1988.226.32

Guo YM, Zeng LJ, Liang SY, Deng RQ, Ye HN, Liao SX (2016) Effect of UV-B radiation on growth and main secondary metabolites of *Dendrobium officinale* Kimura et Migo. Northern Hortic 16:154–156. (in Chinese)

Gupta MN (1966) Induction of somatic mutations in some ornamental plants. In: Proceedings of all India symposium on horticulture, pp 107–114

Gupta MN (1970) Use of gamma irradiation in the production of new varieties of *Perennial portulaca*. In: Proceedings of symposium on radiation and radiomimetic substances in mutation breeding. Food and Agricultural Communication Department at Energy, Bombay, pp 206–214

Gupta PP (1977) Cytogenetics of aquatic ornamentals IV. Effects of gamma irradiation on *Nymphaea rubra* Roxb. Cytologia 42:297–304

Gupta MN, Matsuo T (1969) Effects of thermal neutrons and gamma rays on seeds of Antirrhinum majus and their modification by water content and oxygen concentration. Radioisotopes 18 (2):52–55

Gupta MN, Samata Y (1967) The relationship between developmental stages of flower-buds and somatic mutations induced by acute X- and chronic gamma-irradiation in Cosmos bipinnatus. Radiat Bot 7(3):225–240. https://doi.org/10.1016/S0033-7560(67)80022-X

Gustafson A, Gadd J (1965) Mutations and crop improvement. IV. *Poa pratensis* J. (Gramineae). Hereditas 53(1–2):90–103

Haggo A (2006) Tibouchina urvilleana plant named 'TB01'. US13/573,5862012-09-262012-09-26Tibouchina urvilleana plant named 'TB01'

Haigh MV, Howard A (1973) Radiation induced tumorous outgrowth in young gametophytes of *Osmunda regalis*. Radiat Bot 13:111–119

Hajizadeh HS, Mortaza SN, Tohid F, Yildiz H, Helvaci M, Alas T, Okatan V (2022) Effect of mutation induced by gamma-irradiation in ornamental plant Lilium *(Lilium longiflorum* cv. Tresor). Pak J Bot 54(1). https://doi.org/10.30848/PJB2022-1(23)

Hamatani M, Iitsuka Y, Abe T, Miyoshi K, Yamamoto M, Yoshida S (2001) Mutant flowers of dahlia (*Dahlia pinnata* Cav.) induced by heavy-ion beams. RIKEN Accel Prog Rep 34:169

Hansen AA, Juska FV (1959) A 'progressive' mutation in *Poa pratensis* L. by ionizing radiation. Nature (London) 184:1000–1001

Hansen AA, Juska FV (1962) Induced mutations in Kentucky bluegrass. Crop Sci 2:369–371

Haq MS (1983) Colchicine induced mutants in *Salvia coccinia* Juss. Geobios. New Rep 2:149–150

Harris GP, Hart EMH (1964) Regeneration from lef squares of Peperomia sandersii A. DC: a relationship between rooting and budding. Ann Bot (London) 28:509–526

van Harten AM, Bal ECJ (1986) Chemically induced mutations for male sterility in Petunia x hybrida. In: Horn W, Jensen CJ, Odenbach W, Schieder O (eds) Genetic manipulation in plant breeding: proceedings international symposium Organized by EUCARPIA, 8–13 Sept 1985, Berlin (West), Germany. De Gruyter, Berlin, pp 43–48. https://doi.org/10.1515/9783110871944-006

Hase Y, Okamura M, Takeshita D, Narumi I, Tanaka A (2010) Efficient induction of flower-color mutants by ion beam irradiation in petunia seedlings treated with high sucrose concentration. Plant Biotechnol. 27:99–103. https://doi.org/10.5511/plantbiotechnology.27.99

Hase Y, Akita Y, Kitamura S, Narumi I, Tanaka A (2012) Development of an efficient mutagenesis technique using ion beams: Toward more controlled mutation breeding. Plant Biotechnol. 29: 193–200

Hasing T, Rinaldi E, Manrique S, Colombo L, Haak DC, Zaitlin D, Bombarely A (2019) Extensive phenotypic diversity in the cultivated Florist's *Gloxinia, Sinningia speciosa* (Lodd.) Hiern, is derived from the domestication of a single founder population. Plants People Planet 1:363–374. https://doi.org/10.1002/ppp3.10065

Haspolat G, Kunter B, Ozzambak ME (2016) Some changes in mutant *Crocus* plants. In: ISHS Acta Horticulturae 1127: XXIV international horticultural congress on horticulture: sustaining lives, livelihoods and landscapes (IHC2014): international symposium on plant breeding in horticulture. https://doi.org/10.17660/ActaHortic.2016.1127.60

Hassan AA, Ahmad Z, Ariffin S (2010) RnD seminar 2010: research and development seminar 2010, Bangi (Malaysia), 12–15 Oct 2010. Also available in Malaysian Nuclear Agency Document Delivery Center, 8p

Hassan M, Badr M, El-Maksoud BA, El-Shennawy O (2017) Effect of gamma radiation on genetic improvement against salinity in *Catharanthus roseus* plants. J Adv Agric Res 22(3):424–455. https://doi.org/10.21608/jalexu.2017.243819

Hayashi Y, Ishii S, Hirano T, Ichinose K, Kazama Y, Abe T (2019) New ornamental cherry cultivars induced by heavy-ion beam irradiation. Acta Hortic 1235:99–104. https://doi.org/10.17660/ActaHortic.2019.1235.13

Hazbullah NA, Taha RM, Saleh A, Mahmad N (2012) Irradiation effects on *in vitro* organogenesis, callus growth and seedling development of *Garbera jamesomnii*. Hortic Bras 30:252–257

Heffron LM, Korban SS (2022) Mutagenic responses to ethyl methanesulfonate and phenotypic characterization of an M_1 generation of snapdragon, *Antirrhinum majus*. Euphytica 218(6): 76. ISSN: 0014-2336

Heffron L, Blowers A, Korban S (2006) Chemical mutagenesis in snapdragon (*Antirrhinum majus*). HortScience 41(4). https://doi.org/10.21273/HORTSCI.41.4.1021A

Hekstra G, Broertjes C (1968) Mutation breeding in bulbous Iris. Euphytica 17:315–351

Hemalatha K (1998) Induction of mutation in carnation (*Dianthus caryophyllus* L.) through gamma rays and EMS. Ph.D. Thesis, University of Agricultural Sciences, Bangalore, India

Henny RJ (1978) In vitro propagation of *Peperomia* 'Red Ripple' from leaf discs. HortScience 139(2):150–151

Hentrich W, Beger B (1974) Untersuchungen uber die mutagene Effizienz von N-nitroso-N-Methylharnstoff bei *Saintpaulia ionantha* H. Wendl Arch Zuchtungsforsch 4(1):29–43

Hentrich W, Glawe M (1982) Zuchtung von Edelnelken (*Dianthus caryophyllus*) durch AMS – Applikation in die Blattachseln von Jungpflanzen. Archiv Zuchtungsforsch 12(3):197–207

Hernández-Muñoz S, Pedraza-Santos ME, López PA et al (2017a) Estimulación de la germinación y desarrollo *in vitro* de *Laelia autumnalis* con rayos gamma. Revista Fitotecnia Mexicana 40(3): 271–283

Hernández-Muñoz S, Pedraza-Santos ME, López PA et al (2017b) Determinación de la DL_{50} y GR_{50} con rayos gamma (^{60}Co) en protocormos de *Laelia autumnalis in vitro*. Agrociencia 51(5):507–524

Heslot H (1964) L'induction experimentale de mutations chez les plantes floras. In: P.V. Séance 16 decembre. Academy of Agriculture, France, pp 1281–1308

Heursel J (1972) Studie van de Overerving en de Genotypische Variabiliteit van Kleur, Grootte ev Vorm van de Bloemen bij Rhododendron simsii Planch. (*Azalea indica* L). Rijksstn Sierplantenteelt, Melle, Meded 28:196

Heursel J (1981) Diversity of flower colours in *Rhododendron simsii* Planch. and prospects for breeding. Euphytica 30(1):9–14

Hewawasam WDCJ (2003) Creating novel phenotypes of *Crossandra infundibuliformis* var. 'Danica' through tissue culture and induced mutations. M.phil. Thesis. PGIA, University of Peradeniya

Hewawasam WDCJ, Bandara D, Aberathne WM (2004) New phenotypes of *Crossandra infundibuliformis* var. danica through in vitro culture and induced mutations. Trop Agric Res 16:253–270. Biology Corpus ID: 90815640

Honda K, Maeda T, Ichimuru N, Fukuda T, Fujino M, Anami T, Nishizuku M (1980) Characteristics of a new variety of Chinese mat grass 'Toyomidori'. Bull Oita Mat Grass Techn Guid Stn 1:1–18

Hong YP, Lee JS, Kim KK (1980) Studies on improving Hibiscus syriacus in Korea. Research report of the Office of Rural Development, Horticultute and Sericulture, Suwon, vol 22, pp 56–62

Hopper JE, Peloquin ST (1968) X-ray inactivation of the stylar component of the self-incompatibility reaction in *Lilium longiflorum*. Can J Genet Cytol 10:941–944

Horita M, Sakamoto K, Abe T, Yoshida S (2002) Induction of mutation in Sandersonia aurantiaca Hook. by heavy-ion beam irradiation. RIKEN Accel Prog Rep 35:131

Horn W (1984) Treating *invitro* cultures of floriculture crops with mutagens. Mutat Breed Newsl 24:13–15

Horovitz A, Zohary D (1966) Spontaneous variegation for perianth colour in wild *Anemone coronaria*. Heredity 21(3):513–515. https://doi.org/10.1038/hdy.1966.49. Corpus ID: 41825958

Hoskins T, Contreras RN (2019) Exposing seeds of Sarcococca confusa to increased concentrations and durations of ethyl methanesulfonate reduced seed germination, twinning, and plant size. Hortscience 54(11):1902–1906. https://doi.org/10.21273/HORTSCI14428-19

Hosoguchi T, Uchiyama Y, Komazawa H, Yahata M, Shimokawa T, Tominaga A (2021) Effect of three types of ion beam irradiation on Gerbera (*Gerbera hybrida*) in vitro shoots with mutagenesis efficiency. Plants 10(7):1480. https://doi.org/10.3390/plants10071480

Howard A, Haigh MV (1968) Chloroplast aberrations in irradiated fern spores. Mutat Res 6(2): 263–280

Huang P-L (2021) Preliminary studies on mutagenic effects of sodium azide and γ-Ray irradiation on organogenesis in *Guzmania* and *Aechmea* Bromeliads. Res Bull of KDARES 21(1):28–38

IAEA (1972) Induced mutations and plant improvement. IAEA, Vienna, p 526

Ichikawa S (1997) Somatic mutation frequencies in *Tradescantia* stamen hairs treated with relatively low thermal neutron fluxes. Radiat Res 147(1):109–114. https://doi.org/10.2307/3579449

Ichikawa S, Sparrow AH (1978) Influence of radiation exposure rate on somatic mutation frequency and loss of reproductive integrity in *Tradescantia* stamen hairs. Mutat Res 52(2):195

Ichikawa S, Takahashi CS (1977) Somatic mutation frequencies in stamen hairs of stable and mutable clones of *Tradescantia* after acute gamma-ray treatments with small doses. Mutat Res 9(3):195

Ichikawa S, Sparrow AH, Thompson KH (1969) Morphologically abnormal cells, somatic mutations and loss of reproductive integrity in irradiated *Tradescantia* stamen hairs. Radiat Bot 9:195

Ichikawa S, Takahashi CS, Nagashima-Ishii C (1981) Somatic mutation frequency in the stamen hairs of Tradescantia KU 7 and KU 9 clones exposed to low-level gamma rays. Jpn J Genet 56: 409–423

Iizuka M, Kudo O, Kimurai Y, Tanaka A, Shikazono N, Watanabe H (1998) Induction of mutation in strawberry and hydrangea by ion beam irradiation. Effects of ion beams on germination of strawberry and hydrangea seed. JAERI-Review 99-025 2. 3, TIARA Annual Report, pp 30–31

Iizuka M, Kudo N, Kimura Y, Hase Y, Tanaka A (2001) Induction of mutation in Spiraea by ion beam irradiation. In: Saidoh M, Toraishi A, Namba H, Itoh H, Tanaka S, Naramoto H, Sekine T, Tanaka A, Kobayashi Y, Arakawa K, Otsubo M, Tajima S, Tanaka S (eds) Effects of ion beams on germination of Spiraea seed TIARA annual report 2000, No. 2001-039, pp 45–46

Ikegawa S, Shoji K, Hase Y, Nozawa S (2016) Determination of ion beam irradiation conditions for callus of tulip. QST Takasaki Ann Rep 2015:131

Ishii S, Hayashi Y, Ryuto H, Fukunishi N, Abe T (2009) A new cultivar "Nishina Zaou" induced by heavy ion beam irradiation. RIKEN Accel Prog Rep 42:xi

Ishii S, Hayashi Y, Hirani T, Fukunishi N, Abe T (2011) Development of flowering mutant of cherry blossom by heavy-ion irradiation. RIKEN Accel Prog Rep 44:xiii

Ishizaka H (2018) Breeding of fragrant cyclamen by interspecific hybridization and ion-beam irradiation. Breed Sci 68(1). https://doi.org/10.1270/jsbbs.17117

Ishizaka H, Kondo E, Kameari N (2012) Production of novel flower color mutants from the fragrant cyclamen (*Cyclamen persicum* × *C. purpurascens*) by ion-beam irradiation. Plant Biotechnol. 29:201–208

Ito Y, Seki E, Ohkoshi K, Watanabe M, Saito H, Hayashi Y, Abe T (2007) Mutation induction by ion beam irradiation in *Begonia*. Bull Chiba Agric Res Cent 6:75–84. ISSN: 1347-2585

Jack (2009) Chemicals and hormones to induce mutations. Cacti and succulents. https://www.shaman-australis.com/forum/20765-c

Jala Hayati BD, Krisantini SIA (2016) Radiosensitivity levels of *in vitro* cultured *Celosia cristata* planlets by γ - ray irradiation. J Trop Crop Sci 3(2):61–65

Jamdhade VP, Kashid NG (2016) Effect of physical and chemical mutagenesis in sunflower (*Helianthus Annus* L.) on seed germination through induced mutation. Int J Sci Res 5(6). ISSN: 2319-7064 Index Copernicus Value (2013): 6.14

Janick J, Skirvin RM, Janders RB (1977) Comparison of in vitro and in vivo tissue culture systems in scented geranium. J Hered 68:62–64

Jerzy M, Zalewska M (1992) *In vitro* adventitious bud techniques for mutation breeding of *Gerbera jamesonii*. Acta Hortic 314:269–274. https://doi.org/10.17660/ActaHortic.1992.314.32

Jiang L, Dunn BL (2016) Ethyl methanesulfonate and caffeine mutagenetic treatment to four ornamental *Silene* species. J Environ Hortic 34(4):95–100. https://doi.org/10.24266/0738-2898-34.4.95

Jiang P, Chen Y, Wilde HD (2014) Optimization of EMS mutagenesis on petunia for TILLING. Adv Crop Sci Technol 2:141. https://doi.org/10.4172/2329-8863.1000141

Jin Q, Yao Y, Cai YP, Lin Y, Fan HH, Luo Q (2012) Effects of NO on the growth and subculture of Dendrobium protocorms. Pharm Biotechnol 19:517–520

Johnson EL (1948) Response of *Kalanchoe tubiflora* to X-irradiation. Plant Physiol 23:544–556

Johnson RT (1980) Gamma irradiation and *in vitro* induced separation of chimeral genotypes in carnation. HortScience 15(5):605–606

Julen G (1954) Observations on X-rayed *Poa pratensis*. Acta Agric Scand 4:585–593

Julen G (1958) Uber die Effekte der Rontgenbestrahlung bei Poa pratensis. Zuchter 28:37–40

Julen G (1961) The effect of X-rays on the apomixix in *Poa pratensis*. In: Effects of ionizing radiation on seeds (Proc. Symp. Karlsruhe, 1960). IAEA, Vienna, pp 527–532

Jungnickel F (1977) Induktion und Vermehrung von Mutanten bei *Saintpaulia ionantha* H. Wendl In Sterikultur. Biol Zentralblatt 96:335–343

Kaicker US, Singh HP (1979) Role of mutation breeding in *Amaryllis*. Plant Life 35(1/4):66–73

Kameari N, Okubo N, Kondo E, Nakayama M, Akita Y, Hase Y, Tanikawa N, Morita Y, Tanaka A, Ishizaka H (2011) Floral scent compounds in white mutant generated from fragrant purple cyclamen 'Kokou-no-kaori' (*Cyclamen persicum* × *C. purpurascens*) by ion beam irradiation. Hortic Res 10(suppl 2):220

Kameari N, Akita Y, Kitamura S et al (2012) Characteristics of yellow-flowered mutant generated from dihaploid of fragrant cyclamen (*Cyclamen persicum* 'Golden Boy' × C. purpurascens) by ion- beam irradiation. Hortic Res 11(suppl 1):191

Kameya T (1975) Culture of protoplasts from chimera plants tissue of nature. Jpn J Genet 50(5): 417–420

Kanaya T, Saito H, Hayashi Y, Fukunishi N, Ryuto H, Miyazaki K, Kusumi T, Abe T, Suzuki K (2008) Heavy-ion beam-induced sterile mutants of verbena (*Verbena hybrida*) with an improved flowering habit. Plant Biotechnol. 25:91–96. https://doi.org/10.5511/plantbiotechnology.25.91

Kang SY, Kim DS, Lee GJ (2007) Genetic improvement of crop plants by mutation techniques in Korea. Plant Mutat Rep 1(3):7–15

Kannan M, Sathiyamurthy VA, Shanker V (2002) Mutagenic studies on *Jasminum sambac*. In: Misra RL, Misra S (eds) Floriculture research trend in India. Indian Society of Ornamental Horticulture, Division of Floriculture and Landscaping, IARI, New Delhi, pp 209–211

Kannan B, Davila-Olivas NH, Lomba P et al (2015) *In vitro* chemical mutagenesis improves the turf quality of bahiagrass. Plant Cell Tiss Org Cult 120(2):551–561

Kapoor M, Kumar P, Lal S (2014) Gamma radiation induced variations in corn marigold (*Glebionis segetum*) and their RAPD-based genetic relationship. Indian J Agric Sci 84(7):796–801

Karper JJ, Pierik RLM (1981) Gemuteerde alanchoe bloemen kunnen vegetatief worden vermeeederd. Vakbl Bloemisterij 36(44):44–45

Karunananda D, Ranathunga R, Abeysinghe W (2018) 60Co gamma irradiation-induced variations in vegetatively propagated *Philodendron erubescens* 'Gold'. In: FAO/IAEA international symposium on plant mutation breeding and biotechnolgy, Vienna, 27–31 Aug, Abst. No. IAEA-CN-263-42

Kashikar SG, Khalatkar AS (1981) Breeding for flower colour in *Petunia hybrida* Hort. Acta Hortic 111:35–40. https://doi.org/10.17660/ActaHortic.1981.111.3

Kaur S, Khare PB (1982) Effect of gamma irradiation on development of *Hemionitis arifolia* (Burm.) Moore. Aspects Plant Sci 6:25–30

Kaur R, Kapoor M, Kaur R, Kumar A (2017) Effect of gamma irradiation on cyto-morphology, total phenolic content and antioxidant activity of *Calendula*. J Hill Agric 8(4):395–402. https://doi.org/10.5958/2230-7338.2017.00078.7

Kelly JW, Lineberger RD (1981) Thermal neutron induced changes in *Saintpaulia*. Environ Exp Bot 21(1):95–102

Khalaburdin AP (1991) Effects of gamma radiation from ^{60}Co on changes in some biological and morphological traits in hybrid canna. Genetiko-selektsionnye issledovaniya v Turkmenin, pp 158–167. Cited in: Horti Abstr, 1994: no. 1268

Khalatkar AS, Kashikar SG (1980) Sodium azide mutagenicity in *Petunia hybrida*. Mutat Res 79: 81–85

Khan IA (1997) Induction of mutations in saffron (Crocus sativus L.) by gamma irradiation. J Nucl Agric Biol 26(3):155–160. ISSN 0379-5489

Khan IA (2004) Induced mutagenic variability in saffron (*Crocus sativus* L.). Acta Hortic 650:281–283

Khan IA (2007) Development of high yielding saffron mutant. ISHS Acta Horticulturae, 739. In: II International symposium on saffron biology and technology. https://doi.org/10.17660/ActaHortic.2007.739.31

Khan IM, Khan GA, Islam S (1978) Studies on the spontaneous plastid mutations in *Dahlia variabilis* Desf. Pak J For 28(3):137–143

Khan MA, Nagoo S, Naseer S, Nehvi FA, Zargar SN (2011) Induced mutation as a tool for improving corm multiplication in saffron (*Crocus sativus* L.). J Phytology 3(7):8–10. ISSN: 2075-6240. www.scholarjournals.org

Khare PB (1994) Induced aneuploid in *Adiantum capillus-veneris* L. Cytologia 59:31–34

Khare PB, Kaur S (1980) Growth and differentiation of the gametophyte of a fern, *Adiantum lunulatum* Burm. by means of gamma irradiation. New Bot 7:83–90

Kharkwal MC, Pandey RN, Pawar SE (2004) Mutation breeding for crop improvement. In: Jain HK, Kharkwal MC (eds) Plant breeding, vol 26. Springer, Dordrecht, pp 601–645

Khoddamzadeh AA, Sinniah UR, Kadir MA, Kadzimin SB, Maziah M, Sreeramanan S (2010) Detection of somaclonal varia-tion by random amplified polymorphic DNA analysis during micropropagation of *Phalaenopsis bellina* (Rchb. f.) Christenson. Afr J Biotechnol 4:6632–6639. (17) (PDF) The Application of Biotechnology to Orchids. https://www.researchgate.net/publication/262850225_The_Application_of_Biotechnology_to_Orchids. Accessed 3 Jan 2023

Khosravi AR, Kadir MA, Kadzemin SB et al (2009) RAPD analysis of colchicines induced variation of the *Dendrobium Serdang* beauty. Afr J Biotechnol 8:1455–1465

Kim J-K, Lee K-U, Lim Y-T (1997) Studies on mutation breeding of *Hibiscus syriacus*. Research report, p 89

Kim SH, Jo YD, Ryu J et al (2019a) Effects of the total dose and duration of irradiation on the growth responses and induced SNPs of a *Cymbidium hybrid*. Int J Radiat Biol 96:545–551

Kim SH, Kim Y-S, Lee H-J, Jo YD, Kim J-B, Kang S-Y (2019b) Biological effects of three types of ionizing radiation on creeping bentgrass. Int J Radiat Biol 95(9):1295–1300. https://doi.org/10.1080/09553002.2019.1619953

Kim SH, Kim SW, Ryu J et al (2020) Dark/light treatments followed by –irradiation increase the frequency of leaf-color mutants in *Cymbidium*. Plants 9:532. https://doi.org/10.3390/plants9040532

Kim JH, Lee JS, Oh HJ, Kim SY, Kim HC (2021) New cultivar 'Hwangnarae' of leaf color variegated hosta minor developed by EMS treatment. Korean J Plant Resour 34(2):172

Kleffel B, Walther E, Preil W (1986) X-ray induced mutations in embryogenic suspension cultures of *Euphorbia pulcherrima*. In: Nuclear techniques and in vitro culture for plant improvement, Vienna, 1985, pp 113–120

Klekowski EJ (2011) Plant clonality, mutation, diplontic selection and mutational meltdown. Biol J Linn Soc 79(1):61. https://doi.org/10.1046/j.1095-8312.2003.00183.x

Klekowski EJ Jr (1976) Mutational load in a fern population growing in a polluted environment. Am J Bot 63(7):1024–1030

Klekowski EJ Jr (1984) Mutational lead in clonal plants: a study of two fern species. Evolution 38(2):417–426

Klekowski EJ Jr, Klekowski E (1982) Mutation in ferns growing in an environment contaminated with polychlorinated biphenyls. Am J Bot 69(5):721–727

Klekowski EJ, Kazarinova-Fukshansky N (1984) Shoot apical meristems and mutation: fixation of selectively neutral cell genotypes. Am J Bot 71:22–27

Klinkhamer L (2020) A mutation from Hosta 'Lady in Red' from my good friend Arie Blom. https://www.facebook.com/groups/2248595978/posts/10158507814200979/

Knuth M (1962) Mutation durch Rontgenbestrahlung auch bei Grunpflanzen moglich. Dtsch Gartenbau 9(6):162–164

Kobayashi N, Sakamoto S, Nakatsuka A, Hase Y, Narumi I (2007) Mutation induction in *Azalea* using ion beam irradiation and its gene analysis. JAEA-review 2007-060-87

Kobayashi N, Tasaki K, Kano S, Sakamoto S, Nakatsuka A, Hase Y, Narumi I (2008/2007) Mutation induction in *Azalea* seedlings using ion beam irradiation. JAEA-review 2008-055, p 70

Kobayashi N, Tasaki K, Kano S, Sakamoto S, Nakatsuka A, Hase Y, Narumi I (2009) Mutation induction in azalea seedling using ion beam irradiation. JAEA Takasaki annual report 2007, p 70

Kobayashi N, Sasaki S, Tasaki K, Nakatsuka A, Nozawa S, Hase Y, Narumi I (2011) Effect of ion beam irradiation for *Asclepias* species. JAEA Takasaki annual report 2009, p 67

Kodama M, Tanabe Y, Hirano T, Abe T (2015) Development of flower form and colour mutant of *Hydrangea macrophylla* by heavy-ion beam irradiation. Hortic Res 14(suppl 1):419

Koh YC, Davies FT Jr (2000) Mutagenesis and in-vitro culture of *Tillandsia fasciculata* Swartz var. fasciculata (Bromeliaceae). Sci Hortic 87:225–240

Koh YC, Davies T Jr (2001) Mutagenesis and in vitro culture of *Tillandsia fasciculata* Swartz var. fasciculata (Bromeliaceae). Sci Hortic 87:225–240

Kolar F, Pawar N, Dixit G (2011) Induced chlorophyll mutations in *Delphinium malabaricum* (Huth) Munz. J Appl Hortic 13(1):18–24

Kolar FR, Pai SR, Dixit GB (2013) EMS, sodium azide and gamma rays induced meiotic anomalies in *Delphinium malabaricum* (Huth) Munz. Isr J Plant Sci 61(1–4):64. https://doi.org/10.1080/07929978.2014.961395

Kolar F, Ghatge-Patil SR, Nimbalkar M, Dixit GB (2015) Mutational changes in Delphinium malabaricum (Huth.) Munz.: a potential ornamental plant. J Hortic Res 23(2):5–15. https://doi.org/10.2478/johr-2015-0012

Kolar FR, Ghatge SR, Kudale SS et al (2020) RP-HPLC analysis of delphinidin content in flower colour mutants of *Delphinium malabaricum* (Huth) Munz. In: Research interventions and advancements in plant sciences. ISBN: 978-93-88901-14-7

Kole PC, Meher SK (2005) Effect of gamma rays of some quantitative and qualitative characters in zinnia *Zinnia elegans* N.J. Jacguin in M1 generation. J Ornam Hortic 8(4):303–305

Kondo M, Koike Y, Hase Y, Yokota Y, Kobayashi H (2007) Induction of mutations by the ion beam irradiation to the bulb-scales of Lilium cv. Acapulco. JAEA-review 2007-060 3-09, p 75

Kondo E, Hase Y, Narumi I, Ishizaka H (2008a) Mutation induction by ion-beam irradiation in fragrance Cyclamen (*Cyclamen* persicum × *C. purpurascens*). JAEA-review 2008-055, p 79

Kondo M, Koike Y, Okuhara H, Oda M, Hase Y, Yoshihara R, Kobayashi H (2008b) Induction of mutations affecting pollen formation by ion beam irradiation to *Lilium* × formolongi hort (cv. White Aga). JAEA Takasaki annual report 2007, p 67

Kondo E, Nakayama M, Takamura T, Kurihara Y, Saotome T, Hase Y, Tanaka A, Ishizaka H (2009a) Flower pigments and volatile compounds of mutant obtained from fragrant cyclamen (*Cyclamen persicum* × *C. purpurascens*) by ion beam irradiation. Hortic Res 8(suppl 2):273

Kondo E, Nakayama M, Kameari N, Tanikawa N, Morita Y, Akita Y, Hase Y, Tanaka A, Ishizaka H (2009b) Red-purple flower due to delphinidin 3,5-diglucoside, a novel pigment for *Cyclamen* spp., generated by ion-beam irradiation. Plant Biotechnol 26:565–569. https://doi.org/10.5511/plantbiotechnology.26.565

Kondo E, Nakayama M, Kameari N, Kurihara Y, Tanikawa N, Morita Y, Akita Y, Hase Y, Tanaka A, Ishizaka H (2010) Analyses of flower pigments and volatile compounds of white mutants generated by ion beam irradiation from fragrant purple *Cyclamen* 'Kokou-no-kaori' (*Cyclamen persicum* × *C. purpurascens*). Hortic Res 9(suppl 2):255

Kondo E, Nakayama M, Kameari N, Kurihara Y, Tanikawa N, Morita Y, Akita Y, Hase Y, Tanaka A, Ishizaka H (2011) Analyses of flower pigments and volatile compounds of red-purple mutants generated by ion beam irradiation from fragrant purple cyclamen 'Kaori-no-mai' (*Cyclamen persicum* × *C. purpurascens*). Hortic Res 10(suppl 2):219

Konzak CF, Randolph LF (1956) Radiation and iris breeding. Bull Am Iris Soc 142:68–76

Koo FKS, Cuevas-Ruiz J (1974) Induction of somatic mutations in Jerusalem thorn (*Euphorbia splendens* Bojeri). Mutat Breed Newsl 3:13

Kozlowska-kalisz J (1979) Influence of ionizing radiation on biological activity of endogenous growth regulators in orchids/cymbidium/in tissue culture. Acta Hortic 91(1):261–286

Kruczkowska H, Pawlowska H, Skucinska B (1997) Effect of EMS and sodium azide on callus culture and plant regeneration in *Portulaca grandiflora* (Hook). Mutat Breed Newsl 43:12–13

Krumbiegel G (1979) Response of haploid and diploid protoplasts from *Datura innoxia* Mill. and *Petunia hybrida* L. to treatment with X-rays and a chemical mutagen. Environ Exp Bot 19:99–103. https://doi.org/10.1016/0098-8472(79)90014-5

Kuang AIC (1999) Effects of gamma irradiation on in vitro cultures of selected orchid hybrids. M. Sc. Thesis, University Putra Malaysia, Serdang

Kudo N, Iizuka M, Kimura Y, Tanaka Y, Shikazono N, Watanabe H (1998) Induction of mutation in strawberry and hydrangea by ion beam irradiation: effects of ion beam irradiation on shoot regeneration of strawberry callus and germination of hydrangea seed. In: Tanaka R, Saido M, Nashiyama I, Naramoto H, Suwa T, Morita Y, Toraishi A, Watanabe H, Tani M, Tajima S, Okada S (eds) TIARA annual report 1997, no. 98-016: 30-31. JAERI, Tokai, pp 62–64

Kukimura H, Ikeda F, Fujita H et al (1975) Genetic, cytological and physiological studies on the induced mutants with special regard to effective methods for obtaining useful mutants in perennial woody plants. In: Improvement of vegetatively propagated plants through induced mutations, Tokai, 1974. IAEA, Vienna, pp 83–104

Kukimura H, Ikeda F, Fujita H, Maeta T, Nakajima K, Katagiri K, Nakahira K, Somegou M (1976) Genetical, cytological and physiological studies on the induced mutants with special regard to effective methods for obtaining useful mutants in perennial woody plant. In: International Atomic Energy Agency (ed) Improvement of vegetatively propagated plants and tree crops through induced mutations. International Atomic Energy Agency, Vienna, pp 93–137

Kuleshov G, Bekhtin N, Williams WR (1974) Chemical mutagenesis in perennial grass breeding. In: Iglovikov VG, Movsisyants AP (Eds) Proceedings of the XII interantional grassland congress, Moscow III(II), pp 812–816

Kulkarni RN, Baskaran S (2003) Inheritance of resistance to *Pythium* dieback in the medicinal plant, periwinkle. Plant Breed 122:184–187

Kulkarni RN, Baskaran K (2008) Inheritance of pollen-less anthers and "Thrum" and"Pin" flowers in periwinkle. J Hered 99(4):426–431

Kulkarni RN, Suresh N (1999) Inheritance of characters in periwinkle: leaf pubescence and corolla colour. J Herbs Spices Med 6:85–88

Kulkarni RN, Bhaskaran K, Chandrasekhara RS, Kumar S (1999) Inheritance of morphological traits of periwinkle mutants with modified contents and yields of leaf and root alkaloids. Plant Breed 118(1):71–74

Kumar PRR, Ratnam SV (2010) Macro and micro-morphological mutants in varieties of sunflower (*Helianthus annuus* L.) by using gamma-rays, sodium azide and combined treatment. Int J Plant Sci 5(2):528–534. ISSN: 0973-1547. Record Number: 20103327448

Kumar N, Sambandamurthu S, Khader JMA (1983) Sensitivity of *Jasminum grandiflorum* cuttings to gamma irradiation. South Indian Hortic 31(4/5):250–251

Kumar S, Rai SP, Kumar SR, Singh D, Srivastava S, Mishra RK (2007) Plant variety of *Catharanthus roseus* named 'Ili'. United States Patent PP18315;2007

Kumar M, Kumar V, Solanki RB (2008) Induction of mutation in Acalypha (*Acalypha tricolor* L.) through gamma ras. IN: International symposium on induced mutations in plants (ISIM); Vienna (Austria); 12-15 Aug 2008. Joint FAO/IAEA Division of Nuclear Techniques in Food and Agriculture, Plant Breeding Section, Vienna (Austria), p 129, 207p. IAEA-CN--167-331P

Kurniati R (2004) Genetic diversity induction of *Phalaenopsis hinamatsuri* x *Doritaenopsis modern* Beauty and Phalaenopsis Taipei Gold' GS' using gamma ray radiation. Bogor Agriculture Institute, Bogor

Kwon HJ, Lim JH, Woo SM, Hwang MJ, Pyo SH, Woo JS (2009) Mutant breeding of ornamental trees for creating variations with high value using proton beam. KAERI, 44p. KAERI/CM--1111/2008

Lama S, Aeksomtramaet L, Kanchanapoom K (2005) Induced mutation in Nelumbo nucifera Gaertn. by gamma irradiation. In: Conference: 31. Congress on Science and Technology of Thailand, Nakhon Ratchasima (Thailand), 18–20 Oct 2005. Report Number: INIS-TH-100

Land JB, Whittington WJ, Norton G (1971) Environment dependent chlorosis in a mutant plant of Festuca pratensis, Huds. Ann Bot New Ser 35(141):605–613

Laneri U, Franconi R, Altavista P (1990) Somatic mutagenesis of *Gerbera jamesonii* Hybr.: irradiation and in vitro culture. Acta Hortic 280:395–402. https://doi.org/10.17660/ActaHortic.1990.280.64

Lantin B, Decourtye L (1970) Obtention de types nouvxis chez le *Dahlia* par mutations provoquees. Lien Hortic 7. Rev Fr Maraichage Hortic. Pepiniere, No 99:5875–5878

Lapade AG, Veluz AMS, Marbella LJ, Rama MG (2001) Induced mutation and in vitro culture techniques for the genetic improvement of ornamentals. Phil Nucl J 13:23–31

Largon V, Lyakh V (2002) Mutation variability in M2 in *Linum grandiflorum* Desf. Under influence of ethylmethane sulfonate. Mutation spectrum Visnyak Zaporizhzhia State University 34(1):124–129

Lata P, Gupta MN (1971) Morphological and cytological studies on gamma ray induced mutants of perennial *Portulaca grandiflora*. Jpn J Breed 21(3):121–128

Latado RR, Tulman Neto A (1996) Use of gamma rays in mutation breeding of *Calathea louisae* cv Albertii. Braz J Gen 19:197

Latha S, Dharmatti PR (2018) Gamma rays induced mutation studies in marigold Cv. double orange in M1 generation. Int J Pure Appl Biosci 6(3):443–447. ISSN: 2320 – 7051. https://doi.org/10.18782/2320-7051.5380

Lavanya V, Ganga M, Rajamani K, Meenakumari B, Gnanam R (2022) The efficiency of gamma irradiation and ethyl methane sulphonate in inducing variations in *Jasminum auriculatum* Vahl. J Appl Nat Sci 14(4):1163–1168. https://doi.org/10.31018/jans.v14i4.3761

Lawrence WJC (1931) Mutation or segregation in the octoploid *Dahlia variabilis*. J Genet 24:308–325

Lawrence WJC (1942) The origin of the Garden Dahlia, 1942. Hortus Print, Burnley

Lee YM, Jo YD, Lee HJ et al (2015) DNA damage and oxidative stress induced by proton beam in *Cymbidium* hybrid. Hortic Environ Biotechnol 56:240–246

Lee YM, Lee HJ, Kim YS et al (2016) Evaluation of the sensitivity to ionizing -radiation of a *Cymbidium* hybrid. J Hortic Sci Biotech 91:109–116

Lee SR, Kim K, Lee BY et al (2019) Complete chloroplast genomes of all six *Hosta* species occurring in Korea: molecular structures, comparative, and phylogenetic analyses. BMC Genomics 20:833. https://doi.org/10.1186/s12864-019-6215-y

Lenawaty DY, Sukma D, Syukur M, Suprapta DN, Nurcholis W, Aisyah SI (2022) Increasing the diversity of marigold (*Tagetes* sp.) by acute and chronic chemical induced mutation of EMS (Ethyl Methane Sulfonate). Biodiversitas 23(3):1399–1407. E-ISSN: 2085-4722. https://doi.org/10.13057/biodiv/d230326

Lev-Yadun S, Flaishman MA, Atzmon N (2004) Nonchimeric variegated mutation in *Cupressus sempervirens* L. Int J Plant Sci 165(2):257–261

Li XL, An D (2009) Induction and identification of autotetraploids in *Dendrobium*. Hortic J 8: 1239–1242

Li R, Bruneau AH, Qu R (2010) Morphological mutants of St. Augustinegrass induced by gamma ray irradiation. Plant Breed 129(4):412–416. https://doi.org/10.1111/j.1439-0523.2009.01735.x

Li Y, Zhang Z, Wang P, Wang S, Ma L, Li L, Yang R, Ma Y, Wang Q (2015) Comprehensive transcriptome analysis discovers novel candidate genes related to leaf color in a *Lagerstroemia indica* yellow leaf mutant. Genes Genom 37:851–863. https://doi.org/10.1007/s13258-015-0317-y

Li R, Fan L, Lin J, Li M, Liu D, Sui S (2019) In vitro mutagenesis followed by polymorphism detection using start codon targeted markers to engineer brown spot resistance in Kalanchoe. Am Soc Hortic Sci 144(3):193–200. https://doi.org/10.21273/JASHS04571-18

Li F, Mo X, Wu L, Yang C (2020) A novel double-flowered cultivar of *Gypsophila paniculata* mutagenized by ^{60}Co γ-ray. Am Soc Hortic Sci 55(9):1531–1532. https://doi.org/10.21273/HORTSCI15137-20

Li C, Dong N, Zhao Y, Wu S, Liu Z, Zha J (2021a) A review for the breeding of orchids: current achievements and prospects. Hortic Plant J 7(1). https://doi.org/10.1016/j.hpj.2021.02.006

Li Y-r, Liu L, Wang D, Chen L, Chen H (2021b) Biological effects of the electron beam to target turning X-ray (EBTTX) on two freesias (*Freesia hybrida*) cultivars. PeerJ 9:e10742. https://doi.org/10.7717/peerj.10742

Li Y, Chen L, Zhan X, Liu L, Feng F, Guo Z, Wang D, Chen H (2022) Biological effects of gamma-ray radiation on tulip (*Tulipa gesneriana* L.). PeerJ 10:e12792. https://doi.org/10.7717/peerj.12792

Lin WC, Molnar JM (1983) Riger elatior begonia 'Saanred'. Can J Plant Sci 63(2):563–564

Lin T, Zhou R, Bi B et al (2020) Analysis of a radiation-induced dwarf mutant of a warm-season turf grass reveals potential mechanisms involved in the dwarfing mutant. Sci Rep 10:18913. https://doi.org/10.1038/s41598-020-75421-x

Linderman A (1968) Mutationen bei Elatiorbegonien Rasse Rieger. Gartenwelt 68:266–267

Lineberger RD, Druckenbrod M (1985) Chimeral nature of the pinwheel flowering African violets (*Saintpaulia*; Gesneriaceae). Am J Bot 72:1204–1212

Ling L, Dan W, Yirui L, Kelian M, Min C, Jiwen Z, Shanwen D, Qing L, Yi X, Wei LJ (2019) The biological effects of ^{60}Co-γ rays on *Freesia refracta*. J Nanjing For Univ 43(1):186–192. ISSN: 1000-2006. http://njlydxxb.periodicals.net.cn/de

Liu Y, Chaturvedi P, Jinlei F, Qingqing C, Weckwerth W, Yang P (2016) Induction and quantitative proteomic analysis of cell dedifferentiation during callus formation of lotus (*Nelumbo nucifera* Gaertn.spp. *baijianlian*). J Proteomics 131(10):61–70. https://doi.org/10.1016/j.jprot.2015.10.010

Liu Q, Zhang D, Liu F et al (2019) Micropropagation of *Nelumbo nucifera* 'Weishan Hong' through germfree mature embryos. In Vitro Cell Dev Biol Plant 55:305–312. https://doi.org/10.1007/s11627-019-09984-4

Lizuka M, Ikeda A (1963) Effects of x-ray irradiation on *Lilium formosanum*. Proc Am Soc Hortic Sci 82:508–516

Lizuka M, Kimura Y, Hase Y, Tanaka A (2007) Development of commercial variety of Osteospermum by a stepwise mutagenesis by ion beam irradiation. JAEA-review 2007-060 3-05, p 71

Lizuka M, Kimura Y, Okada T, Hase Y, Tanaka A, Sekiguchi M (2011) A new cultivar 'Viento Flamingo' of Osteospermum. Bull Gunma Agric Tech Cent 8:89–92

Loh PT, Cooper DC (1966) Effects of gamma-irradiation on the pollen of *Lilium rugale*. Can J Genet Cytol 8(1):152–164. https://doi.org/10.1139/g66-020

López-Martínez MI, Robledo-Paz A, Flores-Hernández LA, Corona-Torres T, Gutiérrez-Espinosa MA, Hernández-Rodríguez M, García-de los Santos G (2022) *In vitro* mutagenesis in anthurium induced by colchicine. Rev Chapingo Ser Hortic 28(1):Chapingo ene./abr. 2022. https://doi.org/10.5154/r.rchsh.2021.06.011

Love JE (1966) Some effects of fast neutron irradiation on the somatic tissue of poinsettia. Proc Am Soc Hortic Sci 89:672–676

Love JE (1972) Somatic mutation induction in poinsettia and sweet potato. In: Constantin MJ (ed) Mutattion breeding workshop, Knoxville. University of Tennessee, Knoxville

Love JE, Constantin MJ (1965) The response of some ornamental plants to fast neutrons. Tenn Farm Home Sci 56:10–12

Love JE, Constantin MJ (1966) The induction of bud sports in *Coleus blumei* by fast neutrons. Process Am Soc Hortic Sci 88:627–630

Love JE, Malone BB (1967) Anthocyanin pigments in mutant and non-mutant *Coleus* plants. Radiat Bot 7:549–552

Love JE, Mullenax RH (1964) Exposure of coleus plants to gamma irradiation. Tenn Farm Home Sci 50:18–19

Lu G, Zhang XY, Zou YJ, Zou QC, Xiang X, Cao JS (2007) Effect of radiation on regeneration of Chinese narcissus and analysis of genetic variation with AFLP and RAPD markers. Plant Cell Tiss Org Cult 88:319–327

Lu S, Wang Z, Niu Y, Chen Y, Chen H, Fan Z, Lin J, Yan K, Guo Z, Li H (2009) Gamma-ray radiation induced dwarf mutants of turf-type bermudagrass. Plant Breed 128:205–209

Luan LQ, Uyen NHP, Ha VTT (2012) *In vitro* mutation breeding of *Paphipedilum* by ionizing radiation. Sci Hortic 144:1–9

Lyakh V, Lagron V (2005) Induced mutation variability in Linum grandiflorum Desf. Mutat Breed Newsl Rev 1:4–5. https://inis.iaea.org/search/search.aspx?orig_q=RN:36105145

Ma TH, Cabrera GL, Cebulska-Wasilewska A, Chen R, Loarca F, Vandenberg AL, Salamone MF (1994) Tradescantia stamen hair mutation bioassay. Mutat Res 310(2):211–220. https://doi.org/10.1016/0027-5107(94)90114-7

Ma J, Li Z, Lin Y (2023) Integrating multi-omics analysis reveals the regulatory mechanisms of white–violet mutant flowers in grape Hyacinth (*Muscari latifolium*). Int J Mol Sci *24*(5):5044. https://doi.org/10.3390/ijms24055044

Maeta T, Somegou M, Nakahira K, Miyarzaki Y, Kondo T (1982) Studies on cytological, physiological and genetic characteristics in somatic mutant strains of Sugi (*Cryptomeria japonica* D. Don). In: Induced muttions in vegetatively propagated plants II. IAEA, Vienna, pp 281–304

Magd el Din FR (2019) Inducing some morphological variations in *Cyperus alternifolius* L. by using gamma irradiation. J Adv Agric Res 24(2):178–192

Mahna SK, Garg R (1989) Induced mutation in *Petunia nyctaginiflora* Juss. Biol Plant 31:152–155

Majumder J, Singh SK, Verma M (2018a) Assessment of mutation in marigold (*Tagetes erecta* L.) using morphological and molecular markers. Int J Curr Microbiol Appl Sci 7(7):2588–2597. https://doi.org/10.20546/ijcmas.2018.707.303. ISSN: 2319-7706. http://www.ijcmas.com

Majumder J, Singh SK, Kumari M, Verma M (2018b) Variability and correlation studies on induced mutants of marigold (*Tagetes erecta* L.) for different traits and assessing them using molecular markers. Plant Tissue Cult Biotechnol 28(2):223–223. ISSN 1817-3721, E-ISSN 1818-8745

Mangaiyarkarasi R, Girija M, Gnanamurthy S (2014) Mutagenic effectiveness and efficiency of gamma rays and ethyl methane sulphonate in *Catharanthus roseus*. Int J Curr Microbiol App Sci 3(5):881–889

Matsubara H (1982) Mutation breeding in ornamental plants. Techniques used for radiation induced mutants in Begonia, *Chrysanthemum*, Abelia and winter Daphne. Gamma Field Symp 21:65–68

Matsubara H, Iba S, Oka M, et al (1965) Effects of gamma-irradiation on tulip. II. Effects on various stages of development. Tokyo Metropolitan Isotope Centre, Annual Report, vol 2, pp 157–162

Matsubara H, Shigematsu K, Suda H, et al (1971) The isolation of the mutated plants from sectorial chimera induced by irradiation in *Begonia* and *Chrysanthemum*. In: Japan Atomic Industrial Forum (ed) Proceedigns of the 10th Japan conferences on radioisotoper. Tokyo Metropolitan Isotope Research Center, Tokyo, pp 374–376

Matsubara H, Shigenmatsu K, Suda H (1974) The isolation and fixation of wholly mutant plant from sectorial chimera induced by gamma irradiation in *Begonia* rex. J Jpn Soc Hortic Sci 43(1): 63–68. https://doi.org/10.2503/jjshs.43.63

Matsubara H, Suda H, Sawada T et al (1975) An ozone-sensitive strain from the mutants induced by gamma ray irradiation of the *Begonia rex*, variety Winter Queen. Agric Hortic 50(6):811–812

Matsubara H, Shigenmatsu K, Suda H et al (1976) Useful mutant strains obtained by irradiation breeding. Annual report of Tokyo Metropolitan Isotope Research Center 1975, pp 67–69

Matsuda G (1960) Studies on the breeding of the special plants in the cold districts utilizing radioisotopes. I. Effects of gamma irradiation on tulip. Bull Toyama Agric Exp Stn I

Maynard RCI, John MR (2023) Mutation breeding of *Salvia coccinea* with ethyl methanesulfonate. HortScience 58(5):568–572. https://doi.org/10.21273/HORTSCI17092-23

Maynard RCI, Ruter JM (2023) Mutation breeding of Salvia coccinea with Ethyl Methanesulfonate. Am Soc Hortic Sci 58(5):568–572. https://doi.org/10.21273/HORTSCI17092-23

Mehlquist GAL, Sagawa Y (1959) The effect of gamma radiation on carnations. Proc Int Bot Congr 9(11):258

Mehlquist GAL, Sagawa Y (1964) The effect of gamma radiation on carnations. Proc Int Hortic Congr 16(4):10–18

Mehlquist GAL, Ober DE, Sagawa Y (1954) Somatic mutations in the carnations (*Dianthus caryophyllus* l.). Proc Natl Acad Sci U S A 40:432–436

Mekala P (2009) Improvement of Jasminum sambac cv. Gundumalli (Ait.) through mutation breeding. MSc. Thesis submitted to Tamil Nadu Agricultural University

Mekala Kumar PN, Kannan M, Vijayakumar G, Jawaharlal M (2010) Effect of physical and chemical mutagen on floral characters of J. sambac cv. Gundumalli. Abst: national symposium on lifestyle floriculture: challenges and opportunities. Session-2, crop improvement, biotechnology, and biodiversity, March 19-21, 2010 at DR. Y.S. Parmar University of Horticulture and Forestry, Nauni, Solan, India, Abstract No. 2.50: 37

Melati MR, Alaimo MG, Orecchio S, De Vita F (2004) Stress-induced cytological and chemical adapt at ions in Cupressus plant s from an urban area of Palermo (Italy). Acta Bot Gallica 151(3):265–283. https://doi.org/10.1080/12538078.2004.1051542

Mendoza-Gómez RJ, Tapia-Campos E, Barba-Gonzalez R (2020) In vitro mutagenesis efficiency with EMS (ethyl methanesulfonate) on *Eustoma grandiflorum*. Acta Hortic 1288:163–166. https://doi.org/10.17660/ActaHortic.2020.1288.24

Mericle LW, Mericle HP (1967) Genetic nature of somatic mutations for flower color in *Tradescantia*, Clone 02. Radiat Bot 7:449

Meshitsuka G et al (1962) Effects of gamma irradiation on tulip I. Effects of bulb irradiation after completion of the flowerbud differentiation. Tokyo Metropolitan Isotope Center annu report, vol 1, pp 86–96

Mikkelson EP (1976) Histology of adventitious shoot and root formation on leaf petiole cuttings of *Begonia* x hiemalis Fotsch cv. Aphrodite Peach. Thesis, Michigan State University, East Lansing, 36p

Mikkelson JC, Ryan J, Constantin MJ (1975) Mutation breeding of Rieger's elatior begonias. Am Hortic 54(3):18–21

Minisi FA, El-Mahrouk ME, Rida ME et al (2013) Effects of gamma radiation on germination, growth characteristics and morphological variation of *Moluccella laevis* L. Am Eur J Agric Environ Sci 13(5):696–704

Mishra P, Kumar S (2003) Manifestation of heterostyled character by induction of recessive hsf mutation responsible for thrum type herkogamous flowers in *Catharanthus roseus*. JMAPS 28:2

Mistry V, Tiwari P, Patel P, Vishwakarma GS, Lee G-J, Sharma A (2022) Ethyl methane sulfonate and sodium azide-mediated chemical and X-ray-mediated physical mutagenesis positively regulate *Peroxidase 1* gene activity and biosynthesis of antineoplastic vinblastine in *Catharanthus roseus*. Plants 11(2):2885. https://doi.org/10.3390/plants11212885

Mitra S (1958) Effects of x-rays on chromosomes of *Lilium longiflorum* during meiosis. Genetics 43(6):771–789

Miyazaki K, Suzuki K, Abe T, Katsumoto Y, Yoshida S, Kusumi T (2002) Isolation of variegated mutants of Petunia hybrida using heavy-ion beam irradiation. RIKEN Accel Prog Rep 35:130

Miyazaki K, Suzuki K, Iwaki K, Kusumi T, Abe T, Yoshida S, Fukui H (2006) Flower pigment mutations induced by heavy ion beam irradiation in an interspecific hybrid of *Torenia*. Plant Biotechnol 23:163–167. https://doi.org/10.5511/plantbiotechnology.23.163

Moghaddam ZS, Moshtaghi N, Bagheri A (2019) The effect of mutation on micropropagation of Crocus sativus L. Plant Cell Biotechnol Mol Biol 20(23-24):1059–1065

Mohamad A, Halim AA, Ibrahim R (2002) Mutation induction of *Blechnum gibium* a fern species through in vitro propagation. In: Proceedings of the seminar MINT R and D 2002: strengthening R and D culture for technology generation, 2002

Mohammed MMO, Abdul-Kader HA, Esho KB (2022) Induction of mutation by treatment with colchicine and spraying with progesterone for two cultivars of *Zinnia elegans* by using DNA molecular. Int J Health Sci 6(S9):2927–2951. https://doi.org/10.53730/ijhs.v6nS9.13159

Mohd Nazir B, Sakinah A (2001) Molecular techniques as complementary tools in orchid mutagenesis. In: Proceedings of the 2001 FNCA workshop on agriculture: plant mutation breeding & biofertilizer, 20–24 Aug, Bangkok, Thailand. JAERI, pp 91–102

Mohd Nazir B, Sakinah A, Affrida AH, Zaiton A, Tanaka A, Shikazono N, Oono Y, Hasc Y (2003) Mutation induction in orchids using ion beam, TIARA annual report 2003. JAERI-review 2004-025, pp 45–47

Mohd A, Baharun AHB, Juraimi AS, Alam MA et al (2016) Morphological traits alteration of mutant common turf grass (*Cynodon dactylon*) induced by gamma ray irradiation. Res J Biotechnol 11(12):93–105

de Mol van Oud Loosdrecht WE (1956) Der Einfluss der Rontgenstrahlen auf die Entwicklung des pollens und der Sprosse bei Tulpen. Mit Ruckblick auf die Wissenschafrliche Lebensarbeit des Verfassers von K.J.J. Thamm. Bayerischer Landwirtschaftsverlag, Bonn

Molnar JM (1976) Rieger elatior begonia cv. Northern Sunset. Can J Plant Sci 56:1003. https://doi.org/10.4141/cjps76-169

Moore CN, Haskins CP (1935) X-ray induced modifications of flower color in the *Petunia*. J Hered 26:349–355. https://doi.org/10.1093/oxfordjournals.jhered.a104121

Morgan E, Burge GK, Seelye JF (2002) Sandersonia: towards the new generation. Acta Hortic 570. https://doi.org/10.17660/ActaHortic.2002.570.8. ISHS 2002

Morgan E, Seelye JF, Hofman BL, Hare PD (2004) Production of triploid *Sandersonia aurantiaca* plants. South Afr J Bot 70(2):20–214. https://doi.org/10.1016/S0254-6299(15)30237-4

Morimoto H, Narumi-Kawasaki T, Takamura T, Fukai S (2019) Analysis of flower color variation in carnation (*Dianthus caryophyllus* L.) cultivars derived from continuous bud mutations. Hortic J 88:1

Mostafa GG, Alfrmawy AM, El-Mokadem HE (2014) Induction of mutations in *Celosia argentea* using dimethyl sulphate and identification of genetic variation by ISSR markers. Int J Plant Breed Genet 8(2):44–56

Muangsorn S, Te-chato S (2008) Use of EMS to induce mutation in *Dendrobium friedericksianum* Rchb.f. J Agric 24(2):153–15s

Muhallilin I, Aisyah S, Sukma D (2019) The diversity of morphological characteristics and chemical content of *Celosia cristata* plantlets due to gamma ray irradiation. Biodiversitas 20(3):862–866. https://doi.org/10.13057/biodiv/d200333

Mukherjee I, Khoshoo TN (1970) Genetic-evolutionary studies on cultivated cannas IV: Parallelism between natural and induced somatic mutations. Radiat Bot 10(4):351–364

Mutlu S, Djapo H, Tuncel N et al (2015) Gamma-ray irradiation induces useful morphological variation in Bermudagrass. Not Bot Horti Agrobotanici Cluj-napoca. https://doi.org/10.15835/NBHA4329762. Corpus ID: 85681352

Myodo H (1942) Effects of X-rays upon tulip plants when irradiated in different developmental stages of floal organs. J Fac Agric Hokkaido Imp Univ 48:359–382

Nagatomi S, Katsumata K, Nojiri C (1993) Induction of dwarf-type plants derived from *in vitro* cultures by gamma-ray irradiation in *Cytisus* Genus. Tech News 51:1–2

Nagatomi S, Kamiyo M, Narusawa H et al (1996) Three mutant varieties in *Eustoma grandiflorum* induced through *in vitro* culture of chronic irradiated plants. Tech News 53:1–2

Nakano M, Amano J, Watanabe Y, Nomizu T, Suzuki M, Mizunashi K, Mori S, Kuwayama S, Han D, Saito D, Ryuto H, Fukunishi N, Abe T (2010) Morphological variation in *Tricyrtis hirta* plants regenerated from heavy ion beam-irradiated embryogenic calluses. Plant Biotechnol. 27: 155–160. https://doi.org/10.5511/plantbiotechnology.27.155

Nakatsuka T, Nishihara M, Mishiba K et al (2005) Two different mutations are involved in the formation of white-flowered gentian plants. Plant Sci 169(5):949–958

Nakayama M, Tanikawa N, Moritaa Y, Ban Y (2012) Comprehensive analyses of anthocyanin and related compounds to understand flower color change in ion-beam mutants of cyclamen (Cyclamen spp.) and carnation (*Dianthus caryophyllus*). Plant Biotechnol 29:215–221. https://doi.org/10.5511/plantbiotechnology.12.0102a

Nakronthap A (1965) Radiation-induced somatic mutations in the ornamental *Canna*. Radiat Bot 5 (suppl):707–712

Nakronthap A (1974) Radiation-induced somatic mutations in Kalanchoe (*Kalanchoe lacin iata*). Mutat Breed Newsl 3:14

Nambisan KMP, Krishnan BM, Veeraragavathatham D et al (1980) Induced mutants in jasmine (*Jasminum grandiflorum* L.): leaf spot resistant and dwarf mutants. Sci Cult 46:427–428

Nauman CH, Underbrink A, Sparrow A (1975) Influence of radiation dose rate on somatic mutation induction in *Tradescantia* stamen hairs. Radiat Res. https://doi.org/10.2307/3574186. Corpus ID: 28074907

Nauman CH, Sparrow AH, Underbrink AG, Schairer LA (1976) Low-dose mutation-response relationships in Tradescantia; principles and comparison to mutagenesis following low-dose gaseous chemical mutagen exposure. [X radiation, neutrons]. Seminar on dose effect relationship in radiobiology; Orsay, France; 24–26 May 1976

Nauman CH, Schairer LA, Sautkulis RC, Klug EE (1977) Influence of hyperthermia on the spontaneous, radiation and chemical induced mutation frequency in tradescantia stamen hairs. Radiat Bot 70:632

Naveena N, Subramanian S, Jawaharlal M, Iyanar K, Chandrasekhar CN (2020a) Mutagenic effectiveness and efficiency of gamma rays and ethyl methane sulphonate on Hibiscus rosa-sinensis L. cultivar red single. Electron J Plant Breed 11(4):1187–1193. https://doi.org/10.37992/2020.1104.191

Naveena N, Subramanian S, Jawaharlal M, Iyanar K, Chandrasekhar CN (2020b) Determination of mutagenic sensitivity in *Hibiscus rosa-sinensis* L. (Cultivar Red Single) to physical and chemical mutagens. Int J Curr Microbiol Appl Sci 9(10):1756–1763. https://doi.org/10.20546/ijcmas.2020.910.213

Nehvi F, Khan MA, Lone A, Yasmin S (2010) Effect of radiation and chemical mutagen on variability in saffron (*Crocus sativus* L.). Acta Horticu 850(850):67–74. https://doi.org/10.17660/ActaHortic.2010.850.8

Nelka SAP, Vidanapathirana NP, Dahanayake N, Subasinghe S, Silva TD, Weerasinghe S, Anuruddhi HIGK (2021) Effect of gamma irradiation on survivability, growth performances and floral characters of Jasminum officinale (Samanpichcha). J Agrotechnol Rural Sci 1(1): 24–29. https://doi.org/10.4038/atrsj.v1i1.28

Nelson J (1961) Induction of plant tumors by ultraviolet radiation. Proc Natl Acad Sci USA 47: 1165–1169

Neto TA, Latado RR, Tsai SM, Derbyshire MT, Yemma AF, Scarpare Felho JA, Cera Volo L, Rossi AC, Namekata T, Pompeu Jr J, Figueiicedo JO, Pio R, Tobias Domngues E (2001) Mutation breeding in vivo and in vitro in vegetatively propagated crops. Source: Joint FAO/IAEA Division of Nuclear Techniques in Food and Agriculture, Vienna (Austria); 77 p; ISSN 1011-4289; May 2001; pp 55–62; Final research co-ordination meeting on induced mutations in connection with biotechnology for crop improvement in Latin America

Nezu M (1962) The effect of radiation on tulip breeding. Gamma Field Symp 1:43–49

Nezu M (1963a) Studies on the production of bud sports in tulips by ionizing radiation. III. Frequency and mechanism of color formation. Seiken Jiho 15:75–86

Nezu M (1963b) Studies on the production of bud sports in tulips by ionizing radiation. IV. Comparison of color sports by gamma rays with similar virus disease phenomena. Seiken Jiho 15:87–97

Nezu M (1964) Studies on the production of bud sports in tulips by ionizing radiation. V. Effects of total dose, fractionation and temperature on somatic mutation. Jpn J Genet 39(6):440–446

Nezu M (1965) Studies on the production of bud sports in tulips by ionizing radiation. VI. Selection and observation on the mutant progeny. Jpn J Breed 15(2):13–118

Nezu M (1967) Tulip breeding by bud sports induced by gamma rays. Toyama Agric Exp Stn Spec Rep 7:1–74

Nezu M, Obata S (1964a) Studies on the production of bud sports in tulips by ionizing radiation. I. Morphology and development of the bulbs. Idengaku Zasshi 38:386–391. [not available]

Nezu M, Obata S (1964b) Studies on the production of bud sports in tulips by ionizing radiation. II. Acute and chronic radiation effects on plant growth and bulbil yield. Idengaku Zasshi 38: 392–398

Nishiguchi M, Nanjo T, Yoshida K (2012) The effects of gamma irradiation on growth and expression of genes encoding DNA repair related proteins in lombardy poplar (*Populus nigra* var. italica). J Environ Radioact 109:19–28. https://doi.org/10.1016/j.jenvrad.2011.12.024

Nitarska D, Stefanini C, Haselmair-Gosch C et al (2018) The rare orange-red colored *Euphorbia pulcherrima* cultivar 'Harvest Orange' shows a nonsense mutation in a flavonoid 3'- -hydroxylase allele expressed in the bracts. BMC Plant Biol 18:216. https://doi.org/10.1186/s12870-018-1424-0

Niu C, Li Y (1988) The radiation induced mutation of *Canna* L. Acta Agric Nucl Sin 2:33

Norazlina N, Azhar M, Rusli I (2003) Mutation induction and propagation of Asplenium nidus a fern species from spores. In: Challenges in plant productivity and food security in changing environment: proceedings of the 14th Malaysian Society of Plant Physiology Conference 2003

Nybom N (1961) The use of induced mutations for the improvement of vegetatively propagated plants. In: Proceedings of symposium on mutations and plant breeding. Cornell University, Ithaca. NAS-NRC, Publication 891, pp 252–294

Oates KM, Touchell DH, Ranney TG (2013) Induced variation in tetraploid *Rudbeckia subtomentosa* 'Henry Eilers' regenerated from gamma-irradiated callus. HortScience 48:831–834. https://doi.org/10.21273/HORTSCI.48.7.831

Ocokoljic M, Milosevic B (2004) Somatic mutation of birch (Betula verrucosa Ehrh.) trees. AGRIS Since:319–324

Ogawa D, Fujioka T, Hirano T, Abe T (2014) Effect of C-ion beam irradiation on survival rates and flower color mutations in statice (*Limonium sinuatum* Mill.). RIKEN Accel Prog Rep 47:295

Ohba K (1971a) Studies on the radiation breeding of forest trees. Bull Inst Radiat Breed 2. Ohmiya, Ibaraki, 102p

Ohba K (1971b) Segregation of spontaneous mutants from Sugi elite clones of Cryptomeria Japonica D. Don Tech News, No. 9. Institute of Radiation Breeding, Ohmiya, Ibaraki

Ohba K, Maeta T (1973) Induction of somatic mutations and cross fertility of the mutants in Sugi, *Cryptomeria japonica* D. Don. Gamma Field Symp 12:19–36

Ohba K, Murai M (1966) Studies on radiosensitivity of forest trees 2. Growth and somatic mutations of Alnus, Populus and *Betula* species under chronic gamma irradiation in a gamma field. J Jpn For Soc 48(1):12–19

Ohki S, Sakai Y, Chatani K, Koreeda K (2013) Mutation induction by ion beam in ornamental plants -- effects on seed germination and DNA polymorphism in *Eustoma grandiflorum*. AGRIS 522:123–126. https://doi.org/10.17660/ActaHortic.2000.522.12. Paper presented at the Twenty-fifth International Horticultural Congress held August 2-7, 1998, Brussels, Belgium. Part 12. Includes references

Okada T, Iizuka M, Hase Y, Nozawa I, Narumi I, Sekiguchi M (2012) Development of commercial variety of *Osteospermum* by a stepwise mutagenesis by ion beam irradiation. Hortic Res 11 (suppl 1):428. JAEA-review 2009-041, p 73

Okamura M, Yasuno N, Ohtsuka M, Tanaka A, Shikazono N, Hase Y (2003) Wide variety of flower-color and -shape mutants regenerated from leaf cultures irradiated with ion beams. Nucl Instrum Methods Phys Res B 206:574–578. https://doi.org/10.1016/s0168-583x(03)00835-8

Okamura M, Takeshita D, Hase Y, Narumi I, Tanaka A (2009) New type flower colored petunia obtained by ion beam irradiation at JAEA-TIARA. JAEA Takasaki annual report 2008, p 69

Okamura M, Umemoto N, Onishi N (2012) Breeding glittering carnations by an efficient mutagenesis system. Plant Biotechnol 29:209–214

Okamura M, Nakayama M, Umemoto N, Cano EA, Hase Y, Nishizaki Y, Sasaki N, Ozeki Y (2013) Crossbreeding of a metallic color carnation and diversification of the peculiar coloration by ion-beam irradiation. Euphytica 2013(191):45–56

Olimpienko GS, Titov AF (1979) Cytoplasmic temperature dependent chlorophyll deficient mutants in Festuca pratensis Huds. Mutat Breed Newsl 13:15

Omar MS, Yousif DP, Al-Jibrouri JM, Al-Rawi MS, Hameed MK (1993) Effects of gamma rays and sodium chloride on growth and cellular constituents of Sunflower (*Helianthus annus*) callus cultures. J Islamic Acad Sci 6:69–72

Osiecki M (1989) Mutation breeding in *Streptocarpus* x hybridus. AGRIS 281:73–77

Osipova RG, Shevchenko VA (1984) The use of tradescantia (clone 02 and 4430) in the radiation and chemical mutagenesis studies. Zh Obshch Biol 45(2):226

Padmapriya S, Rajamani K (2017) Induced mutagenesis in glory lily (*Gloriosa superba* L.) for economic variability. Int J Phytomed Relat Ind 9(4):231–236. Print ISSN: 0975-4261. Online ISSN: 0975-6892. https://doi.org/10.5958/0975-6892.2017.00037.5

Padmapriya S, Rajamani K (2021) Induced chlorophyll mutations in *Gloriosa superba*. Electron J Plant Breed 12(4):1454–1460. https://www.ejplantbreeding.org/index.php/EJPB/article/view/3824

Pal S, Ajit K, Preeti C, Srivastava R, Shailesh T (2017) Determination of lethal dose for gamma rays induced mutagenesis in different cultivars of dahlia. J Hill Agric 8(3):279–282. https://doi.org/10.5958/2230-7338.2017.00055.6

Palamine MTL, Cureg RGA, Marbella LJ, Lapade AG, Domingo ZB, Deocaris CC (2005) Some biophysical changes in the chloroplasts of a *Dracaena* radiation-mutant. Philip J Sci 134(2): 121–126. ISSN 0031 – 7683

Pallavi B, Nivas S, Hedge S (2017) Gamma rays induced variations in seed germination, growth and phenotypic characteristics of *Zinnia elegans* var. Dreamland. Adv Hortic Sci. https://doi.org/10.13128/AHS-20289

Palta HK, Mehra PN (1973) Radiobiological investigations on *Pteris vittata* L. II. X-rays effect on the gametophytic generation. Radiat Bot 13:155–164

Pan C, Lu Y-F, Wen P-J, Chen Y-M (2019) Using colchicine to create Poinsettia (Euphorbia pulcherrima × Euphorbia cornastra) mutants with various morphological traits. HortScience 54(10):6p. https://doi.org/10.21273/HORTSCI14143-19

Pandey R, Pal MB, Nagar A (1987) Colchitetraploidy in *Crossandra infundibuliformis* Nees. New Bot 14(1-4):73–78

Papanicolaou K, Kokkini S (1983) Homoeotic mutation as well as inheritance of hairy sepals and anthers in *Ranunculus millefoliatus* Vahl. Feddes. Repertorium 94(9/10):631–634. ISSN: 0014-8962. https://agris.fao.org/agris-search/search.do?recordID=US201302136188

Paramesh TH, Chowdhury S (2005) Impact of explants and gamma irradiation dosage on in vitro mutagenesis in carnation (*Dianthus caryophyllus* L.). J Appl Hortic 7:43–45. https://doi.org/10.37855/jah.2005.v07i01.11

Parkhi RD, Khalatkar AS (1988) Induced reproductive alterations in *Ipomoea purpurea* (L) Roth. Plant Cell Incompatibility Newsl 20:8–16

Parliman B, Stushnoff C (1979) Mutant induction through adventitious buds of *Kohleria*. Euphytica 28(2):521–530

Partanen CR (1958) Quantitative technique for analysis of radiation induced tumorization in fern prothalli. Science 12:1006–1007

Pathak CS, Raghuvanshi SS (1980) Occurrence and significance of polypetalous variant in *Phlox drummondii* L. Proc Indian Natl Sci Acad B 46(4):524–527

Patil S (2015) Mutational changes in *Delphinium malabaricum* (Huth.) Munz.: a potential ornamental plant. J Hortic Res 4:11p

Patil UH, Masalkar SD, Patil AH (2019) Effect of chemical mutagens on growth and flowering of *Carnation*. J Pharmacogn Phytochem 8(2):1982–1984

Patwibul W, Kongchuensin S, Therdbarame P, Limpanavech P (2001) Effect of gamma rays on some characteristics of climbing lily (*Gloriosa superba* L.). In: AGRIS. Proceedings of the 12th genetics: gene revolution era. s, Bangkok. ISBN 974-553-922-8, pp 193–196

Pederson IF, Dickens R (1985) Registration of AU Centennial Centipedegrass. Crop Sci 25(2):364

Pei YR, Di W, Dong L, Huai ZX, Lei S (2018) Radiation breeding of Boston fern via 60Co-γ rays. Acta Hortic Sin 45(5):988–996. http://www.ahs.ac.cn/EN/abstract/abst

Pereau-Leroy P (1969) Effect de I'irradiation gamma sur une chimere complexe d'oeillet Sim. In: Induced mutations in plants, Pullman, 1969. IAEA, Vienna, pp 337–344

Pereau-Leroy P (1970) Detection par irradiation gamma de I'influence de la structure en chimere pericline d'une variete d'oeillet sur la coloration de la fleur. C R Acad Sci Ser D 270(6):810–812

Pereau-Leroy P (1974a) Comparison des effects physiologiques et genetiques provoques par des irradiations aux neutrons et aux yayons gamma de boutures d'oeillets, *Dianthus caryophyllus* L. In: Biological effects of neutron irradiation, Munich – Neuherberg, 1973. IAEA, Vienna, pp 441–449

Pereau-Leroy P (1974b) Genetic interaction between the tissues of carnation petals as periclinal chimeras. Radiat Bot 14:109–116

Pereau-Leroy P (1975) Recherche Radiobiologiques sur des chimeres d'oeillet, *Dianthus caryophyllus* L. Thesis, Universite de Clermont-Ferrand, Ser E, No. 199, p 169

Pierik RML (1975) Callus multiplication of *Anthurium andraeanum* Lind. in liquid media. Neth J Agric Sci 23:299–302

Pierik RML (1976) *Anthurium andraeanum* plantlets produced from callus tissues cultivated in vitro. Physiol Plant 37:80–82

Pierik RML, Steeginans HHM (1978) Vegetative propagation of *Anthurium scherzerianum* Schott through callua cultufres. Sci Hortic 4(3):291–293

Pierik RLM, Van Leeuwen P, Righter GCCM (1979) Regeneration of leaf explants of *Anthurium andraeanum* Lind. In vitro. Neth J Agric Sci 27:221–226

Pillai PRU, Verma RC (1992) A comparison of induced chromosomal aberrations in *Vicia faba* and *Phlox drummondii*. J Cytol Genet 3:17–23

Pimonrat P, Suraninpong P (2009) Mutation Induction in S. plicata Blume by gamma irradiation. In: Proceeding of National Horticultural Congress, 6–9 May 2009. The Empress Chiangmai Hotel, Chiang Mai

Pimonrat P, Suraninpong P, Wuthisuthimethavee S (2012) Acute effect of gamma radiation on stable characteristics of *Spathoglottis plicata* Blume. Acta Hortic 953:173–180

Plummer TH, Leopold AC (1957) Chemical treatment for bud formation in *Saintpaulia*. Proc Am Soc Hortic Sci 70:442–444

Pohleim F (1974) Nachweiss von Mischzellen in variegaten Adventivsprossen von Saintpaulia entstanden nach Behandlung isolierter Blatter mit N-nitroso-N-Methylharnstoff. Biol Zentralbl 93:141–148

Pohlheim F (1974) Induced plastom mutations in *Saintpaulia*. Proc Intl Hortic Congr 19(1B):888

Pohlheim F (1977) Umlagerungen an der Trichimare Pelargonium zonale 'Freak of Nature'; ein Beitrag zur Herstellung von Plastommutanten. Wiss Z Padagog Hochsch Potsdam 16:65–70

Pohlheim F (1983) Induktion von Polyploide an Adventivsprossen aus Wurzelstecklingen von Eryngium planum L. Wiss Z Pad Hochsch Potsdam 27(1):157–163

Pohlheim F, Beger B (1974) Erhohung der Mutationsrate im plastom bei Saintpaulia durch N-Nitroso-N-methylharnstoff. Biologische Rundschau 12:204–206

Pohlheim F, Pohlheim E (1976) Herstellung von plastommutanten bei Saintpaulia ionantha H. Wendl. Biochem Physiol Pflanz 169:377–383

Pohlheim E, Pohlheim F, Gunther G (1972) Die haploid *Pelargonium zonale* 'Kleiner Liebling' als Testsystem fur Mutagene. Wiss Z Padagog Hochsch Potsdam 16:65–70

Pohlheim E, Pohlheim F, Gunther G (1976) Mutagenicity testing of herbicides with a haploid *Pelargonium*. In: European environmental mutagenic society, sixth annual meeting, Gernrode, p 131

Polheim F (1980) Periklinalchimarische Anthozyamuster bei *Saintpaulia ionantha* H. Wendl Nach NMH-Behandlung Arch Zuchtungsforsch 10(4):261–269

Polheim F (1981) Genetischer Nachweis einer NMH-induzierten Plastommutation bei *Saintpaulia ionantha* H. Wendl Biol Rundsch 19:47–50

Potsch J (1964) The induction of extramutative radiation effects on periclinal chimeras with known constitution of *Pelargonium zonale* Ait., *Euphorbia pulcherrima* Willd. and *Abutilon hybridum*. Horticulture dissertation, 64-1053, Potsdam. 214p

Potsch J (1966a) Das Verhalten von Abutilon hybridum hort. 'Andenken an Bonn' nach einmaliger und fraktionierter Rontegenbestrahlung. Z Planzenzucht 55:183–200

Potsch J (1966b) On the dissociation of chimera shoot variants by the use of x-rays. Induced mutations and their utilization. Gatersleben 301–304

Potsch J (1967) On the dissociation of chemical shoot-variants by the use of X-rays. In: Stubbe H (ed) Induced mutations and their utilization. Abh. Dtsche. Akad. Wiss. Berlin KI. Medizin 2, pp 301–304

Powell JB (1974) Induced mutations in turfrgrasses as a source of variation for improved cultivars. In: Roberts EC (ed) Proceedings of the 2nd international Turfgrass research conference, Blackburn. American Society of Agronomy and Crop Science Society of America, Madison, pp 3–8

Powell JB (1976) Induced mutations in highly heterozygous vegetatively propagated grasses. In: Induced mutations in cross-breeding, Vienna, 1975. IAEA, Vienna, pp 219–224

Powell JB, Murray JJ (1978) The vegetative bud mutation technique for breeding improved *Poa pratensis* L. Landwirtsch Zentralbl II 23(1):Abstr. 0762

Powell JB, Toler RW (1980) Induced mutants in 'Floratam' St. Augustinegrass Crop Sci 20(5): 644–646. https://doi.org/10.2135/cropsci1980.0011183X002000050025x

Powell JB, Burton GW, Young JR (1974) Induced mutations in vegetatively propagated turf bermudagrass by gamma irradiation. Crop Sci 14:327–330. https://doi.org/10.2135/cropsci1974.0011183X001400020048x

Prabhukumar KM, Thomas VP, Mohanan K (2015) Induced mutation in ornamental gingers (Zingiberaceae) using chemical mutagens viz. colchicine, acridine, and ethyl methanesulphonate. J Hortic For Biotechnol 19:18–27

Pratiwi OD (2010) The diversity of plant traits kembang kertas (*Zinnia elegans* Jacq,) on M2 Generation of X-ray irradiation results. Faculty of Agriculture, Universitas Gadjah Mada, Yogyakarta

Preil W, Engelhardt M (1982) In vitro Entmischung von Chimarenstrukturen durch Suspensions-kulturen bei *Euphorbia pulcherrima* Wild. Bartenbauwissenschaft 47(6):241–244

Preil W, Walther F (1983) Mutationsinduktion durch bestrahlung von Azaleen. Gb + Gw (Gartnerborse und gartenwelt) 30:788–791

Preil W, Engelhardt M, Walther F (1983) Breeding of low temperature tolerant poinsettia (*Euphorbia pulcherrima*) and *Chrysanthemum* by means of mutation induction in in vitro culture. Acta Hortic. 131:345–351

Privalov GF (1965) Somatical radiomutatons by Acer neguundo L. Bjcell Mosk Obsc Isp. Otd Biol 70(1):153–160

Privalov GF (1967) Experimental mutations in woody plants. In: Stubbe H (ed) Induced mutations and their utlization. Erwin Baur Memorial Lectures, 1966, IV. Akademie-Verlag, Berlin, pp 383–386

Privalov GF (1968a) Mutability in box-elder (Acer neguundo L.) induced by gamma- and X-rays. Genetika 4(5):39–42

Privalov GF (1968b) Investigations of experimental mutagenesis in arboreous plants. Genetika 4(6):144–158

Privalov GF (1974) The effect of treating the seeds with auxin on induced mutagenesis in *Acer neguundo* L. (c.f. Broertjes and Van Harten 1988)

Priyanka D, Dhanavel D (2021) Physical and chemical mutagenesis in *Hibiscus sabdariffa* L. to induce variability on seed germination, survival, and lethal dose. Indian J Nat Sci 12(6): 32280–32284

Przybyla A (1992) Mutation breeding of *Alstroemeria* in Poland. Acta Hortic 325:561–566. https://doi.org/10.17660/ActaHortic.1992.325.78. ISHS Acta Horticulturae 325: VI international symposium on flower bulbs

Przybyla AA (2000) Mutagenesis in creation of new *Alstroemeria* genotypes. Acta Hortic. 508: 351–356. https://doi.org/10.17660/ActaHortic.2000.508.68

Puchooa D (2005) In vitro mutation breeding of *Anthurium* by gamma radiation. Int J Agric Biol 7(1):11–20

Puchooa D, Sookun D (2003) Induced mutation and in vitro culture of *Anthurium andreanum*. Food and Agricultural Research Council, pp 17–27

Purshottam DK, Misra P, Srivastava RK (2021) A mutation study on *Gerbera jamesonii*: an important ornamental plant. Int J Plant Environ 6(1):91–93. https://doi.org/10.18811/ijpen.v6i01.10

Qalby N, Sjahril R, Dachlan A, Akae A (2020) Colchicine induced polyploidy in Common Ice plant *Mesembryanthemum crystallinum* L. IOP Conf Ser Earth Environ Sci 486:012097. https://doi.org/10.1088/1755-1315/486/1/012097

Qiang J-y et al (2005) Effect of (60)Co-γ ray irradiation on germination rate of seed and seedling growth of Celosia cristata. J Anhui Agric Sci, 2005-04

Qin W, Wang S, Yang G (1988) The radiation breeding of Freesia. Acta Horticulturae Sinica, 1988-03

Raghuvanshi SS, Singh AK (1979) Gamma-ray induced mutants in diploid and autotetraploid perennial *Portulaca grandiflora* Hook. Indian J Hortic 36(1):84–87. ISSN: 0972-8538. Online ISSN: 0974-0112

Raghuvanshi SS, Singh AK (1980) Autotetraploid radiosensitivity with special reference to *Portulaca grandiflora* Hook. In: Bir SS (ed) Recent researches in plant sciences. Kalyani Publication, New Delhi, pp 347–352

Rai SP, Kumar S (2000) Induced mutation to monocotyledon in periwinkle, *Catharanthus roseus*, and suppression of mutant phenotype by kinetin. Indian Acad Sci 79(3):97–104

Ramesh Kumar PR, Venkat Ratnam S (2009) Chromosomal induced aberrations in Sunflower (*Helianthus annuus*. L) with gamma-irradiation, sodium azide and combined treatments. Cytologia 74(3):253–258

Ramesh Kumar PR, Venkat Ratnam S (2010) Mutagenic effectiveness and efficiency in varieties of sunflower (*Helianthus annul* L.) by separate and combined treatment with gamma-rays and sodium azide. Afr J Biotechnol 9(39):6517–6521. http://www.academicjournals.org/AJB ISSN 1684-5315

Ramesh A, Verma RC (2015) Effect of gamma radiation on chromosome at mitotic division in *Phlox drummondii*. Int Res J Biol Sci 4(1):82–85. ISSN 2278-3202.

Rather ZA, Jhon AQ (1996) Mutational studies in Dutch Iris (*Iris hollandica* cv. Prof Blaauw). Flora Fauna 2(1):19–21. Accession: 002901519

Rather ZA, Jhon AQ (2000) Effect of 60Co gamma rays on Dutch iris. J Ornam Hortic 3(2):71–74. Accession: 003413908

Rather ZA, Jhon AQ, Zargar GH (2002) Effect of ^{60}CO gamma rays on Dutch iris-II. J Ornam Hortic 5(2):1–4. Accession: 003723274

Relichova J (1984) Induction of somatic mutations in some vegetatively propagated plants. Sb UVTI Genet Slechteni 20(2):85–94

Reinert JA, Toler RW, Bruton BD, Busey P (1981) Retention of resistance by mutants of 'Floratam' St. Augustinegrass to the southern chinch bug and St. Augustine decline. Crop Sci 21(3): 464–466

Richter A, Singleton WR (1955) The effect of chronic gamma radiation of the production of somatic mutations in *Crnations*. Proc Natl Acad Sci 41:295–300

Rifnas LM, Nisansala V, Silva T, Chathurika M et al (2022) Effects of gamma irradiation on the performances of *Acalypha hispida* (Cat's tail plant). In: National symposium on floriculture research (NaSFloR) 2022. Royal Botanical Gardens, Peradeniya

Robinson TR (1931) A chimera in the *Poinsettia*. J Hered 22(11):359

Robinson TR, Darrow GM (1929) A pink poinsettia chimera. J Hered 20(7):335–339. https://doi.org/10.1093/oxfordjournals.jhered.a103215. Accession: 029746307

Roest S, Bokelmann GS (1981) Vegetatieve vermeerdering van Muscari in kweekbuizen. Vakbl Bloemisterij 36(5):134–135

Roest S, Van Berkel MAE, Bokelmann GS et al (1981) The use of an in vitro adventitious bud technique for mutation breeding of *Begonia* x hiemalis. Euphytica 30(2):381–388. https://doi.org/10.1007/BF00034000

Romeida A, Sutjahjo SH, Purwito A et al (2012) Induced mutation by gamma-ray irradiation for the development of superior orchid clones *Spathoglottis plicata* Blume accession bengkulu. Dissertation, Institut Pertanian Bogor, Bogor

Rosna NAH, Azani MT, Mahmad SN (2012) Irradiation effect on in vitro organogenesis, callus growth and plantlet development of *Gerbera jamesonii*. Hortic Bras 30(2):252. https://doi.org/10.1590/S0102-05362012000200012

Roy RK, Singh C, Verma TS, Banerji BK (2010) Effects of gamma irradiation on Canna generalis cv. 'Cleopatra' concerning induction of somatic mutation in foliage and flower color. Abst: National symposium on lifestyle floriculture: challenges and opportunities. session-2, crop improvement, biotechnology, and biodiversity, March 19-21, 2010 at DR. Y.S. Parmar University of Horticulture and Forestry, Nauni, Solan, Abstract No. 2.35: 28–29

Roychowdhury R, Tah J (2011a) Chemical mutagenic action on seed germination and related agrometrical traits in M1 *Dianthus* generation. Curr Bot 2(8):19–23

Roychowdhury R, Tah J (2011b) Assessment of chemical mutagenic effects in mutation breeding programme for M1 generation of carnation (*Dianthus caryophyllus*). Res Plant Biol 1:23–32

Roychowdhury R, Alam MJF, Bishnu S et al (2012) Comparative study for chemical mutagenesis on seed germination, survivability and pollen sterility in M1 and M2 generations of *Dianthus*. Plant Breed Seed Sci 65(1):29–38. https://doi.org/10.2478/v10129-011-0044-8

Sadhukhan R, Ganguly A, Singh PK, Sarkar HK (2014) Study of induced polyploidy in African marigold (*Tagetes ecrecta* L.). Environ Ecol 32:1219–1222

Sagawa Y (1957) Nitrogen as a modifying factor in the growth and flowering responses of carnations to radiation. Plant Physiol 32(supply L):1p

Sagawa Y, Mehlquist GAL (1956) The effect of ionizing radiations on carnation *Dianthus caryophyllus* II. The effect of X- rays on the flower color. Q Prog Rep B 388(29):38–39

Sagawa Y, Mehlquist GAL (1957) The mechanism responsible for some X-ray induced changes in flower colour of the carnation, *Dianthus caryophyllus*. Am J Bot 44(5):397–403

Sagawa Y, Mehlquist GAL (1959) Some Xray induced mutants in the carnation: *Dianthus caryophyllus*. J Hered 50:78–80. https://doi.org/10.1093/oxfordjournals.jhered.a106881

Saggoo MIS, Gill A, Walia S (2011) Cytomixis during microsporogenesis in some populations of *Croton bonplandianum* of North India. Cytologia 76(1):67–72

Saito K (1977) Study on induction of irradiated mutants by gamma-rays and their utilization in flower plant breeding. II. On the mutated semi-double flowered wallflower (*Cheiranthus cheiri*) and whie-flowered moss verbena (*Verbena erinoides*) induced by gamma-irradiation. J Jpn Soc Hortic Sci 46(3):331–337

Sakinah A, Mohd Nazir B (2002) Increasing characteristic variations in Dendrobium orchid through acute irradiation. In: 17th World orchid conference and show, Shah Alam, Malaysia, 26 April–2 May 2002

Sakinah A, Affrida AH, Zaiton A et al (2005) Mutation induction of orchids using ion beams. http://www.iaea.org/inis/collection/NCLCcollectionStore/_Public/43/035/43035281.pdf

Sakuramoto F, Ichikawa S (1996) Effects of X-ray dose fractionations with various intervals in inducing somatic mutations in the stamen hairs of *Tradescantia* clone KU 9. J Free Access 71(6):355–361. https://doi.org/10.1266/ggs.71.355

Salwee Y, Nehvi Firdos A (2017) Ethyl methane sulphonate-a potent mutagen for inducing genetic variability in saffron (*Crocus sativus* L.). Appl Biol Res 19(2):164–171. https://doi.org/10.5958/0974-4517.2017.00023.4

Samadi N, Naghavi MR, Moratalla-Lopez N, Alonso GL, Shokrpour M (2022) Morphological, molecular and phytochemical variations induced by colchicine and EMS chemical mutagens in *Crocus sativus* L. Food Chem (Oxf) 30(4):100086. https://doi.org/10.1016/j.fochms.2022.100086

Samata Y, Tsuyuki Y, Inazu K, Lizuka K (1979) Induction of flower colour changes in carnation (*Dianthus caryophyllus*) by gamma-ray irradiation. Bull Fac Agric Tamagawa Univ 19:29–41

Samatadze TE, Zoshchuk SA, Hazieva FM et al (2019) Phenotypic and molecular cytogenetic variability in calendula (*Calendula officinalis* L.) cultivars and mutant lines obtained *via* chemical mutagenesis. Sci Rep 9:9155. https://doi.org/10.1038/s41598-019-45738-3

Saniya P, Kumar P, Cholin S, Patil B, Mahesh YS, Shirol AM (2023) LD50 value of diethyl sulphate (DES) induced mutagenesis in Nerium. Pharma Innov J 12(2):3490–3493. www.thepharmajournal.com

Sarkar J, Singh SK, Singh KP, Guha SK (2016) In-vivo and in-vitro mutagenesis in marigold (*Tagetes erecta*) using 60Co gamma rays. Indian J Agric Sci 86(7):870–875

Sasaki K, Aida R, Niki T, Yamaguchi H, Narumi T, Nishijima T, Hayashi Y, Ryuto H, Fukunishi N, Abe T et al (2008) High efficiency improvement of transgenic torenia flowers by ion beam irradiation. Plant Biotechnol. 25:81–89. https://doi.org/10.5511/plantbiotechnology.25.81

Sawangmee W, Taychasinpitak T, Jompuk P, Kikuchi S (2011) Effects of gamma-ray irradiation in plant morphology of interspecific hybrids between *Torenia fournieri* and *Torenia baillonii*. Kasetsart J (Nat Sci) 45:803–810

Schairer LA, van't Hof J, Hayes CG, Barton RM, de Serres FJ (1978) Exploratory monitoring of air pollutants for mutagenic activity with the *Tradescantia* stamen hair system. Environ Health Perspect 27:51

Scholz E (1957) Röntgenmutationen bei der Birke. Der Züchter 27(1):54–60. https://doi.org/10.1007/BF00711401

Schum A, Preil W (1998) Induced mutations in ornamental plants. In: Jain SM, Brar DS, Ahloowalia BS (eds) Somaclonal variation and induced mutations in crop improvement. Kluwer Academic Publishers, Dordrecht, pp 333–366. (25) (PDF) Mutation Breeding in Ornamentals: An Efficient Breeding Method?. Available from: https://www.researchgate.net/publication/281624466_Mutation_Breeding_in_Ornamentals_An_Efficient_Breeding_Method. Accessed 10 Sep 2023

Schwemmle B, Robbelen G (1962) Haufigkeit von Blattfarbmutationen nach rontgenbestrahlung verschiedener ontogenetischen Stadien von Kalanchoe. Naturwissenschaften 49:64–65

Sekiguchi F, Yamakawa K, Yamaguchi H (1971) Radiation damage in shoot apical meristem of *Antirrhinum majus* and somatic mutations in regenerated buds. Radiat Bot 11(2):157–169

Selvarasu A, Handhasamy R (2013) Analysis of variability, correlation and coefficients in induced mutants of Glory lily (*Gloriosa superb* L.). Int J Plant Breed 7(1):69–75

Selvarasu A, Kandhasamy R (2017) Molecular and agro-morphological genetic diversity assessment of *Gloriosa superba* mutants. Eur J Med Plants 21(1):1–13. Article no.EJMP.37036 ISSN: 2231-0894, NLM ID: 101583475

Seneviratne KACN, Wijesundara DSA (2004) New African violets (*Saintpaulia ionantha*, H. Wendl.) induced by colchicine. Curr Sci 87(2):138–140

Seneviratne KACN, Wijesundara DSA (2007) First African Violets (*Saintpaulia ionantha*, H. Wendl.) with a changing colour pattern induced by mutation. Am J Plant Physiol 2(3): 233–236

Sharma Rao HK (1977) Gamma ray induced mutations in *Kalanchoe daigremontiana* H. and P. Indian J Exot Biol 15(4):326–328

Sharma Rao HK, Singh Y (1976) Gamma ray induced early bulbil formation and growth abnormalities in *Kalanchoe daigremontiana* H. and P. Indian J Exp Biol 14(3):358–360

Sheela VL, Sheena A (2014) Novel trends and achievements in breeding of tropical ornamental crops especially orchids and anthuriums: the mutation breeding approach, chapter 6. In: Tomlekova NB, Kozgar MI, Wani MR (eds) Mutagenesis; exploring genetic diversity of crops. Wageningen Academic Publishers, The Netherlands, pp 141–157

Sheela VL, Anita S, Sarada K et al (2008) *In vitro* mutagenesis in *Dendrobium* cv. Sonia. In: Abstract of FAO/IAEA international symposium on induced mutations in plants, Vienna, Austria, 12–15 Aug 2008, p 118

Sherpa R, Devadas R, Bolbhat SN, Nikam T, Penna S (2022) Gamma radiation induced in-vitro mutagenesis and isolation of mutants for early flowering and phytomorphological variations in *Dendrobium* 'Emma White'. Plants 11:3168. https://doi.org/10.3390/plants11223168

Shigematsu K, Matsubara H (1972) The isolation and propagation of the mutant plant from sectorial chimera induced by irradiation of *Begonia* rex. J Jpn Soc Horic Sci 42(2):196–200. https://doi.org/10.2503/JJSHS.41.196

Shivaswamy C, Patil S, Sindha M, Solanki KS (2022) Impact of chemical mutagens on the morphological traits of China aster. Pharma Innov J 11(12):1820–1822. www.thepharmajournal.com

Shraf-Eldin MA, Alam P, Elkholy SF (2018) Molecular and chemical characterization of mutant and nonmutant genotypes of saffron grown in Saudi Arabia. Food Sci Nutr 7:247. https://doi.org/10.1002/fsn3.875

Shukla R, Nath P, Jugran HM, Gupta MN (1986) Studies on gamma irradiation of Rudbeckia laciniata cultivar 'Golden Glow'. J Nucl Agric Biol 15(4):207–210

Silvy A (1979) Mutation breeding in carnation. In: Eucarpia meeting on carnation and Gerbera, Alasio, 1978, pp 91–102

Silvy A, Mitteau Y (1986) Diversification des varieties d'oeillet (*Dianthus acryophyllus*) par traitement mutagene. In: International symposium on nuclear techniques and in vitro culture for plant improvement, 1985. IAEA, Vienna, pp 395–407

Simard M, Michaux-Ferriere N, Silvy A (1992) Variants of carnation (*Dianthus caryophyllus* L.) obtained by organogenesis from irradiated petals. Plant Cell Tiss Org Cult 29:37–42. https://doi.org/10.1007/BF00036144

Singh B (2014) Effect of gamma rays on vegetative and flowering parameters of Gerbera (*Gerbera jamesonii* Bolus Ex Hooker F.). HortFlora Res Spectr 3:267–270

Singh AP, Mehra KL (1971) Methods for induction and utilization of variability in the improvement of an apomictic grass, *Dichantium annulatum* complex. Theor Appl Genet 41(6):259–263

Singh JP, Arora RS, Dohare SR, Sengupta K (1970) A spontaneous mutant for flower colour and shape in a white flowering *Dahlia*. Euphytica 19:261–262

Singh KP, Choudhary ML, Kumar P, Suchitra A (2002) Characterization of in-vitro induced mutants of carnation by means of electrophoretic protein analysis. Indian J Hortic 59(4): 427–430. ISSN: 0972-8538

Singh VN, Banerji BK, Dwivedi AK, Verma AK (2009) Effect of gamma irradiation on African marigold (*Tagetes erecta* L.) Cv. Pusa Narangi Gainda. J Hortic Sci 4(1):36–40

Singh S, Dhyani D, Kumar A (2011) Expression of floral fasciation in gamma-ray induced *G. jamesonii* mutants. J Cell Plant Sci 2:7–11

Siranut L, Arunee W, Peernaueh J et al (2000) Exotic canna through gamma-rays induced mutations. Ku-Sci J 18(1-3):15–23

Sirisom Y (2008) Induced mutation in gloxinia (*Sinningia speciosa*) with ultraviolet-c (UV-C) and Ethylmethanesulfonate (EMS). Thesis, Plant Science, Sirisom, Y. 2008. A thesis submitted in partial fulfillment of the requirements for the degree of Master of Science in Plant Science, Prince of Songkla University

Sirisom Y, Te-chato S (2008) Effect of ultraviolet-C (UV-C) on mutation of Gloxinia (*Sinningia speciosa*). J Agric 141–151. ISSN 0857-0841

Skirvin RM, Janick J (1974) Calliclones in geranium. HortScience 9(3):270

Skirvin RM, Janick J (1976a) 'Velvet Rose' *Pelargonium*, a scented geranium. HortScience 11(1): 61–62

Skirvin RM, Janick J (1976b) Tissue culture induced variation in scented *Pelargonium* spp. J Am Soc Hort Sci 10(3):281–290

Skirvin RM, Carlson CL, Gorske S (1982) Natural and tissue culture induced variation in a Portulaca hybrid. In: Earle ED, Demarly Y (eds) Variability in plants. Regenerated from tissue culture. Prager, New York, pp 245–267

Smith WA, Brand MH (2012) 'Summer Skies' Buddleja davidii. HortScience 47(1):126–127. https://doi.org/10.21273/HORTSCI.47.1.126

Smith AG, Noyszewski AK (2018) Mutagenesis breeding for seedless varieties of popular landscape plants. Acta Hortic. 1191:43–52. https://doi.org/10.17660/ActaHortic.2018.1191.7

Soedjono S (1988) Effect of the technique of gamma-ray irradiation on plant trait variability in *Begonia semperflorens*. Bull Penelitian Hortic 16:8–15. Cited in: Plant Breed Abstr.: no.8170

Song HS, Kim JK, Lim YT (1999) A new variety of *Hibiscus syriacus* (Rose of Sharon) 'Baekseol' selected from a gamma ray irradiated population. Korean J Breed 31(4):50–51

Song IJ, Kang HG, Kang JY et al (2009) Breeding of four-leaf white clover (*Trifolium repens* L.) through ^{60}Co gamma-ray irradiation. Plant Biotechnol Rep 3:191–197. https://doi.org/10.1007/s11816-009-0091-x

Songsri P, Jogloy S, Holbrook CC, Puangbut D (2019) Determination of lethal dose and effect of gamma rays on growth and tuber yield of Jerusalem artichoke mutant. SABRAO J Breed Genet 51(1):1–11

Soontornyatara S, Sornjai P, Puripunyavanich V, Taywiya P (2017) Effect of gamma rays on morphological character variation of Nelumbo nucifera 'Pathum'. ISHS Acta Horticulturae, 1167: I International symposium on tropical and subtropical ornamentals. https://doi.org/10.17660/ActaHortic.2017.1167.34

Sparnaaij LD (1974) Progress report on clonal selection in carnations after irradiation. In: Broertjes C (ed) Meeting of the Mutation Breeding Contact Group, Wageningen, October 1974. External Rep. No. 23. Association Euratom-ITAL, Wageningen, pp 28–31

Sparnaaij LD (1978) Current research on carnation with special reference to breeding. In: Eucarpia Meeting on *Carnation* and *Gerbera*, Alassio, 1978, pp 47–55

Sparnaaij LD, Demmink JF (1970) Yield and quality of the glasshouse carnation (*Dianthus caryophyllus* L.) after mutagenic radiation with x-rays. Euphytica 19(3):310–317

Sparnaaij LD, Demmink JF (1971) Mutation breeding as a means of improving productivity in commercial carnation cultivars. In: Eucarpia meeting on ornamentals, Wageningen. Wageningen, Institute of Horticultural Plant Breeding, pp 38–44

Sparnaaij LD, Demmink JF, Garretsen F (1974a) Clonal selection in carnation. In: Eucarpia meeting on ornamentals, Frejus. Wageningen, Institute of Horticultural Plant Breeding, pp 39–50

Sparnaaij LD, Demmink JF, Garretsen F (1974b) Selection clonale chez les oeillets. In: Eucarpia meeting on ornamentals, Frejus. Wageningen, Institute of Horticultural Plant Breeding, pp 51–62

Sparrow AH, Schairer IA (1980) The use of rays to induce somatic mutation in *Saintpaulia*. Afr Violet Mag 13:32–37

Sparrow AH, Underbrink AG, Rossia HH (1972) Mutations induced in *Tradescantia* by small doses of X-rays and neutrons: analysis of dose-response curves. Science 176(4937):916–918. https://doi.org/10.1126/science.176.4037.916

Spencer JL (1955) The effect of X-irradiation on flowering of certain cultivated bulbs and corms. Am J Bot 42:917–920. https://doi.org/10.1002/J.1537-2197.1955.TB10442.X

Srivastava A, Mishra R (2005) Gamma ray induced small flower mutant in *Hibiscus rosa-sinensis*. Plant Mutat Rep Rev 1:20–21

Stein OL, Sparrow AH (1963) The effects of chronic gamma irradiation on the growth of *Kalanchoe* cv. Brilliant Star. Radiat Bot 3:207–222

Stein OL, Sparrow AH (1966) The effects of acute irradiation in air, N, and CO2 on the growth of the shoot apex and internodes of *Kalanchoe* cv. Brilliant Star. Radiat Bot 6:187–201

Stewart RN (1965) The origin and transmission of a series of plastogene mutants in *Dianthus* and *Euphorbia*. Genetics 52:925–947

Stewart RN, Arisumi T (1966) Genetic and histogenic determination of pink bract color in poinsettia. J Hered 57(6):217–220

Stewart RN, Semeniuk P, Dermen H (1974) Competition and accommodation between apical layers and their derivatives in the ontogeny of chimera shoots of *Pelargonium* x hortorum. Am J Bot 61(1):54–67

Streitberg H (1965) Teilabschlussbericht zur Forschungsarbeit "Zuchterische Verbesserung volkswirtschaftlich wichtiger Zierpflanzen". Teil a. Zuchtung Azaleen unter Glas. Dtsch. Akad. Landwirtsch, Wiss., Berlin

Streitberg H (1966a) Schaffung von sprossvarianten bei Azaleen durch behandlung mit rontgenstrahlen. Z Pflanzanzucht 56(1):70–87

Streitberg H (1966b) Neue Rosen-und Azaleen-Sorte mit hilfe der Rontgenbestrahlung. Dtsch Gartenbau 13:267–268

Streitberg H (1967a) Production of economically valuable variation in roses and azaleas by means of X-irradiation. In: Stubbe H (ed) Induced mutations and their utilization, Gartersleben, 1966. Abh. Dtsch. Akad. Wiss. Berlin KI. Medizin 2: 359–362

Streitberg H (1967b) Schaffung von sprossvarianten bei Azaleen durch behandlung mit rontgenstrahlen. Arch Gartenbau 15:101–125

Stubbe H, Doring H (1938a) Untersuchungen uber experimentelle Auslosung von Mutationen bei *Antirrhinum majus*. VII. Z Indukt Abstamm VererbLehre 70:125–129

Stubbe H, Doring H (1938b) Untersuchungen uber experimentelle Auslosung von Mutationen bei *Antirrhinum majus*. VII. Uber den Einfluss des Nahrstoffmangels auf die Mutabilitat. Z Ind Abst Vererb 75:341–351

Sugiyama M, Saito H, Ichida H, Hayashi Y, Ryuto H, Fukunishi N, Terakawa T, Abe T (2008a) Biological effects of heavy-ion beam irradiation on cyclamen. Plant Biotechnol. 25:101–104. https://doi.org/10.5511/plantbiotechnology.25.101

Sugiyama M, Hayashi Y, Fukunishi Y, Ryuto H, Terakawa T, Abe T (2008b) Development of flower colour mutant of *Dianthus chinensis* var. Semperflorens by heavy-ion beam irradiation. RIKEN Accel Prog Rep 41:229

Sulistianingsih R (2013) Genetic variation of natural Orchid *Phalaenopsis amabilis* (L.) Blume produce gamma ray irradiation. Dissertation, Universitas Gadjah Mada, Yogyakarta

Suraninpong P, Wuthisuthimethavee S (2015) Mutation induction of *Anthurium andreanum* using gamma radiation. In: Canhoto JM, Correia SI (eds) Proceedings of VIIIth IS on in vitro culture and horticultural breeding. Acta Hort. 1083, ISHS 2015. [Acta horticulturae, Issue 1083: 139–144. ISSN: 0567-7572]

Susrama GK, Yuliadhi KA (2020) Induced mutagenesis in yellow flowering marigold with colchicine in hydrogen peroxide. Adv Trop Biodivers Environ Sci 4(2). https://doi.org/10.24843/ATBES.2020.v04.i02.p04

Suzuki K, Yomo Y, Abe T, Katsumoto Y, Miyazaki K, Yoshida S, Kusumi T (2002) Isolation of sterile mutants of *Verbena hybrida* using heavy-ion beam irradiation. RIKEN Accel Prog Rep 35:129

Swarup V, Raghava SPS (1974) Induced mutation for resistance to leaf-curl virus and its inheritance in garden *Zinnia*. Indian J Genet Plant Breed 34(1):17–21. ISSN: 0019-5200. Online ISSN: 0975-6906

Syakudo K, Yamagata H, Mori S (1964) Several useful mutants induced by X-ray irradiation of seeds in *Cosmos sulphureus* CAV. Jpn J Breed 14(2):82–87. https://doi.org/10.1270/jsbbs1951.14.82

Taek SJ, Kwangsoo C, Mihee Y et al (2005) Characteristics of mutant by colchicine treatment of wild *Lilium leichtlinii* seed native to Korea. Korean J Hortic Sci Technol 23(4):430–432

Takahashi M, Kohama S, Shigeto J, Hase Y, Tanaka A, Morikawa H (2012) Mutants of *Ficus pumila* produced by ion beam irradiation with an improved ability to uptake and assimilate atmospheric nitrogen dioxide. Int J Phytoremediation 14(3):275–281. https://doi.org/10.1080/15226514.2011.604694

Tanaka A, Shikazono N, Hase Y (2010) Studies on biological effects of ion beams on lethality, molecular nature of mutation, mutation rate, and spectrum of mutation phenotype for mutation breeding in higher plants. J Radiat Res 51:223–233

Tangsombatvitchit C, Wongpiyasatid A, Jompuk P et al (2008) Effects of acute gamma irradiation on mutation from Stem cuttings of *Portulaca oleracea* L. Agric Sci J 39(1):55–64

Tarannum T, Rao M, Sahijram L, Rao M (2016) Mutagenesis in crossandra (*Crossandra infundibuliformis* Nees.) using ^{60}Co gamma radiation. Environ Ecol 34(4B):2185–2192

Taychasinpitak T (2009) Induced mutation of gold-dust *Dracaena* var. Friendmanii and var. Bangkok beauty by gamma radiation https://agris.fao.org/agris-search/search.do?recordID=TH2016003663 ISSN: 0970-0420. http://www.environmentandecology.com/

Te-Chato S, Susanon T (2005) Floral mutation in anthurium cv. Valentine after irraduction by ethyl methane sulphonate. Songklanakarin J Sci Technol 27:675–682

Teng ES (2007) Foundations for a long term *Dracaena* breeding program: flower induction, irradiation, and polyploidization. Tropical plant and soil science, Dissertations and Theses. M. S.Q111.H3_4188 MAY 2007_r.pdf

Teng ES, Leonhardt KW (2009a) Optimum irradiation dosage of unrooted *Dracaena* cuttings for mutation. Acta Hortic. 813:517–523. https://doi.org/10.17660/ActaHortic.2009.813.70

Teng ES, Leonhardt KW (2009b) In vitro and in vivo polyploidization of *Dracaena* with oryzalin. Acta Hortic. 813:509–516. https://doi.org/10.17660/ActaHortic.2009.813.69

Thakur PC, Bhagchandani PM (1978) Spontaneous mutant for floer colour in yellow flowering Dahlia (*Dahlia variabilis*). Indian J Hortic 35(1):62

Thamm KJJ (1956) Ruckblick auf die bisherige wissenschaftliche Arbeid des Amsterdamer Biologen Willem Educard De Mol van Oud Loosdrecht. Bayerischer Landwirtschaftsverlag, Bonn, pp 7–30

Thomas H, Stoddart JL (1975) Separation of chlorophyll degradation from other senescence processes in leaves of a mutant genotype of Meadow Fescue (*Festuca pratensis* L.). Plant Physiol 56:438–441

Thomas H, Luthy B, Matile P (1985) Leaf senescence in a non-yellowing mutant of *Festuca pratensis* Huds. Plants 164:400–405

Tisch RE (1974) Zephyrantheae Report: induced mutations with chemicals. Plant Life. Amaryllis Year Book:124–134

Tiwari AK, Kumar V (2010) Effects of colchicine on germination and survival of Phlox (*Phlox drummondi*). Ann Horticu 3(2):202–204

Tiwari AK, Kumar V (2011) Gamma-rays induced morphological changes in pot marigold (*Calendula officinalis*). Prog Agric 11(1):99–102

Tiwari AK, Mishra SK (2012) Effect of colchicine on mitotic polyploidization and morphological characteristics of *Phlox drummondi*. Afr J Biotechnol 11(39):9336–9342. https://doi.org/10.5897/AJB11.2196. ISSN 1684–5315. http://www.academicjournals.org/AJB

Toker C, Shyam S, Yadav IS, Solanki (2007) Mutation breeding. In: Yadav SS, Mcneil DL, Stevenson PC (eds) Lentil. Springer, New York, pp 209–224

Tolar RW, Beard JB, Grisham MP, Crocker RL (1985) Registration of TXSA 82o2 and TXSA 8218 St. Augustinegrass germplasm resistant to *Panicum* mosain virus St. Augustine decline strain. Crop Sci 25(2):371

Toler RW, Grisham MP (1983) Multi-disease resistance in St. Augustinegrass. Phytopathology 75(3):778

Tomo N, Rosa S, Pintana L, Chile S (2012) Gamma irradiation on *Alstroemeria aurea* G. in vitro rhizomes. Rev FCA UNCUYO 44(1):191–197. ISSN impreso 0370-4661. ISSN (en línea) 1853-8665

Tran Thanh Van M (1973) In vitro control of de novo flower, bud, root and callus differentiation from excised epidermal tissues. Nature (London) 246:44–45

Tran Thanh Van M, Drira A (1971) Definition of a simple epidermental system of directed organosis de novo: organ neoformation from epidermal tissues of Nautilocalyx lynchei. In: Les Cultues de Tissus des Plantes, Strasbourg, 1970, Colloq. Int. C.N.R.S., pp 169–176

Tsuji S, Miyamoto M, Okabe T, Hase Y, Yokota Y, Narumi I (2007) Mutation breeding on the ornamental plants of *Gypsophila* and *Gentiana* species. JAEA-review 2007-060 3-18, p 84

Tsuro M, Iwata C, Kataoka R, Yoshihara R, Hase Y (2008) Comparative effect of 12C6+ Beam and gamma-ray irradiation on callus growth and shoot formation from Lavandin isolated cells. JAEA-review, 2008-055, p 72

Tulman Neto A, Latado RR (1996) Mutation breeding in *Stromantha sanguined* by gamma radiation. Braz J Gen 19:197

Underbrink G, Schaires LA, Sparrow AH (1973) *Tradescantia* Stamen Hairs: A radiobiological test system applicable to chemical mutagenesis. In: Chemical mutagenesis: principles and methods for their detections, vol 3, no 4. Plenum Press, New York, p 159

Urrea AI, Ceballos SM (2005) Empleo de las radiaciones gamma en la inducción de variabilidad genética en *Heliconia psittacorum*. Actual Biol 27(82):17–23

Urrea AI, Ceballos SM (2017) Use of gamma radiation in the induction of genetic variability in *Heliconia psittacorum*. Biol Updates 27(82):17–23. https://revistas.udea.edu.co/index.php/actbio/article/view/329425

Uyama Y, Ohya H, Amano Y, Kashimoto K, Hatano S, Nozawa S, Yoshihara R, Hase Y, Narumi I (2011) Production of mutants by ion beam irradiation in *Dahlia* spp. JAEA Takasaki annual report 2011, p 102

Uyama Y, Ohya H, Amano Y, Kashimoto K, Hatano S, Nozawa S, Yoshihara R, Hase Y, Narumi I (2013) Production of mutants by ion beam irradiation in *Dahlia* spp. JAEA Takasaki annual report 2011, p 102

Vaarama A (1970) Induced mutations and polyploidy in Birch, *Betula* spp. Final report (part V)-P. L.480; 9-30-1970. Project no. E8-FS-47, Grant no. FG-Fi-133. Report no. 10, covering the research period: Jan. 1st, 1962-Dec. 31st, 1966, with additional results of later date, 100p

Van Dordrecht EM (1984) Mutatieveredeling bij *Kalanchoe* veelbeloved. Vakkbl Bloemiemisterij 39(25):50–51

Van Eijk JP, Eikelboom W (1981a) Selecti bij veredeling van tulpen. Hobaho 54(50):6–8

Van Eijk JP, Eikelboom W (1981b) Selecti bij veredeling van de tulp. (II). Bloembollencultuur 91(39):1054–1055

Van Groenestijn JE, Van Tuyl J (1983) Straks geen leliebelichting meer nodig. Hoopvolle resultaten van veredeling op lichtbehoefte. Vakbl Bioemisterij 38(6):30–31

Van Houwelingen A, Souer E, Spelt K, Kloos D, Mol J, Koes R (1998) Analysis of flower pigmentation mutants generated by random transposon mutagenesis in *Petunia hybrida*. Plant J 13:39–50. https://doi.org/10.1046/j.1365-313x.1998.00005.x

Van Raalte D (1980) Nieuwe Hoya carnosa-rassen. Vakbl Bloemisterij 35(5):115. https://nucleus.iaea.org/sites/mvd/SitePages/Search.aspx (Mutant Variety Database Search & Cited Source Material - STEMMA, Hoya Journal, ISSN 2832 417-X, July 2022, V4, N1, Page 37: 'Compacta' variations and seed irradiation)

Van Raatle D, Van Raatle-Wichers D (1974) Toekomst voor Streptocarpus. Vakbl Bloemisterij 29(23):15

Van't Hof J, Schairer A (1982) Tradescantia assay system for gaseous mutagens: A report of the U.S. environmental protection agency Gene-Tox program. Mut Res 99(3):303–315. https://doi.org/10.1016/0165-1110(82)90048-3

Vaughn KC, Wilson KG (1980) A dominant plastome mutation in *Hosta*. J Hered 71(3):203–206

Vaughn KC, Wilson KG, Reibach PH (1980) Ultrastructure and biochemistry of two mutants in *Hosta* (Liliaceae). Cytobios 27(106):71–80. PMID: 7418450

Venkatachalam P, Jayabalan N (1991) Induction of mutants in *Zinnia elegans* Jacq. Mutant Breed Newsl 38:10

Venkatachalam P, Jayabalan N (1992) Analysis of leaf proteins in gamma rays induced mutants of *Zinnia*. Crop Improv 19:97–99

Venkatachalam P, Jayabalan N (1997) Effect of gamma rays on some qualitative and quantitative characters in *Zinnia elegans* Jacq. Indian J Genet Plant Breed 57:255–261

Venkatchalam P, Jayabalan N (1994a) New flower colour mutants evolved through gamma irradiation in *Zinnia elegans* Jacq. J Mendel 11:35

Venkatchalam P, Jayabalan N (1994b) Effect of gamma irradiation on flower structure of *Zinnia elegans* Jacq. J Swamy Bot Cl 11:58

Venkateswarlu M, Kumar P, Mogil T, Kashinath, Ramesh Babu CH (1983) Sterile mutant induced through hydroxylamine in *Catharanthus roseus* (L.) G. Don. by gamma rays and EMS individually and in combination. Bot Rep 2:82

Venkateswarlu M, Susheelamma BN, Kumar P, Subha K (1988) Studies on induced mutation frequency in *Catharanthus roseus* (L.) G. Don. by gamma rays and EMS individually and in combination. Indian J Genet 48:313–316

Venverloo CJ (1974) Exogenous regulation of the formation of roots or shoots in the same culture. In: Street HE (ed) Third international conference of Associations of Plant Tissue and Cell Culture, Leicester. University of Leicester, Leicester, Abstr. No. 31

Venverloo CJ, Koster J, Libbenga KR (1983) The formation of adventitious organs. IV. The ontogeny of shoots and leaves from epidermal cells of Nautilocalyx lynchii. Z Pflanzenphisiol 109(1):55–67

Verma RC, Raina SN (1980) *Phlox drummondii* in the study of radiation induced chromosomal aberrations. Nucleus 23:176–184

Verma RC, Sharma R (2000) Radiation induced cytological change in *Phlox drummondii*. J Cytol Genet 1:33–39

Verma AK, Singh RR (2010) Induced dwarf mutant in *Catharanthus roseus* with enhanced antibacterial activity. Indian J Pharm Sci 72(5):655–657. https://doi.org/10.4103/0250-474X.78541. PMCID: PMC3116317. PMID: 21695004

Verma M, Dwivedi AK, Banerji BK (2010) Effecte of gamma irradiation on African marigold cv. Pusa Basanti Gaida. Abst: National symposium on lifestyle floriculture: challenges and opportunities. Session-2, crop improvement, biotechnology and biodiversity, March 19-21, 2010 at DR. Y.S. Parmar University of Horticulture and Forestry, Nauni, Solan, Abstract No. 2.36: 29

Verma AK, Singh RR, Singh S (2013) Mutation breeding in *Catharanthus roseus* (L.) G. Don: an overview. J Pharmacogn Phytochem 2(1):334–337

Vichai P, Woranuch L, Kanokporn B (2011) Radiation-induced mutant of American yellow lotus and its interspecific crosses with Thai Lotus. Water Garden J 1st Quarter 26(1):5–12, 8p

Vilperte V, Boehm R, Debener T (2011) A highly mutable *GST* is essential for bract colouration in Euphorbia pulcherrima Willd. Ex Klotsch. BMC Genomics 22:208. https://doi.org/10.1186/s12864-021-07527-z

Vinodh S, Kannan M (2020a) Effects of ethyl methane sulphonate on the yield and qualtiy of crossandra (*Crossandra infundibuliformis* (L.) Nees.). Int J Chem Stud 8(1):539–541. P-ISSN: 2349–8528 E-ISSN: 2321–4902. https://doi.org/10.22271/chemi.2020.v8.i1h.8315

Vinodh S, Kannan M (2020b) Effect of gamma irradiation on growth and yield of Crossandra (*Crossandra infundibuliformis* (L.) Nees.). J Pharmacogn Phytochem 9(1):541–542. E-ISSN: 2278-4136 P-ISSN: 2349-8234

Visessuwan R (1985) In vitro mutation breeding of gerbera (*Gerbera jamesonii* Hort.). Thesis, Kasetsart Universty, Bangkok. https://agris.fao.org/agris-search/search.do?recordID=TH8620567

Walther KF (1981) Mutants tolerant to low-temperature conditions induced in suspension culture as a source for improvement of *E. pulcherrima* Will. Ex. IAEA-SM-251/12. Proceedings of an international symposium on induced mutations as a tool for crop plant improvement jointly organized by the International Atomic Energy Agency and The Food and Agriculture Organization of the United Nations and held in Vienna, 9–13 March 1981

Walther F, Sauer A (1986a) Analysis of radiosensitivity – a basic requirement for *in vitro* somatic mutagenesis. II. *Gerbera jamesponii*. In: Nuclear techniques and *in vitro* culture for plant improvement. IAEA, Vienna, Aug 1985, STI/PUB/698, pp 155–159

Walther F, Sauer A (1986b) *In vitro* mutagenesis in *Gerbera jamesonii*. In: Horn W, Jenson CJ, Odenbach W, Schieder O (eds) Genetic manipulation in plant breeding. Proceedings of symposium on Eucarpia, Berlin, 1985. Walter de Gruyter, Berlin, pp 555–562

Walther F, Sauer A (1989a) Increase of genetic variation in 'Blue Daisy' (Brachycome multifida) by *in vitro*-mutagenesis and polyploidization. Mutat Breed Newsl 33:3–4. (Excerpts from a presentation during the XVIth Int. Congress of Genetics, Toronto, Canada, 20-27 August 1988, Source - Joint FAO/IAEA Division of Nuclear Techniques in Food and Agriculture, Vienna (Austria); 36p; ISSN 1011-260X; Jan 1989; pp 3–4; 16. International Congress of Genetics; Toronto (Canada); 20–27 Aug 1988)

Walther F, Sauer A (1989b) Effect of different intervals of X-ray split doses on shoot production of *in vitro* derived explants of *Gerbera jamesonii* Bolus. Vortrage Pflanz, 15. Source - Joint FAO/IAEA Division of Nuclear Techniques in Food and Agriculture, Vienna (Austria); 44p; ISSN 1011-260X; Jan 1990; p 9; 12. EUCARPIA congress on science for plant breeding; Goettingen (Germany); 27 Feb–3 March 1989

Walther F, Sauer A (1990) Influence of acute and fractioned X-ray doses on shoot production of in vitro derived explants of *Gerbera jamesonii* H. Bolus. Plant Breed 105:137–143. https://doi.org/10.1111/j.1439-0523.1990.tb00466.x

Walther F, Sauer A (1991) Split dose irradiation of *in vitro* derived microshoots. An effective procedure for increasing mutability. In: Plant mutation breed for crop improvement. IAEA, Vienna, pp 343–353

Walther F, Sauer A (1992) Interaction between *in vitro* shoot forming capacity and X-ray sensitivity of *Gerbera jamsonii* H. Bolux ex Hook P. Gartenbauwissenschaft 57:68–71

Wang YF, Xi YL, Wie ZC et al (1989) The effect of gamma rays and colchicine on mutagenesis in somaclones of *Lilium davidii* var. Willmottiae. Jiangsu J Agric Sci 5:31–37. Cited in: Plant Breed Abstr., 1990: no. 5930

Wang MG, Zeng RZ, Xie L, Gao XH, Zhang ZS (2011) In vitro polyploid induction and its identification in *Cymbidium sinense*. Chin Agric Sci Bull 27:132–136. (in Chinese)

Wang P, Zhang Y, Zhao L, Mo B, Luo T (2017) Effect of gamma rays on *Sophora davidii* and detection of DNA polymorphism through ISSR marker. Biomed Res Int. https://doi.org/10.1155/2017/8576404

Wannajindapom A, Kativat C, Tantasawat PA (2016) Mutation induction of *Dendrobium* 'Earsakul' using sodium azide. HortScience 51(11):8p. http://repository.unib.ac.id/id/eprint/12851

Warburton J, Gront BWW, Short KC (1984) In vitro selection for cold tolerant cell lines of *Saintpaulia ionantha* (WendL). In: Novak FJ, Havel L, Dolezel J (eds) Proceedings plant tissue and cell culture applications to crop improvement. CSAV, Prague, pp 355–356

Warfield D (1973) Induction of mutations in African Violet (*Saintpaulia ionantha* Wendl.) by ethyl methane sulfonate. Hortic Sci 8:29

Warner RM, Erwin JE (1998) Effect of irradiance on *Hibiscus* L. species flowering. In: 95th annual international conference of the American Society for Horticultural Science Charlotte, North Carolina, USA, 12–15 July 1998, Abstract 015, p 416

Weigle JL, Butler JK (1983) Induced dwarf mutant in *Impatiens platypetala*. J Hered 74:200. https://doi.org/10.1093/oxfordjournals.jhered.a109764

Wellensiek SJ (1960) Mutagenic genes. Proc Kon Ned Akad Wetensch C 63(1):38–42

Wellensiek SJ, van Bren G (1973) A non-specific mutagenic gene in cyclamen. Neth J Agric Sci 21: 99–101

Widiarsih S, Dwimahyani I (2023) Induced mutation on Indonesian black orchid (Coelogyne pandurata Lindley) in-vitro culture by gamma irradiation. IOP Conf Ser Earth Environ Sci 1160:012001. https://doi.org/10.1088/1755-1315/1160/1/012001

Wongpiyasatid A, Hormchan P (2000) New mutants of perennial *Portulaca grandiflora* through gamma radiation. Agric Nat Resour 34:408–416

Wongpiyasatid A, Roongtanakiat N (1992) Effects of gamma radiation on flower colors and types of perennial *Portulaca grandiflora* Hook. In: The 30th Kasetsart University conference proceedings, Bangkok, Thailand, pp 695–704 (In Thai with English abstract)

Wongpiyasatid A, Thinnok T, Taychasinpitak T, Jompuk P, Chusreeaeom K, Lamseejan S (2007) Effects of acute gamma irradiation on adventitious plantlet regeneration and mutation from leaf cuttings of African violet (*Saintpaulia ionantha*). Agric Nat Resour 41:633–640

Wosinska A (1980a) Effect of various doses of gamma ^{60}Co radiation upon some morphological features of China aster (*Callistephus chinensis* Nees) in M_1 and M_2 generations. Acta Agrobot 33(1):5–29

Wosinska A (1980b) Effect of various doses of gamma ^{60}Co radiation upon pollen vitality of China aster (*Callistephus chinensis* Nees) in M_1 and M_2 generations. Acta Agrobot 33(1):31–39

Wosinska A (1982) V. Induction of variability and mutation in China aster (*Callistephus chinensis* Nees) with 60Co gamma rays. Acta Agrobot 35(2):285–301. Accession: 001505627

Wosinska A (1986) A radiomutant with bi-coloured ligulate flowers acquired from the China aster (*Calistephus chinensis* Nees) cultivar Dukat. Acta Agrobot 37:117–121. Cited in: Hort. Abstr. 1987: no.1255

Wosinska A (2013) Induction with 60Co gamma rays of modification variability and mutation of China aster (*Callistephus chinensis* Nees). Acta Agrobot 35(2):285–301. https://doi.org/10.5586/aa.1982.029

Wu J-Z, Zheng Y-B, Chen T-Q, Yi J, Qin L-P, Rahman K, Lin W-X (2007) Evaluation of the quality of lotus seed of *Nelumbo nucifera* Gaertn from outer space mutation. Food Chem 105(2): 540–547. https://doi.org/10.1016/j.foodchem.2007.04.011

Wu DL, Hou SW, Qian PP et al (2009) Flower color chimera and abnormal leaf mutants induced by 12C6+ heavy ions in *Salvia splendens* Ker-Gawl. Sci Hortic 121:462–467

Xia X (2020) Effects of 60Co-γand electron beam irradiation on the germination and seedling growth of *Lagerstroemia indica*. J Trop Biol 11(2):210–216. https://doi.org/10.15886/j.cnki.rdswxb.2020.02.011

Xicun D, Shuang M, Wenjian L, Lixia Y, Libin Z, Hongmei X (2007) Preliminary study on flower color mutant induced by 80 MeV/u $^{12}C^{+6}$ ions in *Dahlia pinnata* Cav using RAPD technique. J Radiat Res Radiat Process 25(1):62–64. ISSN 1000-3436

Yakovleva EV, Beznosikov VA, Kondratenok BM et al (2011) Genotoxic effects in *Tradescantia* plant (clone 2) induced by benzo(a)pyrene. Contemp Probl Ecol 4:594–599. https://doi.org/10.1134/S1995425511060051

Yamaguchi H (2018) Mutation breeding of ornamental plants using ion beams. Breed Sci 68:71–78

Yamaguchi E, Miyatani M, Kawai T, Atsumi H, Hase Y (2017) Re-development of new variety of *Salvia* by ion beam breeding. QST Takasaki annual report 2015, p 132

Yang Y, Chen X, Xu B, Li Y, Ma Y, Wang G (2015) Phenotype and transcriptome analysis reveals chloroplast development and pigment biosynthesis together influenced the leaf color formation in mutants of *Anthurium andraeanum* 'Sonate'. Front Plant Sci 6:139. https://doi.org/10.3389/fpls.2015.00139

Yena AV (2022) New cultivar Hedera helix 'Peregreenus' and some features of bud mutations developing in ivy. Plant Biol Hortic Theory Innov 2(163):36–44. https://doi.org/10.36305/2712-7788-2022-2-163-36-44

Yim JH, Woo SM, Hwang ML, Pyo SH, Kwon HL, Woo JS (2010) Mutant breeding of ornamental trees for creating variations with high value using proton beam. KAERI, 56p

Yin CC, Zhang Y, Zhang JH, Chen YY, Wang GD (2010) Identification of hybrid tetraploid and ploidy induced by colchicine. J Nucl Sci 24:518–521. (in Chinese)

Yu L, Li W, Du Y, Chen G, Luo S, Liu R, Feng H, Zhou L (2016) Flower color mutants induced by carbon ion beam irradiation of geranium (*Pelargonium hortorum*, Bailey). Nucl Sci Tech 27: 112. https://doi.org/10.1007/s41365-016-0117-3

Yuki S, Araki S, Suzuki T, Ohsuga T, Katayama K, Hase Y, Yokota Y (2007) Breeding new varieties of miniature *Cymbidium* using ion beam irradiation. JAEA-review 2007-060 3-19, p 85

Yunus MF, Azizi MA, Kadir MA, Daud SK, Rashid AA (2013) In vitro mutagenesis of *Etlingera elatior* (Jack) and early detection of mutation using RAPD markers. Turk J Biol 37:716–725. https://doi.org/10.3906/biy-1303-19

Zahra N, Narges AM, Farah F (2020) Effect of gamma radiation on morphological and genetic variation in regenerated plantlets *Catharanthus roseus* L. (G) Don. Genetika 52(1):15–28. https://doi.org/10.2298/GENSR2001015N

Zandbergen M (1964) Why should daffodils not have to split cup? In: Daffodil tulip year book, vol 29, pp 82–87

Zeven AC (1972) Inheritance of functional male sterility in *Streptocarpus* 'Constant Nymph' and its mutants. Euphytica 21:265–270

Zeven AC (1973) *Streptocarpus* 'Constant Nymph'. De oorzaak van de populariteit in Nederland. Vakbl Bloemisterij 28(35):14–15

Zhang QH, Li ZL, Tang M, Xu CC, Xi HP (2011) Study on the use of colchicine to induce polyploidy of *Dendrobium candidum* Wall ex Lindl. J Yunnan Agric Univ 26:678–682. (in Chinese)

Zhang ZY, Wang P, Li Y, Ma LL, Li LF, Yang RT, Ma YZ, Wang SA, Wang Q (2014) Global transcriptome analysis and identification of the flowering regulatory genes expressed in leaves of *Lagerstroemia indica*. DNA Cell Biol 33:680–688

Zhou D, Shen S, Wang Y (1990) Induction of mutation with gamma irradiation in Camellia (Camellia japonica L.). Acta Agriculturae Nucleatae Sinica, 1990-01

Zhou LB, Li WJ, Ma S, Dong XC, Yu LX, Li Q, Zhou GM, Gao QX (2006a) Effects of ion beam irradiation on adventitious shoot regeneration from in vitro leaf explants of Saintpaulia ionahta. Nucl Instrum Methods Phys Res B 244:349–353. https://doi.org/10.1016/j.nimb.2005.10.034

Zhou LB, Li WJ, Yu LX et al (2006b) Linear energy transfer dependence of the effects of carbon ion beams on adventitious shoot regeneration from in vitro leaf explants of *Saintpaulia ionahta*. Int J Radiat Biol 82:473–481

Part IV

Genotoxic Effects of Mutagens

13 Characterization of Mutants and Cause of Flower Color Mutations

Abstract

A large number of mutant varieties have been reported in different ornamental plants, but reports on a comparative evaluation of original and mutant varieties are very restricted. This chapter covers the results of the comparative analysis of original cultivars and their respective induced mutants on cytomorphological, radiosensitivity, anatomical, palynological, and biochemical characters for a better and clear understanding of the exact mechanism involved in the origin and evolution of somatic flower color mutations at the molecular level.

Keywords

Original variety · Mutant variety · Characterization of varieties

Many new ornamental cultivars have evolved by induced mutations. The majority of mutants have been selected based on mainly one desirable character, i.e. flower color and/or flower shape. Mutants with other desirable characters have also been selected in many ornamentals. A large number of mutant varieties have been reported in different ornamental plants but reports on a comparative evaluation of original and mutant varieties are very restricted. Comparative characterization reports on original and mutant cultivars are negligible. Analysis of the possible genotoxic effects of different mutagens on experimental crop plants is very important in mutagenesis work. The author gave special efforts to characterize a large number of mutant varieties and their respective control varieties. The objective of characterization was to find out the extent of radiation effects on different quantitative characters. Specific attention was paid to examining the mutagen effects on other characters along with flower color. The general question is how do these flower color mutations arise? According to available literature, the radiation-induced flower color changes may be due to chromosomal aberrations, changes in chromosome number, gene

S. Datta, *Role of Mutation Breeding in Floriculture Industry*,
https://doi.org/10.1007/978-981-99-5675-3_13

mutation, rearrangement of different histogenic layers, and mutation occurring in the biochemical pathway leading to pigment formation. This has been reviewed earlier (Datta 1990a). Based on these interpretations the exact mechanism of induction of somatic mutations cannot be explained with certainty. Therefore, much attention was paid to the comparative analysis of original cultivars and their respective induced mutants on cytomorphological, radiosensitivity, anatomical, palynological, and biochemical characters for a better and clear understanding of the exact mechanism involved in the origin and evolution of somatic flower color mutations at the molecular level. Maximum mutants of chrysanthemum rose and bougainvillea was selected as the target materials for characterization.

Molecular techniques are now very powerful for the characterization of mutants. But there is no universal parameter and rules for characterization. Characterization depends mainly on the type of crop and more importantly on the objective of characterization. Cultivar identification and cultivar relatedness are important issues for horticultural breeders. In mutation breeding, characterization is needed not only to identify mutants but also to explore the sensitive other targets changed due to mutation. A good number of classical and moderately advanced techniques were used to characterize a wide range of parameters like morphological (vegetative and floral characters), cytological (chromosome number, chromosomal behavior, karyotype, etc.), anatomical (number of stomata and size, number of chloroplasts per guard cell, hair structure, etc.), palynological (pollen grain sterility, size, exine ornamentation pattern, etc.), biochemical (phenolic compounds in leaves and petals, pigment composition studies using TLC, spectrophotometer, etc.), and molecular (RAPD) to distinguish new varieties developed through sport and induced mutagenesis from their parental varieties (Figs. 13.1 and 13.2). Characterization reports have been arranged systematically and characterization reports of other scientists have been mentioned as and when required.

The author made concentrated efforts for comparative analysis of original cultivars and their respective mutants to find out the mechanism involved in the origin and evolution of somatic flower color mutations and other changes associated with the mutant trait. Characterization reports from different specialized scientists pointed out that chromosomal abnormalities, alteration in chromosome number, gene mutation, and reformation of histogenic layers are accountable to induce flower color changes (Dowrick 1951, 1952, 1953, 1958; Dowrick and El-Bayoumi 1966a, b; Walker 1955; Ichikawa et al. 1970). The author realized further comprehensive studies are very important on point by point at one place for a precise understanding of somatic mutations. Chrysanthemum and rose mutants were selected as target materials as a large number of mutants were induced by the author. Biochemical markers have received more attention as the data reflect more truly the genetic variability because they are the direct products of genes (Perry and McIntosh 1991). The objective of characterization was to find out the molecular basis of somatic flower color mutations, i.e. how gamma ray induced morphological mutants (flower color/shape) can be identified by molecular markers. A total number of mutants developed in different ornamental plants at CSIR-NBRI, Lucknow, India has been reported earlier (Datta 1988a, b, 2015, 2019). Details of original varieties

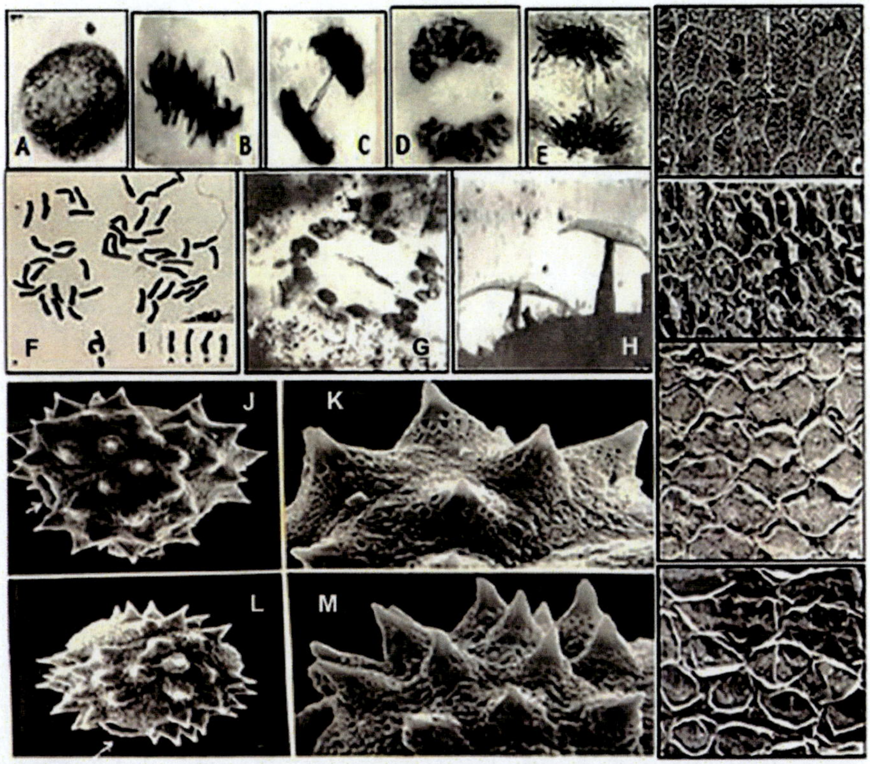

Fig. 13.1 (a–e) Chromosomal abnormalities. (f) Karyotype analysis. (g) Stomata with chloroplast. (h) Epidermal hair. (i) Micromorphology of ray florets showing cell boundaries, cell surface, striations, papillae, etc. in original and mutant varieties. (j–m) Pollen morphological features: apertural character and exine ornamentation. (j) Pollen grain (polar view) of Chrysanthemum cv. "Kingsford Smith" showing lalongate endocolpium (arrow marked). (k) Reticulate undulated surface. (l) Chrysanthemum mutant "Rohit"—polar view showing lalongate endocolpium (arrow marked). (m) Undulated tuberculate surface

and their respective mutants used for present characterization are given in Table 13.1.

In all experimental chrysanthemum materials, gamma irradiation induced various types of chromosomal abnormalities but no change in flower color confirming that flower color changes were not sensible to chromosomal aberrations. It was also noticed that all experimental chrysanthemum cultivars had the same chromosome number ($2n = 54$) but all the cultivars had different flower colors and shapes. It has also been noticed that two chromosome numbers ($2n = 54$ or 52) were present in root tip cells of many control plants but no flower color mutant was detected (Dowrick and El-Bayoumi 1966a, b; Nazeer 1981; Nazeer and Khoshoo 1982, 1983). Sampson et al. (1958) detected that change in flower color is not associated with chromosome number. There is also a different judgment of opinion that

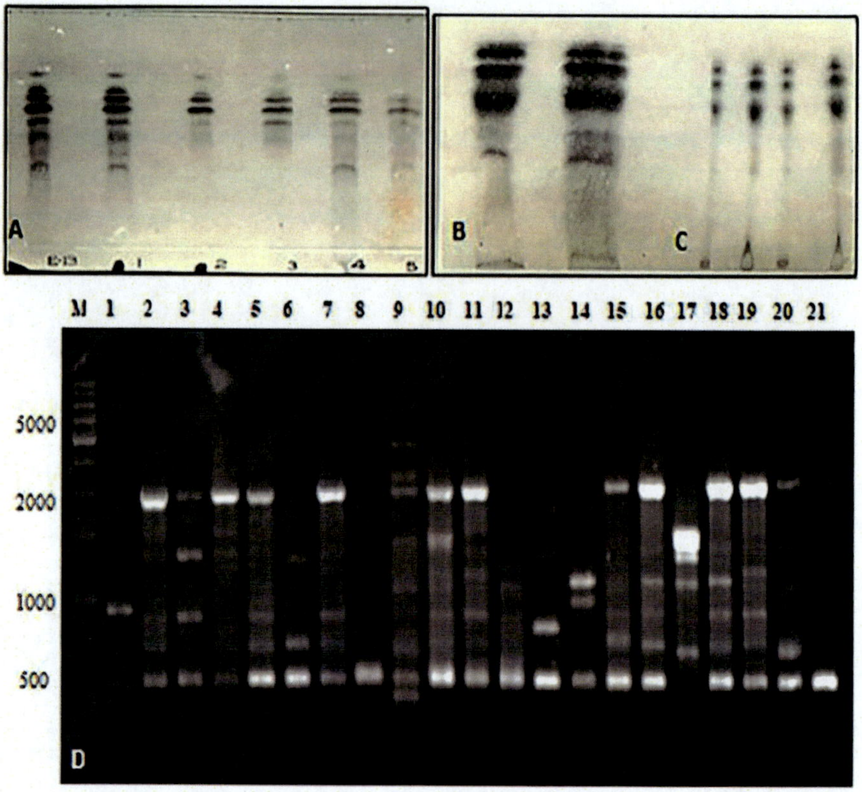

Fig. 13.2 (**a–c**) Thin-layer chromatograms of floret pigments. (**a**) TLC of Chrysanthemum cv. "E-13" and its gamma ray induced mutants (1–5). (**b**) TLC of Chrysanthemum cv. "Maghi" (left) and its mutant (right). (**c**) TLC of Chrysanthemum cv. "Kingsford Smith" and its three mutants. (**d**) RAPD profile of original and mutant cultivars of chrysanthemum

chromosome numbers are found from root cells which develop from L-III (Streanglar 1956) and flowers develop from L-I (Bowen 1965; Sampson et al. 1958; Weaver 1963; Stewart and Dermen 1970). Therefore, it is not wise to connect induced chromosomal abnormalities/chromosome numbers with flower color change. For cytological analysis 18 original chrysanthemum cultivars and their 38 gamma ray induced flower color/shape mutants were selected. All mutants had similar chromosome numbers like parents and no mutant-specific chromosomal abnormalities could be detected. Original and mutant cultivars showed no significant differences in Interphase Chromosome Number (ICV), Interphase Nuclear Volume (INV), and DNA content. The role of alterations in chromosome morphology in different groups of plants can be assured by a study of their karyotypes. The karyotypes of several original cultivars and their induced mutants of chrysanthemum were examined. A comparison of ideograms showed that the mutants did not differ from their respective original cultivars in the number of types of chromosomes and

Table 13.1 Original and mutant cultivars selected for comparative characterization

Original Cultivar	Mutant Cultivar	Original Cultivar	Mutant Cultivar
Bougainvillea			
"Partha"	"Arjun"	–	–
Leaves green	Leaves variegated		
Chrysanthemum			
"E-13"	"Kapish"	"E-13"	"Himani"
Mauve	Brownish		White
"E-13"	"Sheela"	"E-13"	"Basanti"
	Yellow		Terracotta
"E-13"	"Lohit"	"E-13"	"Anamika"
	Dark Reddish		Light reddish
"E-13"	"Basanti"	"Flirt"	"Batik"
	Yellow	Red	Yellow stripe on Red background
"D-5"	"Alankar"	"D-5"	"Agnisikha"
Magnolia Purple	Spanish Orange		Erythrite Red
"Kingsford Smith"	"Rohit"	"Kingsford Smith"	"Taruni"
Purplish Mauve	Rhodonite Red	"Kingsford Smith"	Lighter Fuschia Purple"
"Kingsford Smith"	"Erythrite Red"	"Flirt"	"Man Bhawan"
		Red	Yellow and red
"Himani"	"Sheela"	"Sharad Bahar"	"Colchi Bahar"
White	Canary Yellow	Purple	Terracotta Red
"Megami"	"Hemanti"	"M-24"	"Tulika"
Amaranth Pin	Chinese Yellow	Purple	Purple, paint brush
Rose			
"Arjun"	"Creamish White"	"Arjun"	"Striped Arjun"
Pink			White stripe on light
			Pink background
"First Prize"	"Light Pink Prize"	"Mrinalini"	"Light Pink Mutant"
Blend of light red and deep pink		Phlox Pink	
"Raja of Nalagarh"	"Light Pink Mutant"	"Summer Holiday"	"Pink Summer Holiday"
Dazzling orange		Scarlet Vermillion	
"America's Junior Miss"	"Sukumai"	"Contempo"	"Yellow Contempo"
Coral Pink	Light Pink	Copper orange with Yellow eyes	Empire Yellow
"Contempo"	"Tangerine Contempo"	"Contempo"	"Striped Contempo"
	Tangerine Orange		Bicolored
"Contempo"	"Pink Contempo"	"Imperator"	"Pink Imperetor"

(continued)

Table 13.1 (continued)

		Cherry Red	
"Imperator"	"Twinkle"	"Salmon Beauty"	"Creamish white
	Pink streak on Original background	Salmon	"Mutant"
"Zorina"	"Pink Zorina"	"Queen Elizabeth"	"Sharada"
Grenadine Red		Carmine Rose	Blossom Pink
"Windy City" Pink	"Light Pink Mutant"	"Mrinalini"	"Mrinalini Stripe"
"Sylvia"	"Sylvia White"		

the number of each type represented in them. But a comparison of ideograms showed varietal differences. The karyotype in the original and mutant cultivars was reasonably symmetrical (Fig. 13.1e). This strengthens the opinion that flower color changes have taken place through gene mutation but neither through a change in chromosome number, chromosomal abnormalities nor due to alterations in karyomorphology (Datta and Gupta 1981a, b, c; Datta 1994a).

From a wide range of observations in different ornamentals at CSIR-NBRI, it was very clear that the reshuffling of histogenic layers applied to the development of chlorophyll variegation in leaves. As both, the phenomenon (flower color mutation and chlorophyll variegation) is found very frequently in different vegetative propagated ornamentals (both spontaneously and after mutagen treatment), different workers are biased to explain that the development of new flower color is also due to reshuffling of histogenic layers. The author interpreted from his investigations on several mutants from different ornamentals that rearrangements of histogenic layers do not have any clear-cut capacity in the development of somatic flower color mutation. It is noteworthy to understand how reshuffling of histogenic layers theory can elucidate when a series of new flower color mutants develop from a single starting cultivar not only in the first generation but also in later generations. Similarly, this theory cannot be supported for striped (bicolored or spotted) mutants. The author developed several striped flower color mutants in chrysanthemum ("Batik"), rose ("Contempo," "Imperator," "Mrinalini"), and chlorophyll variegations in bougainvillea and *Lantana depressa* (Datta 1994a, b, 1995, 1997, 2015). It was interesting to note that stripe nature in "Contempo" and "Imperator" and chlorophyll variegations in *Lantana depressa* were variable which strengthen the possibility of involvement of transposable elements. It has been established that reshuffling of histogenic layers is very relevant to develop chlorophyll variegations in leaves. Variations in the comparative ratio of three different pigments in upper and lower epidermal layers of ray florets determine the chrysanthemum flower color (Dowrick 1953; Bowen et al. 1962). Fujii and Mabuchi (1961) presumed that mutation induced in the biochemical pathway leads to the formation of new pigment. Heslot (1968) and Love and Malone (1967) found an increase or decrease of one or several pigments in mutants of rose and coleus and also detected peonin in some mutants which were absent in control. Kaicker et al. (1991) studied chromosome and morphological characters of rose cv. "Folklore" and its 12 induced mutants observed euploidy as most common with $2n = 28$ both in the control and its 12 mutants. Univalents, heteromorphic bivalents, quadri to octavalents, fragments, and micronuclei, in addition to the euploid number in flower mutants and chromosomal mosaicism were also of common occurrence. They have explained that all these factors are perhaps responsible for creating new colors and other morphological changes through the use of a low dose of gamma irradiation (2.5–5 krad) and EMS as a chemical mutagen.

Different other parameters selected for comparative studies between the original and mutant cultivars were plant height; branch, leaf, and flower-head number per plant, floret number per flower-head; the size of leaf, floret, and petals, flower-head and floret; the weight of flower-head and moisture content of the floret. Extensive efforts were made by the author to analyze the floret/petal pigments of a large number of original and mutant cultivars of chrysanthemum and rose by thin layer chromatographic and spectrophotometric methods (Datta 1986a, b, 1987a, b, 1990a, b, 1994a, b; Datta and Gupta 1983a, b, c).

Morphological features: Leaf width was significantly reduced in chrysanthemum mutant "Kapish" than the original cultivar "E-13" (Datta and Gupta 1981a, b, c). Significant increases in floret length and flower-head diameter and a reduction in the total number of florets per head were observed in "Himani" and "Sheela," mutants of "E-13" (Datta and Banerji 1986). Petiole length, flower-head diameter, and height, floret breadth were significantly decreased in "Hemanti," a gamma ray induced mutant of "Megami" (Datta and Gupta 1982a). Both increase and decrease in petiole size were recorded in some of the mutants (Datta and Gupta 1981a, 1982a). Floret number/head, the weight of the flower head, and floret width significantly decreased in one mutant ("Alankar") of "D-5" (Datta and Gupta 1981b). Floret number/head decreased significantly in "Tulika," a mutant of "M-24" (Datta et al. 1985). Floret number per head significantly reduced in "Kapish" and significantly more in "Maghi Yellow" over their respective control (Datta and Gupta 1981a, 1982a). Floret length significantly reduced in mutants "Subarna," "Flirt Mosaic," and "Sonali," but significantly increased in mutants "Alankar," "Agnisikha" "Himani," "Kapish," "Maghi Yellow," and "Sheela" over their respective control (Datta and Banerji 1986; Datta 1986a, b, 1988b, 2015). Floret width increased in mutants "Himani," "Kapish," "Man Bhawan," "Sheela," and "Subarna" and reduced in "Shabnam" over the control (Datta 1986a, b, 1987a, b, 2015). In another mutant analysis program, five original cultivars and 12 of their gamma ray induced mutants were selected and significant change in floret size was observed in some of the mutants over the original cultivar (Datta and Tandon 1994).

In rose, ten original and 15 gamma rays induced mutants were selected to analyze different morphological characters (spine and petal) and found that flower color changes were also associated with significant changes of some other biological characters in some cultivars. The spine character was found to be very stable and the size of the petals was sensitive. This study will be very helpful for the taxonomic characterization and identification of horticultural ornamental varieties (Datta

1986a, b, 1990a, b, 1999; Datta and Gupta 1983a, b, c; Datta and Singh 1999; Gupta et al. 1990a, b, 1993).

Anatomical features: Stomata size reduced significantly in some mutants (Fig. 13.1f). The length of palisade cells in "Basanti" (mutant of "E-13") reduced significantly ($P < 0.05$) (Datta and Gupta 1981a, b, 1982b, c).

Micromorphological features: Petal and leaf micromorphological studies showed noticeable differences among the control and mutant cultivars of chrysanthemum, rose, and *Lantana depressa* in cell shape, cell boundaries, cell wall surface, and striations in the petals (Fig. 13.1h). It indicated that flower color changes due to mutation were also associated with some changed micromorphology of the petal surface. This study indicated that the petal micromorphological characters can be utilized not only for identifying mutants but also for a correlation study that will help in the proper identification of different present-day cultivars of chrysanthemum and rose and their origin (Datta and Shome 1994).

Epidermal features: A very simple epidermal study on flower petals was conducted to focus on how petal color and epidermal pigmented cells can be correlated. Induced mutations in flower color can be easily noticed from its phenotypic expression. A simple experiment was conducted to note how epidermal peeling studies can assist to recognize mutations in flower color. Pigmentation form was studied from petal epidermal peelings of original and gamma ray induced mutants under a microscope. Flavonoids are dissolved in the cell sap and found in the epidermal cells and carotenoids are confined to plastids. Following original rose cultivars and their respective mutants (in parenthesis) were selected for such studies: "Arjun" ("Creamish white" and "Striped Arjun"), "First Prize" ("Light Pink Prize"), "Mrinalini" (Light Pink mutant), "Raja of Nalagarh" (Light Pink mutant), "Summer Holiday" (Pink Summer Holiday), "America's Junior Miss" ("Sukumari") "Contempo" ("Yellow Contempo," "Tangerine Contempo," "Striped Contempo," "Pink Contempo"), "Imperator" ("Pink Imperator," "Twinkle"), "Salmon Beauty" (Creamish yellow mutant), "Zorina" (Pink Zorina), "Queen Elizabeth" ("Sharada"), "Windy City" (Light Pink Mutant) (Table). Pigment features were examined and pigmented cells were counted from peelings/thin sections of petals from all the above-mentioned cultivars. Original and mutant cultivars had similar cell sizes and structures. All epidermal cells were deep orange in "Arjun" and greenish yellow in "Creamish White mutant." The striped mutant showed light orange cells in the pink portion and greenish-yellow cells in the striped sector. Pink cells were present in "First Prize" and "Mrinalini" and the cells were light pink in both their light pink mutants. 0.76% of cells were reddish violet and the remaining cells were deep orange in "Raja of Nalagarh" and all cells were very light pink in the light color mutant. "Summer Holiday" and its mutant had all cells orange except 2.48% and 2.54% of cells were reddish violets in the original and the mutant, respectively. The color intenseness was lighter in both types of mutant cells. Violet cells were mostly detected as a single unit or occasionally in a chain. All the cells were light pink in "America's Junior Miss" and mutant "Sukumari" had all nearly whitish cells. "Contempo" had all orange cells except 1.5% of cells were red. Yellow pigments induced in both the mutants "Yellow Contempo" and "Tangerine Contempo" were intracellular and no pigmented cells were found. Orange pigmented (but lighter than control) cells along with 0.5% red pigmented cells were recorded in "Striped Contempo." "Pink Contempo" mutant showed all dark orange cells along with 0.22% cells with red pigment. Cultivar "Imperator" had mostly deep orange cells, 3.5% of cells were violet, and 13.36% were cells where a portion of the cells was violet. The pink mutant had 1.14% complete light violet cells and the rest orange pigmented cells. Curious composition of pigmented cells was noticed in the mutant "Twinkle." Maximum deep orange cells were present in the original cherry red portion along with partly violet 3.68% cells and 9.10% complete violet cells. No violet cells were noticed in the stripe portion where all cells were light orange. Maximum violet-pigmented cells were noticed around the stripe portion. All cells in "Salmon Beauty" contained light orange pigment except 1.21% of cells with reddish pigment. In the "Creamish yellow mutant" all cells were creamish yellow. "Zorina" showed orange and 0.53% red pigmented cells. Mutant "Pink Zorina" had lighter orange cells and 0.71% red cells. All cells in "Queen Elizabeth" were creamish yellow and 1.57% reddish pigmented cells. Mutant "Sharada" had all lighter creamish yellow and lighter reddish (0.39%) pigmented cells. Epidermal cells were pink in miniature rose "Windy City" with 0.85% cells with dark red pigment. In its mutant, all the cells were light pink, and no red cells. No new pigmented cells were noticed in any mutant except in "Yellow Contempo" and "Tangerine Contempo." These easy and manageable epidermal studies will be very beneficial for an informative perspective on flower coloring before doing chemical analysis. The genetic makeup and inheritance of present-day rose pigments have not yet been resolved due to their reticulate evolution. All modern roses preserve the genes of many ancestors. Epidermal studies will give the initial intimation regarding the genetic composition of pigments in modern roses. This study will also be very useful for the taxonomic characterization of other horticultural ornamental varieties and in clarifying phonetic and evolutionary relationships and variation patterns, hybridization situations, evolutionary trends, and so on (Datta 1993, 2015).

Pollen grain features: Pollen grain sterility significantly increased in some of the chrysanthemum mutants ("Alankar"). Pollen grains were uniform ("normal") in most of the control ("D-5," "Himani," "Flirt") and mutant cultivars. Dimorphic ("small"/"big" and "normal") pollen grains were observed in some original chrysanthemum cultivars ("Megami," "Kingsford Smith," "Sharad Bahar") and in some of the mutant cultivars ("Hemanti," "Alankar," "Anamika," "Basanti," "Lohit," "Colchi Bahar," etc.; Datta and Banerji 1988; Datta and Gupta 1981c, 1982a). Pollen grain size, spine, and wall thickness significantly increased in some of the mutants (Datta and Gupta 1981a, b).

Mutagen treatment created the highest variations in exine surface ornamentations. Changes varied from mutant to mutant. The original cultivar ("D-5") had an exine surface a fosso-reticulate pattern with narrow muri and irregularly shaped lumina. These were modified to reticulate exine with broad muri and uniformly circular lumina in its mutant cvs. "Alankar" and "Agnishikha." The spine character also changed from straight to bent. The Scrobiculate wrinkled exine surface of "Kingsford Smith" altered to a scrobiculate wrinkled surface in its mutant "Rohit" along with a noticeable change in spine shape from a broad base to a bulbous base (Fig. 13.1i–l). Scrobiculate exine was recorded in the mutant ("Sheela") whereas its parent variety ("Himani) had a punctuate surface pattern. "Sharad Bahar" had reticulate exine with narrow muri, closely packed lumina, and broad base spine which abruptly taper to form acute tips. Its mutant ("Colchi Bahar") developed foveolate, broad muri, and undulated exine and conical-shaped spines (Datta and Datta 1998).

Changes in pollen grain characters were noticed after seed treatment with gamma rays in *Mesembryanthemum criniflorum* L.f. (Aizoaceae). The exine pattern changed from spinulose to microverrucate in 10 krad treatment (Chaturvedi et al. 1997). Such a study will provide additional information for understanding genetic control over pollen aperture and exine surface ornamentation which are of potential markers value in plant biosystematics. The changes due to mutation were perhaps induced at several independent loci controlling different characters and the loci are differentially sensitive to mutagenic treatment (Datta and Datta 1998; Datta et al. 2003).

Biochemical features: A good number of experiments have been carried out for the comparative study of pigments by thin-layer chromatographic (TLC) and spectrophotometric methods. Attempts were made to find out the biochemical differences in the distribution of spots for pigments in florets of different control and their respective induced mutants by thin-layer chromatographic methods. From the TLC analysis, it has been found that some of the spots for pigments were lighter/darker in the original cultivar than that of the mutant and vice versa. Some spots were found to be absent and some new spots were developed in some of the mutants indicating both qualitative and/or quantitative differences in pigments of original and mutant varieties (Datta and Gupta 1981b, c, 1982a, 1983a, b, c; Datta 1986a, b, 1987a, b, 1988b, 1997, 2000; Datta and Banerji 1991). These analyses indicated that somatic flower color changes in chrysanthemum and rose are due to both qualitative and/or quantitative changes in pigments as a result of mutation induced by gamma rays during the pigments biosynthesis pathway (Datta and Gupta 1983b).

The interpretations were drawn from the study of the following original and mutant cultivars:

Chrysanthemum: "Otome Zakura" and its mutants ("Purnima" and "Pitambar") (Datta and Gupta 1981c), "D-5" and its mutants ("Alankar" and "Agnisikha" (Datta and Gupta 1981b, c), "E-13" and five of its mutants ("Anamika," "Basanti," "Himani," "Kapish," "Lohit") (Fig. 13.2a) (Datta and Gupta 1981a), "Megami" and its mutant "Hemanti" (Fig. 13.2b) (Datta and Gupta 1982a), "Himani" and its mutant ("Sheela") (Datta 1985a), "Kingsford Smith" and its four mutants ("Taruni," "Rohit," "Lighter Fuschia purple" and "Erythrite red") (Fig. 13.2c) (Datta 1986a, b), "Flirt" and its "Man Bhawan" (Datta 1987b), "Sharad Bahar" and its colchicines induced mutant ("Colchi Bahar") (Datta 1987a; Datta and Banerji 1988).

Bougainvillea: "Partha" and its mutant "Arjuna" (Datta 1985b).

Rose: "Contempo" and its three flower color mutants ("Yellow Contempo," "Tangerine Contempo," "Pink Contempo") (Datta and Gupta 1982b, 1983a, b),

"Montezuma" and its mutant "Angara," "Pink Parfait" and its mutants ("Pink Parfait Lighter," "Pink Parfait Darker") (Datta and Gupta 1983a), "Queen Elizabeth" and mutant "Saroda," "America's Junior Miss" and its mutant "Sukumari" (Datta and Gupta 1982c, 1984), *Rosa damascene* and "Light Pink Mutant" (Datta 1986a, b). As an illustration petal pigments of the original rose cv. "America's Junior Miss" (pink) and its gamma ray induced mutant "Sukumari" (almost white) were analyzed by thin-layer chromatography and a total of nine spots were observed in the original cultivar, whereas spot numbers 3 and 4 were absent in the mutant. Spot number 9 was darker (due to more concentration) in the mutant in comparison to the control indicating both qualitative and quantitative differences in petal pigments between original and mutant cultivars (Datta and Gupta 1982c). Datta (1994b) studied the petal pigments of the original rose cultivar "Salmon Beauty" and its gamma ray induced mutant by spectrophotometric and TLC methods. TLC preparations of "Salmon Beauty" showed seven spots. In the mutant, spot number 1 was absent and also all the spots in the mutant were lighter in concentration than those of its original cultivar. Scanning of petal extracts of "Salmon Beauty" and its mutant in spectrophotometer showed three peaks. The concentration (i.e. O.D. values) of all the peaks in the mutant was very high over the original variety.

Extensive spectrophotometric studies on pigment analysis in original and gamma ray induced flower color mutants of chrysanthemum and rose indicated that somatic flower color mutations are due to qualitative and/or quantitative changes in the pigment/s as a result of mutation during pigment biosynthesis pathway (Datta 2015). Heslot (1968) studied the pigments of induced mutants and that of original cultivars of rose and found that usually the nature of pigments did not alter but the mutants showed either an increase or decrease of one or several of the pigments found in the control. The anthocyanin pigments in mutant and nonmutant Coleus plants have been studied by Love and Malone (1967) and they have reported that color differences between mutant and nonmutant plants are due to a variation in the amount of one anthocyanin pigment rather than a change in the structure of the pigment molecule itself.

Electrophoresis of crude proteins and enzyme extracts has been successfully used to establish control and mutant identities. Kuhns and Fretz (1978) have shown how by combining the results for several isozyme systems, a clear separation could be achieved between the rose cultivar "peace" and three of its sports "Chicago Peace," "Flaming Peace," and "Climbing Peace," a related seedling "Pink Peace" and its mutant "Candy Stripe." Protein electrophoresis has provided a new approach to the problems of species relationships. Fiebich and Henning (1992) used successfully isozyme analysis in the breeding of Chrysanthemum. Analysis of the electrophoretic pattern of soluble proteins in the original and mutant cultivars of chrysanthemum rose and *Lantana depressa* revealed the existence of variability for the number and intensity of protein bands between the original and mutants (Datta 1997).

Effects of mutation after treatment with gamma rays and chemical mutagens (dimethyl sulfate, ethyl methane sulphonate, hydroxylamine) were studied on rose pigments and found gradual loss of pigmentation (mostly anthocyanins) with an increase in irradiation dose of mutagens (Lata 1973). Dohare and Mathew (1991) studied pigment variations in four rose cultivars and their induced flower color mutants. Among the cultivars, "Raktagandha" contained the highest cyanidin content at all stages of development followed by "Jantar Mantar," "Raja Surendra Singh of Nalagarh" and "Neelambari." Cyanidin content was maximum at the bud stage and minimum in the petals of fully opened flowers in all the cultivars. A lighter color mutant was always associated with a reduction in cyanidin content. Kaicker (1990) analyzed anthocyanidin contents in rose cv. "Folklore" and its five gamma ray induced mutants. The amount of cyanidin was maximum in control and decreased considerably in induced mutants. Light flower color mutant associated with reduced cyanidin content was detected in roses (Dohare and Mathew 1991; Kaicker 1990; Kaicker et al. 1991).

RAPD: The author and his colleagues selected many gamma ray induced mutants of chrysanthemum and rose for molecular characterization. The objective was to find out the genetic diversity present in the mutant and original varieties. The special aim was also to trace out the molecular basis of somatic flower color mutations, i.e. how gamma ray induced morphological mutants (flower color/shape) can be identified by molecular markers. 21 mutant varieties of chrysanthemum were selected for this analysis along with their parental varieties. Studies distinguished parent and mutant cultivars and indicated that RAPDs could be useful to identify radio mutants of chrysanthemum (Fig. 13.2d) (Datta and Chatterjee 2006; Chatterjee et al. 2006). Six original rose varieties and their gamma ray induced mutants (in parenthesis) were used in RAPD analysis: "Contempo" ("Contempo New," "Contempo Pink," "Contempo Stripe," "Tangerine Contempo, "Yellow Contempo"), "Imperator" ("Twinkle," "Pink Imperator"), "First Prize" ("Light Pink Prize"), "America's Junior Miss" ("Sukumari"), "Sylvia" ("Sylvia White"), "Mrinalini" ("Mrinalini Lighter," "Mrinalini Stripe"). Most of the mutants appeared to be phenotypically the same as their parents, except for flower color. To clarify to what extent this phenotypical variation was related at the genetic level, the present RAPD analysis showed noticeable differences between parents and their mutants. This indicated that somatic flower color changes could have resulted due to some sort of genomic rearrangements rather than point mutations. The present RAPD analysis can be used not only for estimating genetic diversity present in different floricultural crops but also for the correct identification of mutant/new varieties for their legal protection under plant variety rights (Chakrabarty and Datta 2010). Trigiano et al. (1998) studied the relationship between induced mutants of the chrysanthemum "Charm" family. The cultivars differed only in flower color and were very difficult to distinguish by DAF but could be easily distinguished by ASAP (Arbitrary Signatures from Amplification Profiles). Lema-Ruminska et al. (2004) characterized ten radio mutants of chrysanthemum using RAPD. They studied and confirmed the utility of RAPD markers to identify chrysanthemum cultivars as well as to distinguish the radio mutants from the parents. Kumar et al. (2006) applied the RAPD technique to characterize 11 gamma ray induced radio mutants of two chrysanthemum cultivars. Their study revealed that RAPD molecular markers can be used to assess polymorphism among the radio mutants and can be a useful tool to supplement the distinctness, uniformity, and stability analysis for plant variety protection in the future. Teng et al. (2008) studied the genetic variation of regenerated plantlets in chrysanthemums following in vitro mutation using RAPD methods. Results showed that the genetic variation of generated plantlet was proportional to the dosage of gamma ray, while the 15 and 20 Gy treatments were not significantly different, which was consistent with the common conception that genetic variation of radio mutants was usually proportional to the dosage of mutagen within a certain range. They concluded that RAPD is a useful technique for the rapid and easy assessment of the genetic variation of mutants and may become a potential tool for the quick selection of mutants with great genetic variation during early growth stages. Flower doubleness as a breeding characteristic is of major importance in carnation (*Dianthus caryophyllus*) since flower architecture is of the utmost value in ornamentals. Based on the number of petals per flower, carnations are grouped into "single," "semi-double," and "double" flower types. These flower types are not easily distinguishable due to phenotypic overlaps. Scovel et al. (1998) identified an RAPD marker that was tightly linked to this recessive allele. The RAPD marker was cloned and used to generate an RFLP marker. This RFLP marker could discriminate with 100% accuracy between the semi-double and double flower phenotypes in carnations. The RAPD analysis can be used for the correct identification of mutant/new varieties for their legal protection under plant variety rights.

Cause of flower color mutation: Datta (1986a, b, 1997) from his extensive analysis of florets/petal pigment by TLC and spectrophotometric methods has suggested a schematic representation that explains the probable manner in which differences in the composition of pigments of original and mutant cultivars may arise. It has been clearly shown from analysis of a large number of chrysanthemum and rose mutants that somatic flower color mutation in ornamental plants will arise in four major directions, i.e. when there is any somatic flower color mutation, it has to follow any one pathway among the four represented in the scheme: (1) mutation may be due to either increase or decrease or both in the concentration of one or more existing pigments, (2) mutation may be due to blockage of one or more pigments synthesis and this may also be associated with increase or decrease in the concentration of one or more existing pigment/s, (3) mutation may be due to the origin of a new pigment which may be associated with increase or decrease in the concentration of one or more existing pigment/s, and (4) mutation may result in both synthesis of a new pigment and blocking of development of one or more existing pigment/s. This may be associated with either increase or decrease or both in the concentration of one or more existing pigment/s (Fig. 13.3) (Datta 1990a, b).

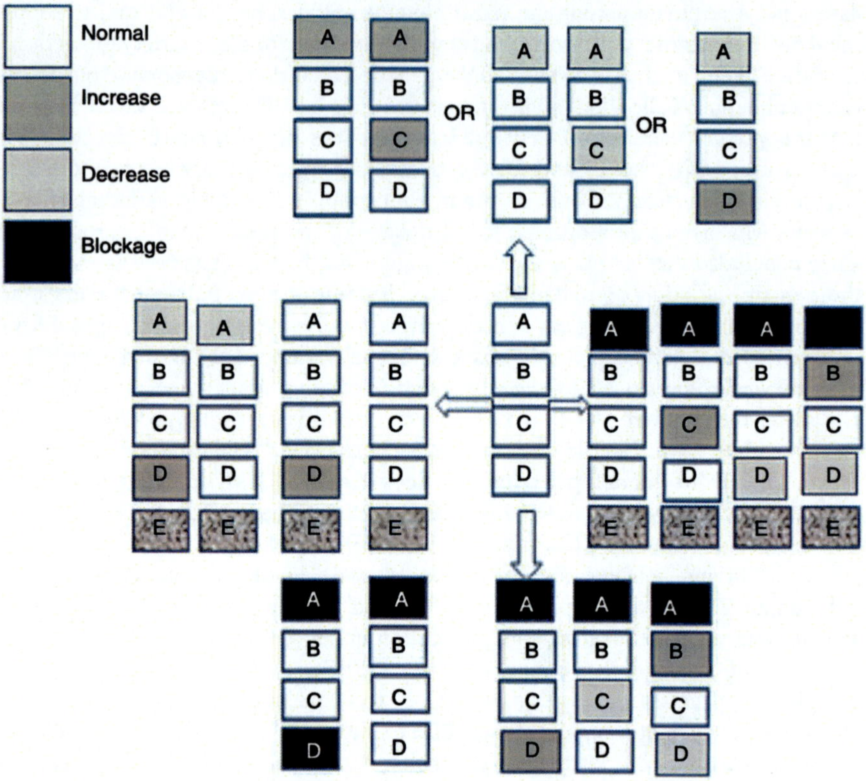

Fig. 13.3 Schematic representation of role of pigments in origin of mutants

References

Bowen HJM (1965) Mutations in horticultural chrysanthemum. Radiat Bot Suppl 5:695–700

Bowen HJM, Cawse PA, Dick MJ (1962) The induction of sports in chrysanthemum by gamma irradiation. Radiat Bot 1:297–303

Chakrabarty D, Datta SK (2010) Application of RAPD markers for characterization of gamma-ray-induced rose mutants and assessment of genetic diversity. Plant Biotechnol Rep 4:237–242

Chatterjee J, Mandal AKA, Ranade SA, Jaime A, da Silva T, Datta SK (2006) Molecular systematics in chrysanthemum x grandiflorum (Ramat.) Kitamura. Sci Hortic 110:373–378

Chaturvedi M, Datta K, Datta SK (1997) Pollen variation in gamma-irradiated Mesembryanthemum criniflorum L.F. (Aizoaceae). Taiwania 42(4):289–295

Datta SK (1985a) Gamma-ray induced mutant of a mutant chrysanthemum. J Nucl Agric Biol 14(4):131–133

Datta SK (1985b) Thin layer chromatographic study of bougainvillea cv. 'Partha' and it's gamma-ray induced mutant 'Arjuna'. J Nucl Agric Biol 13:140–141

Datta SK (1986a) New chrysanthemum cultivars and their spectrophotometric and TLC studies. Sci Cult 52:205–206

Datta SK (1986b) Thin layer chromatographic studies and evolution of somatic mutations in rose. The Indian Rose Annual V:41–48

Datta SK (1987a) 'Colchi Bahar'—a new chrysanthemum cultivar evolved by colchi mutation. The Chrysanthemum 43(1):40

Datta SK (1987b) 'Man Bhawan'—new chrysanthemum cultivar induced by gamma irradiation. J Nucl Agric Biol 16(4):217–218

Datta SK (1988a) Chrysanthemum cultivars evolved by induced mutations at National Botanical Research Institute, Lucknow. The Chrysanthemum 44(1):72–75

Datta SK (1988b) 'Agnishikha'—a new chrysanthemum cultivar evolved by gamma irradiation. Floriculture 9(11):10

Datta SK (1990a) Induction and analysis of somatic mutations in garden chrysanthemum. Adv Hortic For 1:241–254

Datta SK (1990b) Flower colour mutation. In: Plant mutation breeding for crop improvement, vol I. Proc of a Symp., Vienna, 18–22 June 1990, jointly organized by IAEA and FAO, pp 479–489

Datta SK (1993) Epidermal studies for detection of somatic flower colour mutation. J Ornam Hortic 1(2):6–11

Datta SK (1994a) Cytological interpretation of induced somatic flower colour mutation in garden chrysanthemum. In: Ghosh S (ed) Plant cytogenetics in India. Association for Cell and Chromosome Research, Dept. of Botany, University of Calcutta, Calcutta, pp 107–114

Datta SK (1994b) Induction and analysis of somatic mutation in Rose cultivar 'Salmon Beauty'. The Indian Rose Annual XII:79–85

Datta SK (1995) Induced mutation for plant domestication: *Lantana depressa*. Proc Indian Sci Acad B61(1):73–78

Datta SK (1997) Ornamental plants—role of mutation. Daya Publishing House, Delhi, p 219

Datta SK (1999) Flower colour analysis in garden roses: carotenoids. Sci Hortic 6:151–156

Datta SK (2000) Mutation studies on garden chrysanthemum—a review. In: Singh SP (ed) Scientific horticulture, vol 7. Scientific Publisher, Jodhpur, pp 159–199

Datta SK (2015) Indian floriculture: role of CSIR. Regency Publications, A Division of Astral International (P) Ltd., New Delhi

Datta SK (2019) Floriculture work at CSIR-National Botanical Research Institute, Lucknow. Sci Cult 85(7–8):274–283

Datta SK, Banerji BK (1986) 'Sheela'—a new chrysanthemum mutant of a mutant 'Himani'. Sci Cult 52(4):129–130

Datta SK, Banerji BK (1988) Analysis of 'Colchi Bahar'—a new chrysanthemum evolved by colchi mutation. J Indian Bot Soc 67(3&4):275–277

Datta SK, Banerji BK (1991) Analysis of induced somatic mutations in chrysanthemum. J Indian Bot Soc 70(I–IV):59–62

Datta SK, Chatterjee J (2006) Cytological and molecular analysis of gamma ray induced mutants of garden chrysanthemum. In: RKD, SC, GCS (eds) Perspectives in cytology and genetics. AICCG Publ. 12, pp 121–134

Datta K, Datta SK (1998) Palynological interpretation of gamma-ray and colchicine induced mutation in chrysanthemum cultivars. Israel J Plant Sci 46:199–207

Datta SK, Gupta MN (1981a) Cytomorphological, palynological and biochemical studies on control and gamma induced mutant of chrysanthemum cultivar 'E-13'. SABRAO J 134(2): 136–148

Datta SK, Gupta MN (1981b) Cytomorphological, palynological and biochemical studies on control and gamma induced mutant of chrysanthemum cultivar 'D-5'. The Chrysanthemum 34(4):193–200

Datta SK, Gupta MN (1981c) Studies on chrysanthemum cultivar 'Otome Zakura' and its mutants. Botanical Progress 4:88–92

Datta SK, Gupta MN (1982a) Hemanti—a new chrysanthemum cultivar evolved by gamma irradiation. Prog Hort 14(1):33–37

Datta SK, Gupta MN (1982b) Gamma-ray induced yellow flower mutants in rose cv. 'Contempo'. American Rose Annual, pp 36–37
Datta SK, Gupta MN (1982c) Gamma-ray induced white flower mutant in rose cv. Junior Miss. J Nucl Agric Biol 11(2):32–33
Datta SK, Gupta MN (1983a) Effect of gamma rays on pigment synthesis of rose cv. 'Contempo'. J Nucl Agric Biol 12(3):79–80
Datta SK, Gupta MN (1983b) Thin layer chromatographic and spectrophotometric analysis of flower colour mutations in roses. American Rose Annual, pp 102–106
Datta SK, Gupta MN (1983c) Soamtic flower colour mutation in chrysanthemum cv. 'D-5'. J Nucl Agric Biol 12(1):22–23
Datta SK, Gupta MN (1984) 'Sarda' and 'Sukumari'—new rose cultivars evolved by gamma irradiation. Sci Cult 50:200–201
Datta SK, Shome U (1994) Micromorphological studies of original and mutant cultivars of ornamentals. Feddes Repert 105(3–4):167–174
Datta SK, Singh MS (1999) Survey of phenolic compounds in leaves of garden roses: miniature cultivars. Sci Hortic 6:157–164
Datta SK, Tandon S (1994) Analysis of floret characters in chrysanthemum mutants. J Nucl Agric Biol 23(3):146–147
Datta SK, Banerji BK, Gupta MN (1985) 'Tulika'—a new chrysanthemum cultivar evolved by gamma irradiation. J Nucl Agric Biol 14(4):160
Datta K, Shukla R, Datta SK (2003) Effects of gamma irradiation in the context of palynological and cytological parameters on Narcissus tazetta cv. Cicily white. Cytologia 68(3):225–230
Dohare SR, Mathew V (1991) Mutation studies in garden roses. In: Golden Jubilee Symposium on Genetic Research and Education: current trends and the next fifty years, Feb. 12–15, New Delhi, pp 767–768
Dowrick GJ (1951) Sporting in chrysanthemums. Chrysanthemum:152–155
Dowrick GJ (1952) The chromosomes of the chrysanthemum. I. Heredity 6:365–376
Dowrick GJ (1953) The chromosomes of the chrysanthemum. III. Heredity 76:219–226
Dowrick GJ (1958) Chromosome numbers and the origin and nature of sports in the garden chrysanthemum. Natl. Chrysanthemum Soc, pp 60–69
Dowrick GJ, El-Bayoumi A (1966a) The origin of new forms of the garden chrysanthemum. Euphytica 15:32–38
Dowrick GJ, El-Bayoumi A (1966b) The induction of mutations in chrysanthemum using X- and gamma radiation. Euphytica 15:204–210
Fiebich D, Henning F (1992) Use of isozyme analysis inbreeding of chrysanthemum. Gartenbauwissenschaft 57:212–218
Fujii T, Mabuchi T (1961) Irradiation experiments with chrysanthemum. Seiken Hiho 12:40–44
Gupta MN, Nath P, Datta SK (1990a) Morphological analysis of induced mutant. The Indian Rose Annual VIII:56–62
Gupta MN, Nath P, Datta SK (1990b) Role of induced mutation on rose breeding. The American Rose XXX:17–18
Gupta MN, Nath P, Datta SK (1993) Induction and analysis of somatic mutations in rose cv. 'Zorina'. The Indian Rose Annual XI:123–128
Heslot H (1968) Mutation research done in barley, rose and marigolds. A progress report, mutation in plant breeding. IAEA, Vienna, pp 153–159
Ichikawa S, Yamakawa K, Sekiguchi F, Tatsuno T (1970) Variation in somatic chromosome number found in radiation-induced mutants in Chrusanhemum morifolium Hemsl. Cv. Yellow Delaware and Delaware. Radiat Bot 10:557–562
Kaicker US (1990) Anthocyanins in cv. Folklore and its mutants. The Indian Rose Annual VIII:63–69
Kaicker US, Kumar S, Singh SP (1991) Cytogenetics and rose breeding in India. In: Golden Jubilee Symp. on genetic research and education: current trends and the next fifty years, Feb. 12–15, New Delhi, pp 765–766

Kuhns LK, Fretz TA (1978) Distinguishing rose cultivar by polyacrylamide gel electrophoresis II. Isozyme variation among cultivars. J Am Soc Hort Sci 103:509–516

Kumar S, Prasad KV, Choudhary ML (2006) Detection of genetic variability among chrysanthemum radio mutants using RAPD markers. Curr Sci 90:1108–1113

Lata P (1973) Effects of ionizing radiation on roses with special reference to induction of mutations. Ph.D. thesis. Kanpur Univ., Kanpur, p 245

Lema-Ruminska J, Zalewska M, Sadoch Z (2004) Radiomutants of chrysanthemum (Dendranthema grandiflora Tzvelev) of the Lady group: RAPD analysis of the genetic diversity. Plant Breed 123(3):290–293

Love JE, Malone BB (1967) Anthocyanin pigments in mutants and nonmutants *Coleus* plants. Plant Radiat Bot 7:549–552

Nazeer MA (1981) Karyological studies in two species of chrysanthemum L. Cell Chrom Newsl 4: 33–35

Nazeer MA, Khoshoo TN (1982) Cytogenetical evaluation of garden chrysanthemum. Curr Sci 51(12):583–585

Nazeer MA, Khoshoo TN (1983) Variation in the chromosome complement of *Chrysanthemum morifolium* complex. Nucleus 26:22–29

Perry MC, McIntosh MS (1991) Geographical patterns of variation in the USDA soybean germplasm collections.1. Morphological traits. Crop Sci 31:1350–1355

Sampson DR, Walker GWR, Hunter AWS, Bragdo M (1958) Investigations on the sporting process in greenhouse chrysanthemums. Can J Plant Sci 38:346–356

Scovel G, Ben-Meir H, Ovadis M, Itzhaki H, Vainstein A (1998) RAPD and RFLP markers tightly linked to the locus controlling carnation (*Dianthus caryophyllus*) flower type. Theor Appl Genet 96(1):117–122. https://doi.org/10.1007/s001220050717

Stewart RN, Dermen H (1970) Determination of the number and mitotic activity of shoot apical initial cells by analysis of mericlinal chimeras. Am J Bot 57:816–826

Streanglar BB (1956) Origin and development of adventitious roots in stem cuttings of chrysanthemum, carnation and rose, vol 342. Cornell Exp. Stn, Ithaca, NY, pp 3–24

Teng N, Chen F, Jiang Z, Fang W, Chen T (2008) Detection of genetic variation by RAPD among chrysanthemum plantlets regenerated from irradiated calli. Acta Hortic 766:413–420. https://doi.org/10.17660/ActaHortic.2008.766.5

Trigiano RN, Scott MC, Caetano-Anolles G (1998) Genetic signatures from amplification profiles characterize DNA mutation in somatic and radiation-induced sports of chrysanthemum. J Am Soc Hort Sci 123(4):642–646

Walker GWR (1955) Chromosome numbers in chrysanthemum sports. In: Dep. Agric. Hortic. Div., Progr. Rep. 1949-1953, pp 69–70

Weaver GM (1963) The effect of Cesium 137 gamma radiation on plant growth and flower color of greenhouse chrysanthemum cultivars. Can J Genet Cytol 5:73–82

Part V

Conclusion

Conclusion of Mutation Work on Ornamentals in a Nutshell

14

Abstract

Mutation breeding at its present status appears to be well standardized, efficient, and cost-effective. Classical mutagenesis combined with management of chimera and in vitro mutagenesis are most promising and standardized techniques for developing new and novel varieties.

Keywords

Classical mutagenesis · In vitro mutagenesis · Biotechnology · TILLING · CRISPR

The book provides a synthesis of relevant information on the entire mutation technology which will be a source of reference for practical mutation breeding in vegetatively propagated ornamentals crops related to the floriculture industry.

Conventional breeding is a little problematic in vegetatively propagated crops. These group of plants are highly heterozygous and polyploid in nature which cause complexity in inheritance and genetic analysis. The mutation technique is very suitable and appropriate for this group of crops, especially for ornamentals.

There have been many remarkable improvements in mutation experimental procedures for the creation of new desired ornamental cultivars. Mutation breeding is now an important part of any crop improvement program. Mutation technology has already been justified by creating diversity in different ornamentals to meet consumer demand. It takes less time to create variability over conventional breeding technology. The floriculture industry is expecting the most effective technology package to get direct feedback at the grassroots level. Generally, breeding for ornamentals is focused on developing something new which is good for commercialization. In floriculture, it is very important to develop more and more new forms as quickly as possible. The variety which is in good demand in the market today will

S. Datta, *Role of Mutation Breeding in Floriculture Industry*,
https://doi.org/10.1007/978-981-99-5675-3_14

be old fashioned shortly. There is constant change in the floriculture trade for consumer demand. Consumers expect faster and better new cultivars. Consumers' expectations always motivate breeders to develop new forms in desired genera, species, and even specific cultivars. Therefore, breeders should be ready with the right technology to meet the changing demand. Now floriculture economy is more related to consumer tastes which shift from minute to minute. There is a need for coordination between nursery breeders and the market. For selective breeding, one can see that mutation technology plays an important role in developing new forms and thus remain an important component in industrial development.

Technology is a very important pillar in the floriculture market. Technology has a great role in maximum productivity if it can be used in the right way. Mutation technology can play a very crucial role in the development of the floriculture industry. It is very relevant to the floriculture industry since it can diversify both foliage and flower ornamentals. The efficiency of mutation technology is the most important component of industrial development. Earlier there had many drawbacks to mutation technology. Classical mutation technology is now well-equipped. The present status of mutation technology can help the floriculture industry to move forward as it has progressed a lot in the last few years. The mutation is now a very attractive tool to create genetic variability in ornamentals. Commercial breeding companies are using this technique for many specific ornamental plants. All accumulated knowledge on mutation work on different ornamentals will help to achieve targeted results more easily and in less time.

Ornamental plants are very suitable for mutation breeding work as the range of economic characteristics is wide which can be easily supervised after mutagen treatment. The high heterozygous nature of many ornamental species helps to detect the mutation in M1v1 and the isolation of mutant character by vegetative propagation. The heterozygous nature of many of the cultivars offers high mutation frequency. Mutation breeding has been most successful due to some additional advantages. Changes in any phenotypic characteristics like color, shape, or size of the flower and chlorophyll variegation in leaves can be easily detected. The main advantage of mutation work in vegetatively propagated ornamentals is the chances of changing one or a few characteristics of an otherwise outstanding cultivar without altering the remaining part of the genotype. The greatest advantage of mutation breeding lies in its ability to improve one or a few specific traits of the preferred variety. The mutated organ can be easily isolated through vegetative propagation.

Mutation breeding at its present status appears to be well standardized, efficient, and cost-effective. Classical mutagenesis combined with the management of chimera and in vitro, mutagenesis are the most promising and standardized techniques for developing new and novel varieties. At this stage it is possible to increase the rate of mutant development by combining the classical mutation breeding and in vitro technique. In vitro, mutagenesis experiments were started in chrysanthemums for inducing an increased number of mutants and solid mutants. The main advantage of this technique is to overcome chimera. Now, all spontaneous chimeras can be established as new varieties using in vitro chimera management technology. This will be a very helpful technique in floriculture nursery as a routine procedure for creating novelties.

Extensive work on chrysanthemums generated interesting knowledge and a huge number of mutants which may be followed by mutation researchers as an ideal model plant. It covers the entire mutation technology package.

All recent publications cover mostly routine research parameters which do not provide supportive data for the improvement of technology. These research results indicate that such efforts are good for individual scientists but far from the technical refinements. We must be sincere enough to sensitize ourselves about our research results, appropriate methodological perspectives, possibilities, and limitations of mutation technology. We must now realize the limitations and technically feasible measures should be taken with the best available mutation protocol for commercial exploitation. We should enrich ourselves with all basic knowledge and prepare a need base plan and apply the most suitable methods to improve a specific species considering ongoing and oncoming market demand.

The possibilities for creating different forms and improving ornamentals are infinite. Although the induction of mutation is uncertain, it has been very successful to develop desirable changes as per need in some ornamentals (Datta 2020, 2021, 2023a, b, c). Experimental results signify beyond doubt that induced mutation techniques can be utilized for the creation of new and novel cultivars of commercial importance through genetic variation.

It is admitted that all mutation-related technological assets have not been suitably documented and utilized in the right direction by all mutation breeders. Results are spread in different international journals, regional journals, local society journals, and also in some journals which are inaccessible due to language barrier. It is also difficult to estimate the exact contribution of classical mutation breeding in the development of new varieties from published review articles. It has been observed that the same example is cited in all reviews to highlight the prospect of mutation breeding. The actual number of mutants is more than the reported number. Some meaningful observations on routine mutation breeding work are cited. Many scientists work on important regional crops and all such experimental results are published in regional crop-specific literature which is not accessible to international forums. It is one of the notable handicaps that such regional work does not reflect total mutation work. Unfortunately, much regional technical information on the practical approach has not been properly documented, and also regional documented information could not get proper exposure to the world platform due to non-accessibility. One very significant example is many newly selected mutants are not released due to many reasons. The author isolated a good number of new varieties but not released them (Fig. 14.1). Such unaccounted unreleased mutants in different ornamentals throughout the world are maximum. Mutagenesis breeding for horticultural crops is valuable not only for creating new cultivars but also expanding the genetic pool for hybridization. Such mutants may play a very important role in a breeding program (Anter 2022). Mostly all sanctioned projects are for 3 years. The most vital point and serious drawback of any mutation breeding work is the fate of ongoing experiments/variants/mutants after the project tenure is over. Variants/

Fig. 14.1 New selections after gamma irradiation

mutants or even promising lines are lost after tenure is over because organizations do not have the provision of financial support for the maintenance of such important materials. Proper screening of regional literature is also very important to get the exact contribution of mutation technology. There is a need to mention that a wide range of circumstances that be partly responsible for the project failure—>inconsistent communication >poor planning >too small populations >not effective or too effective mutagen treatment >false selection >unsatisfactory selection techniques >false claims based upon improper starting material >out crossing >unnoticed contamination >wrong interpretation of results >premature publication of unconfirmed observations >afraid to publish some unexpected results—which is very unfortunate >unrealistic expectations >lack of cohesion between team members >poor monitoring and risk management, etc. In addition, ill-defined, unrealistic, or poorly communicated success criteria can lead to disappointing technology outcomes (Datta 2023a, b, c). Sincere efforts have been made to provide a synthesis of relevant information on the subject in this book. The mutation technique now has been more perfect and refined in the case of vegetatively propagated crops, especially in ornamentals. All minor technical details have been incorporated in this book to develop a complete mutation technology toolkit. The purpose of this "white paper" is to provide all technological advancements for mutation breeding project teams for more productivity. The success of mutation technology in the long term needs a technology management system so that innovations are not left to chance. Innovation management shapes the structures and framework conditions so that innovation potentials can be systematically identified, ideas generated, and then successfully implemented. It is very relevant to mention that despite the continued creation of novelties, research has declined due to limited funding projects on mutation breeding. Scientists have developed an interest to work on molecular techniques and therefore the development of mutant varieties has declined.

While highlighting and recommending the application of mutation breeding for the development of new ornamental varieties for the floriculture industry, a quick appraisal of other modern techniques is necessary.

Biotechnology has modernized target-oriented technology and research activities in the field of agriculture. Demanding and in-depth research in the area of molecular breeding has resulted in the development of efficient gene transfer methodology which has recovered engineered plants in different ornamental species. It provides a more direct approach to manipulating colors in the most desirable genetic background. The first genetically engineered ornamental crop having commercially desirable traits was reported in petunia (Meyer et al. 1987) and chrysanthemum (Courtney-Gutterson et al. 1993, 1994), both with changed flower color. Recent advances in molecular technology resulted in manipulating the concentration of flower color pigments in a highly controlled direction. Several important genes have been characterized and cloned for flower color, post-harvest attributes, flower shape, disease resistance, regulation of flowering, abiotic stress, etc. Interesting work on genetic engineering has been reported in *Alstroemeria, Antirrhinum*, *Dianthus, Chrysanthemum*, *Cyclamen*, *Gerbera*, *Gladiolus*, orchid, *Lilium, Petunia*, *Tulipa, Rose* (synthesis of blue gene), *Torenia* etc. which have been reviewed from time to time (Aida et al. 1999; Azadi et al. 2016; Bashandy and Teeri 2017; Benetka and Pavingerova 1995; Bolagam 2022; Burchi et al. 1996; Caldwell et al. 2004; Chandler and Sanchez 2012; Chandler and Tribe 2022; Christensen et al. 2008; Courtney-Gutterson et al. 1993; Da Silva et al. 2011; Firoozabady et al. 1994; Fukui et al. 2003; Datta 2005; Datta and Chakrabarty 2005; Handa 1992; Hossain et al. 2013; Jin et al. 2023; Li and Pei 2006; Martin and Gerats 1993; Matzke et al. 1989; Meyer 1991; Katsumoto et al. 2007; Kuehnle et al. 1993; Meyer et al. 1987; Mol et al. 1999; Norman et al. 2017; Ohtsubo 2011; Oud et al. 1995; Ono et al. 2006; Pellegrineschi et al. 1994; Robinson and Firoozabady 1993; Seiichi et al. 1995; Shibata 2008; Ochatt et al. 2022; Shimada et al. 2001; Su et al. 2019; Suzuki et al. 2000; Takatsu et al. 1999; Tanaka et al. 2005, 2009; Teixeira da Silva 2004; Tsuda et al. 1999, 2004, Van der Krol et al. 1988; Van der Krol and Vorst 1997; Vantuyl et al. 2018; Winefield et al. 1994; Woodson 1991; Woodson et al. 1990; Wu et al. 2021; Yagi 2018; Zuker et al. 1998, 2001; Huylenbroeck and Krishna Bhattarai 2022).

Other recent technology (TILLING, EcoTILLING, and EMAIL) made notable progress in mutation technology, especially in seed-propagated plants. It helps early screening of plants sheltering chemically (EMS, NMU) mutagenized genes of interest from a small population. Such technology with novel approaches made radical changes in plant genomics as it has greatly reduced the cost of sequencing the genome of an organism. Plants carrying mutations in known genes can now be quickly identified by reverse genetics methodologies (Wang et al. 2009; Ostergaard and Yanofsky 2004; Alonso and Ecker 2006). TILLING tool with its technical sophistication can only improve the efficiency of using induced mutations to develop crops with random mutagenized improved traits (Ibiza et al. 2010; Nieto et al. 2007; Till et al. 2003, 2004, 2006, 2007a, b, 2009, 2010; Wilde et al. 2012; Amri et al. 2018; McCallum et al. 2000a, b; Coe et al. 2018; Colbert et al. 2001; Greene et al. 2003; Gilchrist and Haughn 2005, 2010; Talame et al. 2008, 2009; Minoia et al. 2010; Guo et al. 2018; Henry et al. 2014; Laouar et al. 2018; Tsai et al. 2011; Uauy et al. 2009; Oleykowsky et al. 1998; Comai and Henikoff 2006; Caldwell et al. 2004; Comai et al. 2004; Slade and Knauf 2005; Cordeiro et al. 2006; Gilchrist et al. 2006a, b; Mejlhede et al. 2006; Sato et al. 2006; Cross et al. 2008; Lee et al. 2009; Martin et al. 2009; Yi et al. 2009; Szarejko and Maluszynski 2011; Chawade et al. 2010; Stemple 2004; Perry et al. 2003; Ritchie and Nielsen 2006; Xin et al. 2008; Gady et al. 2009; Kurowska et al. 2011; Bradley et al. 2007; Cooper et al. 2008; Suzuki et al. 2008; Wang et al. 2008; Gottwald et al. 2009; Ritchie and Nielsen 2006; Schmitt et al. 2012; Simsek and Kacar 2010; Slade et al. 2005; Stephenson et al. 2010; Szurman-Zubrzycka et al. 2018; Szarejko et al. 2017).

Another very important technique is the CRISPR/Cas9 system which is a very impressive and encouraging tool for targeted gene mutagenesis. CRISPR/Cas9 technology will be suitable to develop desirable traits in ornamental crops through genome editing. This will help to increase the diversity in the economic characteristics of ornamental plants. The work reports at present are limited. It has generated basic knowledge on different aspects like color modification, self-life of flowers, flower development, flower qualities, etc. It will be a popular tool for the improvement of ornamental plants through modification. Nitarska et al. (2021) reported for the first time successful targeted mutagenesis with the CRISPR/Cas9 system in the commercially important poinsettias. They applied CRISPR/Cas9 in the red flowering poinsettia (*Euphorbia pulcherrima*) cultivar "Christmas Eve" to knock out the *flavonoid 3′-hydroxylase* (*F3′H*) and to develop plants with orange bract color. Even though F3′H was not completely inactivated, the bract color of transgenic plants changed from vivid red (RHS 45B) to vivid reddish orange (RHS 33A), and cyanidin levels decreased significantly compared with the wild type. The CRISPR-Cas9 system has created a lot of adventure in the scientific community because it is faster, cheaper, more accurate, and more efficient than other genome editing methods (Ahn et al. 2020; Cong et al. 2013; Deveau et al. 2008; Doudna and Charpentier 2014; Giovannini et al. 2021; Gupta and Musunuru 2014; Hahne et al. 2019; Hischler et al. 2016; Hsu et al. 2014; Jansen et al. 2002; Kishi-Kaboshi et al. 2017; Komor et al. 2017; Kui et al. 2017; Laforest and Nadakuduti 2022; Li et al. 2020, 2021; Mali et al. 2013; Mostafa 2021; Nishihara et al. 2018; Nitarska et al. 2021; Sarmast 2020; Sarmast and Janati 2019; Sirohi et al. 2022; Schaeffer and Nakat 2015; Semiarti et al. 2020; Shan et al. 2020; Subburaj et al. 2016; Shibuya et al. 2018; Sun and Kao 2018; Tasaki et al. 2019, 2020; Tong et al. 2020; Watanabe et al. 2017, 2018; Xie and Yang 2013; Xu et al. 2020, 2021; Yan et al. 2019; Yu et al. 2021; Zhang et al. 2015, 2016).

The molecular technique is expected to develop variety as per desire but it will take more time to develop variety-specific technology. The exact understanding of the function and interaction of individual genes is a major prerequisite for the successful application of genetic engineering in the development of new flower colors. Like other molecular breeding techniques, CRISPR/Cas9-based genome editing technique has some drawbacks at present. In CRISPR/Cas9 system plant tissue culture-based gene transformation is a crucial step. *Agrobacterium*-mediated gene transformation has not yet been standardized for all important ornamentals. For the routine application of genetic engineering to develop desirable ornamental varieties for the floriculture trade the most essential steps are to be standardized—suitable reproducible regeneration system not only for genus and species but also at each variety level; powerful transformation system; selection of desirable genes. Management of polygenic characters is another constraint for different ornamental characters. There are many reports of the variation and instability of genes in transgenic plants. Transferred gene can lose its transcriptional activity after crossing into the next generation or after the introduction of another foreign construct into such a transgenic line (c.f. Datta and Chakrabarty 2005; Matzke et al. 1989; Ohtsubo 2011). All molecular process is lengthy. The future scope of improving ornamentals using molecular breeding is unlimited and exciting. But it has just enabled researchers to start unraveling the molecular details of genes of commercial novelties. Technology has not yet been simplified so that general floriculture scientists, florists, and nurserymen can utilize it to create new desired ornamental varieties. High costs, sophisticated instruments, and overall highly skilled manpower are some other barriers that restrict the economic viability for routine application on different ornamentals which contribute differentially to the floriculture industry.

The mutation technique is quicker than other molecular techniques. Both basic and applied mutation work have now established beyond doubt that mutation breeding will constitute an excellent supplement to conventional methods in practice. The mutation technique now has been more perfect and refined in the case of vegetatively propagated crops, especially in ornamentals. The mutation technology can be redesigned based on various success stories such as professional knowledge and technology skills. All minor technical details may be assimilated into the mutation technology package. The present mutation breeding package covers both traditional and biotechnological components (in vitro management). One should start mutation work with the full package. At this stage, the classical mutation technique combined with the management of chimera and in vitro mutagenesis is the most promising and standardized technique for developing new, novel, and desired ornamental varieties (Datta 1995, 2005, 2006, 2009a, b, 2010, 2012, 2014, 2023a, b, c).

No doubt we have generated voluminous information, but considering the past time frame, scientific manpower, and money we need a changed mindset, management with vision, right research topics that would help the scientists to unlock the creative potential in the right direction. We can stop some of our routine activities and we must identify our routine gap-filling activities. Because the majority of such activities have already been explored and are not in synchronization with their practical application.

We must introspect and review our approach and policies carefully and implement whatever correction is needed to redeem mutation technology into a leadership position in the floriculture industry. Most of the recent publications in the fields of S&T research seem to be deficient in innovative components for the improvement of technology. Shortcomings in this quality factor should be a matter of concern to all mutation breeders. If we change some of our routine activities and concentrate on the full technology package we can achieve our goal of developing desirable varieties. A proper approach is required which will shorten the research and help the development of the correct path which may allow mutation breeders to leapfrog to pull mutation technology to industrial status within a reasonable time.

Many simple technical modifications of technology can improve efficiency. Various outcomes of the technology should be reviewed and examined at various stages to reschedule the future technology package. Floriculture-based technology does not have a life cycle. It can always be re-structured based on the enrichment of ideas/experience and need base. Knowledge sharing is perhaps the most important aspect of mutation breeding technology.

We must agree that every innovation has the probability of failure. There are numerous examples of innovations that started with great potential and wound up as dismal failures. Mutation technique for crop improvement is a never-ending science. Mutation technology has yet to achieve its full capability and contribution to inclusive global productivity.

References

Ahn CH, Ramya M, An HR, Park PM, Kim Y-J, Lee SY, Jang S (2020) Progress and challenges in the improvement of ornamental plants by genome editing. Plan Theory 9:687. https://doi.org/10.3390/plants9060687

Aida R, Hirose Y, Kishimoto S, Shibata M (1999) *Agrobacterium tumefaciens*-mediated transformation of *Cyclamen persicum* Mill. Plant Sci 148:1–7

Alonso J, Ecker J (2006) Moving forward in reverse: genetic technologies to enable genome-wide phenomic screens in *Arabidopsis*. Nat Rev Genet 7:524

Amri WT, Laouar M, Abdelguerfi A, Jankowicz-Cieslak J, Till BJ (2018) Genetic variability induced by gamma rays and preliminary results of low-cost TILLING on M2 generation of chickpea (*Cicer arietinum* L.). Front Plant Sci Sect Plant Breed 9. https://doi.org/10.3389/fpls.2018.01568

Anter AS (2022) Induced mutation to enhance plant biodiversity and genetic resources for intensification of crop production to mitigate climatic changes. In: Çalişkan M, Aydin S (eds) Genetic diversity. https://doi.org/10.5772/intechopen.108117

Azadi P, Bagheri H, Nalousi AM, Nazari F, Chandler SF (2016) Current status and biotechnological advances in genetic engineering of ornamental plants. Biotechnol Adv 34:1073–1090

Bashandy H, Teeri TH (2017) Genetically engineered orange petunias on the market. Planta 246: 277–280. https://doi.org/10.1007/s00425-017-2722-8

Benetka V, Pavingerova D (1995) Phenotypic difference in transgenic plants of chrysanthemum. Plant Breed 114:169–173

Bolagam RK (2022) Biotechnological interventions in floriculture. Pharma Innov J 11(7): 1260–1266. ISSN (E): 2277-7695 ISSN (P): 2349-8242

Bradley T, Till J, Cooper J, Tai TH, Colowit P, Greene EA, Henikoff S, Comai L (2007) Discovery of chemically induced mutations in rice by TILLING. BMC Plant Biol 7:19

Burchi G, Mercuri A, De Benedetti L, Glovannini A (1996) Transformation methods applicable to ornamental plants. Plant Tissue Cult Biotechnol 2:94–104

Caldwell DG, McCallum N, Shaw P, Muehlbauer GJ, Marshall DF, Waugh R (2004) A structured mutant population for forward and reverse genetics in barley (*Hordeum vulgare* L.). Plant J 40: 143–150

Chandler SF, Sanchez C (2012) Genetic modification; the development of transgenic ornamental plant varieties. Plant Biotechnol J 10:891–903

Chandler SF, Tribe D (2022) Modern techniques for plant breeding in ornamentals (Chapter 17). In: Datta SK, Gupta YC (eds) Floriculture and ornamental plants, handbooks of crop diversity: conservation and use of plant genetic resources, pp 523–555. https://doi.org/10.1007/978-981-15-3518-5_19

Chawade A, Sikora P, Brautigam M, Larsson M, Vivekanand V, Nakash MALL et al (2010) Development and characterization of an oat TILLING-population and identification of mutations in lignin and beta-glucan biosynthesis genes. BMC Plant Biol 10:86

Christensen B, Sriskandarajah S, Serek M, Muller RPB (2008) Transformation of *Kalanchoe blossfeldiana* with rol-genes is useful in molecular breeding towards compact growth. Plant Cell Rep 27(9):1485–1495. https://doi.org/10.1007/s00299-008-0575-0

Coe RA, Chatterjee J, Acebron K, Dionora J, Mogul R, Lin HC, Yin X, Bandyopadhyay A, Sirault XRR, Furbank RT et al (2018) High-throughput chlorophyll fluorescence screening of *Setaria viridis* for mutants with altered CO_2 compensation points. Funct Plant Biol 45:1017–1025

Colbert T, Till BJ, Tompa R, Reynolds S, Steine MN, Yeung AT, McCallum CM, Comai L, Henikoff S (2001) High-throughput screening for induced point mutations. Plant Physiol 126: 480–484

Comai L, Henikoff S (2006) TILLING : practical single-nucleotide mutation discovery. Plant J 45: 684–694

Comai L, Young K, Till BJ, Reynolds SH, Greene EA, Codoma CA, Enns LC, Johnson JE, Burtner C, Odden AR, Henikoff S (2004) Efficient discovery of DNA polymorphism in natural populations by Ecotilling. Plant J 37:778–786

Cong L, Ran FA, Cox D et al (2013) Multiplex genome engineering using CRISPR/Cas systems. Science 339(6121):819–823

Cooper JL, Till BJ, Laport RG, Darlow MC, Kleffner JM, Jamai A et al (2008) TILLING to detect induced mutations in soybean. BMC Plant Biol 8(1):9

Cordeiro G, Eliott FG, Henry RJ (2006) An optimized ecotilling protocol for polyploids or pooled samples using a capillary electrophoresis system. Anal Biochem 355:145–147

Courtney-Gutterson N, Firoozabady E, Lemiex C, Nicholas J, Morgan A, Robinson K, Otten A, Akerboom M (1993) Production of genetically engineered colour modified chrysanthemum plants carrying a homologous chalecone synthase gene and their field performance. Acta Hortic 336:57–62

Courtney-Gutterson N, Napoli C, Lemieux C, Morgan A, Firoozababy E, Robinson KEP (1994) Modification of flower color in Florist's *Chrysanthemum*: production of a white-flowering variety through molecular genetics. Biotechnology 12:268–271

Cross MJ, Waters DLE, Lee LS, Henry RJ (2008) Endonucleolytic mutation analysis by internal labeling (EMAIL). Electrophoresis 29:1291–1301

Da Silva JAT, Chin DP, Van PT, Mii M (2011) Transgenic orchids. Sci Hortic 130:673–680

Datta SK (1995) Role of mutation breeding in inducing desired genetic diversity in ornamentals for the floriculture trade. In: Proc. Symp. induced mutations and molecular techniques for crop improvement, jointly organized by IAEA and FAO, 19–23, pp 670–672

Datta SK (2005) Technological advances for the development of new ornamental varieties. In: Trivedi PC (ed) Advances in biotechnology. Agrobios, India, pp 285–304

Datta SK (2006) Application of biotechnology for improvement of ornamentals. In: Bhattacherjee SK (ed) Advances in ornamental horticulture. Pointer Publishers, Jaipur, pp 96–104

Datta SK (2009a) A report on 36 years of practical work on crop improvement through induced mutagenesis. In: Shu QY (ed) Induced plant mutations in the genomics era. Joint FAO/IAEA Division of Nuclear Techniques in Food and Agriculture, International Atomic Energy Agency, Vienna, pp 253–256

Datta SK (2009b) Role of classical mutagenesis for development of new ornamental varieties. In: Shu QY (ed) Induced plant mutations in the genomics era. Joint FAO/IAEA Division of Nuclear Techniques in Food and Agriculture, International Atomic Energy Agency, Vienna, pp 300–302

Datta SK (2010) Induced mutagenesis for the development of new varieties in vegetatively propagated ornamentals. In: NAARRI international conference—2010, isotope technologies and applications—new horizons, December 13–15, Mumbai, pp 227–230

Datta SK (2012) Success story of induced mutagenesis for development of new ornamental varieties. In: Bioremediation, biodiversity, and bioavailability 6 (Special Issue I). Global Science Books, pp 15–26

Datta SK (2014) Induced mutagenesis: basic knowledge for technological success. In: Tomlekova NB, Kozgar ML, Wani MR (eds) Mutagenesis: exploring genetic diversity of crops. Wageningen Academic Publishers, pp 95–137

Datta SK (2020) Induced mutations: technological advancement for development of new ornamental varieties. Nucleus 63:119–129

Datta SK (2021) Breeding of ornamentals: success and technological status. Nucleus 65:107. https://doi.org/10.1007/s13237-021-00368-x

Datta SK (2023a) Induced mutation breeding. Springer. https://doi.org/10.1007/978-981-19-9489-0. ISBN 978-981-19-9488-3

Datta SK (2023b) Technology package for induced mutagenesis. J Biol Nat 15(1):70–88. Article no.JOBAN.11268 ISSN: 2395-5376 (P), ISSN: 2395-5384 (O), (NLM ID: 101679666)

Datta SK (2023c) Induced mutagenesis in chrysanthemum. In: Penna S, Jain SM (eds) Mutation breeding for sustainable food production and climate resilience. Springer, Singapore. https://doi.org/10.1007/978-981-16-9720-3_19

Datta SK, Chakrabarty D (2005) Classical mutation breeding and molecular methods for genetic improvement of ornamentals. In: Datta SK (ed) Role of classical mutation breeding in crop improvement. Daya Publishing House, Delhi, pp 260–303

Deveau H, Barrangou R, Garneau JE, Labonte J, Fremaux C, Boyaval P (2008) Phage response to CRISPR-encoded resistance in *Streptococcus thermophilus*. J Bacteriol 190:1390–1400. https://doi.org/10.1128/JB.01412-07

Doudna JA, Charpentier E (2014) Genome editing. The new frontier of genome engineering with CRISPR-Cas9. Science 346:1258096. https://doi.org/10.1126/science.1258096

Firoozabady E, Moy Y, Courtney-Gutterson N, Robinson K (1994) Regeneration of transgenic rose *(Rosa hybrida)* plants from embryogenic tissue. Biotechnology 12:609–613

Fukui Y, Tanaka Y, Kusumi T, Washita I, Nomoto K (2003) A rationale for the shift in colour towards blue in transgenic carnation flowers expressing the flavonoid 3′ 5′ hydroxylase gene. Phytochemistry 63:15–23

Gady ALF, Hermans FWK, de Wal MHBJV, van Loo EN, Visser RGF, Bachem CWB (2009) Implementation of two high through-put techniques in a novel application: detecting point mutations in large EMS mutated plant populations. Plant Methods 5:13

Gilchrist EJ, Haughn GW (2005) TILLING without a plough: a new method with applications for reverse genetics. Curr Opin Plant Biol 8:211–215

Gilchrist E, Haughn G (2010) Reverse genetics techniques: engineering loss and gain of gene function in plants. Brief Funct Genomics 9(2):103–110

Gilchrist EJ, O'Neil NJ, Rose AM, Zetka MC, Haughn GW (2006a) TILLING is an effective reverse genetics technique for Caenorhabditis elegans. BMC Genomics 7:262

Gilchrist EJ, Haughn GW, Ying CC, Otto SP, Zhuang J, Cheung D et al (2006b) Use of Ecotilling as an efficient SNP discovery tool to survey genetic variation in wild populations of Populus trichocarpa. Mol Ecol 15(5):1367–1378

Giovannini A, Laura M, Nesi B, Savona M, Cardi T (2021) Genes and genome editing tools for breeding desirable phenotypes in ornamentals. Plant Cell Rep 40:461–478. https://doi.org/10.1007/s00299-020-02632-x

Gottwald S, Bauer P, Komatsuda T, Lundqvist U, Stein N (2009) TILLING in the two-rowed barley cultivar 'Barke' reveals preferred sites of functional diversity in the gene HvHox1. BMC Res Notes 2(1):258

Greene EA, Codomo CA, Taylor NE, Henikoff JG, Till BJ, Reynolds SH, Enns LC, Burtner C, Johnson JE, Odden AR, Comai L, Henikoff S (2003) Spectrum of chemically induced mutations from a large scale reverse-genetic screen in Arabidopsis. Genetics 164:731–740

Guo T, Wang H, Luo W, Chen Z (2018) Analysing the spectrum of spaceflight-induced mutagenesis via high throughput sequencing. In: FAO/IAEA international symposium on plant mutation breeding and biotechnology, 27–31 August, Vienna, Austria, Abstract No. IAEA-CN-263-46

Gupta RM, Musunuru K (2014) Expanding the genetic editing tool kit: ZFNs, TALENs, and CRISPR-Cas9. J Clin Invest 124(10):4154–4161. https://doi.org/10.1172/JCI72992

Hahne G, Tomlinson L, Nogué F (2019) Precision genetic engineering tools for next-generation plant breeding. Plant Cell Rep 38:435–436. https://doi.org/10.1007/s00299-019-02400-6

Handa T (1992) Genetic transformation of *Antirrhinum majus* L. and inheritance of altered phenotype induced by Ri- T-DNA. Plant Sci 81:199–206

Henry IM, Nagalakshmi U, Lieberman MC, Ngo KJ, Krasileva KV, Vasquez-Gross H, Akhunova A, Akhunov E, Dubeovsky J, Tai TH, Comai L (2014) Efficient genomic-wide detection and cataloging of EMS-induced mutations using exome capture and next-generation sequencing. Plant Cell 26:1382. https://doi.org/10.1105/tpc.113.121590

Hischler J, Bürstmayr H, Stoger E (2016) Targeted modification of plant genomes for precision crop breeding. Biotechnol J 11:1–14. https://doi.org/10.1002/biot.201600173

Hossain MM, Kant R, Van PT, Winarto B, Zeng S, Jaime A, da Silva T (2013) The application of biotechnology to orchids. Crit Rev Plant Sci 32(2):69–139. https://doi.org/10.1080/07352689.2012.715984

Hsu PD, Lander ES, Zhang F (2014) Development and applications of CRISPR-Cas9 for genome engineering. Cell 157(6):1262–1278. https://doi.org/10.1016/j.cell.2014.05.010

Huylenbroeck JV, Krishna Bhattarai K (2022) Ornamental plant breeding: entering a new era? Ornam Hortic 28(3):297–305. https://doi.org/10.1590/2447-536X.v28i3.2516

Ibiza VP, Cañizares J, Nuez F (2010) EcoTILLING in capsicum species: searching for new virus resistances. BMC Genomics 11:631

Jansen R, van Embden JDA, Gaastra W, Schouls LM (2002) Identification of genes that are associated with DNA repeats in prokaryotes. Mol Microbiol 43(6):1565–1575

Jin C, Dong L, Wei C, Wani MA, Yang C, Li S, Li F (2023) Creating novel ornamentals via new strategies in the era of genome editing. Front Plant Sci 14:1142866. https://doi.org/10.3389/fpls.2023.1142866

Katsumoto Y, Mizutani MF, Fukui Y, Brugliera F, Holton TA, Karan M et al (2007) Engineering of the rose flavonoid biosynthetic pathway successfully generated blue-hued flowers accumulating delphinidin. Plant Cell Physiol 48(11):1589–1600

Kishi-Kaboshi M, Aida R, Sasaki K (2017) Generation of gene-edited *Chrysanthemum morifolium* using multicopy transgenes as targets and markers. Plant Cell Physiol 58:216–226. https://doi.org/10.1093/pcp/pcw222

Komor AC, Badran AH, Liu DR (2017) CRISPR-based technologies for the manipulation of eukaryotic genomes. Cell 169(3):559. https://doi.org/10.1016/j.cell.2017.04.005

Kuehnle AR, Chen FC, Jaynes JM (1993) Engineering bacterial blight resistance into Anthurium. In: Schiva T, Mercuri A (eds) Creating genetic variation in ornamentals. Proc. XVIIth Eucarpia Symposium, Sanremo, Italy, pp 127–129

Kui L, Chen H, Zhang W, He S, Xiong Z, Zhang Y, Yan L, Zhong C, He F, Chen J (2017) Building a genetic manipulation tool box for orchid biology: identification of constitutive promoters and application of CRISPR/Cas9 in the orchid, Dendrobium officinale. Front Plant Sci 7:2036. https://doi.org/10.3389/fpls.2016.02036

Kurowska M, Daszkowska-Golec A, Gruszka D, Marzec M, Szurman M, Szarejko I, Maluszynski M (2011) TILLING—a shortcut in functional genomics. J Appl Genet 52(4):371

Laforest LC, Nadakuduti SS (2022) Advances in delivery mechanisms of CRISPR gene-editing reagents in plants. Front Genome Ed 4:830178. https://doi.org/10.3389/fgeed.2022.830178

Laouar M, Tabti D, Jankowicz-Cieslak J, Till BJ, Tilioulne W (2018) Early assessment of lentil and chickpea mutant and evaluation of low cost TILLING on M2 chickpea. In: FAO/IAEA international symposium on plant mutation breeding and biotechnology, 27–31 August, Vienna, Austria, Abstract No. IAEA-CN-263-269

Lee LS, Cross MJ, Henry RJ (2009) EMAIL—a highly sensitive tool for specific mutation detection in plant improvement programmes. In: Shu QY (ed) Induced plant mutations in the genomics era. Food and Agriculture Organization of the United Nations, Rome, pp 243–244

Li Y, Pei Y (2006) Plant biotechnology in ornamental horticulture. Haworth Food and Agricultural Products Press, New York

Li J, Li H, Chen J, Yan L, Xia L (2020) Toward precision genome editing in crop plants. Mol Plant 13:811–813. https://doi.org/10.1016/j.molp.2020.04.008

Li C, Brant E, Budak H, Zhang B (2021) CRISPR/Cas: a Nobel Prize award-winning precise genome editing technology for gene therapy and crop improvement. J Zhejiang Univ Sci B 22: 253–284. https://doi.org/10.1631/jzus.B2100009

Mali P, Esvelt KM, Church GM (2013) Cas9 as a versatile tool for engineering biology. Nat Methods 10:957–963

Martin C, Gerats T (1993) The control of flower colouration. In: Jordan BR (ed) The molecular biology of flowering. CAB International, Wallingford, pp 219–256

Martin B, Ramiro M, Martinex-Zapater JM, Alonso-Blanco C (2009) A high-density collection of EMS-induced mutations for TILLING in Landsberg erecta genetic background of *Arabidopsis*. BMC Plant Biol 9:147

Matzke MA, Primig M, Trnovsky J, Matzke AJM (1989) Reversible methylation and inactivation of marker genes in sequentially transformed tobacco plants. EMBO J 8(3):643–649

McCallum CM, Comai L, Greene EA, Henikoff S (2000a) Targeting induced local lesions in genomes (TILLING) for plant functional genomics. Plant Physiol 123:439

McCallum CM, Comai L, Greene EA, Henikoff S (2000b) Targeted screening for induced mutations. Nat Biotechnol 18(4):455–457

Mejlhede N, Kyjovska Z, Backes G, Burhenne K, Rasmussen SK, Jahoor A (2006) EcoTILLING for the identification of allelic variation in the powdery mildew resistance genes mlo and Mla of barley. Plant Breed 125(5):461–467

Meyer P (1991) Engineering of novel flower colours. In: Singh F, Mol JNM (eds) Genetics and breeding of ornamental species. Kluwer Academic Publishers, Dordrecht, pp 285–307

Meyer P, Heidmann I, Forkmann G, Saedler H (1987) A new petunia flower colour generated by transformation of a mutant with a maize gene. Nature 330:677–678

Minoia S, Petrozza A, D'Onofrio O, Piron F, Mosca G, Sozio G, Cellini F, Bendahmane A, Carriero F (2010) A new mutant genetic resource for tomato crop improved by TILLING technology. BMC Res Notes 3(1):69

Mol J, Grotewold E, Koes R (1999) How genes paint flowers. In: Altman A et al (eds) Plant biotechnology and in vitro biology in the 21st century. Kluwer Academic Publishers, Dordrecht, pp 597–600

Mostafa KS (2021) CRISPR/Cas, a powerful tool for genome editing in ornamental plants. Flower Ornam Plants 6(1):37–48

Nieto C, Piron F, Dalmais M, Marco CF, Moriones E, Gómez-Guillamón ML, Truniger V, Gómez P, Garcia-Mas J, Aranda MA, Bendahmane A (2007) EcoTILLING for the identification of allelic variants of melon eIF4E, a factor that controls virus susceptibility. BMC Plant Biol 7: 34

Nishihara M, Higuchi A, Watanabe A, Tasaki K (2018) Application of the CRISPR/Cas9 system for modification of flower color in *Torenia fournieri*. BMC Plant Biol 18:331. https://doi.org/10.1186/s12870-018-1539-3

Nitarska D, Boehm R, Debener T, Lucaciu RC, Halbwirth H (2021) First genome edited poinsettias: targeted mutagenesis of flavonoid 3′-hydroxylase using CRISPR/Cas9 results in a colour shift. Plant Cell Tissue Organ Cult 147(1):49–60. https://doi.org/10.1007/s11240-021-02103-5

Norman A, Aqeel M, Deng J, Khalid N, Sanaullah T, Shuilin H (2017) Biotechnological advancements for improving floral attributes in ornamental plants. Front Plant Sci 8:530. https://doi.org/10.3389/fpls.2017.00530

Ochatt SJ, Beruto MI, Chan M-T, Prestwich BMD, Eimert K, Lambardi M, Winkelman T (2022) Biotechnology of ornamental plants: when beauty joins science—preface from the editors. Plant Cell Tissue Organ Cult (PCTOC) 149:497–502. https://doi.org/10.1007/s11240-022-02330-4

Ohtsubo N (2011) Beyond the blue rose: modification of floral architecture with plant-specific chimeric repressors. Plant Biotechnol 28:113–121. https://doi.org/10.5511/plantbiotechnology.11.0120

Oleykowsky CA, Bronson Mullins CR, Godwin AK, Yeung AT (1998) Mutation detection using a novel plant endonuclease. Nucleic Acids Res 26:4597–4602

Ono E, Mizutani MF, Nakamura N, Fukui Y, Sakakibara KY, Yamaguchi M (2006) Yellow flowers generated by expression of the aurone biosynthetic pathway. Proc Natl Acad Sci U S A 103(29): 11075–11080

Ostergaard L, Yanofsky M (2004) Establishing gene function by mutagenesis in *Arabidopsis thaliana*. Plant J 39:682

Oud JSN, Schneiders H, Kool AJ, Grinsven MQJM (1995) Breeding of transgenic orange Petunia hybrida varieties. Euphytica 84:175–181

Pellegrineschi A, Damon JP, Valtorta N, Paillard N, Tepfer D (1994) Improvement of ornamental characters and fragrance production in lemon scented geranium through genetic transformation by *Agrobacterium* rhizogens. Bio/Technology 12:64–68

Perry JA, Wang TL, Welham TJ, Gardner S, Pike JM, Yoshida S, Parniske M (2003) TILLING reverse genetics tool and a web-accessible collection of mutants of the legume *Lotus japonicas*. Plant Physiol 131:866–871

Ritchie RD, Nielsen NC (2006) TILLING and its potential for plant agriculture. CAB Rev 1(66):5. http://www.cabi.org/cabreviews14

Robinson K, Firoozabady E (1993) Transformation of floriculture crops. Sci Hortic 55:83–99

Sarmast MK (2020) Principles and application of CRISPR/Cas technology in genetic modification of plants. Gorgan University of Agricultural Sciences and Natural Resources Publication. 124p. (In Persian)

Sarmast MK, Janati M (2019) Advances in new technology for targeted modification of plant genomes. Gorgan University of Agricultural Sciences and Natural Resources Publication. 216p. (In Persian)

Sato Y, Shirasawa K, Takahashi Y, Nishimura M, Nishio T (2006) Mutant selection from progeny of gamma-ray-irradiated rice by DNA heteroduplex cleavage using Brassica petiole extract. Breed Sci 56:179–183

Schaeffer SM, Nakat PA (2015) CRISPR/Cas9-mediated genome editing and gene replacement in plants: transitioning from lab to field. Plant Sci 240:130–142. https://doi.org/10.1016/j.plantsci.2015.09.011

Schmitt MW, Kennedy SR, Salk JJ, Fox EJ, Hiatt JB, Loeb LA (2012) Detection of ultra-rare mutations by next-generation sequencing. Proc Natl Acad Sci U S A 109:14508–14513. PMID: 22853953

Seiichi F, Jong JD, Rademaker W (1995) Efficient genetic transformation of chrysanthemum (D. grandiflorum Ramat. Cv. Kitamura) using stem segments. Breed Sci 45(2):179–184

Semiarti E, Nopitasari S, Setiawati Y, Lawrie MD, Purwantoro A, Widada J, Asano Y (2020) Application of CRISPR/Cas9 genome editing system for molecular breeding of orchids. Indian J Biotechnol 25:61–68. https://doi.org/10.22146/ijbiotech.39485

Shan S, Soltis PS, Soltis DE, Yang B (2020) Considerations in adapting CRISPR/Cas9 in nongenetic model plant systems. Appl Plant Sci 8(1):e11314. https://doi.org/10.1002/aps3.11314

Shibata M (2008) Importance of genetic transformation in ornamental plant breeding. Plant Biotechnol 25:3–8

Shibuya K, Watanabeb K, Ono M (2018) CRISPR/Cas9-mediated mutagenesis of the EPHEMERAL1 locus that 1 regulates petal 2 senescence in Japanese morning glory. Plant Physiol Biochem 131:53–57. https://doi.org/10.1016/j.plaphy.2018.04.036

Shimada Y, Ohbayashi M, Nakano-Shimada R, Okinaka Y, Kiyokawa S, Kikuchi Y (2001) Genetic engineering of the anthocyanin biosynthetic pathway with flavonoid-3′, 5′-hydroxylase : specific switching of the pathway in petunia. Plant Cell Rep 20:456–462

Simsek O, Kacar YA (2010) Discovery of mutations with TILLING and ECOTILLING in plant genomes. Sci Res Essays 5(24):3799–3802

Sirohi U, Kumar M, Sharma VK, Teotia S, Singh D, Chaudhary V, Priya, Yadav MK (2022) CRISPR/Cas9 system: a potential tool for genetic improvement in floricultural crops. Mol Biotechnol 64(12):1303–1318. https://doi.org/10.1007/s12033-022-00523-y

Slade AJ, Knauf VC (2005) TILLING moves beyond functional genomics into crop improvement. Transgenic Res 14:109–115

Slade AJ, Fuerstenberg SI, Loeffler D, Steine MN, Facciotti D (2005) A reverse genetic, nontransgenic approach to wheat crop improvement by TILLING. Nat Biotechnol 23:75–81

Stemple DL (2004) TILLING—a high throughput harvest for functional genomics. Nat Rev Genet 5:145–150

Stephenson P, Baker D, Girin T, Perez A, Amoah S, King GJ, Østergaard L (2010) A rich TILLING resource for studying gene function in Brassica rapa. BMC Plant Biol 10(1):62

Su J, Jiang J, Zhang F, Liu Y, Ding L, Chen S, Chen F (2019) Current achievements and future prospects in the genetic breeding of chrysanthemum: a review. Hortic Res 6:109. https://doi.org/10.1038/s41438-019-0193-8

Subburaj S, Chung SJ, Lee C, Ryu SM, Kim DH, Kim JS, Bae S, Lee GJ (2016) Site-directed mutagenesis in *Petunia hybrida* protoplast system using direct delivery of purified recombinant Cas9 ribonucleoproteins. Plant Cell Rep 35:1535–1544. https://doi.org/10.1007/s00299-016-1937-7

Sun L, Kao TH (2018) CRISPR/Cas9-mediated knockout of PiSSK1 reveals essential role of S-locus F-box protein-containing SCF complexes in recognition of non-self S-RNases during cross-compatible pollination in self-incompatible Petunia inflata. Plant Reprod 31:129–143. https://doi.org/10.1007/s00497-017-0314-1

Suzuki K, Zue H, Tanaka Y, Fukui Y, Fukuchi-Mizutani M, Murakami Y, Katsumoto Y, Tsuda S, Kusumi T (2000) Flower color modifications of *Torenia hybrida* by cosuppression of anthocyanin biosynthesis genes. Mol Breed 6:239–246

Suzuki T, Eiguchi M, Kumanaru T, Matsusaka H, Moriguchi K, Nagato Y, Kureta N (2008) MNU-induced mutant pools and high-performance TILLING enable finding of any gene mutation in rice. Mol Gen Genomics 279:213–223

Szarejko I, Maluszynski M (2011) TILLING—a shortcut in functional genomics. J Appl Genet 52(4):371

Szarejko I, Szurman-Zubrzycka M, Nawrot M, Marzec M, Gruszka D, Kurowska M, Chmielewska B, Zbieszezyk J, Jelonek J, Maluszynski M (2017) Creation of a TILING population in barley after chemical mutagenesis with sodium azide and MNU. In: Biotechnologies for plant mutation breeding. Springer, pp 91–111

Szurman-Zubrzycka M, Daszkowska-Golec A, Chmielewska B, Todorovska E, Szarejko I, Bedkowska I, Jelonek J, Nawrot M (2018) Development of barley TILING mutants tolerant to aluminium stress. In: FAO/IAEA international symposium on plant mutation breeding and biotechnology, 27–31 August, Vienna, Austria, Abstract No. IAEA-CN-263-301

Takatsu Y, Nishizawa Y, Hibi T, Akutsu K (1999) Transgenic chrysanthemum (*Dendranthema grandiflorum* (Ramat.) Kitamura) expressing a rice chitinase gene shows enhanced resistance to gray mold (*Botrytis cinerea*). Sci Hortic 82:113–123

Talame V, Bovina R, Sanguineti MC, Tuberosa R, Lundqvist U, Salvi S (2008) TILLMore, a resource for the discovery of chemically induced mutants in barley. Plant Biotechnol J 6(5): 477–485

Talame V, Rivina R, Salvi S, Sanguineti MC, Piffanelli P, Lundquist U, Tuberosa R (2009) TILLING with TILLMore. In: Shu QY (ed) Induced plant mutations in the genomic era. Food and Agriculture Organization of the United Nations, Rome, pp 240–242

Tanaka Y, Katsumoto Y, Brugliera F, Mason J (2005) Genetic engineering in floriculture. Plant Cell Tissue Organ Cult 80:1–24

Tanaka Y, Brugliera F, Chandler S (2009) Recent progress of flower colour modification by biotechnology. Int J Mol Sci 10:5350–5369

Tasaki K, Higuchi A, Watanabe A, Sasaki N, Nishihara M (2019) Effects of knocking out three anthocyanin modification genes on the blue pigmentation of gentian flowers. Sci Rep 9:15831. https://doi.org/10.1038/s41598-019-51808-3

Tasaki T, Yoshida M, Nakajim M, Higuchi A, Watanabe A, Nishihara M (2020) Molecular characterization of an anthocyanin-related glutathione S transferase gene in Japanese gentian with the CRISPR/Cas9 system. BMC Plant Biol 20:370. https://doi.org/10.1186/s12870-020-02565-3

Teixeira da Silva J (2004) Ornamental chrysanthemum: improvement by biotechnology. Plant Cell Tissue Organ Cult 79:1–18

Till BJ, Reynolds SH, Greene EA, Codomo CA, Enns LC, Johnson JE, Burtner C, Odden AR, Young K, Taylor NE, Henikoff JG, Comai L, Henikoff S (2003) Large-scale discovery of induced point mutations with high-throughput TILLING. Genome Res 13:524–530

Till BJ, Reynolds SH, Well C, Springer N, Burtner C, Young K, Bowers E, Codomo CA, Enns LC, Odden AR, Greene EA, Comai L, Henikoff S (2004) Discovery of induced point mutations in maize genes by TILLING. BMC Plant Biol 4:12

Till BJ, Zerr T, Comai L, Henikoff S (2006) A protocol for TILLING and Ecotilling in plants and animals. Nat Protoc 1(5):2465

Till BJ, Comai L, Henikoff S (2007a) Tilling and Ecotilling for crop improvement. In: Varshney RK, Tuberosa R (eds) Genomics-assisted crop improvement, vol 1: genomics approaches and platforms. Springer, Dordrecht, pp 333–349

Till BJ, Cooper J, Tai TH, Colowit P, Greene EA, Henikoff S, Comai L (2007b) Discovery of chemically induced mutations in rice by TILLING. BMC Plant Biol 7:19

Till BJ, Afza R, Bado S, Huynh OA, Jankowicz-Cieslak J, Matijevic M, Mba C (2009) Global TILLING projects. In: Shu QY (ed) Induced plant mutations in the genomics era. Food and Agriculture Organization of the United Nations, Rome, pp 237–239

Till B, Jankowicz-Cieslak J, Sagi L, Huynh O, Utsushi H, Swennen R et al (2010) Discovery of nucleotide polymorphisms in the Musa gene pool by Ecotilling. Theor Appl Genet 121(7): 1381–1389

Tong CG, Wu FH, Yuan YH, Chen YR, Lin CS (2020) High efficiency CRISPR/Cas-based editing of Phalaenopsis orchid MADS genes. Plant Biotechnol J 18:889–891. https://doi.org/10.1111/pbi.13264

Tsai H, Howell T, Nitcher R, Missirian V, Watson B, Ngo KJ, Lieberman M, Fass J, Uauy C, Tran RK, Khan AA, Filkov V, Tai TH, Dubcovsky J, Comai L (2011) Discovery of rare mutations in populations: TILLING by sequencing. Plant Physiol 156:1257–1268. https://doi.org/10.1104/pp.110.169748

Tsuda S, Suzuki K, Xue H, Tanaka Y, Fukui Y, Mizutani MF, Katsumoto Y, Kusumi T (1999) Molecular breeding of flower colour of *Torenia hybrida*. In: Altman A (ed) Plant biotechnology and in vitro biology in the 21st century, pp 613–616

Tsuda S, Fukui Y, Makamura N, Katsumoto Y, Sakakibara YK, Mizutani FM (2004) Flower color modification of *Petunia hybrida* commercial varieties by metabolic engineering. Plant Biotechnol 21:377–386

Uauy C, Paraiso F, Colasuonno P, Tran RK, Tsai H, Berardi S, Comai L, Dubcovsky J (2009) A modified TILLING approach to detect induced mutations in tetraploid and hexaploid wheat. BMC Plant Biol 9(1):115

Van der Krol AR, Vorst O (1997) Manipulation of flower shape. In: Geneve KL, Preece JE, Merkle SA (eds) Biotechnology of ornamental plants. CAB International, London, pp 237–256

van der Krol AR, Lenting PE, Veenstra J, van der Meer IM, Koes RE, Gerats AGM, Mol JNM, Stuitje AR (1988) An antisense chalcone synthase gene in transgenic plants inhibits flower pigmentation. Nature 333:866–869

Vantuyl JM, Arens P, Shahin A, Marasekciolakowska A, Barba-Gonzalez R, Kim HT, Lim KB (2018) Lilium. In: Van Huylenbroeck J (ed) Ornamental crops. Handbook of plant breeding, vol 11. Springer, Cham, pp 481–512. https://doi.org/10.1007/978-3-319-90698-0_20

Wang N, Wang YJ, Tian F, King GJ, Zhang CY, Long Y et al (2008) A functional genomics resource for *Brassica napus* :development of an EMS mutagenized population and discovery of FAE1 point mutations by TILLING. New Phytol 180(4):751–765

Wang TL, Perry J, Welham T, Pike J, Parniske M, Rogers C, Oldroyd GED, Stephenson P, Ostergaard L, Mccullagh B, Baker D, Walsh S, Clarke J (2009) Driving forward in reverse. In: Shu QY (ed) Induced plant mutations in the genomics era. Food and Agriculture Organization of the United Nations, Rome, pp 247–250

Watanabe K, Kobayashi A, Endo M, Sage-Ono K, Toki S, Ono M (2017) CRISPR/Cas9-mediated mutagenesis of the dihydrofavonol4-reductase-B (DFR-B) locus in the Japanese morning glory Ipomoea (Pharbitis) nil. Sci Rep 7:10028. https://doi.org/10.1038/s41598-017-10715-1

Watanabe K, Oda-Yamamizo C, Sage-Ono K, Ohmiya A, Ono M (2018) Alteration of flower color in Ipomoea nil through CRISPR/Cas9-mediated mutagenesis of carotenoid cleavage dioxygenase 4. Transgenic Res 27:25–38. https://doi.org/10.1007/s11248-017-0051-0

Wilde HD, Chen Y, Jiang P, Bhattacharya A (2012) Targeted mutation breeding of horticultural plants. Emir J Food Agric 24(1):31–41

Winefield C, Davis K, Schwinn K, Deroles S, Boulton G, Miller R, Boase M, Konezac II (1994) Genetic engineering for pelargonidin production in chrysanthemum. In: IVth Annual Queenston molecular biology meeting, Queenston, New Zealand, Abs. 072

Woodson WR (1991) Biotechnology of floricultural crops. Hort Sci 26(8):1029–1033

Woodson WR, Lawton KA, Meyer RC, Ragnothama KG, Goldbrough PB (1990) Regulation of gene expression in sensing carnation petals. In: Berunett AB, O'Neill SD (eds) Horticultural biotechnology, pp 203–212

Wu X, Hulse-Kemp AM, Waldl PA, Smith Z, Mockaitis K, Staton ME, Rinehart T, Alexander LW (2021) Genomic resource development for hydrangea (*Hydrangea macrophylla* (Thunb.) Ser.)—a transcriptome assembly and a high-density genetic linkage map. Horticulturae 7(2): 25. https://doi.org/10.3390/horticulturae7020025

Xie K, Yang Y (2013) RNA-guided genome editing in plants using a CRISPR–Cas system. Mol Plant 6:1975–1983. https://doi.org/10.1093/mp/sst119

Xin ZG, Wang ML, Barkley NA, Burow G, Franks C, Pederson G et al (2008) Applying genotyping (TILLING) and phenotyping analyses to elucidate gene function in a chemically induced sorghum mutant population. BMC Plant Biol 8:103

Xu J, Kang BC, Naing AH, Bae SJ, Kim JS, Kim H, Kim CK (2020) CRISPR/Cas9-mediated editing of 1-aminocyclopropane-1-carboxylate oxidase1 enhances Petunia flower longevity. Plant Biotechnol J 18:287–297. https://doi.org/10.1111/pbi.13197

Xu J, Naing AH, Bunch H, Jeong J, Kim H, Kim CK (2021) Enhancement of the flower longevity of petunia by CRISPR/Cas9-mediated targeted editing of ethylene biosynthesis genes. Postharvest Biol Technol 174:111460. https://doi.org/10.1016/j.postharvbio.2020.111460

Yagi M (2018) Recent progress in whole genome sequencing, high-density linkage maps, and genomic databases of ornamental plants. Breed Sci 68:62–70. https://doi.org/10.1270/jsbbs.17080

Yan R, Wang Z, Ren Y, Li H, Liu N, Sun H (2019) Establishment of efficient genetic transformation systems and application of CRISPR/Cas9 genome editing technology in *Lilium pumilum*

DC. Fisch. and *Lilium longiflorum* White Heaven. Int J Mol Sci 20:2920. https://doi.org/10.3390/ijms20122920

Yi G, Luth D, Goodman TD, Lawrence CJ, Becraft PW (2009) High-throughput linkage analysis of mutator insertion sites in maize. Plant J 58(5):883–892

Yu J, Tu L, Subburaj S, Bae S, Lee GJ (2021) Simultaneous targeting of duplicated genes in Petunia protoplasts for flower color modification via CRISPR-Cas9 ribonucleoproteins. Plant Cell Rep 40:1037–1045. https://doi.org/10.1007/s00299-020-02593-1

Zhang F, Puchta H, Thomson JG (2015) Advances in new technology for targeted modification of plant genomes. Springer. 166p, New York. https://doi.org/10.1007/978-1-4939-2556-8

Zhang B, Yang X, Yang C, Li M, Guo Y (2016) Exploiting the CRISPR/Cas9 system for targeted genome mutagenesis in *Petunia*. Sci Rep 6:20315. https://doi.org/10.1038/srep20315

Zuker A, Tzfira T, Vainstein A (1998) Genetic engineering for cut-flower improvement. Biotechnol Adv 16:33–79

Zuker A, Tzfira T, Scovel G, Ovadis M, Shklarman E, Itzhaki H, Vainstein A (2001) RolC-transgenic carnation with improved horticultural traits: quantitative and qualitative analysis of green-grown plants. J Am Soc Hort Sci 126(1):13–18

Cropwise Mutation References

Mutation research results on different ornamental plants have been mentioned in different chapters. An attempt has been made in this chapter to give a summary list of all ornamental species alphabetically along with the author references. One can quickly experience almost entire mutation breeding work on ornamentals from following and cross-references.

Abelia grandiflora (Matsubara et al. 1976); **Abutilion** (Potsch 1966a, b, 1967); **Acalypha** (Das et al. 1977; Rifnas et al. 2022); ***Acer negundo*** (Broertjes and Van Harten1988; Privalov 1965, 1967, 1968a, b, 1974); ***Acer ginnala*** (Smith and Noyszewski 2018); **Achimenes** (Broertjes 1971, 1972, 1973, 1974, 1976, 1977; Broertjes et al. 1983); ***Aechmea fasciata*** (Broertjes and Van Harten 1988; Huang 2021); ***Alnus glutinosa*** (Ohba and Murai 1966); **African violet** (Alstroemeria) (Anastassopoulos and Keil 1996; Broertjes 1971; Broertjes and Verboom 1974; Przybyla 1992, 2000; Tomo et al. 2012); **Amaryllis** (Hippeastrum) (Kaicker and Singh 1979; Broertjes and Van Harten 1988; Affrida et al. 2009; Ariffin et al. 2016); **Anemone** (Horovitz and Zohary 1966; Alkema 1974a, b; Boertjes and Van Harten 1988); **Antirrhinum** (Broertjes and Van Harten 1988; Pierik 1975, 1976; Pierik and Steegmans 1976; Pierik et al. 1979; Heffron et al. 2005, 2006; Gupta and Matsuo 1969; Dyer et al. 2007; Eltayeb and Ei-Metainy 2012; Cuany et al. 1985a, b; Sekiguchi et al. 1971; Stube and Doring 1938a, b; Sheela and Sheena 2014; Doring and Stube 1938); **Anthurium** (Broertjes and Van Harten 1988; Puchooa and Sookun 2003; Puchooa 2005; Te-Chato and Susanon 2005; Bin et al. 2006; Chen 2009; Sheela and Sheena 2014; Suraninpong and Wuthisuthimethavee 2015; Yang et al. 2015; López-Martínez et al. 2022); Asclepias species (Kobayashi et al. 2011); **Asclepias species** (Kobayashi et al. 2011); **Begonia** (Dorenbos 1973a, b; Doorenbos and Karper 1975; Shigematsu and Matsubara 1972; Matsubara et al. 1974; Molnar 1976; Linderman 1968; IAEA 1972; Mikkelson 1976; Mikkelson et al. 1975; Anonymous 1977a, b; Lin and Molnar 1983; Benetka 1987; Soedjono 1988; Roest et al. 1981; Broertjes 1982; Brown and Harney 1974; Matsubara 1982; Masubara et al. 1971; Matsubara et al. 1975; Roest et al. 1981; Ito et al. 2007; Chen et al. 2014); ***Berberis thunbergii*** (Smith and Noyszewski 2018); **Betula** (Ohba and Murai 1966; Scholz 1957; Vaarama 1970; Ocokoljic and Milosevic 2004); **Blue Daisy** (*Brachycome multifida*) (Walther and Stauer 1989a); **Bougainvillea** (Datta 1992,

S. Datta, *Role of Mutation Breeding in Floriculture Industry*,
https://doi.org/10.1007/978-981-99-5675-3

2004; Gupta and Shukla 1974; Gupta and Nath 1977; Abraham and Desai 1977; Hong and Shaode 1990; Deng and Liu 1990; Datta and Banerji 1997; Swaroop et al. 2015; Swaroop and Jain 2016; Swaroop and Janakiram 2015; Chen et al. 2012a, b; Judsri et al. 2016; Anonymous 1982; Banerji and Datta 1987, 1993; Banerji et al. 1987a, b; Datta 1992; Datta and Banerji 1990, 1997; Sharma et al. 2002; Srivastava et al. 2002; Jayanthi et al. 1999, 2000; Jayanthi and Datta 2006); **Bouvardia** (Broertjes and Van Harten 1988); ***Buddleja davidii*** (Smith and Brand 2012); **Cactus and Succuents** (Milica Filipovic 2019; Anonymous 2022); ***Calendula officinalis*** (Kaur et al. 2017; Tiwari and Kumar 2011; El-Nashar 2012); **Calathea** (Broertjes and Van Harten 1988; Anonymous 1988; Latado and Neto 1996; Tulmann et al. 2001); ***Camellia japonica*** (Zhou et al. (1990); **Canna** (Broertjes and Van Harten 1988; Nakornthap 1965; Gupta 1966; Mukherjee and Khoshoo 1970; Chuantang and Li 1989; Fucui et al. 2011, 2012; Chemarin et al. 1973; Deng and Liu 1987, 1890; Khalaburdin 1991; Niu and Li 1988; Siranut et al. 2000); ***Catharanthus roseus*** (Verma et al. 2013; Vekaateswarlu et al. 1983, 1988; Bhattacharjee 1998; Chatacharjee 1978; Kulkarni and Baskaran 2003, 2008; Kulkarni and Suresh 1999; Kulakarni et al. 1999; Mishra and Kumar 2003; Rai and Kumar 2000; Anjalica et al. 2005; Kumar et al. 2007; Baskaran et al. 2013; Verma and Singh 2010; Mangaiyarkarasi et al. 2014; Hassan et al. 2017); ***Celosia cristata*** **L** (Qing et al. 2005; Mostafa et al. 2014); ***Cheiranthus cheiri*** (Saito 1977); **China aster (*Callistephus chinensis* Nees)** (Wosinska 1980a, b, 1982, 1986, 2013; Anwar et al. 2019, 2020; Shivaswamy et al. 2022); **Chionodoxa** (Alkema 1974a, b; Broertjes and Van Harten 1988); ***Chlorophytum tuberosum*** (Chowta and Dnyansagar 1974); **Chrysanthemum** (Jank 1957a, b; Sheehan and Sagawa 1959; Fuji and Mabuchi 1961; Jain et al. 1961; Rana 1964a, b, c, d, 1965a, b; Bowen et al. 1962; Bowen 1965; Cawase 1965; Brock 1966; Dowrick and El-Bayoumi 1966a, b; Broertjes 1966a, b, 1971, 1976; Broertjes et al. 1980, 1983; Broertjes and Jong 1984; Yamkawa and Sekiguchi 1968; Matsubara 1982; Matsubara et al. 1971; Ichikawa et al. 1970; Das et al. 1974; Solanki and Sharma 2001; Sarkar and Sharma 1988; Sneepe 1977; Usenbaev and Imankulova 1974; Buiatti and Tesi 1968; Jong 1978, 1984; Jong and Custers 1986; Jung-Heiliger and Horn 1980; De Jong and Custers 1986; Huitema et al. 1986, 1989; Preil et al. 1983; Broertjes and Van Harten 1988; Jerzy 1990; Jerzy and Zalewska 1996; Dalsou and Short 1987; Schum and Preil 1998; Lema-Rumińska and Zalewska 2002; Latado et al. 1996, 2004; Nagatomi et al. 2003; Zalewska et al. 2007; Yamaguchi et al. 2008; Lema-Rumińska and Sliwinska 2009, 2015; Dwimahyani and Widiarsih 2010; Barakat et al. 2010; Salleh et al. 2010, 2012; Kaul et al. 2011; Kapadiya 2014; Kapadiya et al. 2014; Verma et al. 2012; Kumari et al. 2013; Singh and Bala 2015; Telem et al. 2015; Patil et al. 2015, 2017; Kapadiya et al. 2016; Mile and Kulus 2018; Bajpay and Dwivedi 2019; Haspolat et al. 2019; Ghormade et al. 2020; Miller et al. 2020; Din et al. 2020, 2021; Anitha et al. 2021; Haspolat 2022; Nasri et al. 2022; Castillo-Martinez et al. 2015; Nagatomi and Degi 2009; Fujii and Matsumura 1967; Broertjes 1972; Kovalchuk et al. 2000; Mergen and Thielges 1966; Nagatomi 1991, 1992, 2002; Nagatomi et al. 1993a, b, 1996a, b, c, 2000; Natarajan and Maric 1961; Richter and Singleton 1995; Du et al. 2017; Kim et al. 2018; Tanaka et al. 2010; Bolon et al. 2011, 2014; Love and Constantin 1965; Smith and Noyszewski 2018; Nakagawa 2009; Tanaka et al. 2010; Wu et al. 2005; Nagatomi et al. 1997, 1998, 2003; Suzuki et al. 2005; Ikegami et al. 2006; Shirao et al. 2007; Ueno et al. 2005; Furutani et al. 2008; Watanabe et al. 2008; Wakita et al. 2009; Sato et al. 2006; Matsumura et al. 2007, 2009, 2010; Toyoda et al. 2007; Okamura et al. 2007, 2008; Furutani et al. 2008; Wakita et al. 2009; Yamaguchi et al. 2009; Asami et al. 2010, 2011; Tanaka et al. 2010; Yamaguchi et al. 2010; Shirao et al. 2013; Ueno et al. 2013; Okamura et al. 2015; Sakamoto et al. 2016; Tanokashira et al. 2015, 2016; Tamaki et al. 2017; Tamari et al. 2017; Suryawati et al. 2022; CSIR-NBRI—Datta 1985a, 1987a, 1990, 1991, 1992, 1994, 1997, 2012, 2015, 2019; Datta and da Silva 2006; Gupta 1966, 1979; Gupta and Shukla 1971; Gupta and Datta 1978; Gupta and Jugran 1978; Datta and Gupta 1980, 1981a, b, 1983, 1984a, b, 1987; Datta et al. 1985; Banerji and Datta 1990, 1993; Datta 1992a, 2005a, b; Shukla and Datta 1993; Datta and Banerji 1991, 1993, 1995a, b; Datta and Mandal 2005; Mishra et al. 2003; Chakrabarty et al. 1999, 2000; Mandal et al. 2000a, b; Dwivedi et al. 2000; Datta et al. 2001; Misra and Datta 2007; Hossain et al. 2004, 2006; Mandal and Datta 2005); ***Cosmos sulphureus*** (Syakudo et al. 1964; Gupta and Samata 1967; Ayudhaya et al. 2022); **Codiaeum (Croton)** (Broertjes and Van Harten 1988; Alabi et al. 2010; Saggoo et al. 2011); **Crocus** (Akhund-Zade and Muzaferova 1975; Khan 1997, 2004, 2007; Khan et al. 2011; Nehvi et al. 2010; Haspolat et al. 2016; Salwee and Nehvi Firdos 2017; Magd el Din F. Rida 2019; Mahmoud A. Shraf-Eldin et al. 2018; Moghaddam et al. 2019; Samadi et al. 2022); **Crossandra** (Broertes and Van Harten 1988; Tarannum et al. 2016; Anonymous 2017; Hewawasam et al. 2004; Hewawasam 2003; Pandey et al. 1987; Girija et al. 1999; Almeida et al. 2008; Dhivya et al. 2015; Vinodh and Kannan 2020a, b); **Cupressus** (Broertjes and Van Harten 1988; Lev-Yadun et al. 2004; Maria et al. 2004); ***Cryptotneria japonica*** (Kuimura et al. 1975, 1976; Ohba 1971a, b; Ohba and Maeta 1973; Maeta et al. 1982; Kukimura et al. 1976); **Cyclamen** (Breider 1959; Broertjes and Van Harten 1988; Wellensiek 1960; Wellensiek and van Bren 1973; Sugiyama et al. 2008a; Kondo et al. 2008, 2009a, b, 2010, 2011; Hase et al. 2012; Kameri et al. 2011, 2012; Ishizaka et al. 2012; Ishizaka 2018; Nakayama et al. 2012); ***Cynodon dactylon*** (Lu et al. 2009); **Cyperus** (Denisova and Shurshikova 1976); **Cytisus** (Nagatomi et al. 1993); ***Dahlia variabilis*** (Broertjes and Ballego 1967, 1968, 1969; Das et al. 1974, 1975, 1977, 1978; Asahira et al. 1975; Khan et al. 1978; Lawrence 1931a, b; Singh et al. 1970; Thakur and Bhagchandani 1978; Grabowska and Mynett 1964; Lantin and Decourtye 1970; Dube et al. 1980; Xicun et al. 2007; Hamatani et al. 2001; Uyama et al. 2011, 2013; Pal et al. 2017); ***Datura innoxia*** Mill. (Krumbiegel 1979); ***Delphinium malabaricum*** (Huth) Munz. (Chinone et al. 2008; Firdose et al. 2011; Kolar et al. 2020); ***Dianthus caryophyllus*** **L. (Carnation)** (Richter and Singleton 1955; Sagawa and Mehlquist 1956, 1959; Buiatti and Ragazzini 1965; Dommergues et al. 1966; Simard et al. 1992; Cassells et al. 1993; Hemlata 1998; Singh et al. 2002; Bhattacharaya 2003; Okamura et al. 2003, 2012, 2013; Paramesh and Chowdhury 2005; Roychowdhury and Tah 2011a, b; Roychowdhury et al. 2012; Sugiyama et al. 2008b; Tanaka et al. 2010; Morimoto et al. 2019; Patil et al. 2019; Richter and Singleton 1955; Heslot 1964; Mehlquist et al. 1954; Sagawa 1957; Sagawa and Mehlquist 1956, 1957, 1959; Mehlquist and

Sagawa 1959, 1964; Johnson 1980; Samata et al. 1979; Bugnon et al. 1965; Dulieu 1968, 1969; Gaufillier 1965; Farestveit and Klougart 1966; Farestveit 1969; Badr and Etman 1977; Buiatti and Ragazzini 1965; Buiatti et al. 1965; D'Amato et al. 1964; Pereau-Leroy 1969, 1970, 1974a, b, 1975; c.f. Broertjes and Van Harten 1988; Dommergues and Gillot 1965, 1973; Broertjes 1982; Custers 1978; Silvy 1979; Sparnanji 1974, 1978; Sparnanji and Demmink 1970, 1971; Sparanji et al. 1974a, b, Carrier 1983; Silvy 1979; Silvy and Mitteau 1986; Custers et al. 1977; Hentrich and Glawe 1982); ***Digitalis obscura*** (Gavidia and Perez-Bermudez 1999); **Endymion** (Broertjes and Alkema 1970); ***Etlingera elatior*** (Yunus et al. 2013); **Eryngium** (Broertjes and Van Harten 1988; Pohlheim 1983); ***Euphorbia pulcherrima*** (Love 1966, 1972; Preil and Engelhardt 1982; Preil et al. 1983; Robinson and Darrow 1929; Robinson 1931; Stewart 1965; Stewart and Arisumi 1966; Kleffel et al. 1986; Koo and Cuevas-Ruiz 1974; Canul-Ku et al. 2012; Vilperte et al. 2011; Walther 1981; Nitarska et al. 2018; Pan et al. 2019); ***Eustoma grandiflorum*** (Nagtomi et al. 1996; Ohki et al. 2013; Abou-Daheb et al. 2017a, b, 2019; Mendoza-Gómez et al. 2020); **Ferns** (Broertjes and Van Harten 1988; Partanen 1958; Nelson 1961; Carlson 1969; Haigh and Howard 1973; Howard and Haigh 1968; Khare and Kaur 1980; Kaur and Khare 1982; Palta and Mehra 1973; Khare 1994; Klekowski 1976, 1984, 2011; Klekowski Jr. and Klekowski 1982; Klekowski and Kazarinova-Fukshansky 1984; Mohamad et al. 2002; Norazlina et al. 2003; Pei et al. 2018); **Ficus** (De Loose 1981; Takahashi et al. 2012); **Forsythia** (Dai and Magnusson 2012); **Freesia** (Broertjes and Van Harten 1988; Qin et al. 1988; Ling et al. 2019); **Fuchsia** (Broertjes and Van Harten 1988; Bouharmont and Dabin 1986a, b; Dabin and Bouharmont 1984); **Gentiana** (Nakatsuka et al. 2005; Tsuji et al. 2007); **Gerbera** (Laneri et al. 1990; Walther and Sauer 1990; Ghani et al. 2014; Ghani and Sharma 2019; Walther and Sauer 1986a, b, 1989, 1990, 1991, 1992; Jerzy and Zalewska 1992; Singh et al. 2011; Hazbullah et al. 2012; Hosoguchi et al. 2021; Purshottam et al. 2021); **Gloxinia (*Sinningia speciosa*)** (Sirisom 2008; Sirisom and Te-chato 2008; Radiasi 2011; Dong et al. 2018; Hasing et al. 2019); ***Gloriosa superba*** L. (Patwibul et al. 2001; Anandhi et al. 2013a, b; Selvarasu and Handhasamy 2013; Selvarasu and Kandhasamy 2017; Padmapriya and Rajamani 2017, 2021); **Grasses** (Bashaw and Hoff 1962; Burton 1972, 1974, 1975, 1976, 1979, 1981, 1985; Burton and Hanna 1977; Burton and Jackson 1962; Burton et al. 1980, 1982; Busey 1980; De Loose 1964; Hansen and Juska 1959, 1962; Julen 1954, 1958, 1961; Kuleshov et al. 1974; Powell 1974, 1976; Powell and Murray 1978; Powell et al. 1974, 1980; Lu et al. 2009; Li et al. 2010; Chen et al. 2011, 2016; Mutlu et al. 2015; Mohd et al. 2016; Kim et al. 2019; Merr. Lin et al. 2020); ***Glebionis segetum*** (corn marigold) (Kapoor et al. 2014); ***Gypsophila paniculata*** (Tsuji et al. 2007; Barakat and El-Sammak 2011a, b; Li et al. 2020); **Gladiolus** (Annonymous 1967, 1977, 1978, 1979; Dryagina 1962a, b, 1964, 1968, 1970, 1971, 1975a, b, 1977; Dryagina et al. 1967; Dryagina and Akhramova 1962, 1963, 1966a, b; Dryagina and Kazarinov 1963, 1965, 1966, 1972; Cantor et al. 1967; Abraham and Desai 1976b; Sazonova and Syrovatko 1974; Banerji 1982; Banerji et al. 1981, 1994, 2000, 2001; Banerji and Datta 1986a, 1987a, b, 1988, 2000, 2001; Banerji and Gupta 1982, 1985; Dhaduk 1992; Dryagina and Kazarinov 1965, 1966; Dryagina and Akhramona 1962; Grabowaska 1971, 1972, 1973, 1975, 1978; Grabowaska and Mynett 1970, 1974; Grabowaska and Chemagin 1974; Grabowaska et al. 1971; Dhara and Bhattacharya 1972; Gupta and Banerji 1977, 1984; Pandey and Gaur 1984; Iba et al. 1964, 1965; Isaev et al. 1960; Kaicker 1975; Kaicker and Singh 1983; Meshitsuka et al. 1963; Moes 1966, 1969; Raghava et al. 1988; Rao et al. 1990; Sheehan 1959; Sheehen and Lorz 1960; Mynett 1971; c.f. Grabowaska and Mynett 1970; Misra 1975, 1976a, b, 1977, 1978a, b, 1983, 1984, 1996; Misra and Bajpai 1978, 1982, 1983a, b, c; Misra and Choudhary 1979; Misra and Mahesh 1993; Uzenbaev and Nazanenko 1970; Buatti et al. 1965, 1967, 1969; Buiatti and Tesi 1968; Buiatti et al. 1967, 1969, 1970; Marek 1954; Broertjes and van Harten 1978; Drust 1973; Hubpard 1966; Jenkins 1961; Sax 1955; Sheehan and Sagawa 1959–1960; Sheehen and Lorz 1960; Sheehan and Sagawa 1959–1960; Jenkins 1961; Rodrinquez 1965; Hubbard 1966; Patil 2014; Patil and Dhaduk 2009; Srivastava et al. 2007; Kasumi 2001; Bhajantri and Patil 2013; Tanabe and Dohino 1993; Kasumi et al. 1999; Rao et al. 1990; Yamoto et al. 1958; Methews 1943; Mol 1939, 1952a, b; Spencer 1955; Sax 1955; Tirkey and Singh 2019; Patel et al. 2018; Dogra et al. 2017; Patil et al. 2017; Seilleur 1975, 1977a, b; Scarascia-Mugnazza 1968; Negi et al. 1983; Raghava 2000; Raghava et al. 1981, 1988; Raikov and Bulykov 1975; Glazurina and Chemagin 1974; Drust 1973, 1975; Broertjes and Van Harten 1978; Awad and Harried 1985); ***Glebionis segetum*** (corn marigold) (Kapoor et al. 2014); **Guzmania peacockii** (De Loose 1966, 1969a, 1973a; Huang 2011); **Hebe** (Gallone et al. 2012); **Hedera helix** (Knuth 1962); **Hibiscus** (Das et al. 1974, 1977; Broertjes and Van Harten 1988; Banerji and Datta 1986, 1988; Srivastava and Mishra 2005; Anonymous 1989; Warner and Erwin 1998; Naveena et al. 2020a, b; Hong et al. 1980; Song et al. 1999; Kang et al. 2007; Kim et al. 1997); ***Helianthus tuberosus*** **(Jerusalem artichoke)** (Coppola 1986; Songsri et al. 2019); ***Helianthus annus*** (Omar et al. 1993; Cecconi and Durante 2000; Barakat et al. 2002; Ramesh Kumar and Venkat Ratnam 2009, 2010; Kumar and Ratnam 2010; Jamdhade and Kashid 2016; Feng et al. 2016; Diaz et al. 2018); ***Heliconia psittacorum*** (Urrea and Ceballos 2005, 2017); **Hosta** (Vaughn and Wilson 1980; Vaughn et al. 1980; Vaugh and Wilson 1980); **Hoya** (Van Raatle 1980; Broertje and van Harten 1988); ***Hoya carnosa*** (Broertjes and Van Harten 1988; Anonymous 1988; Van Raatle 1980); **Hyacinthus** (De Mol 1926, 1931, 1933, 1934, 1937a, b, 1940, 1953; Broertjes and Van Harten 1988; Broertjes and Alkema 1970); **Hydrangea** (Iizuka et al. 1998; Kodama et al. 2015; Kudo et al. 1998); ***Impatiens platypetala*** (Alston and Sparrow 1962; Arisumi 1973, 1978; Weigle and Butler 1983); **Iresine** (Das et al. 1977); **Iris** (Konzak and Randolph 1956; Hekstra and Broertjes 1968; Rather and Jhon 1996, 2000; Rather et al. 2002); ***Ipomoea purpurea*** **(L) Roth.** (Parkhi and Khalatkar 1988; Bhate 1999, 2001); **Jasmine** (Nambisan et al. 1980; Kumar et al. 1983; Ghosh et al. 2018a, b, c, 2019; Kannan et al. 2002; Ghosh and Ganga 2019; Nelka et al. 2021; Gopitha et al. 2022); **Kalanchoe** (Broertjes and Leffring 1972; Nakoranthap 1974; Sharma Rao and Singh 1976; Sharma Rao 1977; Van Dordrecht 1984; Horn 1984; Karper and Pierik 1981; Johnson 1948; Schwemmle and Robbelen 1962; Stein and Sparrow 1963, 1966); **Kohleria** (Parliman and Stushnoff 1979; Geier 1983, 1988, 1989,1994, 2012); ***Lagerestroemia indica*** (Boddie and Whitcomb 1978; Li et al.

2015; Zhang et al. 2014; Yim et at. 2010; Xia 2020); ***Lavandula intermedia*** **Emeric** (Tsuro et al. 2008); **Lilium** (Brown and Cave 1953, 1954a, b; Brown and Zohary 1953, 1955; Bowen and Sparrow 1954; Crouse 1954; Mitra 1958; Lizuka and Ikeda 1963; Taek et al. 2005; De Mol 1926; Broertjes and Van Harten 1988; De Mol 1949; Custers et al. 1977; Van Eijk and Eikelboom 1981a, b; Hopper and Peloquin 1968; Brown and Cave 1953, 1954a, b; Loh and Cooper 1966; VanGroenestijn and Van Tuyl 1983; Largon and Lyakh 2002; Lyakh and Largon 2005; Wang et al. 1989; Cuany et al. 1958; Broertjes 1969, 1972a; Broertjes and Alkema 1970; Anonymous 1991a, b; Chinone et al. 2008; Grassotti et al. 1987; Chiba et al. 2007; Chinone et a. 2007; Kondo et al. 2008); ***Limonium sinuatum*** Mill (Chinone et al. 2008; Ogawa et al. 2014); ***Lonicera japonica*** (Boyarskikh et al. 2016; Cambecedes et al. 1992); **Lotus (*Nelumbo nucifera*** Gaertn.) (Arunyanart and Soontronyatara 2002; Wu et al. 2007; Arunyanart and Chaitrayagun 2005; Vichai et al. 2011; Liu et al. 2016; Soontornyatara et al. 2017; Liu et al. 2019); **Mesembryanthemum** (Chaturvedi et al. 1997; Qalby et al. 2020); ***Moluccella laevis*** (Minisi et al. 2013); **Muscari** (*M. armeniacum*) (Broertjes and Alkema 1970; Roest and Bokelmann 1981; Broertjes and Van Harten 1988); **Narcissus** (De Mol 1926; Zandbergen 1964; Broertjes and Van Harten 1988; Alkema 1974a, b; Datta et al. 2003; Lu et al. 2007); **Nautilocalyx** (Broertjes and Van Harten 1988; Tran Thanj Van 1973; Tran Thanj Van and Drira 1971; Venverloo 1974; Venverloo et al. 1983; Chouard 1938); ***Nertera granadensis*** (Broertjes and Van Harten's 1988); ***Nymphaea rubra*** **Roxb.** (Gupta 1977); **Orchid** (Kozlowska-kalisz 1979; Kuang 1999; Yuki et al. 2007; Affrida et al. 2008; Aurigue et al. 2008; Chen et al. 2008; Gonzales et al. 2008; Muangsorn and Te-chato 2008; Khostravi et al. 2009; Pimonrat and Suraninpong 2009; Khoddamzadeh et al. 2010; Ariffin et al. 2012; Luan et al. 2012; Atra et al. 2014; Lee et al. 2015, 2016; Wannajindapom et al. 2016; Dehgahi and Joniyasa 2017; Hernández-Muñoz et al. 2017a, b; Kim et al. 2019, 2020; Pimonrat et al. 2012; Sherpa et al. 2022; Affrida et al. 2008; Romeida et al. 2012; Sulistianingsih 2013; Sheela and Sheena 2014; Sheela et al. 2008; Sakinah et al. 2005; Mohd. Nazir and Sakinah 2001; Mohd. Nazir et al. 2003; Sakinah and Mohd Nazir 2002; Hassan et al. 2010; Kurniati 2004; Khoddamzadeh et al. 2010); ***Ornithogalum virens*** L. (Broertjes and Van Harden 1988; Biswas and Biswas 2006; Contreras and Shearer 2020); **Osteospermum** (Lizuka et al. 2007, 2011; Okada et al. 2012); **Paspalum notatum Flugge** (Kannan et al. 2015); **Pelargonium** (Bergann 1967a, b; Bergann and Bergann 1959; Pohlheim 1977; Potsch 1964; Daker 1966, 1967; Pohlheim et al. 1972, 1976; Stewart et al. 1974; Craig 1963; Skirvin and Janick 1974, 1976a, b; Grunewaldt 1983; Janick et al. 1977; Kameya 1975; Yu et al. 2016); **Peperomia** (Harris and Hart 1964; Henny 1978; Broertjes and Van Harten 1988); **Petunia** (Moore and Haskins 1935; Colijn et al. 1979; Mahana and Garg 1989; Khalatkar and Kashikar 1980; Kashikar and Khalatkar 1981; Gerats 1991; Van Houwellingen et al. 1998; Berenschot et al. 2008; Okamura et al. 2009; Hase et al. 2010; Dona et al. 2013; Jiang et al. 2014); ***Philodendron erubescens*** (Karunananda et al. 2018); **Phlox** (Pathak and Raghuvanshi 1980; Verma and Raina 1980; Pillai and Verma 1992; Verma and Sharma 2000; Ramesh and Verma 2015; Tiwari and Kumar 2010; Tiwari and Mishra 2012); ***Plectranthus (Coleus)*** (Love and Mullenax 1964; Love and Constantin 1965, 1966; Love and Malone 1967; Aisyah et al. 2015; Aisyah et al. (2017); **Populus** (Biswas et al. 2013; Nishiguchi et al. 2012); ***Portulaca grandiflora*** Hook. (Gupta 1966, 1970; Lata and Gupta 1971; Cotter 1963; Banerjee 1967; Desai 1973, 1974; Abraham and Desai 1977, 1978; Skirvin et al. 1982; Raghuvanshi and Singh 1979, 1980; Kruczkowska et al. 1997; Wongpiyasatid and Roongtanakiat 1992; Wongpiyasatid and Hormchan 2000; Tangsombatvitchit et al. 2008; Wongpiyasatid and Hormchan 2000; Bennani and Rossi-Hassani 2001); ***Prunus lannesiana*** (Ishii et al. 2009, 2011; Hayashi et al. 2019); **Ranunculus** (Alkema 1974; Broertjes and Van Harten 1988; Dorion et al. 1975; Papanicolaou and Kokkini 1983); **Rhododendron simsii** (Broertjes and Van Harten 1988; De Loose 1968, 1969a, b, 1970a, b, c, d, 1971a, b, 1973a, b, 1974a, b, d, e, 1979; Heursel 1972, 1981; Streitberg 1965, 1966a, b, 1967a, b; Preil and Walther 1983; Akabane et al. 1973; Kobayashi et al. 2007, 2009); **Ribes** (Contreras and Friddle 2015); **Rose** (Gelin 1965; Chan 1966; Heslot 1966, 1968; Walther and Sauer 1986a, b; Kaicker and Swarup 1972, 1978, 1985; Dommergues 1976, 1980; Heslot 1966, 1968; Gupta 1966; Gupta and Shukla 1970, 1971a, b; Swarup et al. 1971, 1973; Lata 1973, 1975; Lata and Gupta 1971a, b, 1975; Streitburg 1964, 1966a, b, c, 1967; Nakajima 1965, 1970, 1973; Usenbaev and Imankulova 1974; Desai and Abraham 1978; James 1983; Smilansky and Zieslin 1986; Benetka 1985; Kaicker 1982, 1983, 1985, 1987, 1988, 1995; Kaicker et al. 1991; Kaicker and Dhyani 1985a, b, 1986a, b, 1987; Khalatkar 1986; Datta and Gupta 1983a, b; Gupta and Datta 1984; Datta 1994b, 1986a, b, 1987a, b, 1988, 1990a, b, 1989a, b, c, d, 1991a, 1992, 1993; Datta and Gupta 1985a, b; Dohare and Mathew 1991; Chan 1966; Dommergues 1976, 1980; Swarup et al. 1971, 1973; Gupta and Datta 1983; Kaicker and Kumar 1992; Gupta et al. 1982b; Klimenko et al. 1977; Guo et al. 1983; Arinshtein and Krapivenko 1980; Lata et al. 1977; Huang and Chen 1986; Walther and Sauer 1986a, b; Hara et al. 2003; Yamaguchi et al. 2003; Canli and Skirvin 2008; Ryu et al. 2019); **Rudbeckia** (Shukla et al. 1986; Oates et al. 2013); ***Saintpaulia ionantha*** (African Violet) (Broertjes 1972b; Warfield 1973; Sparrow and Schairer 1980; Grunewaldt 1980, 1983, 1988; Espino and Vazquez 1981; Kelly and Lineberger 1981; Geier 1983; Warburton et al. 1984; Lineberger and Druckenbrod 1985; Ando et al. 1986; Seneviratne and Wijesundara 2004; Zhou et al. 2006; Seneviratne and Wijesundara 2007; Fang and Traore 2011; Wongpiyasatid et al. 2007; Craig and Hampson 1979; Espino and Vazquez 1981; Kelly and Lineberger 1981; Relichova 1984; Eyerdom 1981; Hentrich and Beger 1974; Jungnickel 1977; Plummer and Leopold 1957; Polheim 1974a, b, 1977, 1980, 1981; Pohlheim and Beger 1974; Pohlheim and Pohlheim 1976); ***Sarcococca confuse*** (Hoskins and Contreras 2019); **Silene species** (Jiang and Dunn 2016); ***Salvia coccinea*, *S. splendens*** (Feiglova 1968; Haq 1983; Wu et al. 2009; Yamaguchi et al. 2017; Bugallo 2021; Rebekah C.I. Maynard and John M. Ruter 2023); ***Sandersonia aurantiaca* Hook.** (Horita et al. 2002; Morgan et al. 2002, 2004; Davisel et al. 2002); ***Sarcococca confuse*** (Hoskins and Contreras 2019); **Sansevieria** (now under **Dracaena**) (Lapade et al. 2001; Palamine et al. 2005; Teng and Emily Shih-wen 2007; Fernando Aurigue 2019; Taychasinpitak 2009; Teng and Leonhardt 2009a, b); **Schefflera** (c.f. Broertjes and Van Harten 1988); **Scilla** (Akema 1974a, b; Broertjes and Van Harten 1988; Chakravarty and

Sen 1987, 2001); ***Sophora davidii*** (Franch.) Kom. ex Pavol (Wang et al. 2017); ***Spiraea thunbergii*** (Iizuka et al. 2001); ***Streptocarpus*** (Anonymous 1975; Broertjes 1968, 1969a, 1970, 1973, 1974; Broertjes et al. 1969; Brown 1971, 1973, 1974; Brown and Davies 1971; Choudhary 1976; Davis 1971; Davies and Hedley 1975; Van Raatle and Van Raatle-Wichers 1974; Zevan 1972, 1973; Osiecki 1989); ***Stromantha sanguinea*** (Tulman Neto and Latado 1996; Neto et al. 2001); ***Tagetes erecta*** L. (African Marigold) (Singh et al. 2009; Sarkar et al. 2016; Majumder et al. 2018a, b; Latha and Dharmatt 2018; Aravind and Dhanavel 2021; Susrama and Yuliadhi 2020; Lenawaty et al. 2022); ***Tigridia pavonia*** (Díaz-López et al. 2003); ***Tillandsis fasciculate*** (Koh and Davies Jr. 2001); ***Torenia fournieri*** (Miyazaki et al. 2006; Abe and Ohtsub 2008; Sasaki et al. 2008; Sawangmee et al. 2011); **Tradescantia** (Sparrow et al. 1972; Nauman et al. 1975; Dennis 1976; Nauman et al. 1976; Ichikawa and Takahashi 1977; Schairer et al. 1978; Ichikawa and Sparrow 1978; Ichikawa et al. 1981; Van't Hof and Schairer 1982; Osipova and Shevchenko 1984; Sakuramoto and Ichikawa 1996; Ichikawa 1997; Yakovleva et al. 2011; Evseeva and Zainullin 2000; Ichikawa et al. 1969; Nauman et al. 1977; Mericle and Mericle 1967; Underbrink et al. 1973; Evseeva and Geras'kin 2001); ***Tricyrtis hirta*** (Nakano et al. 2010); **Tuberose** (Younis and Borham 1975; Abraham and Desai 1976a, b; Patil et al. 1975; Gupta et al. 1974; Krasaechai 1976, 1992; Sambandamurthi 1983; Datta and Shukla 1996; Ali 2002; Krishnan et al. 2003; Guo et al. 2009; Estrada-Basaldua et al. 2011; Singh et al. 2011, 2013, 2015, 2016; Navabi et al. 2016; Kainthura and Srivastava 2015; Panigrah and Saiyad 2013; Pohare et al. 2012, 2013; Wani et al. 2014; Navabi et al. 2016; Kaintura et al. 2015, 2016, 2018; Kayalvizhi et al. 2016a, b, 2017a, b; Pooja et al. 2016; Kayalvizhi et al. 2017; Sah et al. 2017; Yadav et al. 2018; Abhangrao 2019; Abhangrao et al. 2019, 2020; Singh and Sadhukhan 2019; Jyoti et al. 2019; Sharavani et al. 2019; Kutty et al. 2020; Kumar et al. 2021); **Tulip** (De Mol 1933, 1949; Thamm 1956; Mol van Oud Loosdrecht 1956; Nybom 1961; Graboska and Mynett 1970; Matsubara et al. 1965; Matsuda 1960; Myodo 1942; Nezu 1962, 1963a, b, 1964, 1965, 1967; Nezu and Obata 1964a, b; Van Eijk and Eikelboom 1981a, b; Custers et al. 1977; Broertjes and Van Harten 1988; Anonymous 1988; Meshitsuka et al. 1962; Ikegawa et al. 2016; Li et al. 2022); **Verbena** (Kanaya et al. 2008; Suzuki et al. 2002; Saito 1977); ***Vitex agnus-castus*** (Ari et al. 2015); ***Weigela*** (Duron 1992); ***Zephyranthes*** (Banerjee 1967; Spencer 1955; Tisch 1974; Broertjes and Van Harten 1988); ***Zinnia elegans*** Jacq. (Swarup and Raghava 1974; Venkatachalam and Jayabalan 1991, 1992, 1994a, b, 1997; Pallvi et al. 2017; Kole and Meher 2005; Pratiwi 2010).